INTERPRETATION OF CARBON-13 NMR SPECTRA

INTERPRETATION OF CARBON-13 NMR SPECTRA

SECOND EDITION

F. W. Wehrli, A. P. Marchand, and S. Wehrli

JOHN WILEY & SONS

Chichester · New York · Brisbane · Toronto · Singapore

Copyright © 1983 Wiley Heyden Ltd;
© 1988 John Wiley & Sons Ltd.

Library of Congress Cataloging-in-Publication Data:

Wehrli, F. W.
  Interpretation of carbon-13 NMR spectra.

  1. Carbon—Isotopes—Spectra.   2. Nuclear
magnetic resonance spectroscopy.   I. Marchand, Alan P.
II. Wehrli, S.   III. Title.
QC462. C1W43   1988   547.3'0877   87–19020

ISBN   0 471 91742 7

British Library Cataloguing in Publication Data:

Wehrli, F. W.
  Interpretation of carbon-13 NMR spectra
  —2nd ed.
  1. Chemistry, Physical organic   2. Nuclear
  magnetic resonance spectroscopy   3. Carbon
  —Isotopes
  I. Title   II. Marchand, A. P.   III. Wehrli, S.
  547.3'0877   QD476

ISBN   0 471 91742 7

Phototypesetting at Thomson Press (India) Limited, New Delhi
Printed by Anchor Brendon Ltd, Tiptree, Essex

Contents

Single-frequency off-resonance decoupling (SFORD)—Evaluation of off-resonance data with the aid of graphical methods—Selective decoupling—NOE-enhanced proton-coupled carbon spectra—Analysis of proton-coupled carbon spectra—Selective low-power irradiation—Stereospecificity of the three-bond coupling constant $^3J_{CH}$

Preface

Carbon-13 n.m.r. still is and remains one of the most powerful structural tools in organic chemistry, in particular when used in conjunction with its closest counterpart, proton n.m.r. After having been reprinted three times, *Interpretation of Carbon-13 NMR Spectra*, which first appeared in 1976, was felt to be in need of a thorough revision, necessitated primarily by the instrumental and methodological advances made during the past 10 years. Although today n.m.r. spectroscopy is taught routinely at an undergraduate level, the practicing chemist is increasingly overwhelmed by the unusual richness of this spectroscopic technique. Contrary to other predictions and unlike any other spectroscopic method, n.m.r. continues to evolve rapidly. The technique received a new impetus with the advent of two-dimensional spectroscopy, just barely missed by the first edition. Along with this, instrumentation evolved to ever higher levels of sophistication. Whereas in the mid-70s, the superconducting spectrometer was an expensive research option, today virtually all routine spectrometers used in chemical and analytical practice are based on superconducting magnets. The concomitant increase in spectral dispersion, albeit less critical in carbon-13 than in proton spectroscopy, has greatly expanded the scope of applications. Along with the higher magnetic fields came a significant increase in signal-to-noise, making possible acquisition of carbon-13 spectra of a few milligrams of a typical organic compound in minutes rather than hours.

Spurred by these events, the authors completely rewrote and significantly expanded the five chapters while leaving the basic concept intact.

Chapter 1 briefly reviews the basic physics of carbon-13 n.m.r., expanding in particular on changes that occurred in instrumentation.

Chapter 2 (Spectral Parameters) was in least need of refurbishing. Updates were primarily necessary in the area of carbon–proton coupling constants as well as coupling constants involving other nuclei which, owing to improved instrumentation, have been studied very broadly during the past 10 years. The known stereospecificity of many of these coupling constants, together with increased instrument sensitivity for obtaining proton-coupled spectra, have considerably expanded the scope of such applications.

The most dramatic innovations, however, have occurred in the area of new experimental techniques. In the light of these developments, Chapter 3 was renamed 'Experimental Techniques for Spectral Assignment'. It is virtually impossible, and completely beyond the scope of this book, to do complete justice to the monumental advances in this area. The authors therefore attempted to make a judicious selection of the most relevant new techniques not available 10 years ago. These comprise new polarization transfer experiments like INEPT, DEPT, APT and some of their two-dimensional counterparts. Perhaps the most powerful among the latter category of experiments is the chemical shift correlation 2D technique, permitting a new and straightforward way for cross-assigning proton and carbon spectra. Also new are carbon–carbon connectivity experiments in which spectra are obtained from naturally occurring isotopomers containing two C-13 isotopes, in principle making it possible to perform a complete assignment of an unknown molecule *ab initio*, i.e. without the need of additional structural information. Nevertheless, some of the classical approaches to structure determination such as chemical shift correlation and selective double resonance, will maintain their utility and are therefore covered in much detail. The increased spectral dispersion, achievable at high field, further makes it possible to study subtle isotope effects, as they are induced by deuterium and oxygen-18. Finally, Chapter 3 also accommodates a section on solid-state carbon-13 n.m.r., by providing a cursory overview of the basic experiments and a discussion of solid-state spectral phenomenology.

Chapter 4 (Nuclear Spin Relaxation) was expanded, with a more detailed discussion of electron-nuclear relaxation in terms of the Bloembergen–Purcell–Pound model and a brief discussion of anisotropic rotational diffusion. The field dependence of relaxation is also covered in some more detail. Overall, however, it was felt that the area of relaxation measurements, in particular from the point of view of structure determination, has been relatively quiescent.

Chapter 5 (Applications) reflects, by and large, the much increased scope of new techniques available to the practicing applied spectroscopist. Concomitantly, the algorithmic approach to structure determination has changed significantly during the past 10 years. Whereas at the time the first edition of this book was written, single frequency off-resonance decoupling was the only practical adjunct to proton noise-decoupled carbon-13 spectra, proton-coupled spectroscopy with polarization transfer, 2DFT (notably proton–carbon correlated 2D spectroscopy) are now used routinely. Some of the new techniques benefit a variety of applications in the macromolecular field, permitting assignment of polymer microstructure at high magnetic field. Quantitative analysis, a rapidly growing application of industrial applications, and the experimental requirements for accurate quantitation are treated in some detail. Finally, two new sections have been added, one on solid-state applications and a discussion of *in-vivo* spectroscopy. The latter allows us, for the first time, to study metabolic processes from intact cells, organs, animals and even humans and thus add a totally new dimension to carbon-13 n.m.r.

From the feedback received during the past 10 years, the authors concluded that it would be worthwhile to expand on a particularly popular feature of the previous version of the book, i.e. the problem section. While some of the more esoteric and less useful problems were removed, a number of new problems were added, notably relating to structure determination of organic and biomolecules, as well as novel mechanistic applications.

Finally, the authors would like to thank all those who have contributed to the book, in terms of advice and constructive suggestions: Professor Barry Shapiro, for critically reviewing the first four chapters of the book; Professor James Cook from the University of Wisconsin in Milwaukee for providing data prior to publication and, last but not least, Dr T. Wirthlin from Varian AG, who coauthored the first edition of the book, for encouragement and advice.

1

Basic Principles

1.1 HISTORY OF CARBON-13 NUCLEAR MAGNETIC RESONANCE

Following hydrogen, carbon is the most abundant element in organic chemistry. Whereas hydrogen occurs at the periphery, carbon constitutes the backbone of molecules. On these premises the impetus to extend n.m.r. to carbon was an obvious one: in analogy to proton n.m.r. carbon chemical shifts and coupling constants could be assumed to provide information on functional groups, structure, and stereochemistry, and might ultimately give the chemist a tool to trace out the carbon skeleton of completely unknown molecules. Today we know that these farsighted anticipations have been fulfilled. It may therefore appear surprising that it took more than a decade following the first carbon-13 n.m.r. experiments in 1957, performed on molecules containing the isotope in natural abundance,[1] until carbon n.m.r. spectra could be obtained routinely.

The reason for this long evolution phase is a technological one, and has to do with the intrinsically low receptivity of carbon-13, coupled with a low natural abundance of this only magnetic isotope of carbon. Swept spectrometers of the kind they were—and to some extent still are—utilized in proton n.m.r. therefore turned out to be inappropriate for natural-abundance carbon-13. These instrumental inadequacies were finally overcome thanks to a breakthrough in computer technology and digital data processing, along with instrumental developments such as broadbanded proton decoupling[2] and pulsed excitation techniques.[3]

Most problems that early spectroscopists were struggling with were in some way associated with spectrometer sensitivity. N.m.r. is known to be an intrinsically insensitive technique when compared, for example, with methods such as optical or mass spectroscopy. This is so for basically two reasons. The first is related to the energy difference ΔE between ground and excited state, which is relatively small and so are therefore the Boltzmann population differences:

$$N_2/N_1 = \exp(-\Delta E/kT) \simeq 1 - \Delta E/kT \qquad (1.1)$$

In Eqn. (1.1) N_1 and N_2 represent the spin populations on the ground and excited state, respectively, k is the Boltzmann constant, and T the absolute temperature. The linearized approximation on the right of the exponential is justified since $\Delta E \ll kT$. Insertion of the difference in Zeeman energy

$$\Delta E = \hbar\gamma H_0 \tag{1.2}$$

into Eqn. (1.1) gives

$$N_2/N_1 \simeq 1 - \hbar\gamma H_0/kT \tag{1.3}$$

with $\hbar$ being the modified Planck constant $(h/2\pi)$, γ the magnetogyric ratio, and H_0 the field strength of the polarizing field. At a field of 2.3 Tesla (25.2 MHz carbon frequency) one obtains for the relative excess population $(N_1 - N_2)/N_1 = 4 \times 10^{-6}$ for carbon nuclei at 25°C. Hence there is only a tiny fraction of spins available to be promoted from the ground to the excited state.

The second cause for the inherent low sensitivity of the n.m.r. experiment lies in the long lifetimes of excited states (typically on the order of $10^{-3} - 10^3$ s). At constant excitation power this severely limits the repetition rate for successive excitations since sufficient time has to be allowed for restoration of the equilibrium populations. In contrast to optical spectroscopy, spontaneous emission is negligibly small in magnetic resonance and the only mechanism counteracting radiofrequency (rf) perturbation is relaxation, more precisely, spin–lattice relaxation. The two time-dependent processes affecting the spin populations are diagrammatically illustrated in Fig. 1.1.

Let us now briefly look into the nucleus-specific factors contributing to sensitivity. The three non-instrumental quantities are the magnetogyric ratio γ, the spin number I, and the natural abundance a, which can be shown to be related to the signal strength S at a given field H_0 as follows:[4]

$$S \propto I(I + 1)\gamma^3 a \tag{1.4}$$

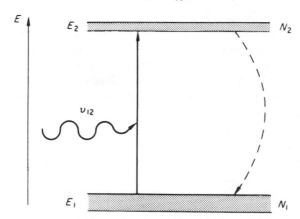

Fig. 1.1 Nuclear Zeeman levels E_1 and E_2 with equilibrium spin populations N_1 and N_2. Irradiation at the Larmor resonance frequency ν_{12} promotes spins from the E_1 to the E_2 level whilst spin–lattice relaxation (dashed line) tends to reestablish Boltzmann populations.

TABLE 1.1   Physical properties of isotopes ^{13}C and ^{1}H

	^{13}C	^{1}H
Relative abundance (%)	1.11	99.98
Spin I (in multiples of $\hbar$)	$\frac{1}{2}$	$\frac{1}{2}$
Magnetogyric ratio γ (in $s^{-1}T^{-1}$)	6.741×10^{7}	2.675×10^{8}
Larmor frequency at 2.34 T (MHz)	25.16	100.0
Relative receptivity	1.78×10^{-4}	1

The quantity $I(I + 1)\gamma^{3}a$ has also been termed *receptivity*[5] to distinguish it from the sensitivity which involves instrumental factors. When we compare the receptivity of ^{13}C with that of the most common n.m.r. isotope, ^{1}H, we can further ignore the spin term since both have $I = \frac{1}{2}$. However, because $\gamma(^{13}C)/\gamma(^{1}H) \simeq \frac{1}{4}$ and $a = 1.11 \times 10^{-2}$ we realize that ^{13}C is about 5600 times less receptive than ^{1}H. Some of the relevant physical properties relating ^{13}C and the proton are compiled in Table 1.1.

Until major advances were made to overcome the crucial signal-to-noise (S/N) hurdle, ^{13}C n.m.r. on molecules of practical analytical interest were not amenable to the experiment. The early work was therefore mostly concerned with low-molecular-weight compounds, usually recorded as neat liquids in the so-called rapid-passage dispersion mode.[1] The fast sweep rates implicit to this technique resulted in fairly low-resolution spectra, as exemplified in Fig. 1.2(a).[6]

Another technique alleviating the sensitivity limitation is based upon hetero-nuclear double resonance. In the case where ^{13}C is coupled to protons, for example, it may be possible to monitor a proton transition (^{13}C satellite) while simultaneously sweeping through the ^{13}C resonance region with a weakly perturbing double-resonance field.[7,8] The concomitant alterations of the ob-served proton line results in what has been termed an *INDOR* (internuclear double resonance) spectrum.

Major progress was made in the mid 1960s with the realization of proton noise decoupling[2] which resulted in a sensitivity gain of one to two orders of magnitude and which henceforth permitted recording of slow-sweep high-resolution spectra. The advent of electronic storage devices falls roughly into the same period. Signal-averaging techniques[9] are based upon the stochastic properties of white noise. While coherent signals add up linearly, noise only increases with the square root of the number of passages. Signal averaging therefore led to a further substantial improvement in S/N, thereby raising the detection limit to *ca* 1 mode/liter.[10] Figures 1.2(b) and (c) show the effect of proton noise decoupling and signal averaging on a swept spectrometer. By correcting for the different number of scans recorded for each trace one realizes that broadband proton decoupling augments S/N by a factor of *ca* 20. This is primarily a consequence of the collapse of the spin coupling multiplets into single lines.

Paralleling such developments, instrumental research was also aimed at

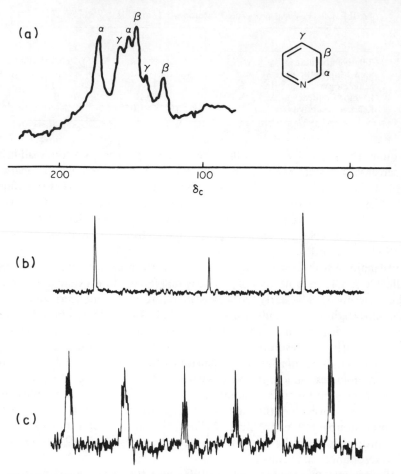

Fig. 1.2   (a)[13]C spectrum of pyridine recorded at 8.5 kG in the adiabatic rapid passage dispersion mode;[1] (b) 25.2 MHz absorption mode ^{13}C spectrum of pyridine single passage with simultaneous noise modulated proton decoupling; (c) as (b) but with no proton irradiation; 16 scans were accumulated in a time-averaging device.

increasing basic signal strength and attenuating noise sources. The former was mainly achieved by increasing the magnetic field and sample diameters, whereas the latter can be attributed to advances in probe and receiver design. The one decisive breakthrough, however, without which [13]C n.m.r. would never have attained its current analytical importance is the development of broadband excitation and Fourier transform techniques.[3,9] Most of the remainder of this chapter will be devoted to the description of this technique and its instrumental requirements.

1.2 RELAXATION AND NUCLEAR OVERHAUSER EFFECT

Even though relaxation and its various applications to chemistry will be treated in detail in Chapter 4, it is felt that a brief introduction of the subject is necessary at this stage because of the profound impact the phenomenon has on the appearance of ^{13}C n.m.r. spectra.

Phenomenologically, one differentiates between spin–lattice and spin–spin relaxation.[11] The former term implies a process involving interaction between spins and their surroundings, commonly referred to as the *lattice*. When transitions are induced by application of a rf field at resonance the equilibrium populations are perturbed in that spins are promoted from the lower to the higher energy levels, leading to a net increase in the spin system's energy. In order to return to thermal equilibrium the excess energy has to be dissipated to the lattice. Since the spins are isolated from the lattice the mutual interaction is weak and consequently the time constant for this process is long.

Another way of visualizing spin–lattice relaxation involves the *macroscopic spin magnetization* $\mathbf{M}$, which results from the polarization of the individual magnetic moments $\boldsymbol{\mu} = \hbar \mathbf{I}$ by the static magnetic field $\mathbf{H}_0$. In the absence of a magnetic field $\mathbf{M} = 0$ the spins are randomly oriented. The magnetization $\mathbf{M}$ can be calculated from the Boltzmann populations. In thermal equilibrium the magnetization $\mathbf{M}$ for a spin-$\frac{1}{2}$ nucleus is given by[12]

$$\mathbf{M} = \frac{3N\hbar^2 \gamma^2 \mathbf{H}_0}{4kT} \tag{1.5}$$

In Eqn. (1.5) N represents the number of spins per unit volume; the remaining quantities have previously been defined. In the absence of an external perturbation $\mathbf{M}$ points along the axis of $\mathbf{H}_0$.

The polarization of the individual magnetic moments does not take place instantaneously when the sample is placed in the field, and $\mathbf{M}$ evolves with a characteristic time constant equal to the *spin–lattice* or *longitudinal relaxation time*, T_1, as illustrated in Fig. 1.3(a). Likewise, when the spin system has been subjected to a strong saturating rf field, the return towards thermal equilibrium is governed by the same time constant T_1. ^{13}C spin–lattice relaxation times in isotropic liquid phase are found to be of the order of 10^{-2}–10^3 s. Even within the same sample they may differ by up to two orders of magnitude.

Spin–spin relaxation, on the other hand, results from interactions among the spins themselves. In the absence of an oscillating rf field the spins precess about the external field axis with their phases at random; there is consequently no net magnetization perpendicular to the field (transverse magnetization). Let us suppose now that a rf field is applied at resonance with the precessing spins. This rf field, as we will see later, has the effect of tipping the magnetization away from the equilibrium position, thereby generating transverse magnetization M_{xy}. We could also say that the individual magnetic moments are polarized along the axis

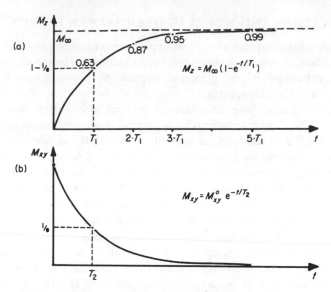

Fig. 1.3   (a) Recovery of the longitudinal magnetization M_z towards its thermal equilibrium value M_∞. (b) Decay of an initial transverse magnetization towards its equilibrium value zero.

of the rf field. M_{xy} does not persist indefinitely, however. It decays to zero because of a second relaxation process, transverse or spin–spin relaxation. Even in an ideally homogeneous magnetic field there is a spread of Larmor frequencies across the sample. As a consequence of this the spins dephase, and they do so with a time constant given by the *transverse* or *spin–spin relaxation time*, T_2. The exponential decay of an initial transverse magnetization M_{xy}^0 as a function of time is illustrated in Fig. 1.3(b). In a perfectly homogeneous field the width of a resonance line at half height is inversely proportional to T_2: $\Delta v_{1/2} = (\pi T_2)^{-1}$. However, the magnetic field is never perfectly homogeneous, i.e. there are always gradients across the sample, which means that the transverse magnetization decays faster than predicted by T_2. This is usually expressed in terms of the effective transverse relaxation time, T_2^*, which is always shorter than T_2. It is obvious that the full magnetization along the magnetic field axis can only be attained after M_{xy} has decayed to zero. From this it follows that $T_2 \leqslant T_1$. In ^{13}C n.m.r. it is found that in the majority of cases $T_1 = T_2$.

Another phenomenon which plays an important role in ^{13}C n.m.r. is the *nuclear Overhauser effect* (*NOE*).[13] This topic will be discussed more thoroughly in connection with the treatment of dipolar relaxation in Chapter 4.

It is known that relaxation of a ^{13}C nucleus in most organic molecules originates from dipolar interaction with the magnetic moments of neighbouring protons. Since the majority of ^{13}C experiments are carried out under simultaneous proton decoupling, protons are saturated, which leads to a redistribution of spin populations on the ^{13}C energy levels. Provided the ^{13}C–proton dipolar relaxation mechanism dominates, it can be shown that the population difference between ground and excited states of the nuclear Zeeman levels increases above that at thermal equilibrium. This is equivalent to increased magnetization and therefore enhanced carbon signal. Apart from very small molecules, proton-bearing carbons are known to be predominantly relaxed by this mechanism.[11] Under these conditions the NOE, as defined by the quotient M_z/M_0, where M_z represents the ^{13}C magnetization with and M_0 without irradiation of protons, is given by

$$NOE = M_z(^{13}C)/M_0(^{13}C) = 1 + \frac{\gamma(^1H)}{2\gamma(^{13}C)} \simeq 3 \qquad (1.6)$$

In Eqn. (1.6) $\gamma(^1H)$ and $\gamma(^{13}C)$ are the magnetogyric ratios of the proton and ^{13}C, respectively.

Apart from the increased S/N brought about by the collapse of spin multiplets when protons are decoupled, the signals of most carbons experience an additional threefold enhancement due to the NOE.

Quaternary carbons do not necessarily give the full NOE because mechanisms other than ^{13}C—^{1}H dipolar relaxation may be significant. However, in larger molecules even quaternary carbons are found to exhibit the full theoretical NOE.[14]

Under standard operating techniques ^{13}C n.m.r. therefore does not provide quantitative information because it is always possible that not all carbons within the molecule have equal NOE. A second cause for the non-quantitative behaviour lies in the differences of the spin–lattice relaxation times, causing enhanced saturation for more slowly relaxing carbons.

1.3 INSTRUMENTAL REQUIREMENTS

With the exception of low-field permanent-magnet-based proton n.m.r. spectrometers, which are almost extinct, all currently available commercial n.m.r. spectrometers are based on pulsed excitation and Fourier transform (FT) methods. The authors therefore feel that any comparison between continuous-wave and pulsed excitation has become redundant.

For an in-depth treatment of Fourier transform n.m.r. the interested reader is referred to the pertinent specialized texts.[15,16] However, a basic understanding of the experimental aspects of n.m.r. becomes increasingly important as many of the n.m.r. facilities in academic and industrial research are open access, i.e. recording of the spectra is often done by the originator of the sample rather than by a dedicated operator. Our experience with practical ^{13}C n.m.r. has shown us that erroneous data interpretation is often compounded by a lack of experimental understanding. Although today's spectrometers are easier to operate than their predecessors the operator is still not—and will never be—relieved from making certain parameter choices. Instrumental settings are related to fundamental molecular properties (such as, for example, molecular correlation times) and, of course, also to the sample state (solvent, sample quantity, temperature, etc.). They further depend on the objective of the experiment.

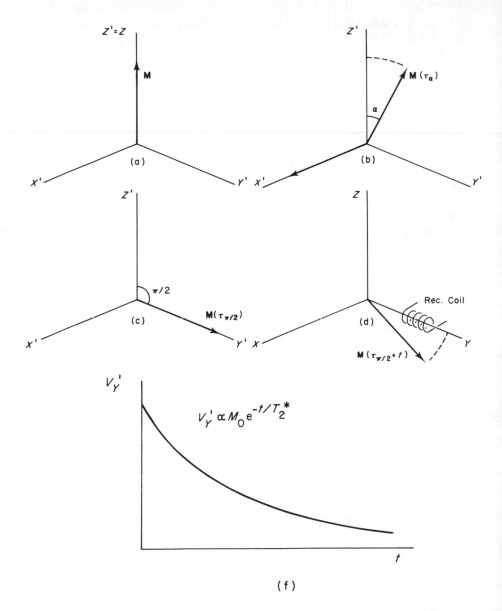

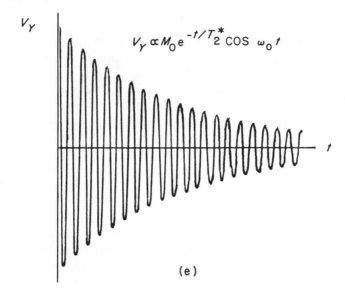

$$V_Y \propto M_0 e^{-t/T_2^*} \cos \omega_0 t$$

(e)

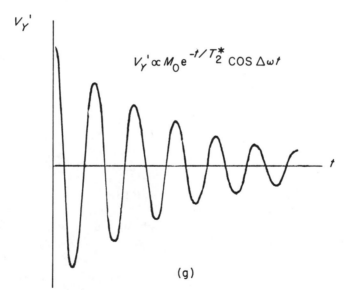

$$V_Y' \propto M_0 e^{-t/T_2^*} \cos \Delta\omega t$$

(g)

Fig. 1.4 Magnetization in the rotating frame, subjected to rf excitation. (a) Equilibrium; (b) effect of radio frequency field H_1, causing nutation of magnetization by an angle α; (c) magnetization at the end of a $\pi/2$ rf pulse; (d) precession of the magnetization around the z axis of the laboratory frame, following $\pi/2$ pulse, induces an a.c. voltage in the receiver coil situated along the y axis of the laboratory frame; (e) out-of-phase component of the free induction signal in the laboratory frame, following $\pi/2$ rf excitation; (f) out-of-phase component of the free induction signal after phase detection, assuming an rf pulse has been applied on resonance; (g) same as (f) for the case where the rf pulse was applied off resonance by an amount $\Delta\omega$ rad/s.

Behaviour of the magnetization subjected to rf excitation[15-17]

The resonance phenomenon is most conveniently described in terms of the previously introduced macroscopic spin magnetization $\mathbf{M}$. We have seen that in thermal equilibrium $\mathbf{M} = (0, 0, M_0)$, i.e. the magnetization has no transverse components. It is straightforward to show that rf absorption can only ensue following generation of transverse magnetization. For this purpose a rf field $\mathbf{H}_1$ has to be applied perpendicular to $\mathbf{H}_0$ in such a way that it rotates at a frequency near the Larmor frequency of the precessing magnetic moments. If the two are in synchronism, $\mathbf{M}$ experiences a continued *torque*. It is readily seen that the motion of $\mathbf{M}$ is quite complex, as it is composed of a precession around $\mathbf{H}_0$ and one around $\mathbf{H}_1$ (this latter is also denoted *nutation*). In order to simplify the description of this motion it is practical to convert the static co-ordinate system into a rotating one, such that the x and y axes rotate in phase with $\mathbf{H}_1$. In this so-called *rotating frame of reference* the $\mathbf{H}_1$ field obviously becomes static. In the special case, where the excitation frequency is exactly equal to the resonance frequency of the precessing spins, the only motion occurring is a nutation at angular frequency $\omega_N = 2\pi \cdot \nu_N = \gamma H_1$. The nutation persists as long as $\mathbf{H}_1$ is acting on the spin system. The resulting nutation or *pulse flip angle* α is given by

$$\alpha = \gamma H_1 \tau \tag{1.7}$$

where τ stands for the duration or width of the rf pulse. The behaviour of $\mathbf{M}$ before and during the pulse is shown in Figs 1.4(a) and (b). The voltage induced in the receiver coil which is oriented perpendicular to $\mathbf{H}_0$ is maximum when all initial magnetization has become transverse following the pulse. This is the case for $\alpha = \pi/2$ (Fig. 1.4(c)).

It is appropriate at this stage to return to the fixed reference frame in which the transverse magnetization precesses at the Larmor frequency ω_0 around $\mathbf{H}_0$, thus inducing a voltage

$$V_y \propto [M_0 \cos \omega_0 t] e^{-t/T_2^*} \tag{1.8}$$

in the receiver coil, as shown in Figs 1.4(d) and (e). In the rotating frame, however, the signal is detected as a simple exponential (Fig. 1.4(f)). In effect, it turns out that the rotating frame representation has some physical justification as well, since in the receiver coil the signal is mixed with the transmitter frequency so that actually a difference frequency is detected which, at resonance, is zero. This means that for a transmitter operating slightly off resonance the detected signal V_y' is a low-frequency damped cosine

$$V_y' \propto [M_0 \cos \Delta\omega t] e^{-t/T_2^*} \tag{1.9}$$

with $\Delta\omega$ being the difference between transmitter and nuclear resonance frequency (Fig. 1.4(g)).

Because of the spread of resonance frequencies caused by the chemical shift the transmitter frequency will always differ from the nuclear resonance frequencies.

In a real situation the detected signal therefore is a complex interference pattern resulting from the superposition of the individual precession signals. It is therefore sometimes termed 'interferogram' in analogy to infra-red interferometry. The generally adopted term, however, is *free induction decay (FID)*, implying the origin of the signal (induction) as well as its transient character.

The FID contains all relevant information, i.e. frequency, linewidth, and intensity for each of the spectral lines. However, instead of relating intensity to frequency, as is common in spectroscopy, the FID can be considered the time evolution of the magnetization.

Time and frequency domain[9,15,17]

Although the FID contains the desired spectral information the individual frequencies cannot directly be extracted save in very simple systems. Figure 1.5, for example, shows an expanded section of an experimental FID for a two-line system. The two frequencies present are clearly discernable, a high-frequency $(v_1 + v_2)/2$ and a low-frequency $(v_1 - v_2)/2$.*

It has been known for some time that FID and frequency-domain spectrum form a pair of Fourier transforms, i.e.

$$F(\omega) \leftarrow \mathscr{F} \rightarrow f(\tau) \tag{1.10}$$

or, explicitly for the sine and cosine transforms:

$$F(\omega) = \int_0^\infty f(t) \cos \omega t \, dt \tag{1.11a}$$

$$F(\omega) = \int_0^\infty f(t) \sin \omega t \, dt \tag{1.11b}$$

It is readily verified that the cosine transform for a decaying exponential $\exp(-t/T_2^*)$, as shown in Fig. 1.4(e), is equal to a Lorentzian

$$F(\omega) = T_2^*/[1 + \omega^2(T_2^*)^2] \tag{1.12}$$

of half-width $1/T_2^*$ (in rad s^{-1}) and a maximum at $\omega = 0$. The FID in Fig. 1.4(g) likewise transforms into a Lorentzian with the sole difference that its position is displaced by an amount $\Delta\omega$. The sine transform (Eqn. (1.11b)) yields the respective dispersion spectrum.

It is important to realize that only if the FID is sampled until it has completely decayed (theoretically until $t = \infty$), it transforms into a Lorentzian. There are clearly practical impediments to this, but it turns out that the line distortions caused by the finite sampling period are normally not too serious.

For the real case of a multicomponent spectrum, Fourier transformation is

$$*\cos 2\pi v_1 t + \cos 2\pi v_2 t = 2\cos\left(2\pi\frac{v_1 + v_2}{2}t\right)\cos\left(2\pi\frac{v_1 - v_2}{2}t\right)$$

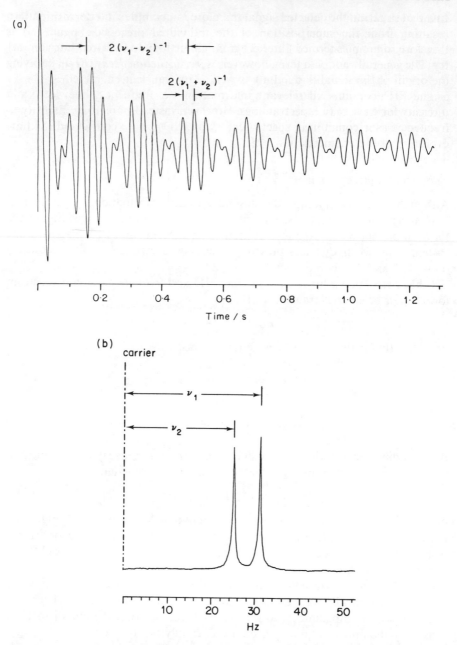

Fig. 1.5 (a) Experimental free induction signal for two spectral components resonating at frequencies v_1 and v_2 relative to the carrier frequency. Note the characteristic beat pattern from which frequencies v_1 and v_2 can be derived from a measurement of the periods of the respective oscillations. (b) Spectrum obtained after Fourier transformation of the free induction decay signal in (a).

carried out digitally by the minicomputer controlling also data acquisition and most of the spectrometer functions. The result of such a discrete Fourier transformation is shown in Fig. 1.5(b) for the FID in Fig. 1.5(a). The stringent requirement of a computer for this task becomes more apparent from the FID in Fig. 1.6(a), which fails to allow even a crude guess as to the frequencies encoded in it.

So far, we have only been concerned with time and frequency domain of the response to an rf signal. The rf field $H_1(t)$, of course, is also a periodic signal which can be represented in either domain. The duration of the pulse, i.e. the time during which H_1 acts on the spin system, turns out to be directly related to the bandwidth of excitation. In order to uniformly nutate the magnetic moments of the various chemically shifted nuclei it is important that the pulse be short. An equal amplitude excitation, theory shows, requires an infinitely short pulse.

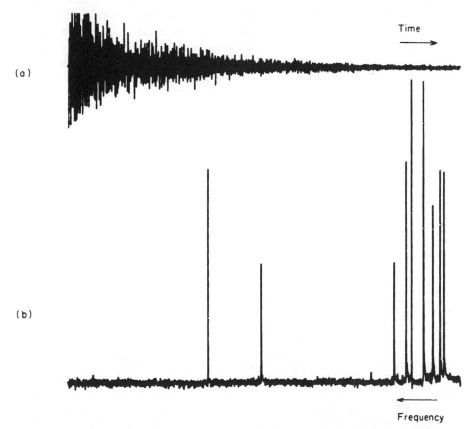

Fig. 1.6 (a) 25.2 MHz FID of 80% α-pinene, 200 transients, 0.8 s acquisition time, 8 k data points. (b) Frequency domain spectrum obtained by Fourier transforming the FID shown in (a).

Although this can in principle be achieved to almost any degree of approximation, there is a second requirement to be fulfilled, namely that the amplitude H_1 be sufficient to nutate the magnetization by an appreciable angle. One therefore realizes that the shorter the pulse, the larger the $\mathbf{H}_1$ amplitude has to be in order to flip the magnetization by a given angle of say 90°. A realistic pulse is therefore rectangular at best, and the resulting frequency-domain spectrum has a sin x/x distribution with $x = \pi v \tau$, in which v is the frequency relative to that of $\mathbf{H}_1$.[2,9]

In ^{13}C n.m.r. a single-pulse FID seldom gives sufficient S/N, so that in practice the sample is subjected to a train of equally spaced pulses, each generating a FID that is received and consecutively added up in the computer's memory. The periodic gating of the transmitter field $\mathbf{H}_1(t)$ introduces a modulation so that its frequency-domain spectrum consists of a series of equidistant sidebands, $1/T$ s apart, when T is the pulse repetition time. The envelope of the sideband amplitudes is determined by the width τ of the pulses. Figure 1.7 shows such a pulse train together with its Fourier inverse. From the latter it is seen that the amplitudes decrease with increasing distance from the $\mathbf{H}_1$ (rf carrier) frequency, v_0, passing through zero at $v_0 \pm 1/\tau$. Both the discrete nature of the excitation spectrum and also the finite width of the pulses impose limitations with respect to achievable resolution and quantitative nature of the spectra. We will see later that the choice of pulse width and interval are critical operator-adjustable parameters.

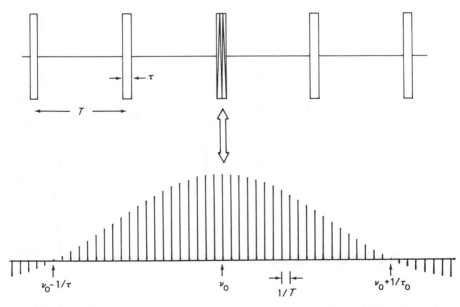

Fig. 1.7   Top: Repetitive RF pulses of width τ and interpulse spacing T and their Fourier transform (bottom).

Elements of a pulsed n.m.r. spectrometer

In this section we shall briefly elaborate on the various components making up an n.m.r. spectrometer together with a discussion of their critical requirements.

The magnet

The heart of every n.m.r. spectrometer unquestionably is the magnet. It largely determines spectral dispersion, resolution, line shape, and therefore to a large extent also sensitivity. Among the three types of magnets* employed in ^{13}C n.m.r. spectrometry, all current commercial systems use either resistive electromagnets with an iron core or superconducting cryomagnets. The former typically operate between 1.4 and 2.1 T with a practical upper limit of 2.3 T. High-resolution superconducting magnets span a range from 2.3 to 14.0 T, although the highest field at which ^{13}C has been observed so far is 11.7 T (125 MHz).

It had not been generally realized until a few years ago that, besides the gain in sensitivity, magnetic fields beyond 4.7 T are desirable even for nuclei such as carbon having a large intrinsic shielding range. Nevertheless, it is fair to say that the overwhelming majority of analytical problems in organic chemistry, amenable to ^{13}C n.m.r., can be solved at field strengths between 1.4 and 2.3 T, whereas fields below 4.7 T are usually inadequate to fully analyse the proton spectra of even small-size molecules.

Superconducting magnets are likely to supersede electromagnets even on low-field spectrometers (< 4.7 T); provided the cost for cryogens (liquid He and N_2) does not grow at a much faster rate than that of electricity. At the current cost of power, an electromagnet operating at 2.3 T is at least twice as expensive to run than a 4.7 T cryomagnet. This trend is further favoured by the improved helium economy of cryostats (up to six months' refill interval) and decreasing production cost of solenoids due to improvements in manufacturing technology.

Another criterion is the gap size (electromagnet) or bore diameter (super-conducting magnet). More than 90% of the carbon spectra are recorded with the objective of deriving structural information on organic molecules, a situation which is usually characterized by sample limitation. The ideal probe size is therefore 5 or 10 mm (see below), which can easily be accommodated by a 50 mm bore. Such magnets are considerably less expensive than the so-called wide-bore magnets (⩾ 90 mm). In fact, a 30 mm bore magnet would be adquate to accommodate a 5 mm dewared probe, although, surprisingly, no manufacturer currently offers such a magnet.

Power amplifier and probe

We have previously seen that the excitation bandwidth is critically dependent upon the duration of the pulse. We may now next inquire about the amplitude of

*Electro-, permanent, and superconducting magnets.

H_1 required to achieve a 90° nutation with a tolerable fall-off in rf power from the centre to the edges of the spectrum. We herewith assume that the rf carrier is placed at the centre of the spectrum, which is permissible on spectrometers operating in the quadrature detection mode.[18,19] (See below.)

Let us now return once again to the rotating frame. Suppose that the nucleus farthest from the carrier resonates at angular frequency ω_i while the carrier's frequency is ω_0. The effective field of the nucleus experiences is then given as

$$\mathbf{H}_{eff} = \frac{1}{\gamma}[(\omega_i - \omega_0)^2 + (\gamma \mathbf{H}_1)^2]^{1/2} \tag{1.13}$$

as illustrated in Fig. 1.8, which also shows that $\mathbf{H}_{eff}$ is tilted with respect to $\mathbf{H}_1$. Ideally, we would like $\mathbf{H}_{eff} \sim \mathbf{H}_1$, which requires that

$$\gamma H_1 / 2\pi \gg \Delta F / 2 \tag{1.14}$$

with ΔF expressing the width of the spectrum in Hertz. By combining Eqns (1.7) and (1.14) for $\alpha = \pi/2$ one obtains the following inequality for the 90° pulse width:*

$$\tau_{\pi/2} \ll 1/2\Delta F \tag{1.15}$$

At 4.7 T (50.3 MHz ^{13}C frequency) and 250 ppm spectral width $\tau_{\pi/2} \ll 40 \,\mu$s. In practice, it turns out that the less stringent condition $\tau_{\pi/2} \lesssim 1/2\Delta F$ is adequate for most experiments.

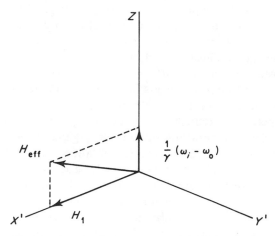

Fig. 1.8   Rotating frame representation of the effective field H_{eff}, resulting from application of a radio frequency field H_1, applied at frequency ω_i relative to the carrier frequency ω_0.

*In single-phase detection where the carrier is placed at one end of the spectrum, Eqn. (1.15) becomes $\tau_{\pi/2} \ll 1/(4\Delta F)$.

The next question regarding the power requirements of the rf amplifier is more difficult to answer. Basically, the H_1 amplitude depends not only on the output power P but also on the quality factor Q of the coil, the resonance frequency ω_0, and the coil volume V.[20]

$$H_1 \propto (PQ/\omega_0 V)^{1/2} \tag{1.16}$$

Equation (1.16) conveys that at high fields and large sample (coil) diameters generation of a short 90° pulse is more difficult to achieve.

The currently adopted probe designs are of the single-coil type with separate orthogonal decoupler coils, the latter usually being doubly tuned to ^{1}H and ^{2}H (decoupling and field/frequency lock). Whereas for electromagnet systems a solenoidal coil concentric with the sample axis is used, this design is difficult to realize on superconducting systems, where the sample axis is collinear with the direction of the (vertical) field. In order to maintain the convenience of vertical sample positioning and axial symmetry of the field, saddle-shaped Helmholtz-type coils are typically used on high-field spectrometers. Unfortunately, the latter design turns out to be less efficient from a sensitivity point of view by a factor of 2 to 3.[21,22] A hybrid design has been applied with some success for the measurement of very small samples. This consists essentially of a solenoidal coil which is first concentric with the external field axis and only after insertion of the sample is rotated into a 45° tilted position.[23]

Receiver

The rf signal induced by the magnetization in the receiver coil subsequently enters the receiver where it is amplified in two stages. The band of nuclear precession frequencies $v_0 + \Delta v$ following a first amplification in the preamplifier is then mixed with the local oscillator frequency $v_0 + v_{IF}$ and converted down to the lower frequency $v_{IF} + \Delta v$, where v_{IF} denotes the intermediate frequency (typically around 10 MHz). This so-called heterodyne detection scheme has the advantage that it is nucleus-independent as the outcoming frequency always is v_{IF}, irrespective of the Larmor frequency of the observed nucleus. Following this first conversion the signal is amplified and mixed in the second stage of the receiver. In the IF receiver mixing with v_{IF} ensues, converting the signal to a band of audiofrequencies Δv which are subsequently digitized and stored in the computer's memory. We can now see that positive and negative frequencies may be produced, depending on whether a particular nucleus resonates above or below the carrier frequency. A single-phase detector is not capable of discriminating f_i against $-f_i$, therefore treating both alike. This can result in artifacts denoted *carrier folding*. By contrast, the *quadrature receiver*, which simultaneously detects both in- and out-of-phase components of the FID, correctly reproduces signals on either side of the carrier. The rf carrier can therefore be placed at the centre of the spectrum in this detection mode. Most modern

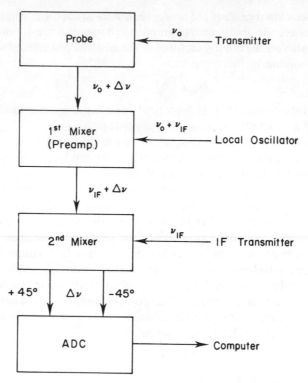

Fig. 1.9 Signal path between probe and ADC. The transmitter pulse excites a band of frequencies Δv centered around the carrier frequency v_0. In the first mixer stage of the preamp, mixing occurs with a local oscillator frequency, resulting in the intermediate frequency v_{if} carrying the spectral frequencies Δv. In the second mixer, the IF transmitter frequency v_{if} is subtracted, resulting in a band of audio frequencies Δv. In the quadrature receiver, the second mixer has two outputs which are 90° out-of-phase relative to one another. This signal is digitized in the ADC and processed in the computer.

spectrometers use quadrature defection while maintaining the option of operating in single-phase detection. A simplified block diagram of the signal path between probe and analogue-to-digital converter (ADC) is shown in Fig. 1.9.

Analogue-to-digital converter (ADC)

Since the process of Fourier transformation is carried out on a digital computer the FID has to be converted from its original analogue into digital form before it can be stored and further processed. This is accomplished by means of the ADC.

During digitization the amplitude of the FID is sampled at regular time intervals and discrete numbers are assigned to each sample. Each sampling value or data point is subsequently stored as a computer word at a particular memory location. We may now ask at what intervals this has to occur or, in other words, what the rate of digitization has to be. Let us, for the sake of simplicity, first consider single-phase detection, and suppose that the carrier has been placed at the low-frequency end of the spectrum. If one assumes a spectral width of ΔF Hz, ΔF is the maximum frequency to be digitized. Information theory states that a frequency can unambiguously be represented in digital form provided there are at least two points per cycle.[15] Sampling for a period A_t, the *data acquisition time*, therefore gives $2 \cdot \Delta F \cdot A_t$ digitization points, each requiring one memory location for storage, or a total of N data points:

$$2 \cdot \Delta F \cdot A_t = N \qquad (1.17)$$

In quadrature detection, where the carrier is placed at the centre of the spectrum, the largest frequency to be digitized is $\Delta F/2$ and the maximum conversion rate consequently is ΔF. Any frequency which falls outside the chosen spectral window, will, however, not just be ignored but converted down to a frequency within the chosen spectral range. A nucleus resonating at a frequency $\Delta F + \delta v$, for example, will give rise to a frequency $\Delta F - \delta v$. This folding phenomenon, also denoted *aliasing*,[15] is a consequence of violating the sampling theorem. It occurs irrespective of the detection scheme used.* Although most carbons resonate within a 200 ppm window, there are situations where a larger spectral width may be needed (cf. Chapter 2).

In addition, the better utilization of pulse power in quadrature detection enhances S/N by a factor of 2 because carrier folding of noise is avoided. However, it can be shown that in both detection modes the limiting resolution is $1/A_t$, the inverse of the acquisition time. The same applies to the memory requirements which, in either case, are dictated by Eqn. (1.17). In our previous example of a 250 ppm carbon spectrum, recorded at 4.7 T, a 1 Hz digital resolution therefore requires $2 \times 12\,500 \times 1 = 25\,000$ storage locations. Since the fast Fourier transform (FFT) algorithm[24] uses data points in increments of powers of 2, at least 32 762 computer words have to be available for data storage. As ΔF is proportional to the external field strength, one realizes that the demands put on data memory space increase at high field.

The last question relating to the ADC concerns the accuracy of digitization. In ^{13}C n.m.r. this is uncritical since the S/N ratio of the incoming signal is normally low. The precision of digitization is determined by the word length of the ADC. Typical digitizers have a maximum resolution of 12–13 bits. If the maximum admissible peak-to-peak voltage of the FID is V volts the quanta of digitization are $V/2^{12}$, for example, for a 12-bit ADC. A signal which falls below this value can therefore not be sampled.

*In single-phase detection one distinguishes carrier folding besides digitizer folding. The manifestation of the two effects is identical although their cause is different.

The ADC resolution can become a critical limitation when one of the incoming signals is very intense (for example, a solvent resonance), whereas the signal of analytical interest is very weak. In order to avoid signal truncation, i.e. clipping of the initial sections of the FID, the receiver gain may have to be turned down to the extent that the small signal, which rides on top of the FID, may no longer trigger the least significant bit.

The data system

Although the data system's prime task is to convert the accumulated time-domain data into the frequency domain, the computer in a modern spectrometer has been assigned a number of additional tasks. Among these are data acquisition, signal storage and averaging, and execution of pulse programs as well as a variety of processing functions prior and after the data have been transformed. These include digital filtering, phasing, integration, base line correction, etc.

An important characteristic of an n.m.r. data system is the computer word length, expressed in bits. This quantity is analogous to ADC resolution and it determines the precision with which a signal can be stored in the computer's memory and therefore the maximum S/N that can ultimately be attained. As long as all signals are of comparable magnitude this is of little concern, since there is no need for a S/N of more than, say, 200/1. However, if our spectrum covers a large dynamic range, computer word length may become critical. In ^{13}C n.m.r. the classical case is the study of branching or terminal groups in synthetic polymers. If we want to detect a single branching site in a polymer chain consisting of 5000 repeating units with a S/N of 20/1, the intense degenerate peak will have S/N = 100 000/1. Assuming that one bit is allocated to the sign, a 16-bit word can store numbers up to 32 000. Hence the previously described experiment would not be feasible. A simple trick to augment dynamic range consists of combining two computer words for the storage of a number (double-precision arithmetics). While this method greatly increases dynamic range, its drawback is a reduction of memory space by a factor of 2. The latter, however, is alleviated by the fact that the cost for computer memory has been rapidly decreasing over the past years, a trend which will certainly continue. Whether a data system uses a 16-, 20-, or 24-bit word (all three are offered by instrument manufacturers) is therefore no longer a criterion of significance.

The computer word length has, however, yet another implication. At given ADC resolution it determines the maximum number of scans that can be acquired. During data acquisition the FID of each successive scan is coherently added, so that a particular data point is always stored at the same assigned memory location. During signal averaging the data memory content following each scan is replaced by the sum of the incoming digitized signal and the current content of the computer word. One realizes that after a certain number of scans the computer word reaches its capacity. In order to avoid overflow, downscaling is necessary, i.e. division by two of the memory content. This situation should be avoided since in this manner dynamic range is lost. The maximum number of scans, $(n_s)_{max}$, possible before overflow occurs, is given by[24]

$$(n_s)_{max} = 2^{b-d} \qquad (1.18)$$

In Eqn. (1.18) b and d represent the word lengths of the computer and digitizer, respectively. Thus for a 12-bit ADC and a 24-bit word, $(n_s)_{max} = 4096$. The above is strictly true for coherent singals only. However, since real FIDs are composed of signal and noise and since the latter, as we have seen, increases only with $n_s^{1/2}$, the maximum number of scans is in fact larger than implied by Eqn. (1.18). Furthermore there is no need to utilize the full resolution to digitize such weak signals. If the ADC resolution is lowered to, say, 6 bits, the maximum number of scans becomes 262 144 according to Eqn. (1.18), or about twice this number for a single-pulse S/N of 1.[25]

The ultimately achievable S/N is not only instrument and data system dependent. In fact, the operator has considerable influence by massaging the data. Although in principle this can be done after Fourier transformation, the mathematics of digital filtering is simpler when carried out in the time domain. Enhancing the sensitivity at the expense of resolution—a trade-off which cannot be avoided—can be accomplished in various ways. In order to maintain the Lorentzian line shape, however, a so-called *matched filter* ought to be used.[9] The mathematical equivalent of the matched filter in the time domain is a decaying exponential. It is obvious that the tail of the FID contains less S/N than its beginning. Multiplying the FID by $\exp(-t/t')$, with a positive value of the time constant t', de-emphasizes the end of the signal, thus enhancing S/N. Since the effective time constant for signal decay is artificially shortened, the lines broaden, the extent of broadening being $(\pi t')^{-1}$ Hz. The optimum choice for t' can be shown to be $t' = T_2^*$. The effect of exponential multiplication for sensitivity enhancement in time and frequency domains is exemplified by Figs 1.10(a) and (b).

Often multiplet structure is not resolvable, be it for instrumental reasons or because the natural linewidths exceed the magnitude of the splittings. In such a situation it may be desirable to artificially enhance resolution. Instead of shortening the effective time constant for the FID, it is prolonged, which is achieved by choosing $t' < 0$. This may result in gross signal distortions following transformation, since the thus-modified FID may now have a finite value at the end of the acquisition period, therefore introducing a discontinuity. We have previously seen that in order for the lines to be Lorentzian the FID need be decayed completely. The truncation effect can be alleviated by multiplication with a continuously decaying function such as a Gaussian one. The complete weighting function then becomes

$$f_{\mathrm{RE}}(t) = \exp\left(-t/t' - ct^2\right) \tag{1.19}$$

with $t' < 0$ and $c > 0$.[26] From Fig. 1.10(c), illustrating the application of this function, one notices that the lines still show some distortion, their leading and trailing edges being accompanied by negative excursions.

The computer's next following task is the Fourier transformation. One prime criterion for this operation obviously is speed, which is related to the computer's

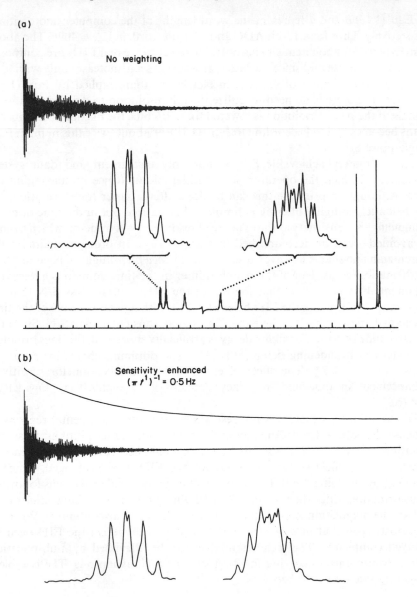

(a)   No weighting

(b)   Sensitivity - enhanced
$(\pi\, t')^{-1} = 0.5\,\mathrm{Hz}$

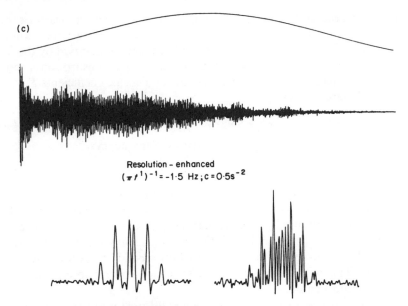

(c)

Resolution – enhanced
$(\pi / ^1)^{-1} = -1.5$ Hz; $c = 0.5 s^{-2}$

Fig. 1.10   Effect of digital filtering: (a) free induction decay and frequency spectrum, without application of a weighting function; (b) free induction decay following convolution with an exponential line broadening function causing a 0.5 Hz line broadening. Note improved signal-to-noise for the two multiplets expanded in (a), occurring at the expense of resolution; (c) resolution-enhanced FID following convolution with a Lorentz Gauss function (Eqn. 1.19), using the time constant as indicated. Note the greatly enhanced resolution in the multiplet structure, achieved at the expense of S/N.

cycle time, the computational algorithms used, and, of course, the size of the data block to be transformed. The integrals in Eqns (1.11a) and (1.11b) are computed as discrete sums. Before the FFT algorithm[24,26] was known, computation of the $N/2$ sine and cosine transform coefficients required N^2 multiplications, whereas the Cooley–Tukey algorithm reduces this number to $2N \log N/\log 2$. This corresponds to a time saving of about 300 for an 8 k FID. On a state-of-the-art n.m.r. data system a 16 k transform requires ca 10–20 s. Since most n.m.r. software systems are designed such as to time-share the computer between various tasks (for example, acquisition, processing, and plotting), the time requirements for processing are normally not very critical. However, when it comes to processing of very large data arrays as, for example, in two-dimensional experiments (cf. Chapter 3), processing time may become prohibitively long. The new generation of n.m.r. spectrometers is therefore fitted with hardwired array processors, which have just recently become affordable and which lend promise to lower transform times by about two orders of magnitude.[27]

Usually the last step before plotting is phase correction, so that all peaks of the spectrum appear in absorption. One might think that by correctly offsetting the receiver phase relative to that of the transmitter the cosine transform should afford the pure absorption spectrum. There are several reasons why this is not so. The principal cause for phase shifts is delayed acquisition, imposed by the finite recovery time of the receiver following the rf pulse. The pure absorption spectrum is obtained as a linear combination of the sine and cosine coefficients $F_s(k)$ and $F_c(k)$ for each data point k:

$$F_a(k) = \cos \phi_k F_c(k) + \sin \phi_k F_s(k) \tag{1.20}$$

The phase angle ϕ is usually a linear function of frequency:

$$\phi_k = \phi_a + a.v_k \tag{1.21}$$

Most spectrometers provide knob control of the frequency-independent (ϕ_a) and frequency-dependent (a) contribution to the phase angle. Turning one of the knobs makes the computer recalculate the phases across the whole spectrum and display the result on the screen. In practice, it is sufficient to zoom in on one line, correct the phase, and repeat this procedure on a second line at the other end of the spectrum. This procedure has become simple on current spectrometers and has now been automated.

On the previous generation of spectrometers the operator, unless prepared to expend considerable programming efforts (in machine code or assembler at best), had to content himself with a few preprogrammed experiments, such as standard FT, inversion-recovery T_1, gated decoupling, etc. By contrast, today's systems offer user-programmable pulse sequence generators which allow high-level programming of almost any pulse sequence the experimenter can conceive. Besides the transmitter, decoupler, and receiver gates, the phases of the three channels are also under program control. This is an important prerequisite for many of the novel experiments described in Chapter 3 of this book. One of the criteria of a pulse sequence generator relates to the accuracy and reproducibility of the gated circuits. The typical time resolution of such devices is 10–100 ns. For experiments in liquid isotropic phase 100 ns appears to be sufficient.

Peripheral devices

A discussion of the n.m.r. data system would be incomplete without mention of peripherals. The term refers to modules operating at the periphery of the data system. These can be grouped into two categories: (1) input/output (I/0) devices and (2) storage devices.

Most spectrometers use as an input device either an alphanumeric keyboard or a light-pen system. It is through these modules that the operator communicates with the data system and the spectrometer. In the case of the more common keyboard the operator enters the computer-controlled experimental parameters (for example, acqusition time, pulsewidth, spectral width, carrier offset, etc.), usually in the form of two- or multicharacter mnemonics. It also serves for command entry to activate execution of an experiment (for example, start of data acquisition, plotting or printing of data, etc.). In addition, most spectrometers provide a set of user-assignable knob routines to control, for example, display modification (zoom, left and right shift), cursor movement, phasing, baseline correction, etc.

Upon completion of an experiment the data are routed to one of the output devices, usually the display monitor first. This allows the operator to inspect the data and to put them into an appropriate format. In this area considerable progress has been made in recent years, and the new generation of spectrometers offer raster-scan scopes which allow simultaneous display of entire spectral arrays such as relaxation data and two-dimensional experiments, the latter in either multiple-trace or contour format.

Once the data have been suitably modified they are output on an XY recorder or digital plotter. The latter have the advantage that the spectrum can be plotted along with the relevant instrumental parameters and listings of chemical shifts, integrals, etc. The printer therefore becomes redundant as a separate digital output device. Digital plotters with continuous paper feed are a perquisite for output automation such as execution of plotting queues. However, even state-of-the-art digital plotters are not apt to cope economically with the vast amount of data generated in array experiments. They are therefore likely to be superseded by fast-memory plotters, which are capable of printing a dot matrix of data one row at a time, therefore enormously speeding up data output.

The task of the second category of peripherals is to store and retrieve programs and data. One distinguishes between serial access devices such as magnetic tapes and random-access devices whose most prominent representative is the high-speed magnetic disc. Random-access storage devices are superior because every location can be accessed equally fast. The space requirements for advanced n.m.r. programs are often beyond the computer's memory capacity, and software is therefore designed in such a way that only certain sections are co-resident at a given point in time. In order to effect disc–computer and computer–disc memory transfers without excessively slowing down execution of the program it is important that the transfer rate be sufficiently high. Another performance criterion is disc storage capacity, usually expressed in megabytes (Mbytes).* Considering the fact that a single two-dimensional spectrum requires 1–4 megawords (Mwords) of storage space (2–12 Mbytes), one realizes that disc storage capacities of 96 Mbytes, as they have become standard, are by no means luxurious. Due to the larger spectral widths demanded at high magnetic field, even more storage space may be needed on high-field (> 4.7 T) spectrometers.

---

*Depending on computer word length, 2–4 bytes are required to store the content of one word.

Choice of instrumental parameters

This section is intended to give the novice in ^{13}C n.m.r. some guidance on how to set up an experiment. Although the instrument manufacturers' operating manuals provide recipes which allow the user to reproduce the performance test spectra, they usually omit discussion of the general criteria for the choice of such critical parameters as pulsewidth and acquisition time. The beginner may therefore have to go through an unnecessarily long learning period until he or she can make efficient usage of the spectrometer.

The sequence of steps discussed below is meant for recording straightforward proton broadband-decoupled spectra. The requirements for more elaborate experiments are more severe, and for a detailed discussion the reader is referred to the more specialized literature.[16]

Sample preparation, choice of probe/sample diameter

In organic structural applications the amount of available sample is usually limited, typical quantities ranging from 5 to 50 mg. It has been shown recently for solenoidal coils that for a given sample amount the sensitivity can be optimized by constraining the sample in a minimum volume and reducing the size of the receiver coil accordingly.[28] Some attention should therefore be paid to finding a solvent system ensuring dissolution in a minimum volume of solvent, typically 300–500 μl, depending on the length of the receiver coil. If the sample dissolves in this amount of solvent the optimum probe diameter is 5 mm.* Filling the tube beyond the height of the receiver coil is not only unnecessary but detrimental to sensitivity, since molecules outside the active volume of the coil do not contribute to the signal. Increased dilution also increases T_1, therefore further impairing sensitivity. Too low a solution volume, on the other hand, should also be avoided, as this will spoil resolution and therefore indirectly also sensitivity. Recourse to a 10 mm or larger diameter probe should only be made in the case of insufficient solubility. Since $H_1 \propto V^{1/2}$ (cf. Eqn. (1.16)), smaller transmitter coils have the added advantage of providing shorter 90° pulses. A final benefit of the 5 mm probe is one of convenience. Most proton spectra are run in 5 mm sample tubes, and previously made up (possibly even sealed) samples may already exist. It should also be mentioned in this context that most current-design fixed-frequency ^{13}C probes are of the dual-observe type, permitting observation of the proton spectrum via the decoupler coil.

Field/frequency stabilization (Lock)

On all pulse FT spectrometers provisions are made to hold the field/frequency ratio constant. This is achieved by means of a second independent resonance

*A microprobe may become advantageous for solution volumes of < 100 μl.

circuit based on the deuterium solvent signal. We will not elaborate on this operation, which is described in the spectrometer's operating manual. It suffices to say that, in order to establish resonance, the field has to be shifted so as to match the chemical shift of the deuterated solvent, since the deuterium lock frequency is fixed. Once resonance is attained, the lock circuit is closed and rf amplitude and gain are adjusted to ensure a stable signal while avoiding saturation. On some spectrometers the essential first two steps have been automated.

Homogeneity adjustment

All n.m.r. magnets are fitted with *shims*, correction coils, designed to produce compensating field gradients for eliminating field imperfections. The non-uniformity of the external magnetic field at the sample has a variety of causes; for superconducting magnets, for example, the non-concentricity of the windings, variations in conducter thickness, etc. Inhomogeneities in diamagnetic suscepti-bility are further produced by the probe and, last but not least, by the sample itself. Fine adjustment of the homogeneity has therefore to be done individually for each sample. This is an operation which, depending on the quality of the basic field homogeneity, may require some operator skill. For this purpose, the lock signal is displayed on the oscilloscope and, while monitoring signal amplitude, the shim currents are systematically varied until a maximum is obtained. Homogeneity adjustment is again an operation which can be computerized to a large extent, and automatic shimming has become standard on most modern spectrometers.

This concludes the initial set-up operations common to all experiments. All subsequent operator interactions occur via I/0 devices.

Spectral width

The spectral width has to be chosen so as to encompass all possible lines in the spectrum. In the absence of allenes and saturated ketones and aldehydes all ^{13}C resonances in diamagnetic neutral molecules resonate within a 200 ppm window (cf. Chapter 2). If the nature of the functional groups is not known it is therefore prudent to work with a spectral width of 250 ppm.

Acquisition time

^{13}C n.m.r. lines in ^{1}H broadband decoupled spectra of organic molecules are typically 0.5–1 Hz wide. The determining factor dictating linewidth is usually the efficiency of decoupling, rather than spin–spin relaxation or field inhomogene-ities. This is not necessarily the case for quaternary carbons, which are narrower because the splittings which have to be removed by decoupling are smaller. From the foregoing it can be inferred that, for most practical purposes, an acquisition time of 1–2 s is sufficient. According to Eqn. (1.7), this requires 25–50 k of storage space for a 250 ppm spectrum at 4.7 T. Because the FFT algorithm demands that the number of time domain data points be a power of 2, the actual minimum data memory needed under these conditions is 32 or 64 k, respectively. Shorter acquisition times are usually not harmful except that they may evoke digitization errors, which manifest themselves in inaccurate peak heights and areas.

Pulse flip angle

In standard $(\alpha - A_t)$ experiments it is advisable to set the acquisition time A_t equal to the pulse repetition time t_p and to adjust the flip angle α such that attenuation of the more slowly relaxing quaternary carbons is not too severe. The optimum pulse flip angle α_{opt} at a given pulse repetition time t_p and relaxation time T_1 is given by[15]

$$\alpha_{opt} = \arccos e^{-t_p/T_1} \tag{1.22}$$

Since $T_1(C_{quat}) \gg T_1(CH_n)(n \geq 1)$, it is clear that a compromise has to be sought.

Let us assume average values of 1 and 10 s for proton-bearing and quaternary carbons, respectively, values which are characteristic of medium-sized molecules. At $t_p = 1$ s the sensitivity therefore is optimum at $\alpha = 68°$ for protonated carbons, whereas the quaternary carbons demand $\alpha = 25°$. Selection of a value closer to the latter assures recovery of all quaternary carbons with a tolerable loss of sensitivity for the protonated carbons (Fig. 1.11).

Transmitter and decoupler frequency setting

In quadrature phase detection the transmitter frequency carrier is set at the centre of the chosen spectral window. The absolute frequency is, of course, instrument-specific, and reference values are usually given in the respective operating manuals. Likewise, the decoupler frequency is centred with respect to the width of the proton spectrum (typically 4–5 ppm to high frequency from the tetramethylsilane reference signal). It should further be noted that these frequency settings are dependent upon the deuterium chemical shift in the solvent used for field/frequency lock. A change of solvent from chloroform to acetone, for example, requires both transmitter and decoupler frequency to be increased by 5.2 ppm (at 4.7 T this corresponds to 260 and 1040 Hz, respectively). Failure to offset the transmitter frequency is not harmful as long as the spectral width has been chosen well beyond the high-frequency end of the spectrum. The effect of incorrect decoupler frequency, however, is incomplete decoupling and therefore loss of signal-to-noise.

Decoupler power

In order to minimize heating effects it is desirable to carry out the experiment with a minimum of decoupler power. The power required to obtain a given amplitude

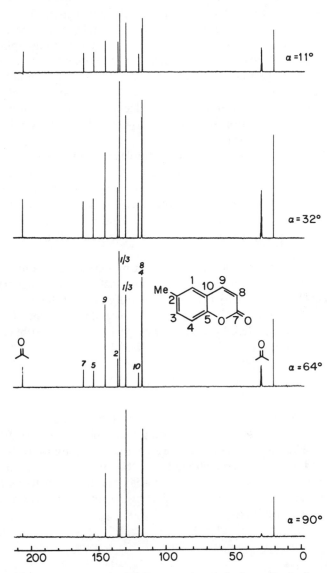

Fig. 1.11 Effect of pulse flip angle on the relative intensities in the proton noise-decoupled spectrum of a small organic molecule. The pulse repetition time was held constant ($t_p = 1$ s). Note the nearly complete suppression of quaternary carbon resonances at $\alpha = 90°$, whereas at smaller flip angles (for example, $\alpha = 32°$, $11°$) the relative intensities for proton-bearing and quaternary carbon signals become comparable. However, there is a loss in overall S/N for proton bearing carbons at the smaller flip angle, (*Note*: the spectra are displayed at constant noise level.)

of H_2 depends on the characteristics of the probe, the magnetic field, and, to some extent, on the solvent. The amplitude of H_2 is usually measured as $\gamma H_2/2\pi$ in Hertz, and can conveniently be determined from the residual splitting in the single-frequency off-resonance decoupled (SFORD) spectrum (cf. Chapter 3). Since the width of the proton spectrum increases linearly with field, high-field spectrometers require larger decoupler field amplitudes. At 4.7T $\gamma H_2/2\pi \sim 4\,kHz$ is about sufficient. The completeness of decoupling further depends on the efficiency of the modulation scheme, i.e. how evenly the power is distributed across the proton spectrum.

In solvents with an appreciable electric dipole moment or conducting solutions the rf field is attenuated.[28] At 4.7 T (200 MHz ^{1}H frequency) the rf field is reduced by 27% at the centre of a 10 mm tube and almost halved at the centre of a 20 mm tube filled with a 0.5 N aqueous NaCl solution. At high field the effect is even more dramatic. A change in solvent dielectric constant also affects probe tuning, therefore requiring adjustment of the decoupler (as well as observer) tuning capacitor.

Decreased decoupling efficiency which, at moderate fields and small sample diameter, usually does not cause too much grief, is always accompanied by heating effects.[28] One remedy is to keep the duty cycle of the decoupler low. In order not to lose the NOE a two-level decoupling scheme, in which the decoupling power is lowered during a delay period and turned back to normal during the acquisition period, has proved useful.

Receiver gain

In order to avoid artificats the receiver gain has to be set such that the FID is not truncated. This is best done by pulsing and displaying the FID on the oscilloscope. The gain is then adjusted so that the amplitude is maximized without the initial parts being clipped. In [13]C n.m.r. this is seldom a problem, since the single-pulse FID is usually very weak and it suffices to set the receiver noise to the requested level . Too low a receiver gain should also be avoided, as this will impair sensitivity.

Accumulation time

The total accumulation time determines the S/N ratio of the spectrum for a given sample and given spectrometer settings. A proton noise-decoupled ^{13}C spectrum, in order to be of practical use, should show a S/N of at least 30/1 (signal divided by peak-to-peak noise times 2.5) for proton-bearing carbons. This assures that the weaker quaternary carbon resonances can clearly be discriminated against the noise. Assuming $T_1(CH) \sim 1\,s$ (MW ~ 300 Daltons), 0.1 m mole of sample, run under optimum conditions on a state-of-the-art midfield spectrometer, should afford above S/N in less than 10 min accumulation time. For the inexperienced operator it may be advisable to do a trial run and calculate the time requirements by extrapolation. If, for example, after 5 min the signals are still not visible, a 1 h run will afford S/N of 3.5/1 at best and an overnight S/N of 13/1. In this way the waste of valuable spectrometer time can be avoided.

References

1 P. L. Lauterbur, *J. Chem. Phys.* **21**, 217 (1957); C. H. Holm, *J. Chem. Phys.* **21**, 707 (1957).
2 R. R. Ernst, *J. Chem. Phys.* **45**, 3845 (1966).
3 R. R. Ernst and W. A. Anderson, *Rev. Sci. Instr.* **37**, 93 (1966).
4 G. H. Fuller, *J. Phys. Chem. Ref. Data* (National Bureau of Standards, Washington, DC) **5**, 835 (1976).
5 R. K. Harris, in *NMR and the Periodic Table*, (Eds R. K. Harris and B. E. Mann), Academic Press, New York, 1978.
6 P. C. Lauterbur, *Ann. N. Y. Acad. Sci.* **70**, 841 (1958).
7 E. B. Baker, *J. Chem. Phys.* **37**, 911 (1962).
8 R. Freeman and W. A. Anderson, *J. Chem. Phys.* **39**, 806 (1963).
9 R. R. Ernst, *Advanc. Magn. Res.* **2**, 1 (1966).
10 (a) F. J. Weigert, M. Jautelat, and J. D. Roberts, *Proc. Nat. Acad. Sci. US* **60**, 1152 (1969); (b) L. F. Johnson and M. E. Tate, *Can. J. Chem.* **47**, 63 (1969).
11 See, for example, (a) J. R. Lyerla, Jr, and D. M. Grant, *Int. Rev. Science. Phys. Chem. Series* (Ed. C. A. McDowell), Vol. 4, Chap. 5, Medical and Technical Publishing Company, Chicago, 1972; (b) J. R. Lyerla, Jr, and G. C. Levy, in *Topics in* ^{13}C *NMR*, Wiley-Interscience, New York, 1974, Vol. 1.
12 See, for example, E. D. Becker, *High-Resolution NMR, Theory and Chemical Applications*, Academic Press, New York 1980.
13 K. F. Kuhlmann and D. M. Grant, *J. Am. Chem. Soc.* **90**, 7355 (1968).
14 A. Allerhand, D. Doddrell, and R. Komoroski, *J. Chem. Phys.* **55**, 189 (1971).
15 D. Shaw, *Fourier NMR*, Elsevier, New York, 1982.
16 M. L. Martin, J. J. Delpuech, and G. J. Martin, *Practical NMR Spectroscopy*, Heyden, London, 1980.
17 T. C. Farrar and E. D. Becker, *Pulse and Fourier Transform Nuclear Magnetic Resonance*, Academic Press, New York, 1971.
18 A. G. Redfield and R. K. Gupta, *Advanc. Magn. Res.* **5**, 82 (1974).
19 E. O. Stejskal and J. Schaefer, *J. Magn. Res.* **14**, 160 (1974).
20 C. Deverell, *Mol. Phys.* **18**, 319 (1970).
21 D. I. Hoult and R. E. Richards, *J. Magn. Res.* **24**. 71 (1976).
22 D. I. Hoult, in *Topics in Carbon-13 NMR*, Vol. 3, Wiley-Interscience, New York, 1979.
23 See, for example, G. C. Levy, R. L. Lichter, and G. L. Nelson, in "*Carbon-13 Nuclear Magnetic Resonance Spectroscopy*", second edition, p. 24, Wiley Interscience, New York, 1980.
24 C. Deverell, *op. cit.*, and references cited therein.
25 J. Cooper, in *Topics in Carbon-13 NMR*, (Ed. G. C. Levy), Vol. 2, Wiley-Interscience, New York, 1976.
26 J. C. Lindon and A. C. Ferridge, *Progr. NMR Spectr.*, Oxford University Press, 1980.
27 D. I. Hoult, C. -N. Chen, and H. Eden, *J. Magn. Res.* **45**, 181 (1981).
28 J. N. Shoolery, in *Topics in Carbon-13 NMR*, Vol. 3, Wiley-Interscience, New York, 1979.
29 See, for example, J. J. Led and S. B. Pedersen, *J. Magn. Res.* **32**, 1 (1978).

2

The Spectral Parameters

2.1 THE CHEMICAL SHIFT

The *chemical shift* or *nuclear shielding constant* is the spectral parameter which characterizes the chemical environment of a carbon nucleus. More accurately, it reflects the distribution of the bonding electrons and is therefore a sensitive probe for the configurational and conformational characteristics of carbon atoms in a molecule.

Although considerable progress has been made in the recent past towards a refined theory of nuclear shielding (see the pertinent review series[1,2]), the *ab initio* calculation of chemical shifts still constitutes a formidable problem, and satisfactory correlation with experimental data remains limited to simple molecules of little practical interest to the chemist. In this chapter an attempt has been made to discuss the physical properties in chemically relevant terms and to make the reader familiar with empirical substituent rules,[3] permitting a straightforward interpretation of experimental chemical shifts. The material given should enable the user to estimate ^{13}C chemical shifts of a carbon within a given spatial arrangement and to rationalize observed chemical shifts.

As this book focuses on the interpretative and experimental aspects of applied ^{13}C n.m.r. spectroscopy, tabular material has been limited to a strict minimum. Furthermore, the amount of chemical shift data reported during the past ten years has assumed such prodigious proportions that any attempt at comprehensive coverage would probably be doomed to fail. For chemical shift reference data the reader is therefore referred to the specialized literature.[4-13] An in-depth discussion of carbon chemical shifts by compound class and functional group in a large number of organic molecules is provided in the monographs by Levy *et al.*[4] and also by Breitmaier[5] and, covering the pre-1972 literature, by Stothers.[6] Specialized reviews have further appeared, treating polycyclic aromatic compounds,[7] boron compounds,[8] heterocyclic non-aromatic compounds,[9] alkaloids,[10] steroids,[11] saccharides,[12] and natural products in general.[13]

Reference scale and solvent shifts

In naturally occurring uncharged molecules the ^{13}C shielding range is found to be slightly over 200 ppm, i.e. *ca* twenty times that of hydrogen. If one takes into account the four times smaller magnetogyric ratio and assuming comparable linewidths for the two nuclear resonances, the chemical shift dispersion of carbon is about five times that of hydrogen. By denoting Δv and $\delta \Delta v$ the total shielding range and the width of a typical line, respectively, chemical shift dispersion may be defined as the ratio $\Delta v/\delta \Delta v$. Whereas early ^{13}C chemical shift data were referenced to carbon disulphide, which resonates at the high-frequency end of the shielding scale, tetramethylsilane (TMS) has for the past twenty years been the accepted shielding standard. With only few exceptions (for example, iodinated hydrocarbons), TMS is more highly shielded than the carbons of the analytical sample. Throughout this book we will adhere to the generally accepted convention that positive δ-values imply *deshielding* relative to the reference nucleus. Because of the insolubility of TMS in very polar solvents such as water, usually dioxane ($\delta = 67.4$ ppm) is used as a secondary standard. Furthermore, it is often convenient to reference to a fictitious TMS standard via the known chemical shift of the solvent, therefore obviating the need of adding TMS to the solution. A word of caution may be necessary, however, as some solvent shifts are concentration-dependent through solvent–solute interactions. This may be demonstrated with the deuterochloroform chemical shifts in Table 2.1, obtained at various mole fractions

$$x = n(CDCl_3) \bigg/ \sum_i n_i$$

in the ternary system deuterochloroform/benzene/TMS. Clearly, such effects are only of concern in highly concentrated samples. It should further be noted that the shielding in the deutero compound is not the same as in the protio analogue. Generally, the carbon atom in the heavier isotopomer is more highly shielded, the magnitude of the isotope shift depending on both structure and number of isotopic sites in α and β positions (cf. Chapter 3). Carbon chemical shifts for a number of common solvents are collated along with other physical properties in Table 2.2.

TABLE 2.1 Carbon chemical shift of deuterochloroform, measured relative to TMS at various values of the mole fraction $x(CDCl_3) = n(CDCl_3)/\sum n_i$ in a benzene/deuterochloroform/TMS mixture

$\delta_c(CDCl_3)$	77.00	77.04	77.09	77.15	77.22
$x(CDCl_3)$	0.94	0.79	0.64	0.48	0.32

TABLE 2.2   Carbon chemical shifts and physical data for common ^{13}C n.m.r. solvents[14]

Solvent	Density[a]	Melting point[b]	Boiling point[b]	δ_C[c]
Acetic acid-d_4	1.12	17	118	178.4; 20.0
Acetone-d_6	0.87	−94	57	206.0; 29.8
Acetonitrile-d_3	0.84	−45	82	118.2; 1.3
Benzene-d_6	0.95	5	80	128.0
Carbon disulphide	1.27	−112	46	192.8[d]
Carbon tetrachloride	1.58	−23	77	96.0[d]
Chloroform-d	1.50	−64	62	77.0
Cyclohexane-d_{12}	0.89	6	81	26.4
Deuterium oxide	1.11	3.8	101.4	[e]
1, 2-dichloroethane-d_4	1.25	−40	84	43.6
Diethyl-d_{10} ether	0.82	−116	35	65.3; 14.5
Diglyme-d_{14}	0.95	−68	162	70.7; 70.0
Dimethylformamide-d_7	1.04	−61	153	162.7; 35.2
Dimethyl-d_6 sulphoxide	1.18	18	189	39.5
p-Dioxane-d_8	1.13	12	101	66.5
Ethanol-d_6	0.91	< −130	79	56.8; 17.2
Glyme-d_{10}	0.86	−58	83	71.7; 57.8
Methanol-d_4	0.89	−98	65	49.0
Methylene chloride-d_2	1.35	−95	40	53.8
Nitrobenzene-d_5	1.25	6	211	148.6; 134.8; 129.5
Nitromethane-d_3	1.20	−29	101	62.8
Isopropanol-d_8	0.90	−86	83	62.9; 24.2
Pyridine-d_5	1.05	−42	116	149.9; 135.5; 123.5
Tetrahydrofuran-d_8	0.99	−109	66	67.4; 25.3
Toluene-d_8	0.94	−95	111	137.5; 128.9; 128.0
				125.2; 20.4
Trifluoroacetic acid-d	1.50	−15	72	164.2; 116.6

[a] g cm^{-3} at 20°C.
[b] °C for light isotopomer.
[c] Measured in solution containing deuterated solvent and 5% TMS.
[d] Reference 4.
[e] TMS insoluble.

Theoretical aspects of carbon shieldings

Although we will not elaborate on the theory of chemical shifts, it proves useful for a better qualitative understanding to briefly review the physics behind nuclear shielding. Basically, carbon chemical shifts are, with few exceptions, found to follow similar trends as their hydrogen counterparts. A complicating factor for the theoretical understanding of carbon chemical shifts is the circumstance that the relative values observable are made up of at least two, each other opposing contributions. According to Saika and Slichter,[15] the shielding constant can be divided into three additive contributions:

$$\sigma_N = \sigma_N^{dia} + \sigma_N^{para} + \sum_{B \neq N} \sigma_N^{NB} \qquad (2.1)$$

σ_N^{dia} represents the contribution from diamagnetic electron currents at the site of atom N. σ_N^{para} is a contribution, which produces a local field at the nucleus, supporting the external field. It arises from the unbalance of the valence electrons in the p states and, in contrast to the diamagnetic term, involves ground as well as excited electronic states. The σ_N^{NB} terms, finally, describe the magnetic fields at the site N, produced by field-induced electron currents at neighbouring atoms.

The local diamagnetic shielding term, σ_N^{dia}

The local diamagnetic term σ_N^{dia} in Eqn. (2.1) describes the isotropic circulation of local electrons around the nucleus. This circulatory motion is perpendicular to H_0 and, according to Lenz's rule, produces a secondary field $H_N^{loc}(dia)$ opposing the applied field. The greater the electron density at the nucleus, the greater the diamagnetic contribution and the further upfield the resonance occurs. For an isolated spherical atom, σ_N^{dia} represents the only contribution and is given by the Lamb formula:[16]

$$\sigma_N^{dia}(\text{free atom}) = \frac{e^2}{3mc^2} \sum_i \langle r_i^{-1} \rangle \tag{2.2}$$

In Eqn. (2.2) the summation of the ground-state mean inverse distances r_i is over all electrons i. An approximate calculation using this formula indicates that addition of an electron into a 2p orbital of a carbon atom would produce a shielding of 14 ppm. From this result it has been concluded[17] that σ_N^{dia}, though numerically large on an absolute scale, cannot be the dominant factor for ^{13}C shieldings.

It has been pointed out,[18-20] however, that in molecules Eqn. (2.2) must be summed over all electrons in the molecule. This can be achieved in a satisfactory approximation[19] without knowledge of the ground-state molecular wave-function, using the semi-empirical relationship

$$\sigma_N^{dia} = \sigma_N^{dia}(\text{free atom}) + \frac{e^2}{3mc^2} \sum_{K \neq N} Z_K \cdot (R_{NK})^{-1} \tag{2.3}$$

In Eqn. (2.3) Z_K is the atomic number of nucleus K, and R_{NK} represents the internuclear distance between K and N.

Calculations[19,21] based on Eqn. (2.3) indicate that indeed considerably larger variations in the σ_N^{dia} term have to be considered when contributions from adjacent atoms are taken into account. For simple hydrocarbons, replacement of a hydrogen atom by a carbon substituent has been estimated to increase the σ_N^{dia} term by 28 ppm.[19]

The neighbour anisotropy shielding term, σ_N^{NB}

The term σ_N^{NB} in Eqn. (2.1), also referred to as the neighbour anisotropy effect, describes the effect of local electron circulations at neighbour atoms B and of interatomic electron currents (i.e. currents due to the bonding electrons). McConnell[22] and Pople[23] derived a simple expression for σ_N^{NB}, which is based on the assumption that these electronic currents may be approximated by point magnetic dipoles for which atomic magnetic susceptibilities $\chi_B^i (i = x, y, z)$ can be defined. For the frequent case where the susceptibility tensor χ_B is axially symmetric, i.e. $\chi^x = \chi^y \neq \chi^z$, the anisotropy term can be expressed as

$$\sigma_N^{NB} = \tfrac{1}{3} R_{NB}^{-3} \Delta\chi_B (1 - 3\cos^2\theta_B) \qquad (2.4)$$

where R_{NB} is the distance between nucleus N and the dipole B; $\Delta\chi_B = \chi_B^z - \chi_B^{xy}$ is the anisotropy of the magnetic susceptibility of the dipole B; and θ_B is the angle between the symmetry axis of B and the NB distance vector.

It is evident from Eqn. (2.4) that the neighbour anisotropy term only depends on the nature of B and on geometry. It is independent of the nature of the observed nucleus N and is thus of the same order of magnitude in ^{1}H and ^{13}C n.m.r.

The local paramagnetic shielding term, σ_N^{para}

For all nuclei other than hydrogen, excited electronic states have to be taken into account for the discussion of local contributions to σ_N. Field-induced mixing of the electronic ground-state with these excited states causes a paramagnetic contribution to the shielding constant which may be visualized as being due to anisotropic, i.e. non-spherical, local electron circulations around nucleus N. Karplus and Pople[24] have derived the following approximate expression (average excitation energy approximation):

$$\sigma_N^{para} = -\frac{e^2\hbar^2}{2m^2c^2}(\Delta E)^{-1}\langle r^{-3}\rangle_{2pN}\left[Q_{NN} + \sum_{B \neq N} Q_{NB}\right] \qquad (2.5)$$

In this approach the difficulty of knowing all molecular wavefunctions and eigenvalues is circumvented by using a mean electronic excitation energy ΔE. $\langle r^{-3}\rangle_{2p}$ stands for the expectation value of the inverse cube of the distance between a 2p electron and the nucleus. The Q terms represent elements of the charge density and bond order matrix in the MO formalism of the unperturbed molecule. Q_{NN} assumes the value of 2 if the charge density in each 2p orbital is 1, an assumption which holds for carbon atoms in hydrocarbons. The term Q_{NB} represents multiple-bond contributions. It is non-zero only if there are both σ and π bonds between N and B.

The terms ΔE, $\langle r^{-3}\rangle_{2p}$, and Q_{NB} are mutually dependent, i.e. any change in the local electronic structure of nucleus N effects all the terms in Eqn. (2.5). However, for a qualitative estimate of the effects governing σ_N^{para} it has proved useful to discuss these terms separately.

σ_N^{para} depends primarily on the availability of low-lying excited electronic states of the carbon atom. A low value for ΔE causes deshielding. In the ultra-violet

TABLE 2.3   Effects of bond order ($\sum Q_{NB}$) and mean excitation energy (ΔE) upon ^{13}C chemical shifts[2,6,19]

Compound type	Hybridization	$\sum Q_{NB}$	$\Delta E(eV)$ (transition)	δ_C
Alkanes	sp^3	0	$\sim 10(\sigma \to \sigma^*)$	0–50
Alkynes	sp	0	$\sim 8(\pi \to \pi^*)$	50–80
Allenes (terminal)	sp^2	0.4	$\sim 8(\pi \to \pi^*)$	70–100
Alkenes, aromatics	sp^2	0.4–0.6	$\sim 8(\pi \to \pi^*)$	100–150
Allenes (central)	sp	0.8	$\sim 8(\pi \to \pi^*)$	200
Ketones	sp^2	0.4	$\sim 7(\pi \sim \pi^*)$	200

spectra of alkanes it is found that $\sigma \to \sigma^*$ transitions, which may be taken with some caution as a measure for ΔE, move to longer wavelengths (i.e. lower energies) with increasing substitution.[19] This behaviour reflects the delocalization or linear combination of σ electrons. In accordance with this, more heavily substituted carbon atoms are found to resonate at lower fields in the carbon n.m.r. spectrum. The energies of excited π^* states are generally much lower in unsaturated systems, and this explains in part why unsaturated carbons are less shielded than saturated carbons (see Table 2.3).

$\sum Q_{NB}$ gives a contribution in the same sense. It measures the relative importance of σ versus π bond character and is zero if the π bond order is zero. Table 2.3 illustrates the effects of $\sum Q_{NB}$ and ΔE upon the chemical shifts for a few examples. It is interesting to note that ΔE, which is the most difficult parameter to estimate, exerts a very strong influence on σ_N^{para}. A difference of 1 eV makes a shielding contribution of about 30 ppm.

The $\langle r^{-3} \rangle_{2p}$ term is of prominent importance for rationalizing ^{13}C chemical shifts. It depends primarily on the effective nuclear charge at nucleus N. An increase of electron density at a carbon atom tends to expand the 2p orbitals; consequently $\langle r^{-3} \rangle_{2p}$ is lowered. Thus a linear dependence of aromatic carbon chemical shifts on local π electron density has been found,[25] which indicates that addition of an electron to a 2p orbital causes an upfield shift of the corresponding carbon resonance of about 160 ppm (Fig. 2.1). From this finding it can be inferred in a somewhat simplified way that the $\langle r^{-3} \rangle_{2p}$ dependence of the σ_N^{para} term is the main cause for the large ^{13}C chemical shift range.

Changes in the $\langle r^{-3} \rangle_{2p}$ term also account for the influence of inductive effects, bond delocalization, electric field effects, and of steric factors on ^{13}C chemical shifts. It will be shown in the subsequent section how such effects can be understood in terms of 2p orbital dimension changes.

Nevertheless, a word of warning has to be given with respect to attempts to rationalize ^{13}C shieldings solely in terms of local charge densities and ignoring changes in electronic transition energies. The pitfalls of doing so are illustrated with the shieldings of the centre sp-hybridized carbon in allenes and their mono and diaza analogues:[26]

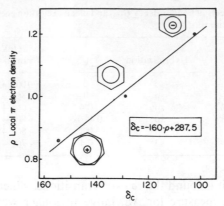

Fig. 2.1 Plot of local π electron densities of simple 6π electron aromatic systems against δ_C, the ^{13}C chemical shift.

Whilst one would intuitively predict the centre carbon (C-2) in **2** and **3** to be progressively deshielded due to electron withdrawal by the electronegative nitrogen, C-2 in reality is more shielded by *ca* 20 and 50 ppm in **2** and **3**, respectively, compared to **1**. Calculated values for $\langle r^{-3} \rangle_{2p}$ for C-2 in systems **1**–**3** do increase in the predicted order (1.455, 1.508, and 1.537 au^{-3}).[26] However, it turns out that the increases in the $\pi - \sigma^*$, $\sigma - \pi^*$, and $\sigma - \sigma^*$ more than counterbalance the charge density effects.

Transmission of shielding effects within molecules

The contributions to the shielding constant σ_N, which have been outlined in the preceding section, are governed by inter- and intramolecular electronic effects. It is common practice to rationalize these very complex electronic effects in terms of quantities which are familiar to the chemist.

In the subsequent discussion emphasis will be placed on intramolecular shielding effects. Medium-induced shifts are usually small in ^{13}C n.m.r. because carbon atoms are buried in the molecular framework (unlike protons, which are located at the periphery of the molecule and which are therefore more sensitive to solvent contact).

Most ^{13}C shieldings can be rationalized in terms of one or several of the following effects:

(1) The *hybridization* state of the observed nucleus;
(2) *Inductive effects* of substituents;
(3) *Van der Waals'* and *steric effects* between closely spaced nuclei;
(4) *Electric fields* originating from molecular dipoles or point charges;
(5) *Hyperconjugation*;
(6) *Mesomeric* interactions in π-electron systems (delocalization effects);
(7) Diamagnetic shielding due to heavy substituents ('*heavy-atom*' *effect*);
(8) *Neighbour anisotropy effects*;
(9) *Isotope effects*.

Hybridization

Inspection of the chemical shift correlation chart at the end of this book clearly reveals the state of hybridization of the observed carbon nucleus to be a dominant factor determining its chemical shift. This is not surprising, since ΔE and $\sum Q_{NB}$ (Eqn. (2.5)) combine to increase the paramagnetic term as illustrated in Table 2.3. The general trend, $\sigma(sp^3) > \sigma(sp) > \sigma(sp^2)$, parallels the order found in ^1H n.m.r. This coincidence is probably fortuitous, since for hydrogen the observed shift sequence seems to largely be a consequence of the neighbour anisotropy term.

Inductive effects

In many series of compounds satisfactory correlations of ^{13}C chemical shifts with substituent electronegativity can be found. A few typical data are listed in Table 2.4 (see also Table 2.7). The effect can be understood as being due to

TABLE 2.4 ^{13}C chemical shifts induced by replacement of a terminal hydrogen atom by electronegative substituents in *n*-alkanes

Substituent		Carbon		
Electronegativity	X ——	$\overset{\alpha}{CH_2}$——	$\overset{\beta}{CH_2}$——	$\overset{\gamma}{CH_2}$——CH_2——
2.1	H	0	0	0
2.5	CH_3	+9	+10	−2
2.5	SH	+11	+12	−6
3.0	NH_2	+29	+11	−5
3.0	Cl	+31	+11	−4
4.0	F	+68	+9	−4

inductive substituents removing electron density from the carbon 2p orbitals; this is associated with an increase in the $\langle r^{-3} \rangle_{2p}$ factor (Eqn. (2.5)) hence causing a deshielding effect. Theory predicts[27] this charge transfer to be propagated along the carbon backbone, producing alternating effects and falling off with the inverse third power of the distance:

$$X \overset{\delta^-}{\underset{}{—}} \overset{\delta^+}{C_\alpha} \overset{\delta\delta^-}{\underset{}{—}} C_\beta \overset{\delta\delta\delta^+}{\underset{}{—}} C_\gamma —\cdots$$

Comparison of the experimental data in Table 2.4 with this prediction clearly reveals that, apart from charge polarization, additional effects must be operative, because β and γ carbons show an opposite shielding behaviour. This implies that the substituent-induced shift of the α carbon, too, can only in part be due to an inductive effect. Other mechanisms which may be involved will be discussed in subsequent section.

Steric effects

^{13}C chemical shifts are sensitive to molecular geometry. Carbons separated by several bonds strongly influence each other if they are spatially close. There are, in principle, two types of short-range nonbonded interactions to explain these effects.

First, one may argue that attractive van der Waals' forces between closely spaced atoms lead to an expansion of orbitals. Such an effect would decrease the $\langle r^{-3} \rangle_{2p}$ term in Eqn. (2.5) and hence produce upfield shifts.[28] Although this model predicts shifts of the correct sign if not only diamagnetic circulations are considered, it is no longer used.

It has instead become a widely accepted practice to rationalize steric ^{13}C shift effects in terms of the repulsive forces between closely spaced atoms. They arise from locally crowded electron distributions within a molecule and they influence the shielding constants of the involved nuclei in two ways. Apart from causing distortions of the electron distribution, these interactions also affect shielding indirectly by producing small changes in molecular geometry. These two components of the steric effect are very difficult to separate and therefore any quantification remains doubtful.

As a qualitative tool, however, the interpretation of the steric effect as a consequence of induced polarization of C—H bonds[29,30] has proved fruitful. According to this concept a steric perturbation of a C—H bond leads to a drift of charge along the bond towards carbon, thus causing orbital expansion and hence increased shielding. In practice, the effect is found to be operative whenever two proton-bearing carbons are in a γ-*gauche* relative orientation. Although C—C bonds have polarizabilities similar to those of C—H bonds, no effect is expected for a carbon with perfect tetrahedral bond symmetry. C—C bond polarization will therefore not produce any change in net charge density on a quaternary

TABLE 2.5 Summary of observed ^{13}C chemical shift increments resulting from *gauche* interactions[31]

Interaction	Shift induced at marked carbon	Fractional shift per H...H interaction	Interaction	Shift induced at marked carbon	Fractional shift per H...H interaction
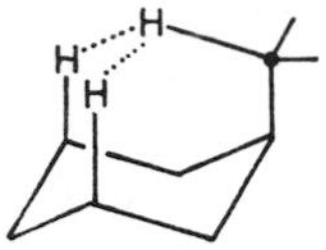	− 3.97	− 1.98	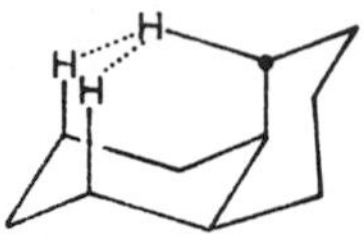	− 8.51	− 4.26
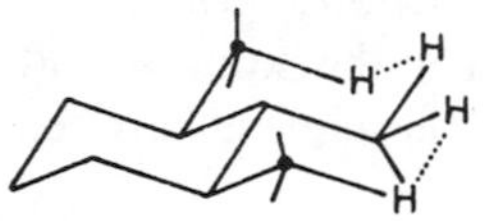	− 2.05	− 2.05	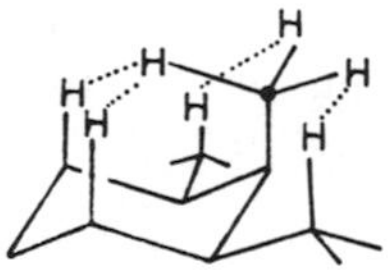	− 17.79	− 4.45
	− 2.58	− 2.58		− 5.10	− 5.10
	− 6.20	− 3.10		− 10.75	− 5.38
	− 11.44	− 3.81			
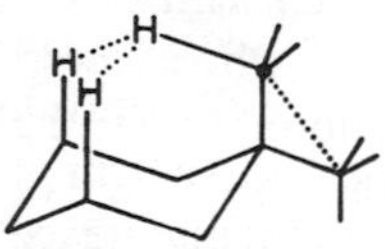	− 8.37	− 4.18	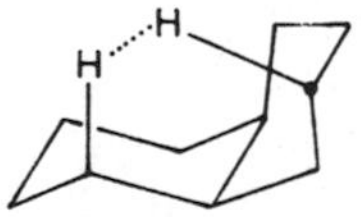	− 5.48	− 5.48
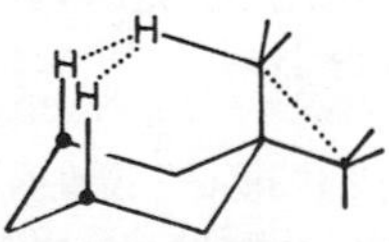	− 4.22	− 4.22	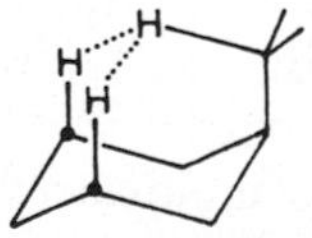	− 6.32	− 6.32

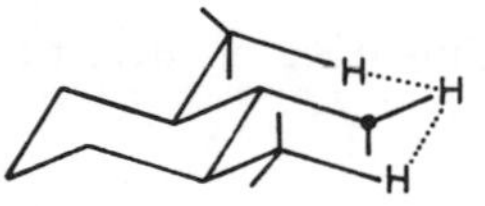

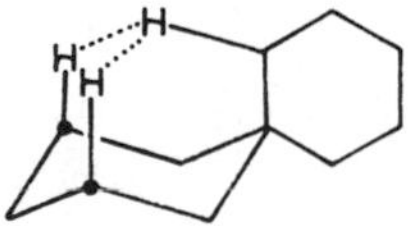

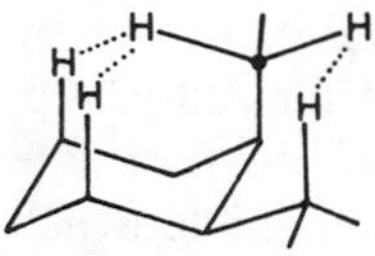

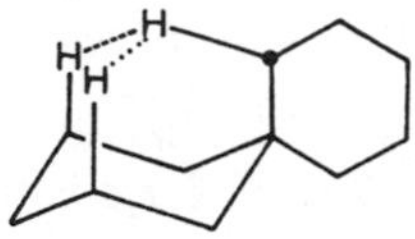

carbon. The flow of charge to a quaternary carbon along a C—C bond will be compensated by a loss of charge along the other C—C bonds. In Table 2.5 some examples of sterically induced upfield shifts in conformationally rigid systems are listed.[30] All interactions shown are of the *γ-gauche* type and exhibit similar geometrical features when molecular models are considered. The large scatter of incremental shifts per H...H interaction can be viewed as reflecting the tendency of the molecule to divert from idealized geometry to avoid steric stress.

The *γ-gauche* effect is also observed, though to a lesser extent, in conformationally mobile systems. In open-chain alkanes (for example, where the *gauche* rotamer population is about 30%) upfield shifts of the order of -2 ppm are observed at the γ carbon upon introduction of a CH_3 group. Other substituents induce similar upfield shifts of the order of -1 to -5 ppm (see Table 2.4).

It would be an oversimplification to associate steric interactions with upfield ^{13}C shifts generally. Grant's model[31] predicts a dependence of the induced shift $\Delta\delta_{st}$ not only on the proton–proton distance r_{HH} but also on the angle Θ between the H...H axis and the perturbed C—H bond:

$$\Delta\delta_{st} = C \cdot F_{HH}(r) \cdot \cos\Theta \qquad (2.6)$$

In Eqn. (2.6), $F_{HH}(r)$ represents the repulsive force between the interacting protons and C is a constant. Depending on Θ, $\Delta\delta_{st}$ can thus have positive or negative values.

Electric field effects

Another type of shielding effect has its origin in a charge polarization caused by electric fields. The latter may, for example, be produced by an ionized group, resulting from protonation or deprotonation. These effects can give rise to sizable shieldings and are believed to be the major cause of ionization shifts. Much smaller shifts of the same origin are induced by electric dipoles, produced, for example, by a polar bond C—X, where X represents a hetero atom. Electric field effects were first postulated for proton shieldings by Buckingham,[32] who showed that the shifts can be expressed as the sum of two terms, proportional to E^2, the square of the electric field, and a linear term, proportional to E_z, the component of the electric field pointing along the C—H bond axis. Whereas E^2 has a fourth power distance dependence and is therefore very rapidly attenuated, E_z falls off only with the square of distance. We will therefore confine ourselves to linear electric field shifts (LEFS) only.[33,34]

A complication arises for multivalent atoms such as carbon, as these possess several polarizable bonds. Assuming a uniform electric field (valid for atoms sufficiently remote from the centre of the field), the induced shift can be expressed as[34]

$$LEFS = \sum_{bonds} A_{bond} \cdot E_{bond} \qquad (2.7)$$

A_{bond} in essence is a function of bond polarizability. It further contains a factor which converts the induced charge density into shielding. The theory therefore relies on the concept of the paramagnetic shielding term, implying a proportionality between electronic charge density and magnetic shielding. Since the polarizabilities of double and triple bonds are about three times those of single bonds, unsaturated carbons are particularly sensitive to electric field shifts.[33] The relatively large shielding differences of the olefinic carbons in mono-unsaturated fatty acids could be rationalized in terms of LFFSs.[33]

The field close to an electric dipole or point charge is, of course, not uniform. In radial direction it decreases with r^2, hence there is a gradient $\partial E/\partial r \propto r^{-3}$. In practical terms this means that for a single bond, for example, placed in radial direction of the point charge field, the two atoms do not experience exactly the same field. The field gradient thus produces a charge separation and therefore a displacement of the resonance. Whereas a sp^3 carbon of perfectly tetrahedral symmetry does not experience a linear electric field shift because the four components, defined by Eqn. (2.7) cancel one another, it is still sensitive to field gradient shifts. The very different shieldings found for proton-bearing and quaternary carbons upon protonation of amines illustrate the effect of the two competing mechanisms. This may be exemplified with the β-carbon protonation shifts in ethylamine (**4a**) and neopentylamine (**4b**):[35]

H   CH$_3$
| |
H—C$_\beta$—CH$_2$NH$_2$ CH$_3$—C$_\beta$—CH$_2$NH$_2$
| |
H   CH$_3$

4a   **4b**

$\Delta\delta_{C_\beta} = -5.18$ ppm $\Delta\delta_{C_\beta} = -1.82$ ppm

The difference $-5.18 - (-1.82) = -3.36$ ppm can thus be ascribed to the LEFs alone. At short distances from the field centre the gradient effect dominates, leading to only small shielding at the carbon (for example, ethylamine: $\Delta\delta = -0.28$ ppm), or even deshielding for tertiary and quaternary carbons (for example, $+2.36$ ppm in isopropylamine and $+5.50$ ppm in t-butylamine).[34]

A particularly simple approach toward predicting LEFs, caused by a polar bond C—X in a neutral molecule, is based on the idea of dissecting the dipole into point charges of opposite sign and thus calculating the fields for each point charge and along each of the bonds extending from carbon C_i.[36] The resulting charge separation ΔQ_{C_iY} in the bond C_i—Y is given as

$$\Delta Q_{C_iY} = P_{C_iY}l_{C_iY}^{-1}qr^{-2}\cos\Theta \qquad (2.8)$$

In Eqn. (2.8) P_{CY} and l_{CY} represent the polarizability and length of the C_i—Y bond, respectively, q is the point charge on either the hetero atom X or its α carbon, and r is the distance between q and the centre of the C_i—Y bond. Finally, Θ is the angle between the electric field vector and the C_i—Y bond, as shown in the diagram below. By summing over all bonds extending from carbon C_i the

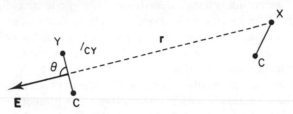

total charge separation ΔQ_{C_i} can be calculated. Electric field effects account for the long-range shielding contributions such as the δ substituent effects in substituted cyclohexanes.[36] This is illustrated with a plot of experimental carbon shieldings for C_δ, plotted versus electric field-induced charge densities, calculated according to Eqn. (2.7) (Fig. 2.2). Such long-range shielding effects have also been observed for the olefinic carbons in steroids.[37]

Hyperconjugative effects

Hyperconjugation has been invoked[38] in an attempt to explain characteristic upfield shifts caused by first-row heteroatoms located at the γ position and antiperiplanar to the ^{13}C nucleus (γ-anti effect). Thus in a system of the type

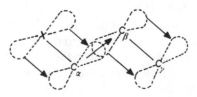

the C_γ resonance moves upfield by -2 to -6 ppm upon replacement of X = H or C by N, O or F. While 'classical' mechanisms such as σ bond polarization (inductive effect) or through-space electric field effects were discounted, the concept of a hyperconjugative interaction of the lone electron pairs of X with the C_α—C_β bond appears to explain the observed effects:

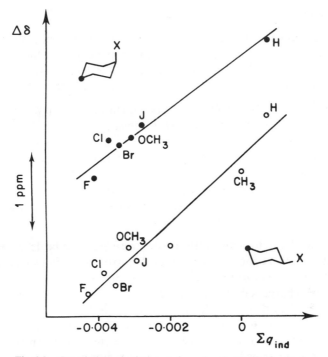

Fig. 2.2   Correlation of relative carbon chemical shifts for C_δ and
local charge densities calculated from linear electric fields induced by
polar substituents.[36]

Such an interaction results in enhanced electron density at C_γ. Partial overlap of
the lone electron pairs of X with the p_π orbitals of C_α is particularly favourable for
X = N, O, F because for these elements the C—X bond is short and the p_π orbital
radii are comparable to that of carbon.

   More recently,[39] the hyperconjugative mechanism has been challenged.
Whereas it was earlier assumed that second-row hetero atoms (P, S, Cl) would
have no effect (because of the larger atomic radii), more recent data[39] indicate
deshielding of the γ carbon in monosubstituted cyclohexanes (5), with decreasing
substituent electronegativity:

5

$$X = C(CH_3)_3, Si(CH_3)_3, Ge(CH_3)_3, Sn(CH_3)_3, Pb(CH_3)_3$$

Furthermore, it was found that the shielding capability of first-row hetero atoms
turns into deshielding, when intervening carbon atoms are heavily substituted.
This behaviour is more consistent with the earlier discussed electric field effect.

Mesomeric effects

In unsaturated frameworks, delocalization of charge across the π electron system produces large shielding changes due to the dependence of the paramagnetic shielding term on the effective nuclear charge. The local charge density of an aromatic carbon is governed by various mechanisms, such as[40]

(1) Overlap of π orbitals with filled or unoccupied substituent p (or d) orbitals of suitable size and symmetry (*mesomeric effect*);

(2) Redistribution of π electron density as a result of repulsive interactions between the π system and a filled orbital or a substituent (*orbital repulsion effect*);

(3) Polarization of the π system by a polar substituent (*π polarization effect*);[41]

(4) Polarization of the σ electron system inducing changes in the π electron density (inducto-electromeric or *π-inductive effect*).

In the valence-bond formalism the mesomeric effects can be described in terms of contributions from canonical structures. Electron-donating substituents such as NH_2, OH, F delocalize their lone electron pairs(s) into the π system, thus increasing the charge density at the β and δ carbons. Electron-attracting substituents, on the other hand (for example, NO_2, CN, COR) can delocalize the π electrons on the oxygen or nitrogen atom, thereby reducing the charge density at the β and δ carbons. These changes in charge density affect the paramagnetic shielding term as described by Eqn. (2.5). Substituents with lone pairs thus shield the β and δ carbons while electron-attracting substituents have a deshielding influence. The observed chemical shifts of the *para* carbons in substituted benzenes correlate well with the total charge density, calculated by CNDO/2.[42] The *ortho* carbons, however, show no correlation because other effects (steric interactions, neighbour group anisotropy) contribute significantly to the shielding.

Good correlation of aromatic carbon chemical shifts has also been obtained with Hammett σ constants. For *para* carbons the best correlation is found[43] for σ_p^+ coefficients (see Fig. 2.3). These coefficients, according to the formalism of

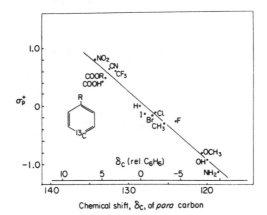

Fig. 2.3   Substituted benzenes. Plot of ^{13}C chemical shift of *para* carbon versus Hammett σ_p^+ parameter.[41,42]

Swain and Lupton,[44] represent a 66% contribution of resonance to the charge density (the remaining 34% being due to inductive effects). This finding may be taken as a measure for the importance of mesomeric effects in governing ^{13}C chemical shifts.

Meta carbon shifts similarly are found to correlate with σ_m^+.[42] The values for substituents with filled non-bonded p (or d) orbitals and those for substituents without lone electron pairs fall on two different lines, suggesting the presence of additional (other than mesomeric) effects. An interpretation in terms of repulsive (exchange) interactions between the occupied substituent orbitals and the π system has been given,[40] but since *meta* carbon chemical shifts vary over no more than 3 ppm (see Table 2.15) the effect is of little practical importance.

Comparison of the mechanisms governing aromatic shieldings reveals the mesomeric effects to be clearly dominant (apart from directly substituted and *ortho* carbons, where inductive and steric effects interfere). Canonical resonance structures as shown on page 46 therefore represent useful aids for estimating relative chemical shifts in such systems.

The chemical shift behaviour of other unsaturated systems can also be described in terms of mesomeric structures. A few examples are shown in Fig. 2.4.

Carbonyl carbons are particularly illustrative in this respect. A carbonyl carbon bears a partial positive charge. This is one reason why carbonyl carbons resonate at the low-field end of the chemical shift scale. The carbonyl carbon in a typical aliphatic ketone resonates near 210 ppm. In unsaturated carbonyl systems and those bearing substituents with lone electron pairs the positive charge can be delocalized from the carbonyl carbon. Consequently the carbonyl resonance moves upfield in these derivatives, as illustrated by the data in Table 2.6. This effect can only be operative if the C=O bond and the substituent π system are

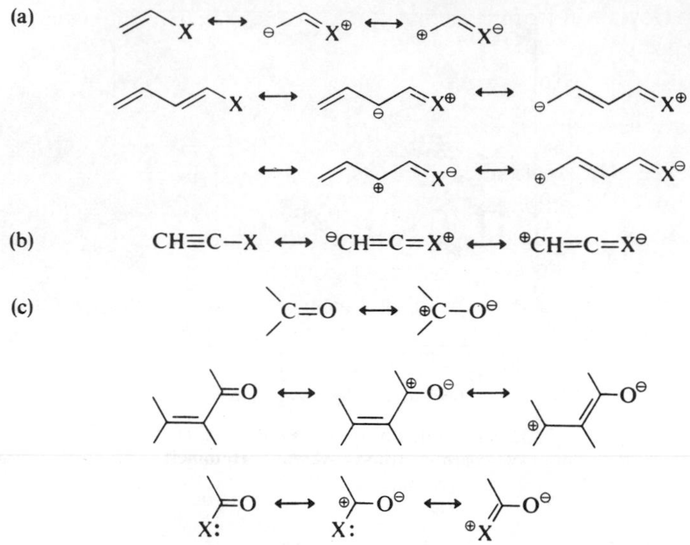

Fig. 2.4   Canonical structures of substituted unsaturated systems.

TABLE 2.6   Carbonyl chemical shifts in various compounds

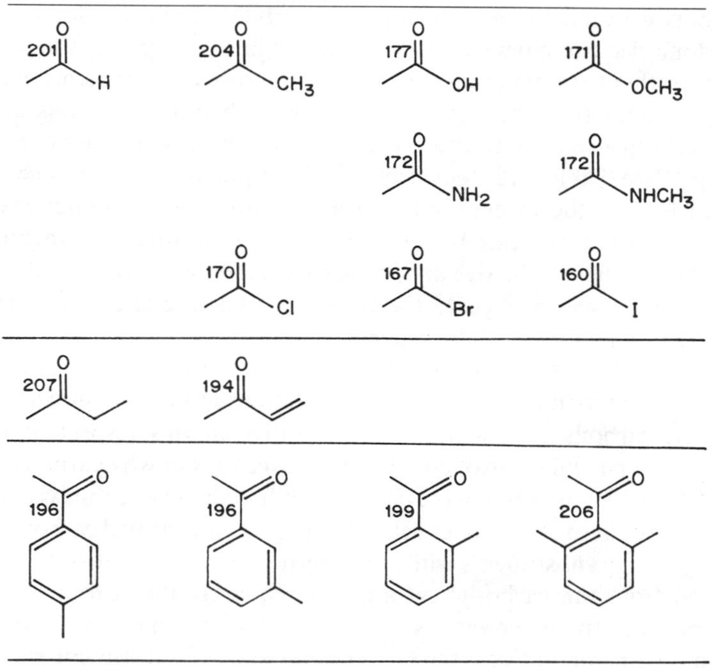

coplanar. Deviation from coplanarity thus causes deshielding ('steric inhibition of resonance').

Neighbour anisotropy effects

The influence of magnetically anisotropic functional groups on nuclear shielding has been discussed on page 35. Since it is generally small in magnitude it cannot be unequivocally distinguished from other shielding contributions. A ring current effect, which is a special case of neighbour group anisotropy, has been demonstrated for [12]-paracyclophane (**6**),[45] whose most shielded ring carbons appear 0.7 ppm upfield from those of cycloheptadecane.

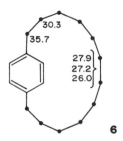

6

'Heavy atom' effect

While the effects of most electronegative substituents upon the substituted aliphatic carbon shieldings have been shown to be mainly inductive (see page 39), this does not apply for the heavier halogens. These substituents exhibit an increasing diamagnetic shielding with increasing atomic number. The observed ^{13}C chemical shifts of compounds $CH_{4-n}X_n(X = F, Cl, Br, I)$ are plotted in Fig. 2.5 against n. In terms of the formalism used in the section on chemical shift, these effects are due to a diamagnetic electron currents produced by the large number of electrons carried by heavy atoms. When the observed shifts are corrected for a diamagnetic contribution using Eqn. (2.3), the usual linear dependence on substituent electronegativity can be found.[19]

Figure 2.5 illustrates a more general problem encountered in ^{13}C n.m.r. chemical shift calculations. Chemical shifts are composed of a sum of two large terms of opposite sign which, to a large extent, cancel one another. Hence relatively small individual errors are magnified upon arithmetic addition.

Isotope effects

Replacement of an atom X by a heavier isotope usually results in a low-frequency shift for the carbons one and two bonds away from X. The heavy isotope lowers

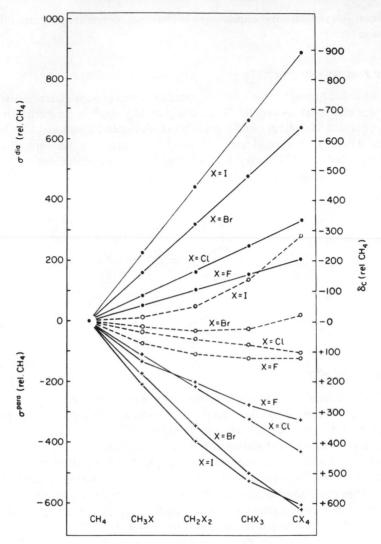

Fig. 2.5   Diamagnetic and paramagnetic shielding contributions in halomethanes $CH_{4-n}X_n$.[19] ●: σ^{dia}, calculated from Eqn. (2.3) and referenced to CH_4. +: Difference in total observed shift, δ, and σ^{dia}, representing essentially σ^{para}. ○: Observed chemical shift, δ, relative to CH_4.[19]

the potential energy of the molecule with a concomitant reduction of bond lengths, thus increasing the diamagnetic shielding term (Eqn. (2.3)). Deuterium and oxygen-18 induced isotope shifts, which are of considerable analytical potential, will be discussed in detail in Chapter 3.

Empirical relationships: substituent effects

The various factors contributing to the shielding of a ^{13}C nucleus generally combine in a complex way. Type and number of substituents as well as the nature of the functional group the carbon under consideration is part of the determining factors. A chemical shift correlation chart is given at the end of this book, allowing the reader to make rough estimates of the chemical shift in various functional groups.

A salient feature of ^{13}C chemical shifts is the empirical finding that substituent contributions are additive. This is particularly so for aliphatic carbons, were practice shows that carbon shieldings can be divided up into a number of additive contributions produced by substituents in α, β, γ, and δ positions. We shall discuss these effects in phenomenological terms and, for the time being, be less concerned with the question about the nature of the shifts.

Unsaturated carbon shifts are generally more straightforward to interpret because they primarily depend on the local π-electron density, which is more easily predictable. Other contributions are often of secondary importance. A brief survey of empirical additivity rules for the prediction of chemical shifts of carbons in various environments is given in the subsequent sections.

The α effect

The empirical data listed in Table 2.7 suggest that the chemical shift of a substituted aliphatic carbon (α carbon) mainly depends on substituent electronegativity. The heavy halogen derivatives deviate from this trend, as discussed on page 49. The α effect is 10–20% smaller for secondary with respect to primary substituents, again with the exception of Cl, Br, and I.

The substituent parameters given in Table 2.7 have been compiled from standard sources.[4-6,46] The data basis is not specified in order to avoid unnecessary limitations in the application of the parameters. As a consequence, the values given are reliable only within a margin of $\pm 5 \div 10\%$.

The β effect

With the exception of carbonyl, cyano, and nitro groups, the substituent effects at the β carbon are found to be fairly constant and independent of the nature of the substituent. Several attempts to interpret this effect have been made. A conclusive answer is lacking, however, since the effect is obviously a sum of competing contributions.

The values for β effects listed in Table 2.7 are generally applicable to situations without exceptional steric stress. Steric congestion reduces the β effect, as illustrated by the slightly smaller parameters for secondary substituents.

TABLE 2.7   Empirical substituent effects upon replacement of H by R in linear and branched alkanes

R	α		β		γ
	n	iso	*n*	iso	
CH_3	+ 9	+ 6	+ 10	+ 8	− 2
COOH	+ 21	+ 16	+ 3	+ 2	− 2
$COO^{\ominus}$	+ 25	+ 20	+ 5	+ 3	− 2
COOR	+ 20	+ 17	+ 3	+ 2	− 2
COCl	+ 33	+ 28		+ 2	
COR	+ 30	+ 24	+ 1	+ 1	− 2
CHO	+ 31		0		− 2
Phenyl	+ 23	+ 17	+ 9	+ 7	− 2
OH	+ 48	+ 41	+ 10	+ 8	− 5
OR	+ 58	+ 51	+ 8	+ 5	− 4
OCOR	+ 51	+ 45	+ 6	+ 5	− 3
NH_2	+ 29	+ 24	+ 11	+ 10	− 5
NH_3^+	+ 26	+ 24	+ 8	+ 6	− 5
NHR	+ 37	+ 31	+ 8	+ 6	− 4
NR_3	+ 42		+ 6		− 3
NO_2	+ 63	+ 57	+ 4	+ 4	
CN	+ 4	+ 1	+ 3	+ 3	− 3
SH	+ 11	+ 11	+ 12	+ 11	− 4
SR	+ 20		+ 7		− 3
F	+ 68	+ 63	+ 9	+ 6	− 4
Cl	+ 31	+ 32	+ 11	+ 10	− 4
Br	+ 20	+ 25	+ 11	+ 10	− 3
I	− 6	+ 4	+ 11	+ 12	− 1

The γ-gauche effect

Carbons three bonds away from a substituent exhibit upfield shifts due to sterically induced polarization of C—H bonds, as discussed on page 40. The values given in Table 2.7 represent the motionally averaged parameters for open-chain alkanes. In rigid, cyclic systems the effect is small for a *trans* configuration and large when the substituent and the γ carbon are in a *gauche* orientation:

trans   gauche

For R = N, O, F other mechanisms may be operative in a *trans* geometry, as discussed earlier.

Typical values for rigid hydrocarbons are listed in Table 2.5. If the substituent is a heteroatom (O, N, F, S, Cl) instead of CH_3 or CH_2, the effect is slightly (1 to 3 ppm) larger,[38] as illustrated by the data listed in Table 2.8. The same trend can be observed for the motionally averaged γ effects listed in Table 2.7.

The γ-trans effect

For first-row heteroatom substituents (R = N, O, F) upfield shifts also occur for *trans*-oriented γ carbons. The effect had been speculated to be due to a

TABLE 2.8 Typical γ-*gauche* effects in rigid systems[38]

Compound	R	Observed shift	Increment
C-6	H	30.1	
	CH_3	22.4	− 7.7
	OH	20.4	− 9.6
	OCH_3	20.5	− 9.5
	NH_2	20.6	− 9.4
C-3.5	H	28.0	
	CH_3		$(− 5.4)^a$
	OH	21.0	− 7.0
	F	21.5	− 6.5
	Cl	21.2	− 6.8
	SCH_3	21.9	− 6.1

^{a}Estimated value.

TABLE 2.9 Typical γ-*trans* effects in rigid systems[38]

Compound	R	Observed shift	Increment
C-6	H	30.1	
	CH_3	29.0	− 1.1
	NH_2	27.0	− 3.1
	OH	24.9	− 5.2
	F	22.6	− 7.5
C-3.5	H	28.0	
	CH_3		$(\sim 0)^a$
	NH_2	26.4	− 1.6
	OH	25.7	− 2.3
	F	25.3	− 2.7
	Cl	27.8	− 0.2
	S	27.6	− 0.4

^{a}Estimated value.

hyperconjugative electron release, although this view has more recently been questioned.[39] It is apparent that CH_3, Cl, and S have relatively little effect, while the values for N, O, and F increase monotonically. Throughout, however, the γ-*trans* effect is smaller than the γ-*gauche* effect.

The δ effects

Substituent effects across four bonds are usually small in aliphatic systems (< 1 ppm), except for the case of highly electronegative substituents, where electric field effects can become substantial (cf. Fig. 2.2). In the energetically favoured conformations (a)–(c) non-bonded interactions due to steric charge polarization should be negligible. For the *syn* axial conformation (d), on the other hand, where the non-bonded internuclear distance is even smaller than in γ-*gauche* conform-

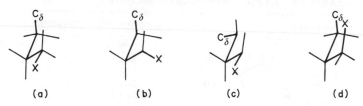

(a)   (b)   (c)   (d)

ations, a strong steric effect is expected. In Table 2.10 a few observed δ-synaxial effects, induced by hydroxyl and methyl groups, are contrasted with other spatial arrangements. In open-chain compounds the synaxial rotamer population is very low due to steric hindrance; therefore usually no δ effects are detected. 2, 2, 4, 4-Tetramethylpentane is an exception; due to heavy substitution C_1 persists for two-thirds of the time in a *gauche*($+$)-*gauche*($-$) interaction with one of the δ methyls. From the observed downfield shift of 3.1 ppm, a δ effect of 4.65 ppm for a rigid synaxial conformation can be calculated (cf. Table 2.10). The steric δ effect is thus of opposite sign to what would be predicted by analogy to the well-known γ effect. This suggests that there must be a competing mechanism which becomes dominant at short distances. It has been shown that second-order electric field effects due to fluctuating magnetic dipoles are large enough to explain the observed downfield shifts.[51]

Empirical additivity rules

Prediction of chemical shifts based on empirical additive substituent parameters has proved to be quite successful in ^{13}C n.m.r. The approach relies on the selection of suitable model compounds and on the availability of a minimal number of substituted compounds for regressional analysis. Thus it was demonstrated in the pioneering work of Grant and Paul[52] that the chemical shift of a paraffinic carbon k in a linear or branched hydrocarbon can be expressed as

$$\delta_C(k) = C + \sum_i n_{ik} A_i \tag{2.9}$$

TABLE 2.10 Typical δ effects in rigid and conformationally mobile systems

No substituent	δ Substituted _gauche–gauche_ or _gauche–trans_	Synaxial	Ref.

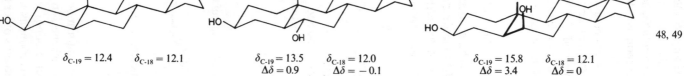

			48, 49

$\delta_{\text{C-19}} = 12.3 \qquad \delta_{\text{C-18}} = 12.6$

$\delta_{\text{C-19}} = 12.9 \qquad \delta_{\text{C-18}} = 13.4$
$\Delta\delta = 0.6 \qquad \Delta\delta = 0.8$

$\delta_{\text{C-19}} = 15.5 \qquad \delta_{\text{C-18}} = 15.1$
$\Delta\delta = 3.2 \qquad \Delta\delta = 2.5$

			48, 49

$\delta_{\text{C-19}} = 12.4 \qquad \delta_{\text{C-18}} = 12.1$

$\delta_{\text{C-19}} = 13.5 \qquad \delta_{\text{C-18}} = 12.0$
$\Delta\delta = 0.9 \qquad \Delta\delta = -0.1$

$\delta_{\text{C-19}} = 15.8 \qquad \delta_{\text{C-18}} = 12.1$
$\Delta\delta = 3.4 \qquad \Delta\delta = 0$

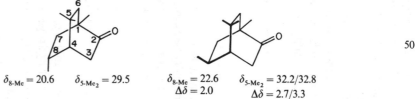

$\delta_{\text{8-Me}} = 20.6 \qquad \delta_{\text{5-Me}_2} = 29.5$

$\delta_{\text{8-Me}} = 22.6 \qquad \delta_{\text{5-Me}_2} = 32.2/32.8$
$\Delta\delta = 2.0 \qquad \Delta\delta = 2.7/3.3$

50

$\delta_{\text{C-1}} = 28.7$

$\delta_{\text{C-1}} = 31.8$
$\Delta\delta = 3.1$

51

where C is a constant corresponding to the chemical shift of the reference compound (methane in this particular case), A_i is the shift increment associated with the ith substituent, and n_{ik} represents the number of substituents in position i relative to carbon k. A_1 (for an α carbon) is $+9.1$ ppm, A_2 (β effect) $= +9.4$ ppm, and A_3 (γ effect) $= -2.5$ ppm.

For more complex compounds where mutual interactions of substituents become important, additional correction terms have to be used, and this sets a limit to the applicability of the method.

Some useful additivity relationships will be discussed subsequently, and a few examples of their application will be given in Chapters 3 and 5.

Open-chain alkanes (the 'Lindeman–Adams rule')[53]

Scope: linear and branched open-chain alkanes. The chemical shifts of substituted alkanes may be estimated (to a lesser degree of confidence) by calculating the shift of the parent alkane and by adding the appropriate substituent parameters from Table 2.7. The Standard error for a set of 59 paraffins is 0.8 ppm.

$$\overset{\displaystyle k}{\underset{\displaystyle\downarrow}{}}\qquad \alpha\quad\ \beta\quad \gamma\quad \delta$$
$$----\overset{}{CH}_n-CH_m-C-C-C-$$

n number of hydrogens at carbon k
m number of hydrogens at the α carbon
N_m^∞ number of CH_m groups at α position ($m = 0, 1, 2$; α-CH_3 groups are ignored)
N^γ number of γ carbons
N^δ number of δ carbons

The chemical shift $\delta_C(k)$ of carbon k is calculated according to Eqn. (2.10).:

$$\delta_C(k) = A_n + \sum_{m=0}^{2} N_m^\alpha \alpha_{nm} + N^\gamma \gamma_n + N^\delta \delta_n \tag{2.10}$$

The values of the empirical parameters A_n, α_{nm}, γ_n, and δ_n are listed in Table 2.11.

Alkenes[54]

Scope: linear and branched open-chain mono-unsaturated hydrocarbons (alkenes), also applicable to alicyclic alkenes. In small rings, carbons are counted twice (from both ends of the double bond) where appropriate. When used for polyenes, count additional alkene carbons as alkane carbons. Table 2.12 also lists a few parameter values for substituents other than carbon; they should be used

TABLE 2.11 Empirical parameters for the calculation of
alkane chemical shifts using Eqn. (2.10)[53]

n	A_n	m	α_{nm}	γ_n	δ_n
3	6.80	2	9.56	-2.99	0.49
		1	17.83		
		0	25.48		
2	15.34	2	9.75	-2.69	0.25
		1	16.70		
		0	21.43		
1	23.46	2	6.60	-2.07	~ 0
		1	11.14		
		0	14.70		
0	27.77	2	2.26	$+0.86$	~ 0
		1	3.96		
		0	7.35		

TABLE 2.12 Empirical parameters for the calculation of alkene chemical shifts using Eqn. (2.11)[6,54]

	Parameters $A_{ki'}(\mathbf{R}_{i'})$				$A_{ki}(\mathbf{R}_i)$	
	$-C$	C	$C-C=C-C$		C	$C-$
$\mathbf{R}_i$	γ'	β'	α'	$\uparrow$ α	β	γ
C	$+1.5$	-1.8	-7.9	$+10.6$	$+7.2$	-1.5
OH		-1	—	—	$+6$	
OR		-1	-39	$+29$	$+2$	
OAc			-27	$+18$		
COCH$_3$			$+6$	$+15$		
CHO			$+13$	$+13$		
COOH			$+9$	$+4$		
COOR			$+7$	$+6$		
CN			$+15$	-16		
Cl		$+2$	-6	$+3$	-1	
Br		$+2$	-1	-8	~ 0	
I			$+7$	-38		
C$_6$H$_5$			-11	$+12$		

Correction terms	
α, α' (*trans*)	0
α, α' (*cis*)	-1.1
α, α	-4.8
α', α'	$+2.5$
β, β	$+2.3$
All other interactions	~ 0

with caution, however, since they are not based on regressional analysis.

$$\overset{\gamma'}{\mathrm{C}}-\overset{\beta'}{\mathrm{C}}-\overset{\alpha'}{\mathrm{C}}-\overset{k'}{\mathrm{C}}=\overset{k}{\mathrm{C}}-\overset{\alpha}{\mathrm{C}}-\overset{\beta}{\mathrm{C}}-\overset{\gamma}{\mathrm{C}}-$$

$$\delta_{\mathrm{C}}(k) = 123.3 + \sum_{i} A_{ki}(\mathrm{R}_i) + \sum_{i'} A_{ki'}(\mathrm{R}_{i'}) + \text{correction terms} \qquad (2.11)$$

In Eqn. (2.11), $A_{ki}(\mathrm{R}_i)$ represents the incremental shift of carbon k upon introduction of substituent R_i at position i, i' referring to substituent positions across the double bond. Values for A_{ki} and $A_{ki'}$ and appropriate correction terms are listed in Table 2.12.

Cycloalkanes

Several sets of shift parameters for cycloalkanes have been published.[31,55-7] Unfortunately, their scope of validity is usually limited to a specific ring skeleton. Because of small changes in molecular geometry (changes of bond lengths and

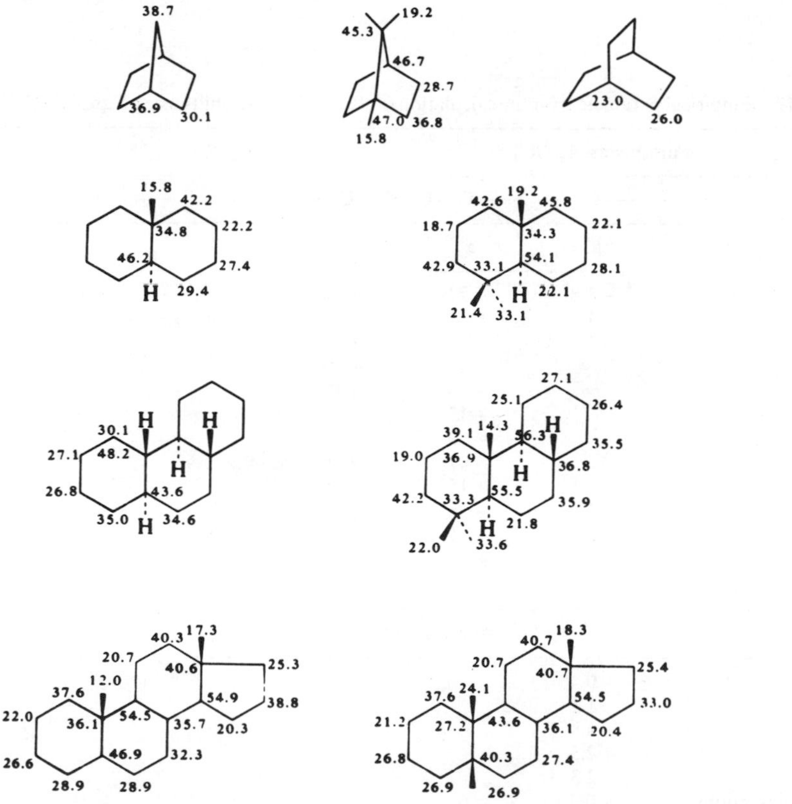

(Refs 4, 6)

(Ref. 60)

(Ref. 61)

(Ref. 5)

Fig. 2.6 Chemical shifts of some important alicyclic hydrocarbon skeletons.

angles not usually detected and determinable from molecular models), different sets of parameters have to be used for cyclohexanes, decalins, perhydrophenanthrenes, gonanes, etc.

As an alternative to these additivity relationships, which are often difficult to use and of limited applicability, the chemical shifts of a number of carbocyclic skeletons are given in Fig. 2.6. Together with the substituent increments compiled in Table 2.7, these values allow rough chemical shift estimates to be made for a vast number of derivatives (e.g. terpenes and steroids).

Substituted cyclohexanes

$$\delta_C(k) = 27.3 + \sum_i^n A_i(R_i) + \text{correction terms} \tag{2.12}$$

Values for shift increments A_i are listed in Table 2.13 for selected substituents R_i. Correction terms are listed for methyl substituents only. The use of Eqn. (2.12) for multiply substituted compounds may therefore be misleading.

Methyldecalins, perhydroanthracenes, perhydrophenanthrenes: Carbon shifts in these classes of compounds have been factorized using identical procedures. Different types of steric interactions occur, however, in the various types of backbones; therefore slightly different sets of correction terms have to be used. Consequently, the factor analysis yields different values even for the basic parameters. The parameter sets listed in Table 2.14 take the following structural features into account:

(1) The number of α and β carbons (α, β), including correction terms for heavily substituted carbons;
(2) Dihedral angles: for each *gauche*, *trans*, or eclipsed interaction between the kth carbon and its neighbours the appropriate V_g, V_t, or V_e increment has to be added. An extra $V_{g,t}$ increment applies for a successive 1, 2-*gauche*, 2, 3-*trans* arrangement;
(3) γ-*gauche* effects (γ_g): special parameters apply for the case of axial CH_x ($x = 1, 2, 3$) groups interacting with the 1, 3-diaxial 3 and 5 hydrogen atoms (γ_g^2) and for the 1, 4-'prow' interaction in boat cyclohexanes (γ_p) (refer also to Table 2.5);
(4) 1, 3-*syn*axial interactions (δ_{sa}): in contrast to the generally observed downfield shifts for synaxial δ effects (page 54), a negative increment is derived from the spectra of the *cis–syn–cis* fused systems (a) and (b). This indicates that these

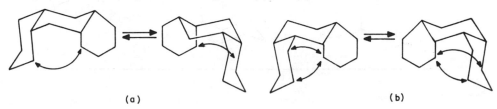

(a)   (b)

TABLE 2.13 Empirical parameters for the calculation of cyclohexane chemical shifts using Eqn. (2.12). (Numbers in parentheses are References)

A_i		CH$_3$ (31)	CN (58)	OH (6, 58)	OCH$_3$ (58)	OAc (58)	NH$_2$ (58)	NC (58)	F (59)	Cl (59)	Br (59)	I (59)
α_e		+6.0	+1	+43	+52	+46	+24	+25	+64	+33	+25	+3
α_a		+1.4	0	+39	+47	+42		+23	+61	+33	+28	+11
β_e		+9.0	+3	+8	+4	+5	+10	+7	+6	+11	+12	+13
β_a		+5.4	−1	+5	+2	+3		+4	+3	+7	+8	+9
γ_e		0	−2	−3	−3	−2	−2	−3	−3	0	+1	+2
γ_a		−6.4	−5	−7	−7	−6		−7	−7	−6	−6	−4
δ_e		−0.2	−2	−2	−2	−2	−1	−2	−3	−2	−1	−2
δ_a		0	−1	−1	−1			−2	−2	−1	−1	−1

Correction terms[31] (for CH$_3$ only)

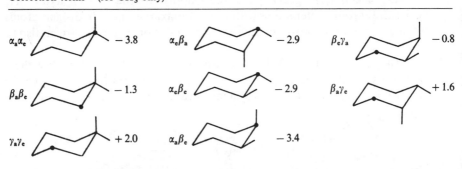

TABLE 2.14 Empirical parameters for the calculation of chemical shifts in methyldecalins, perhydroanthracenes and perhydrophenanthrenes

Parameter		Methyldecalins[56]	Perhydroanthracenes and -phenanthrenes[57]
Constant term		−3	− 2.5
α		+9.9	+ 9.4
β		+8.5	+ 8.8
T		−2.9	− 1.1
Q		−9	
V_g		−3.5	− 3.4
V_t			− 0.8
V_e			− 8.2
$V_{g,t}$		+1.9	
γ_g		−4.6	− 5.5
γ_g^2		−4.0	− 3.0
γ_p			− 11.0
δ_{sa}			− 3.7

molecules probably do not exist in pure chair conformation and that the non-bonded distance between these δ carbons is somewhat larger and falls in the range where shielding effects prevail.[51]

The following example illustrates the use of Table 2.14. A chemical shift calculation of carbon-9 in *trans*-9-methyldecalin[56] (7).

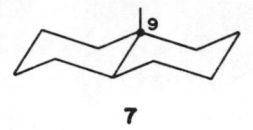

7

TABLE 2.15   Empirical parameters for the calculation of chemical shifts in substituted benzenes[4,6] (Eqn. (2.13))

R	A_i			
	C-1	*ortho*	*meta*	*para*
H	0	0	0	0
CH_3[52]	+ 9.3	+ 0.8	0	− 2.9
CH_2CH_3[52]	+ 15.6	− 0.4	0	− 2.6
$CH(CH_3)_2$[52]	+ 20.2	− 2.5	+ 0.1	− 2.4
$C(CH_3)_3$[52]	+ 22.4	− 3.1	− 0.1	− 2.9
CF_3	− 9.0	− 2.2	+ 0.3	+ 3.2
C_6H_5	+ 13	− 1	+ 0.4	− 1
$CH{=}CH_2$	+ 9.5	− 2.0	+ 0.2	− 0.5
$C{\equiv}CH$	− 6.1	+ 3.8	+ 0.4	− 0.2
CH_2OH	+ 12	− 1	0	− 1
COOH	+ 2.1	+ 1.5	0	+ 5.1
$COO^{\ominus}$	+ 8	+ 1	0	+ 3
$COOCH_3$[53]	+ 2.1	+ 1.1	+ 0.1	+ 4.5
COCl	+ 5	+ 3	+ 1	+ 7
CHO	+ 8.6	+ 1.3	+ 0.6	+ 5.5
$COCH_3$	+ 9.1	+ 0.1	0	+ 4.2
$COCF_3$	− 5.6	+ 1.8	+ 0.7	+ 6.7
COC_6H_5	+ 9.4	+ 1.7	− 0.2	+ 3.6
CN	− 15.4	+ 3.6	+ 0.6	+ 3.9
OH	+ 26.9	− 12.7	+ 1.4	− 7.3
OCH_3	+ 31.4	− 14.4	+ 1.0	− 7.7
$OCOCH_3$	+ 23	− 6	+ 1	− 2
OC_6H_5	+ 29	− 9	+ 2	− 5
NH_2	+ 18.0	− 13.3	+ 0.9	− 9.8
$N(CH_3)_2$	+ 23	− 16	+ 1	− 12
$N(C_6H_5)_2$	+ 19	− 4	+ 1	− 6
$NHCOCH_3$	+ 11	− 10	0	− 6
NO_2	+ 20.0	− 4.8	+ 0.9	+ 5.8
NCO	+ 5.7	− 3.6	+ 1.2	− 2.8
F	+ 34.8	− 12.9	+ 1.4	− 4.5
Cl	+ 6.2	+ 0.4	+ 1.3	− 1.9
Br	− 5.5	+ 3.4	+ 1.7	− 1.6
I	− 32	+ 10	+ 3	+ 1

THE SPECTRAL PARAMETERS

TABLE 2.16 Empirical parameters for the calculation of chemical shifts of substituted pyridines[60] (Eqn. (2.14))

Constant terms C_k

$C_2 = C_6$	149.6
$C_3 = C_5$	124.2
C_4	136.2

Increments A_{ik}

R_i $i=2$	A_{22}	A_{23}	A_{24}	A_{25}	A_{26}
CH_3	+ 9.1	− 1.0	− 0.1	− 3.4	− 0.1
CH_2CH_3	+ 14.0	− 2.1	+ 0.1	− 3.1	+ 0.2
$COCH_3$	+ 4.3	− 2.8	+ 0.7	+ 3.0	+ 0.2
CH	+ 3.5	− 2.6	+ 1.3	+ 4.1	+ 0.7
OH	+ 14.9	− 17.2	+ 0.4	+ 3.1	+ 6.8
OCH_3	+ 15.3	− 13.1	+ 2.1	− 7.5	− 2.2
NH_2	+ 11.3	− 14.7	+ 2.3	− 10.6	− 0.9
NO_2	+ 8.0	+ 5.1	+ 5.5	+ 6.6	+ 0.4
CN	− 15.8	+ 5.0	− 1.7	+ 3.6	+ 1.9
F	+ 14.4	− 14.7	+ 5.1	+ 2.7	− 1.7
Cl	+ 2.3	+ 0.7	+ 3.3	− 1.2	+ 0.6
Br	− 6.7	+ 4.8	+ 3.3	− 0.5	+ 1.4

R_i $i=3$	A_{32}	A_{33}	A_{34}	A_{35}	A_{36}
CH_3	+ 1.3	+ 9.0	+ 0.2	− 0.8	− 2.3
CH_2CH_3	+ 0.3	+ 15.0	− 1.5	− 0.3	− 1.8
$COCH_3$	+ 0.5	+ 0.3	− 3.7	− 2.7	+ 4.2
CHO	+ 2.4	+ 7.9	0	+ 0.6	+ 5.4
OH	− 10.7	+ 31.4	− 12.2	+ 1.3	− 8.6
NH_2	− 11.9	+ 21.5	− 14.2	+ 0.9	− 10.8
CN	+ 3.6	− 13.7	+ 4.4	+ 0.6	+ 4.2
Cl	− 0.3	+ 8.2	− 0.2	+ 0.7	− 1.4
Br	+ 2.1	− 2.6	+ 2.9	+ 1.2	− 0.9
I	+ 7.1	− 28.4	+ 9.1	+ 2.4	+ 0.3

R_i $i=4$	A_{42}	$A_{45} = A_{43}$	A_{44}	$A_{46} = A_{42}$
CH_3	+ 0.5	+ 0.8	+ 10.8	
CH_2CH_3	0	− 0.3	+ 15.9	
$CH=CH_2$	+ 0.3	− 2.9	+ 8.6	
$COCH_3$	+ 1.6	− 2.6	+ 6.8	
CHO	+ 1.7	− 0.6	+ 5.5	
NH_2	+ 0.9	− 13.8	+ 19.6	
CN	+ 2.1	+ 2.2	− 15.7	
Br	+ 3.0	+ 3.4	− 3.0	

yields:

$$\delta_C(\text{C-9}) = \text{Const.} + 4\alpha + 4\beta + Q + 8\,V_g$$
$$= -3 + 39.6 + 34 - 9 - 28 = 33.6$$

The predicted shift of 33.6 ppm is in good agreement with the observed value (34.8 ppm).

Benzenes

Substituent effects in benzenes are approximately additive unless substituents are *ortho* with respect to each other. Therefore, no correction terms are usually required and chemical shifts may be estimated using the simple relationship

$$\delta_C(k) = 128.5 + \sum_i A_i(\text{R}) \tag{2.13}$$

In Eqn. (2.13), $A_i(\text{R})$ represents the chemical shift increment for a substituent R in the ith position (C-1, *ortho, meta,* or *para*). Parameters A_i for a number of common substituents are listed in Table 2.15).

Pyridines

In heteroaromatic systems, substituent effects depend on the relative positions of both the substituent and the heteroatom. As an illustration, substituent additivity parameters for mono-substituted pyridines are given below. They may be used for disubstituted, polyhetero, and polynuclear systems if deviations due to steric and mesomeric effects are allowed for. Heteroaromatic ring chemical shifts are calculated according to Eqn. (2.14):

$$\delta_C(k) = C_k + \sum_i A_{ik}(\text{R}_i) \tag{2.14}$$

where C_k is the constant term for nucleus k (= chemical shift of carbon k in pyridine), and A_{ik} is the shift increment predicted for carbon k upon introduction of substituent R_i at carbon i. The parameters C_k and $A_{ik}(\text{R}_i)$ are listed in Table 2.16.

2.2 SPIN–SPIN COUPLING

The last ten years saw an almost explosive growth of ^{13}C n.m.r. papers reporting spin–spin coupling data. This is largely a result of the more widespread utilization of proton-coupled ^{13}C spectra along with a growing awareness of the potential spin coupling parameters offer as a stereochemical assignment aid. Besides $^{13}\text{C}-^{1}\text{H}$ and $^{13}\text{C}-\text{X}$ (X $= ^{19}\text{F}, ^{31}\text{P}$, etc.) homonuclear spin–spin coupling has aroused the interest of chemist and spectroscopist, in particular since the latter quantity can now be obtained on natural-abundance molecules (cf. Chapter 3). Because a detailed and systematic compilation of coupling constants would be beyond the scope of this book, the reader is again referred to the pertinent reviews.[63-9]

Theoretical aspects

In principle, two different types of magnetic interactions between nuclei are distinguished: (1) the dipole–dipole interaction and (2) the indirect coupling transmitted via the electronic system of the molecule. In high-resolution n.m.r. of liquids (1) can be ignored because Brownian motion averages this interaction to zero. However, in liquid crystals and solids dipolar interactions are the main source of the fine structure observed. The second type of spin–spin coupling (2) is independent of molecular orientation. The mechanism is known to involve the electron spins of the bonding electrons.

The interaction energy has empirically been found to be proportional to the scalar product of the nuclear spins nuclei of $\mathbf{I}_N$ and $\mathbf{I}_K$ according to expression (2.15):

$$E = J_{NK}\mathbf{I}_N.\mathbf{I}_K \tag{2.15}$$

This type of coupling is therefore also referred to as *scalar coupling* in contrast to *dipolar coupling*. In Eqn. (2.15) J_{NK} represents the indirect spin–spin coupling constant, subsequently designated *coupling constant*. J is usually expressed in hertz and is characterized by its magnitude and sign (cf. also Chapter 3). The theory of scalar coupling is probably even more complex than that for nuclear shielding, and computed coupling constants in satisfactory agreement with theory have not been obtained except in the simplest systems.[70]

Theory indicates that J_{NK} is composed of several contributions whose most significant one is believed to be the *Fermi contact term*. The expression originates from the fact the mechanism underlying the interaction stipulates electron density at the site of the nucleus. Similar to the paramagnetic term of the shielding constant, the contact interaction energy involves matrix elements between ground and excited electronic states. Several attempts have been made to calculate J in terms of molecular orbital theory using a mean value for the various excitation energies. The most frequently applied formalism resulting from these considerations is the one developed by Pople and Santry[71] for coupling between directly bonded atoms:

$$J_{NK} = h\frac{4}{9}\frac{e^2\hbar^2}{m^2c^2}\gamma_N\gamma_K S_N^2(0)S_K^2(0)\frac{1}{\Delta E}P_{S_N S_K}^2 \tag{2.16}$$

In Eqn. (2.16) γ_N and γ_K represent the magnetogyric ratios of nuclei N and K, $S_N^2(0)$ and $S_K^2(0)$ are the electron densities in the S valence orbitals of atoms N and K at the nucleus, ΔE is the mean electronic excitation energy, and $P_{S_N S_K}$ is the σ bond order of orbitals S_N and S_K.

Since removal of electron density (for example, brought about by attachment to an electronegative substituent) increases the effective nuclear charge of the carbon in question, this augments the probability of the valence S electron for nuclear contact. Hence an increase of ^1J should result,* which is experimentally observed. If, for example, in methane the hydrogens are successively replaced by chlorine, the one-bond coupling constant $^1J_{CH}$ increases from 125 Hz over 147 Hz in CH_3Cl and 177 Hz in CH_2Cl_2 to 208 Hz in chloroform.

According to Eqn. (2.16), ^1J should also be sensitive to a change in carbon orbital hybridization. In an sp^3-hybridized carbon the probability of an electron occupying the s-orbital is lower than for an sp^2 orbital (25% against 33% s fraction). Consequently $^1J_{CH}$ is expected to be smaller in the former, which is in accordance with experiment.

From Eqn. (2.16) it is readily seen that when replacing nucleus K by a nucleus L (for example, a proton by ^{13}C), and on the assumption that $S_N^2(0)$ is not affected, the coupling constants J_{NK} and J_{NL} are related by

$$\frac{J_{NK}}{J_{NL}} = \frac{\gamma_K}{\gamma_L} \frac{S_K^2(0)}{S_L^2(0)} \frac{\Delta E_L}{\Delta E_K} \tag{2.17}$$

In the special case of isotopic substitution (for example, ^1H by ^2H) the two coupling constants are related through the magnetogyric ratios:

$$J_{NK}/J_{NL} = \gamma_K/\gamma_L \tag{2.18}$$

Equation 2.18 is only approximately valid since isotopic substitution affects the vibrational states of a molecule. However, isotope effects on spin–spin coupling constants are normally negligibly small.

Carbon–proton coupling constants

One-bond C—H couplings

Empirically, it has been found that the one-bond coupling constant $^1J_{CH}$ can be approximated by the following simple relationship:[72]

$$^1J_{CH} = 5 \times (\% s)[Hz] \tag{2.19}$$

where % s represents the percentage s character of the carbon hybrid orbital participating in the C—H bond. This quantity is 25 for sp^3, 33 for sp^2, and 50 for sp hybridized carbon. Accordingly, Eqn. (2.19) predicts 125, 165, and 250 Hz for the CH coupling constant in ethane, ethylene, and acetylene, in good argeement with experiment. Although originally established empirically, Eqn. (2.19) has its theoretical justification (cf. Eqn. (2.16)). Based on INDO-MO calculations of hybridization parameters, the following improved relationship could be derived:[73]

$$^1J_{CH} = 5.7 \times (\% s) - 18.4[Hz] \tag{2.20}$$

*Index n in $^nJ_{NK}$ refers to the number of bonds by which nuclei N and K are separated.

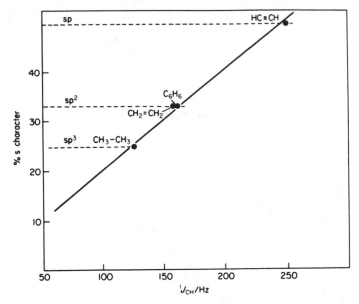

Fig. 2.7 Dependence of $^1J_{CH}$ on s character of carbon. The straight line corresponds to Eqn. (2.17).[72]

A plot of experimental against predicted values of $^1J_{CH}$ is given in Fig. 2.7. Since bond hybridization is related to bond angles, $^1J_{CH}$ provides information on ring size in alicyclic hydrocarbons. A few typical values are listed in Table 2.17.

The second important factor affecting the magnitude of $^1J_{CH}$ is substituent electronegativity. A linear correlation was found between the one-bond CH coupling constant and the Swain–Lupton field effect factor F,[44] which is a measure for the inductive effect exerted by the substituent. The plot in Fig. 2.8 reveals two straight lines. Points on the upper line belong to substituents with no lone pairs (NO_2, CN, etc.) whereas those on the lower line refer to substituents possessing non-bonded orbitals (NH_2, halogens, OH, etc.). Hyperconjugation between the carbon and the substituent is believed to be the cause for different behaviour of the latter type of substituents.[74]

Substituent effects on $^1J_{CH}$ are approximately additive. For a molecule of the type CHXYZ, $^1J_{CH}$ can be expressed as[75]

$$^1J_{CH} = \xi_X + \xi_Y + \xi_Z \tag{2.21}$$

where ξ_X is the contribution of substituent X to the coupling. This contribution is obtained from the monosubstituted methane CH_3X according to

$$\xi_X = {}^1J_{CH}(CH_3X) - 2 \cdot \xi_H \tag{2.22}$$

with $\xi_H = \frac{1}{3} {}^1J_{CH}(CH_4) = 41.7$ Hz. However, additivity is poor for highly electronegative substituents. $^1J_{CH}$ values of the parent compounds CH_3X and the

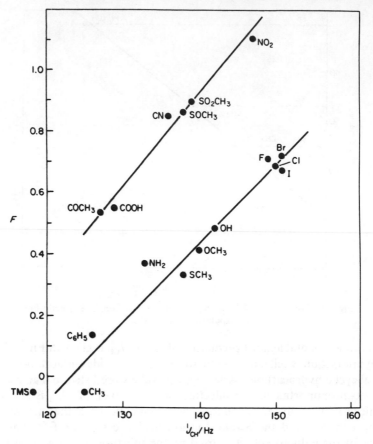

Fig. 2.8 $^1J_{CH}$ in substituted methanes CH$_3$-X versus field effect factors.[74]

corresponding ξ_X increments for various substituents X are listed in Table 2.18.

The effect of remote substituents is generally small, as the pertinent examples in Table 2.18 demonstrate. This is also true for one-bond coupling to sp^2-carbons; therefore substituent effects are of minor importance in olefinic and aromatic systems ($^1J_{CH}$ = 150–180 Hz in monosubstituted benzenes). A few illustrative data for aromatic and heteroaromatic systems are listed in Table 2.19.

Two-bond C, H couplings

Unlike $^3J_{C,H}$ which, like its proton–proton counterpart, has been utilized rather early as a stereochemical probe, $^2J_{C,H}$ exhibits a far more complex behaviour. It can be positive or negative and spans an unusually large range (typically − 10 to + 20 Hz).[64,76] Another conspicuous feature of $^2J_{C,H}$ is its pronounced dependence

TABLE 2.17 $^1J_{CH}$ in hydrocarbons (values from Ref. 6 unless otherwise noted)

sp³		sp²	
$H_3C\!-\!CH_3$	125	$H_2C\!=\!CH_2$	156
	123		157
	128		159
	136		160
	161		170
	130[74]		170
	130		166
	142[74]		173
	142[74]		177
	179		
	205		

TABLE 2.18   Increments ξ_X for the calculation of $^1J_{CH}$ in substituted methanes CHXYZ according to Eqn. (2.19)[75]

Substituent X	ξ_X[Hz]	$^1J_{CH}$ (parent compound CH_3X)
—H	4.17	125
—CH_3	42.6	126
—$CH=CH_2$	38.6	122
—C_6H_5	42.6	126
—$C\equiv CH$	47.6	131
—CHO	43.6	127
—$COCH_3$	43.6	127
—COOH	45.6	129
—$COOCH_3$	46.6	130
—OH	58.6	142
—OCH_3	56.6	140
—OC_6H_5	59.6	143
—NH_2	49.6	133
—$NHCH_3$	48.6	132
—$N(CH_3)_2$	48.6	132
—NO_2	63.6	147
—CN	52.6	136
—F	65.6	149
—Cl	66.6	150
—Br	67.6	151
—I	67.6	151
—CCl_3	50.6	134
—$CHCl_2$	47.6	131
—CH_2Cl	44.6	128
—CH_2Br	44.6	128
—CH_2I	48.6	132
—SCH_3	54.6	138
—$SOCH_3$	54.6	138

on substituent electronegativity. Basically, one can distinguish four different coupling paths, which can be further subdivided according to whether or not there exists a proton–proton analog.[76] This classification turns out to be useful, as there exist many parallels between $^2J_{CCH}$ and $^2J_{HCH}$. In addition, of course, geminal coupling also occurs across hetero atoms such as oxygen and nitrogen. Specific examples of such couplings will be discussed in Chapters 3 and 5.

In saturated systems, which we will address first, the effects of substituent electronegativity and lone-pair electrons are entirely analogous to the corresponding H, H coupling constants. Attachment of electronegative substituents at the central carbon, for example, increases ^2J, as exemplified by structures **8** and **9** in Fig. 2.9,[76] and also by the plot of $^2J_{C,H}$ versus Pauling substituent electronegativities in monosubstituted ethanes and 2-substituted n-propanes,[77] reproduced in Fig. 2.10.

If, however, the electronegative substituent is at the terminal carbon, $^2J_{C,H}$ depends on substituent orientation. In the case of oxygen substitution, for

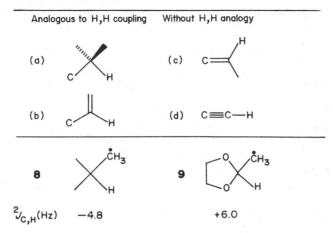

Fig. 2.9 $^2J_{C,H}$ coupling paths

example, it was found that for anti-periplanar orientation (dihedral angle $\phi = 180°$) ^2J is positive, whereas for $\phi = 0°$, ^2J is negative.[78]

$^2J_{C,H} > 0$ $^2J_{C,H} < 0$

An illustration of this behaviour is provided with $^2J_{C(1),H(2)}$ in the anomeric glucopyranoses[78] (**10** and **11**):

10 $^2J_{C(1),H(2)}$ -5.7 Hz   **11**   1 Hz

Among the two possible coupling paths involving sp²-hybridized carbons ((b) and (c) of Fig. 2.9), only (b) has a proton–proton analogue. Both coupling pathways have recently been studied in great detail.[64,76] For type (b) the following trends were ascertained:

(1) Electronegative substituents in β position cause a *negative* contribution to ^2J (structures **12** and **13**):

12   **13**

$^2J_{C,H}$ (Hz) $+4.1$ ±1.9

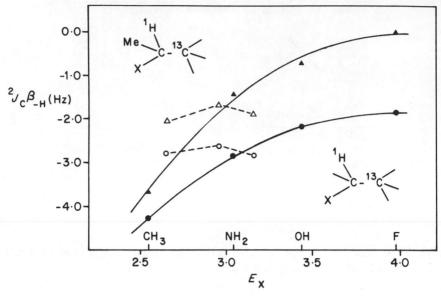

Fig. 2.10 $^2J_{C-,H}$ in ethyl and isopropyl compounds versus Pauling substituent electronegativities.[77] Open circles and triangles refer to I, Br and Cl substituents. Reproduced by permission of the Editorial Office Recueil.

(2) Electronegative substitution in α position or at the coupling carbon causes a *positive* contribution to ^2J (structures **14–16**):

14 $\overset{\bullet}{C}H_3$ ...

15 $\overset{\bullet}{C}H_3$ ...

16 $\overset{\bullet}{C}Cl_3$ ...

$^2J_{C,H}$(Hz) ±0.5 $+26.7$ $+51.6$

(3) With increasing C—C—H bond angle ^2J becomes more positive (structures **17** and **18**):

17   **18**

$^2J_{C,H}$ (Hz) ±0.5 $+5.7$

(4) An electronegative substituent at the coupling carbon invokes a particularly large positive contribution to ^2J if C—X and C—H are *trans*-oriented (compare **17** with **19** and **20**):

	19		20

$^2J_{C,H}$(Hz)   +4.7   +5.3

Of particular importance is type (c) coupling (Fig. 2.9), since it bears the potential as a configurational probe in those cases where other couplings ($^3J_{H,H}$ or $^3J_{C,H}$) are non-existent due to substitution. A large body of data has recently been gathered suggesting that the coupling constant can be expressed as the sum of individual substituent increments:[64]

$$^2J_{C,H} = C + \Delta J^g + \Delta J^c + \Delta J^t \qquad (2.23)$$

In Eqn. (2.23) C is a constant which corresponds to the geminal coupling constant

TABLE 2.19   Substituent increments (Hz) for $^2J_{C,H}$, derived from monosubstituted ethylenes[64]

Substituent	ΔJ^g	ΔJ^c	ΔJ^t
F[a]	+ 17	− 11	+ 10
Cl	+ 9.2	− 5.9	+ 9.5
Br	+ 8.2	− 6.1	+ 9.1
I	+ 6.4	− 5.4	+ 6.5
OC_2H_5	+ 12.1	− 2.9	+ 7.7
$OCH=CH_2$	+ 12.1	—	—
$OCOCH_3$	+ 12.0	− 5.5	+ 10.0
$NDCOCH_3$	—	− 2.9	+ 5.9
	+ 5.9	− 2.3	+ 5.9
$SiCl_3$	− 4.4	+ 1.6	− 0.1
$Si(CH=CH_2)_3$	− 4.6	—	—
CH_3	+ 2.7	− 0.2	+ 1.2
$CH=CH_2$	+ 2.3	—	—
C_6H_5	+ 1.4	− 2.1	+ 2.4
$C=CH$	− 1.6[b]	− 1.3	+ 2.1
CN	− 1.3	− 2.0	+ 2.7
CHO	+ 2.6	− 1.0	+ 1.8
$COCH_3$	+ 1.8	− 2.2	+ 2.2
COOH	+ 1.8	− 2.2	+ 2.2
COOR	+ 1.8	—	—
$CON(CH_3)_2$	+ 1.5[c]	—	—
H	0	0	0

[a] Increments for fluorine were derived from Z- and E-1, 2-difluoroacrylic acid.
[b] From Z-1-methoxy-1-butene-3-yne.
[c] From E-3-chloro-N, N-dimethylacrylamide.

in the parent compound ethylene ($^2J_{C,H} = -2.4\,Hz$) and ΔJ^g, ΔJ^c, and ΔJ^t represent the substituent increments for substituents in *geminal, cis,* and *trans* position, respectively. Substituent increments derived from monosubstituted ethylenes are listed in Table 2.19.[64]

From Table 2.19 we note that an electronegative substituent produces a negative or positive contribution to ^2J depending on whether the substituent at the terminal carbon is *cis* or *trans* with respect to the C—H bond. An electronegative substituent in geminal position causes a positive enhancement of ^2J. Good agreement is generally found between experimental values of multiply substituted ethylenes and those predicted by Eqn. (2.23) and the parameters of Table 2.19.

Recourse to $^2J_{C,H}$ as a configurational assignment aid lends itself for the determination of substitution geometry in trisubstituted ethylenes. In the ethylene derivative **21**, for example, $^2J_{C,H} = 16.3\,Hz$ was found, whereas the values predicted on the basis of the increments in Table 2.19 are 17.4 and 6.5 Hz, respectively, thus unambiguously proving the *cis* geometry indicated by formula **21**.

21

It is interesting to note that the trends delineated by the ethylene substituent increments are also borne out by the behaviour of $^2J_{C,H}$ in aromatics. In monosubstituted benzenes, for example, $^2J_{C(1),H(2)}$ is small and positive for substituents, whose electronegativity does not greatly deviate from that of hydrogen (benzene: $+1.15\,Hz$), but becomes large and negative for highly electronegative substituents such as fluorine ($-4.9\,Hz$) and large and positive for electropositive substituents such as trimethylsilyl ($+4.19\,Hz$).[79] Long-range C, H coupling constants in aromatic systems play a key role in signal assignment (cf. Chapter 3). For reference purposes the two-bond coupling constants $^2J_{C,H}$, together with $^3J_{C,H}$ and $^4J_{C,H}$ are listed in Table 2.20 for a host of monosubstituted benzenes.[79]

In heteroaromatics $^2J_{C,H}$ follows the trend established for the substituent increments in Table 2.19. Accordingly, one predicts a large positive enhancement of $^2J_{C(\alpha),H(\beta)}$ and, even more so, for $^2J_{C(\beta),H(\alpha)}$. For both predictions there is broad experimental evidence. In pyridine, for example, $^2J_{C(\alpha),H(\beta)} = 3.1\,Hz$ and $^2J_{C(\beta),H(\alpha)} = 8.5\,Hz$, compared with the value of 1.15 Hz for benzene. A collection of geminal and longer-range C, H coupling constants in some common heteroaromatic compounds is reproduced in Table 2.21.

TABLE 2.20 $^nJ_{C,H}(n \geq 2)$ in monosubstituted benzenes[a],[79]

Substituent	Carbon 1			Carbon 2				Carbon 3				Carbon 4	
	$^2J_{12}$	$^3J_{13}$	$^4J_{14}$	$^2J_{23}$	$^3J_{24}$	$^3J_{26}$	$^4J_{25}$	$^2J_{32}$	$^2J_{34}$	$^3J_{35}$	$^4J_{36}$	$^2J_{43}$	$^3J_{42}$
F	−4.9	11.0	−1.7	1.1	8.3	4.1	−1.5	−0.6	1.7	9.0	−0.8	0.8	7.6
Cl	−3.4	10.9	−1.8	1.5	8.1	5.1	−1.4	0.3	1.6	8.4	−0.9	0.9	7.5
Br	−3.4	11.2	−1.9	1.5	8.0	5.4	−1.4	0.4	1.6	8.4	−1.1	0.9	7.5
I	−2.5	10.8	−1.9	1.6	8.0	6.1	−1.3	0.6	1.5	8.2	−1.2	0.9	7.5
NH$_2$	−0.9	8.6	−1.4	1.3	7.9	5.4	−1.4	0.2	1.7	8.4	−0.8	0.8	7.4
NO$_2$	−3.6	9.7	−1.8	1.8	8.1	4.5	−1.3	−0.3	1.5	8.2	−0.7	1.3	7.7
H	1.2	7.6	−1.3										
CH$_3$	0.5	7.6	−1.4	1.2	7.8	6.6	−1.4	1.1	1.4	7.9	−1.06	1.1	7.5
CN	0.1	9.0	−1.4	1.8	7.9	6.1	−1.3	0.4	1.3	7.6	−1.1	1.0	7.5
CHO	0.3	7.2	−1.3	1.4	7.8	6.3	−1.3	0.8	1.3	7.6	−1.1	1.5	7.6
OH	−2.8	9.7	−1.6	1.2	8.1	4.7	−1.4	−0.3	1.7	8.7	−0.7	0.8	7.4
OCH$_3$	−2.8	9.2	−1.5	1.4	8.0	4.8	−1.4	−0.3	1.8	8.7	−0.8	0.9	7.5
Si(CH$_3$)$_3$	4.2	6.3	−1.1	1.4	7.5	8.6	−1.2	1.6	1.1	7.3	−1.4	1.2	7.5

[a] All signs were determined by simulation of the experimental spectra. No implicit sign implies the value to be positive.

INTERPRETATION OF CARBON-13 NMR SPECTRA

TABLE 2.21 C,H spin–spin coupling constants (Hz) in benzene[a] and hetero-aromatic compounds

	$^1J_{\text{C,H}}$	$^2J_{\text{C,H}}$	$^3J_{\text{C,H}}$	$^4J_{\text{C,H}}$
Benzene[80]	158.8	+1.1	+7.6	−1.2
Pyridine[81]	2,2: +177.6 3,3: +163.0 4,4: +162.4	2,3: +3.1 3,2: +8.5 3,4: +0.8 4,3: +0.7	2,4: +6.9 2,6: +11.2 3,5: +6.6 4,2: +6.3	2,5: −0.9 3,6: −1.7
Pyrrole[82]	2,2: 182 3,3: 170	2,3: 7.6 3,2: 7.8 3,4: 4.6	2,4: 7.6 2,5: 7.6 3,5: 7.8	
Furan[83]	2,2: 201.7 3,3: 175.1	2,3: 6.9 3,2: 13.8 3,4: 4.1	2,4: 11.1 2,5: 6.9 3,5: 6.0	
Thiophene[83]	2,2: 184.7 3,3: 167.0	2,3: 7.6 3,2: 4.7 3,4: 5.8	2,4: 10.0 2,5: 5.0 3,5: 9.8	
N-Methylpyrazole[84]	3,3: 183.8 4,4: 175 5,5: ∼187	3,4: 5.7 4,3: 9.8 4,5: 9.8 5,4: ?	3,5: 8.4 5,3: ?	
Isoxazole[84]	3,3: 187.6 4,4: 184.6 5,5: 203.5	3,4: 8.2 4,3: 14.2 4,5: 5.4 5,4: 8.0	3,5: 6.2 5,3: 4.4	
Adenosine[85,c]			4,2: 12.8 4,8: 5.5 5,8: 11.7 6,2: 7.7	
Allopurinol riboside[85,c]		5,7: 9.7	4,2: 13.6 4,7: 3.9 6,2: 6.8 6,7: < 1.0	

TABLE 2.21 (Contd.)

	$^1J_{C,H}$	$^2J_{C,H}$		$^3J_{C,H}$		$^4J_{C,H}$
Tubercidin[85,c]		5,7:	8.4	4,2:	12.1	
		7,8:	6.6	4,7:	7.7	
		8,7:	7.0	5,8:	4.0	
		6,2:	7.3			
				6,7:	< 1.0	
Guanosine[85,c]				4,8:	6.6	
				5,8:	7.0	

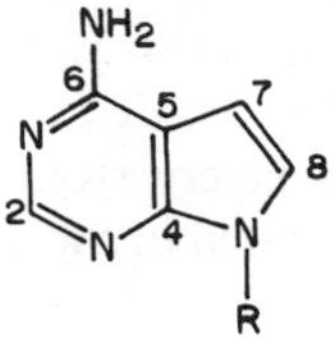

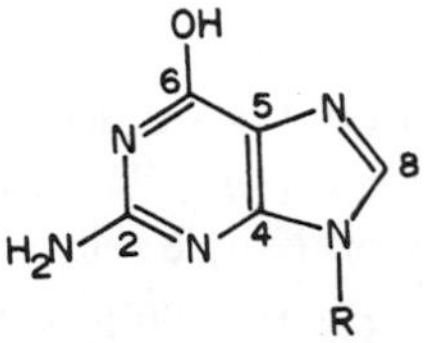

[a] Provided for reference purposes.
[b] The sign is marked in those cases where it had actually been determined; however, ^1J, ^2J, and ^3J can be assumed to be positive in all cases except where $J \sim 0$.

c

R =

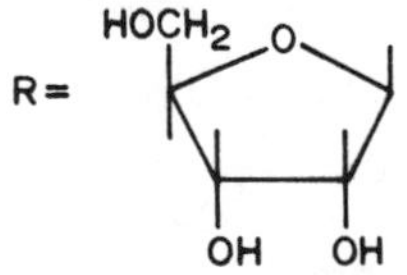

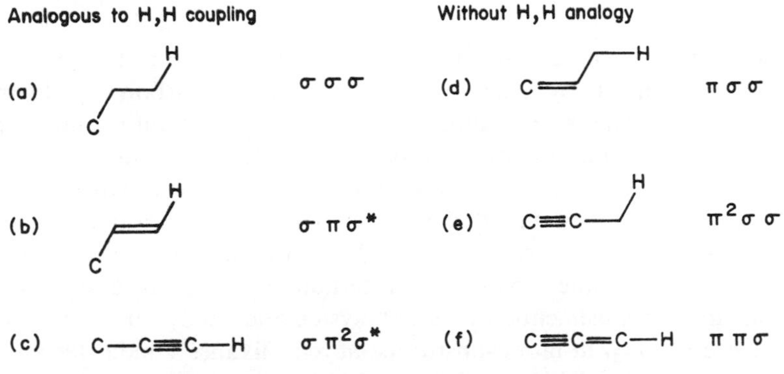

HOCH$_2$

OH OH

Analogous to H,H coupling		Without H,H analogy	
(a)	$\sigma\,\sigma\,\sigma$	(d)	$\pi\,\sigma\,\sigma$
(b)	$\sigma\,\pi\,\sigma^*$	(e) C≡C	$\pi^2\sigma\,\sigma$
(c) C—C≡C—H	$\sigma\,\pi^2\sigma^*$	(f) C≡C=C—H	$\pi\,\pi\,\sigma$

---

*assuming that the terminal(Coupling) Carbon is nonconjugating

Fig. 2.11 $^3J_{C,H}$ Coupling paths.[76a]

Three-bond C,H couplings

Since sign and magnitude of coupling constants are largely a function of the intervening bonds it seems logical to categorize three-bond couplings according to the types of bonds that constitute the coupling path.[76] Further more, we shall again subdivide them into two groups, depending on whether or not a H, H analogue exists (Fig. 2.11). In addition, one would have to consider analogous coupling situations with the central carbons substituted by hetero atoms. Again we will see that electronegative hetero atoms, be they integrated in the coupling network or peripherally substituting, have a significant impact on the magnitude of $^3J_{C,H}$. In contrast to $^2J_{C,H}$, however, the vicinal carbon–proton coupling constant has always been found to be positive, irrespective of the nature of bonding and substitution.

The most widely occurring and therefore most thoroughly studied coupling path is that constituted by three contiguous single bonds (a) ($\sigma\sigma$). There is broad evidence for this coupling to obey a Karplus-type dihedral angle relationship[86] (cf. also Chapter 3). The geometric dependence of $^3J_{C,H}$ is exemplified in **22** and **23**:[87]

22

$$^3J_{C,H_a} = 2.1\,\text{Hz}$$
$$^3J_{C,H_e} = 8.1\,\text{Hz}$$

23

$$^3J_{C,H_{endo}} = +5\,\text{Hz}$$
$$^3J_{C,H_{exo}} = +2.5\,\text{Hz}$$

It is further to be noted that in geometrically analogous HCCH coupling situations (obtained by substituting the terminal carbon by hydrogen), $^3J_{C,H} \approx 0.6\,^3J_{H,H}$. This rule of thumb, though to be used with caution, proves useful to estimate unknown carbon–proton coupling constants..

The sensitivity of $^3J_{C,H}$ to hetero atoms, in particular the extreme values for $\phi = 0°$ and $\phi = 180°$, demand careful model studies if accurate geometric information is to be derived. Figure 2.12 provides an impression of the bandwidth of ^3J values found for a certain dihedral angle ϕ in different substructural arrangements. A recent systematic study of the substituent dependence of $^3J_{C,H}$ in monosubstituted acyclic alkanes[93] indicates an almost linear increase with substituent electronegativity for substitution at the coupling carbon, as opposed to a decrease for substitution at the central or H-terminal carbon.

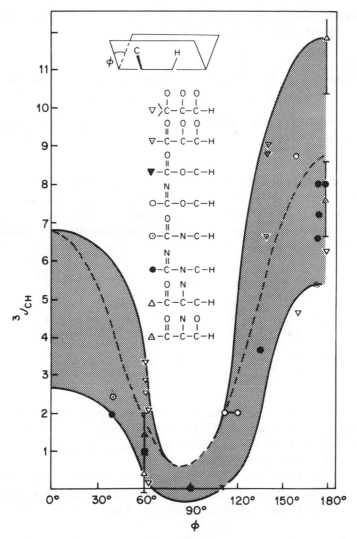

Fig. 2.12 Karplus curve for $^3J_{C,H}$. Dashed line: calculated (INDO) curve.[89] Experimental values from different sources: ▽ ▽ ▼,[88] ○ ⊙ ●,[90,91] ▽ ▽.[92]

Of particular stereochemical significance is coupling path (b) ($\sigma\pi\sigma$), which occurs in substituted olefins where one of the substituents is carbon. These systems have been investigated in considerable detail[63] and, although absolute values of $^3J_{C,H}$ are critically sensitive to the nature of the substituents and steric congestion effects, $^3J^{trans}$ was consistently found to be larger than $^3J^{cis}$ in a given substitution pattern; the typical ratio being $^3J^t/^3J^c = 1.7 \pm 0.1$. Steric perturbation, brought about by substituents in cis position, lowers this ratio, whereas electronegative substituents at the central carbon raise it. This behaviour is illustrated by the data in systems **24–6**:

24

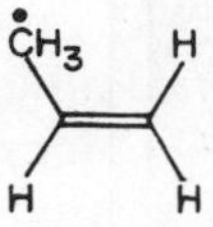

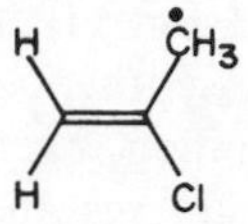

$^3J^t = 12.7\,\text{Hz}$
$^3J^c = 7.6\,\text{Hz}$
$^3J^t/^3J^c = 1.67$

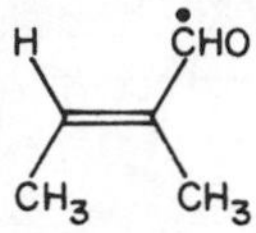

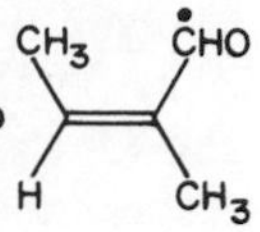

25a

$^3J^c = 9.7\,\text{Hz}$

25b

$^3J^t = 11.0\,\text{Hz}$

1.13

26

$^3J^t = 8.7\,\text{Hz}$
$^3J^c = 4.1\,\text{Hz}$
2.12

With respect to the actual magnitude of the coupling constant we can generally state that an electronegative substituent at the coupling carbon entails a *positive* contribution, whereas substitution at the central or H-terminal carbon invokes a negative contribution to ^3J, as seen by juxtaposing the data for systems **24, 26**, and **27**.

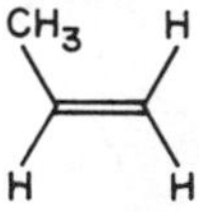

	24	**26**	**27**
$^3J^t\,(\text{Hz})$	12.7	8.7	15.5
$^3J^c\,(\text{Hz})$	7.6	4.1	9.1

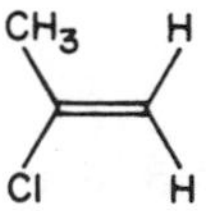

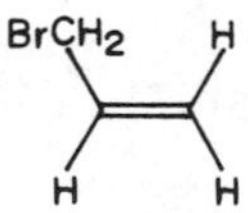

The enhancement of $^3J_{C,H}$ by electron withdrawal from the terminal carbon seems to be a rule of general validity. Further corroboration of this trend is found for substituted aromatics (Table 2.20), where $^3J_{C(1),H(3)}$ increases from 7.6 Hz in benzene to 11.0 Hz in fluorobenzene.

Finally, the implications of π-bond order, which comes into play in conjugated systems, deserves mention. In the methyl-substituted cyclohexadiene derivative **28**, in which the π-electron system is known to be nearly planar,

28

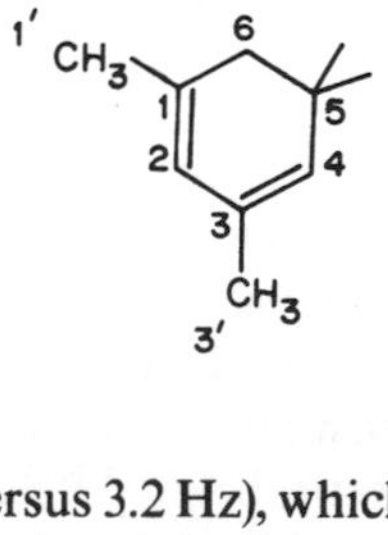

$^3J_{C(1'),H(2)} \gg {}^3J_{C(3'),H(2)}$ (6.3 versus 3.2 Hz), which was ascribed to the shorter bond length of the C(1)—C(2) bond and the larger π-bond order.[63]

Carbon–carbon coupling constants

The rapidly growing interest in C, C coupling constants has a number of reasons: (1) a more widespread utilization of ^{13}C as a tracer in mechanistic and biosynthetic studies; (2) the potential of such couplings to provide insight into binding mechanisms; and (3) the geometrical dependence of C,C long-range couplings and their stereochemical ramifications. Beyond this, the parameter received an additional boost by recent progress in experimental techniques, permitting retrieval of this quantity from natural-abundance samples (cf. Chapter 3).

One-bond C,C couplings

Similar to $^1J_{CH}$, $^1J_{CC}$ in hydrocarbons varies over a wide range depending on the hybridization of the two carbon atoms involved. In terms of the 's character' of the two bonding hybrid orbitals, $^1J_{CC}$ may be described by the expression[94]

$$^1J_{C_xC_y} = 7.3\frac{(\%\,s_x)(\%\,s_y)}{100} - 17[\text{Hz}] \qquad (2.24)$$

Figure 2.13[94] illustrates this behaviour. Equation (2.24) again has its theoretical basis in Eqn. (2.16). Assuming unaltered hybridization for C_x, one expects a linear

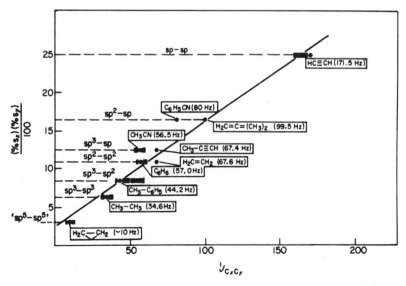

Fig. 2.13 Dependence of $^1J_{CC}$ on product of s characters of coupled carbons. The straight line corresponds to the empirical relationship described by Eqn. (2.24). Observed data from Ref. 94. Horizontal bars indicate range of $^1J_{CC}$ for substituted analogues.[67]

$$\{C^{\delta-} - C^{\delta+}\} - X^{\delta-} \qquad\qquad\qquad - I^+$$
$$\{C^{\delta+} - C^{\delta-}\}^{\delta+} - \{X^{\delta+} - Y^{\delta-}\}^{\delta-} \qquad - I^-$$

Fig. 2.14   Effect of $-I^+$ and $-I^-$ electron-withdrawing substituent on the polarization of a CC bond. Examples: $-I^+$; OH, NH_2, F, Cl, ...; $-I^-$: CN, COOR, NO_2, ...

relationship between $^1J_{C_x,H}$ and $^1J_{C_x,CH_3}$, which is confirmed by the following empirical proportionality:

$$^1J_{C_x,CH_3} = 0.27 \cdot {}^1J_{C_x,H} \tag{2.25}$$

Substituent effects on $^1J_{C,C}$ are comparatively small, as can be seen by the horizontal bars in Fig. 2.13, indicating the spread of experimental values. Exceptional in this respect are coupling constants between sp^2 and sp^3 carbons. It is therefore desirable to be able to predict one-bond carbon–carbon coupling constants.

Inspired by the Pople–Santry formalism, it has recently been suggested[95] to predict $^1J_{C,C}$ as the product of two empirical factors $I(C_i)$ and $I(C_j)$, where the subscripts i and j denote the two bonding carbons. The limitations inherent to this approach is the requirement of a very large set of factors $I(C_i)$. In a more qualitative sense, it is worthwhile pointing out that the so-called $-I^+$ substituents[96] enhance the magnitude of $^1J_{C,C}$ whereas $-I^-$ substituents dehance it. Both types of substituents withdraw electron density from the CC bond. However, they lead to opposite polarizations, as becomes apparent from the scheme in Fig. 2.14.

As an illustration of the effect of a $-I^+$-type substituent on $^1J_{C,C}$ we may compare the various C, C coupling constants in the aromatic ring of codeine (29):[97]

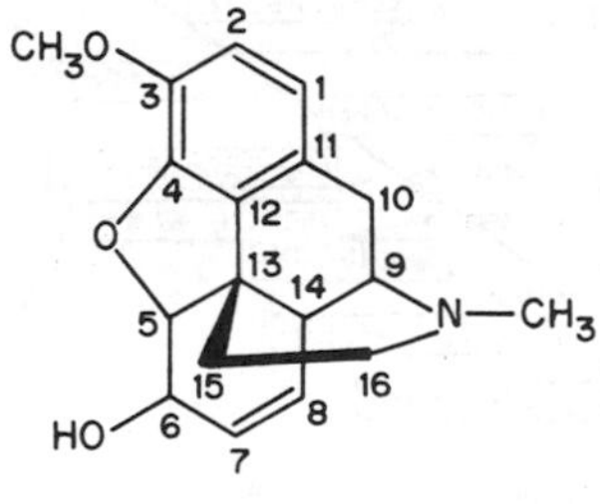

From the data in Table 2.22 we see that each oxygen adds ca 10 Hz to the value of $^1J_{C,C}$.

Two-bond C, C couplings

$^2J_{C,C}$ does not appear to be of great analytical potential. The few sign determinations reported[98] indicate that $^2J_{C,C}$ is usually positive, although

TABLE 2.22 Dependence of $^1J_{C,C}$ on the number of oxygen substituents for some aromatic CC bonds in codeine[97]

Bond	Number of oxygen subst.	$^1J_{C,C}$ (Hz)
1–2	0	60.8
1–11	0	59.1
11–12	0	61.4
2–3	1	69.1
3–4	2	80.0

negative values have been observed as well. Typically, the value of $^2J_{C,C}$ varies from 0.5 to 2.5 Hz, with the following notable exceptions: coupling across (1) carbonyl carbons and (2) CC triple bonds; (3) coupling involving sp-hybridized carbons (compare structures 30–33).[99]

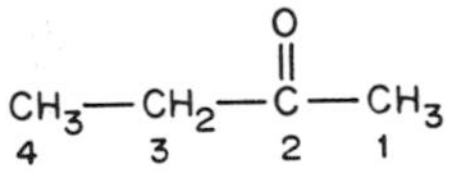

30

$^2J_{C(1),C(3)}$ 15.2 Hz

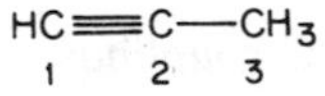

31

$^2J_{C(1),C(3)}$ 11.8 Hz

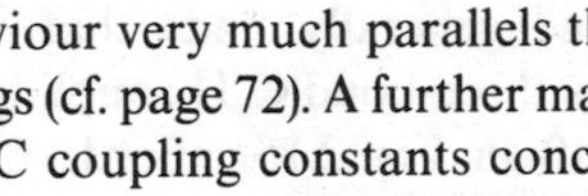

32

$^2J_{C(1),C(\beta)} = +13.1$ Hz

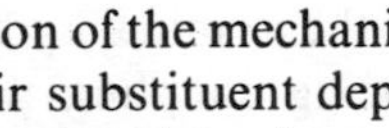

33

$^2J_{C(1),C(3)} = 33.0$ Hz

This behaviour very much parallels that found for the structurally analogous C, H couplings (cf. page 72). A further manifestation of the mechanistic kinship of C, H and C, C coupling constants concerns their substituent dependence. This explains, for example, the largely differing values found for $^2J_{C(1),C(5)}$ in some α and β anomeric pyranoses. Whereas in the β anomers (34) a value of ca 4 Hz was ascertained, this coupling is nearly zero in the α-anomeric series.[100]

34

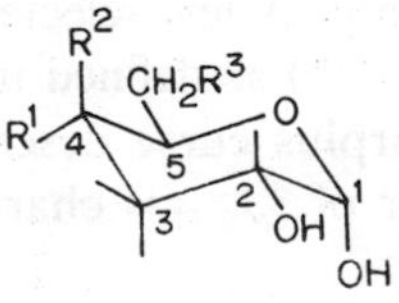

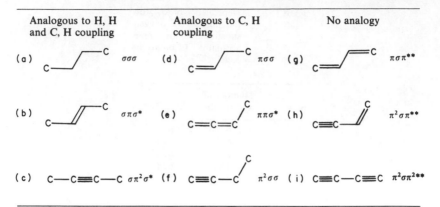

*Classification strictly correct only, provided terminal carbons are nonconjugating;** bond designations ignore π-bond order for the central bond

Fig. 2.15 $^3J_{C,C}$ coupling paths.

Three-bond C, C couplings

Because of the behavioural similarities between C, C and C, H couplings on the one hand, and C, H and H, H couplings on the other, we shall again, as we did for the three-bond C, H coupling constants, categorize $^3J_{C,C}$ according to the different coupling paths and whether or not a C, H or H, H analogue exists (Fig. 2.15). Like its C, H counterpart, $^3J_{C,C}$ can be assumed to be positive throughout,[69] therefore obviating a discussion of the sign.

The major stimulus to the measurement of $^3J_{C,C}$ is the expected dihedral angular dependence, established earlier for H, H and C, H couplings. Before treating this question, however, it is worthwhile to summarize some general characteristics of three-bond C, C coupling constants. Disregarding the exceptions cited in the previous section, one generally finds that $^3J_{C,C} \gg {}^2J_{C,C}$. Substitution of the terminal hydrogen in a three-bond C, H coupling network by carbon (producing coupling paths (a)–(f) in Fig. 2.15) affords C, C coupling constants which are typically 0.6–$0.8 \times {}^3J_{C,H}$.[65] Finally, one finds that a terminal electron-withdrawing substituent raises the value of $^3J_{C,C}$.

The angular dependence of the vicinal C, C coupling constant has been the subject of very detailed studies, and there are numerous manifestations of the existence of such a relationship. A few selected examples are given below for coupling paths of type (a) and (b) as defined in Fig. 2.15.

Inasmuch as no single Karplus curve describes the angular dependence of $^3J_{C,H}$, the Karplus behaviour of $^3J_{C,C}$ is characterized by a family of curves,

35   Ref. 65

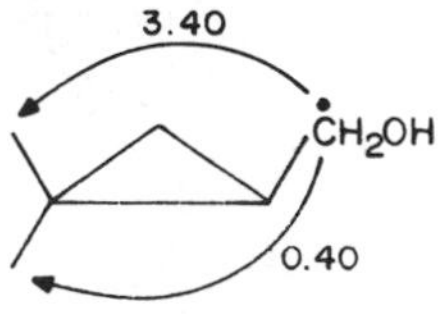

36   Ref. 65

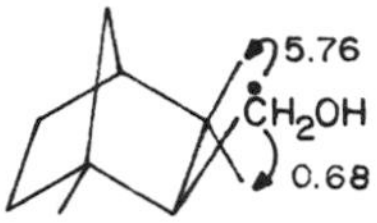

37   Ref. 65

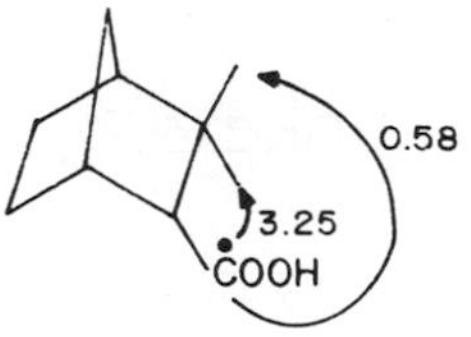

38   Ref. 101

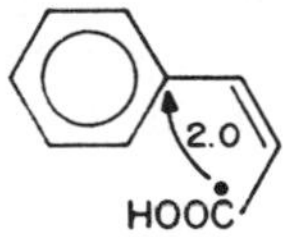

39   Ref. 101

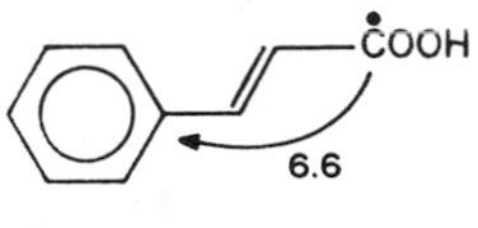

accounting for the effect of substituents. As in the case of $^3J_{C,H}$ the curve is asymmetric, i.e. $\phi = 0°$ and $\phi = 180°$ do not afford the same coupling constants. This is illustrated by the experimental and theoretical curves for a $\sigma\sigma\sigma$ coupling path (Figs 2.16 and 2.17). Whereas in Fig. 2.16[102,103] all carbons are sp^3-hybridized, the data in Fig. 2.17[102,103] were collected from systems where one of the terminal carbons pertains to a carbonyl function. In the case of terminal substitution calculations[65] show that for a *cis–cis* conformation (**40a**) through-space interactions are operative, which are predicted to lower the value of $^3J_{C,C}$. Clearly such non-bonding interactions are absent in the *cis–trans* conformation

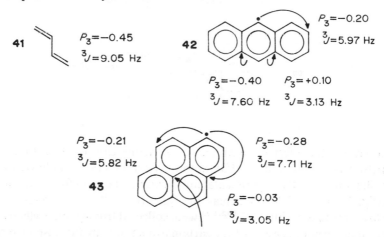

40a   **40b**

40c   **40d**

(**40b**) and, of course, in the *trans–trans* and *trans–cis* conformations (**40c** and **40d**, respectively).

In unsaturated systems of type (g) (Fig. 2.15), $^3J_{C,C}$ appears to depend primarily on π-bond order:[104,105]

$$^3J_{C,C} = 2.7 + 15.2|P_3| \qquad (2.26)$$

The empirical relationship expressed by Eqn. (2.26), in which P_3 represents the π-bond order between the two coupling carbons, implies a σ contribution of 2.7 Hz in such systems. In fact, rather good agreement is obtained for benzenoid hydrocarbons, if simple HMO-derived π bond orders are used, as borne out by the data provided for systems **41–43**:

41 $P_3 = -0.45$ **42** $P_3 = -0.20$
          $^3J = 9.05$ Hz $^3J = 5.97$ Hz

$P_3 = -0.40$ $P_3 = +0.10$
$^3J = 7.60$ Hz $^3J = 3.13$ Hz

$P_3 = -0.21$ $P_3 = -0.28$
$^3J = 5.82$ Hz $^3J = 7.71$ Hz

43

$P_3 = -0.03$
$^3J = 3.05$ Hz

If, by contrast, one of the terminal carbons is exocyclic, there appears again to be a geometric dependence, besides the bond-order effect. It is generally observed that *transoid* couplings are larger than *cisoid* ones, as exemplified by the data given for carboxylic acids (**44** and **45**),[65] suggesting a substituent orientation

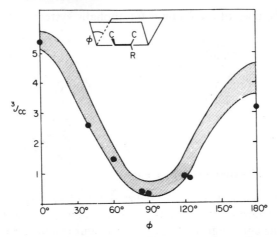

Fig. 2.16   Karplus curve for $^3J_{CC}$ between two sp^3 carbons.[103] The shaded area designates the envelope of the curves calculated (INDO) for 2-butanol and for n-butane. ●: Experimental values.[102,103]

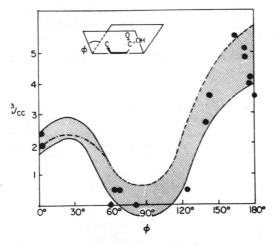

Fig. 2.17   Tentative Karplus curve for $^3J_{CC}$ to carbonyl carbon. Dashed line: values calculated (INDO)[103] for butanoic acid. Shaded area encompasses experimental values.[102]

effect along the lines of the findings for aliphatic systems.

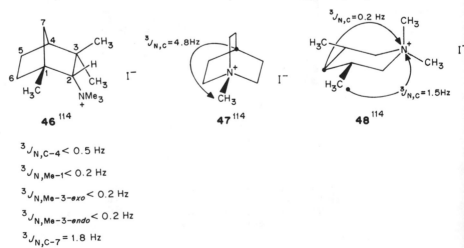

$^3J=0.86$Hz · 4.78 Hz · **44**

2.4Hz · 4.3 Hz · **45**

Coupling between carbon and other nuclei

Among the requirements for facile detectability of spin–spin coupling between carbon and a hetero nucleus X is the existence of a spin-$\frac{1}{2}$ isotope of element X, possessing sufficiently high abundance. Both requirements are ideally met for ^{19}F and ^{31}P. Although nuclei such as ^{14}N, ^{17}O, ^{35,37}Cl, ^{79,81}Br, etc. are magnetic, they are all quadrupolar, and spin–spin coupling is usually not observable because of rapid relaxation of these nuclei, causing effective spin decoupling. Quadrupolar relaxation demands the presence of a net electric field gradient caused by the valence electrons. This field gradient interacts with the quadrupolar moment to cause a fluctuating magnetic field which can be the source of very efficient relaxation, in which case spin–spin coupling to spin-$\frac{1}{2}$ nuclei is not observable.[106-109] However, in the case of quadrupolar nuclei in which the electric field gradients cancel due to symmetry (e.g. ^{14}N nuclei in symmetrically substituted ammonium salts and isonitriles[110-113]), spin–lattice relaxation will be slowed significantly, and coupling to quadrupolar nuclei in such systems generally will be observable. Some examples of quaternary ammonium salts that contain symmetrically substituted ^{14}N$^+$ nuclei wherein $^3J_{N,C}$ couplings have been measured are shown in **46–48**.[114] Three-bond $J_{N,C}$ coupling data obtained

46 [114] $^3J_{N,C}=4.8$Hz   **47** [114] $^3J_{N,C}=0.2$ Hz $^3J_{N,C}=1.5$Hz   **48** [114]

$^3J_{N,C-4}< 0.5$ Hz

$^3J_{N,Me-1}< 0.2$ Hz

$^3J_{N,Me-3-exo}< 0.2$ Hz

$^3J_{N,Me-3-endo}< 0.2$ Hz

$^3J_{N,C-7}= 1.8$ Hz

via examination of the ^{13}C n.m.r. spectra of twelve rigid or anancomeric tetra-alkylammonium iodides were fitted satisfactorily to a Karplus-type curve calculated by the method of Solkan and Bystrov.[115] It was concluded from this study that $^3J_{N,C}$ provides a useful tool for conformational and configurational analysis.[114]

Other spin-$\frac{1}{2}$ isotopes that often play a significant role in organic or in organometallic chemistry are ^{15}N, ^{29}Si, ^{57}Fe, ^{77}Se, ^{103}Rh, ^{119}Sn, ^{195}Pt, ^{199}Hg, ^{203}Tl, ^{205}Tl, and ^{207}Pb. The first three have rather low abundances (i.e. 0.37%, 4.70%, and 2.19%, respectively), and spin–spin couplings that involve these nuclei, therefore, are difficult to detect at natural abundance. The remainder of the isotopes mentioned are sufficiently abundant, so that spin coupling information can be derived from satellite spectra. In the case of rhodium, ^{103}Rh is the only stable isotope, while thallium has two stable spin-$\frac{1}{2}$ isotopes: ^{203}Tl (natural abundance 29.50%) and ^{205}Tl (natural abundance 70.50%). For spin-coupling information regarding the previously listed nuclei, the reader is referred to more specialized literature.[116,117]

Carbon, phosphorus coupling

A discussion of P, C spin–spin coupling is complicated by the enormous versatility of organophosphorus chemistry, caused by the different possible valence states of phosphorus and its ability to form a large number of functionalities. In line with this is a fairly erratic and not yet fully understood coupling behaviour. Whereas the signs of C, H and C, C coupling constants can now be predicted with a high level of confidence, this is not the case for P, C couplings. $^1J_{P,C}$, $^2J_{P,C}$, and $^3J_{P,C}$ can all be positive or negative, and it is therefore clear that by ignoring the sign, erroneous conclusions may be drawn from the data. $^1J_{P,C}$ is found to vary between $ca - 50$ and $+ 300$ Hz,[117,118] depending on phosphorus valence state, substituents, and hybridization of the coupling carbon. Whilst in phosphines $^1J_{P,C}$ is usually small and negative, it is positive and larger in magnitude in phosphonium ions (P^{IV}) and systems with pentavalent phosphorus such as phosphine oxides, phosphonates, etc., as illustrated below (49–53):[117-19].

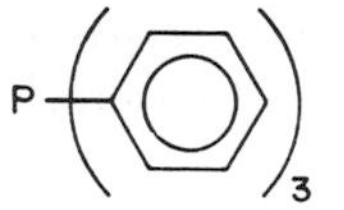

	49	**50**	**51**
$^1J_{P,C}$(Hz)	-12.5^{119}	-50.0^{118}	$+48.5^{117}$

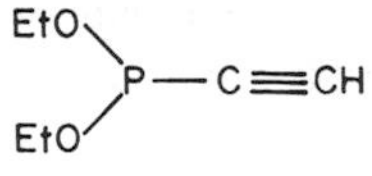

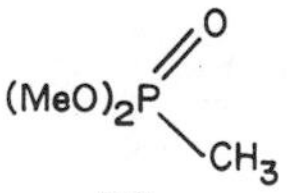

	52	**53**
$^1J_{P,C}$(Hz)	$+142.2^{117}$	$+294.0^{118}$

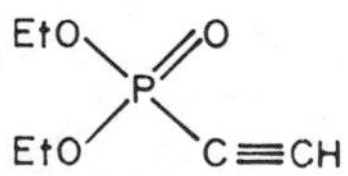

A salient feature of geminal P, C coupling constants in their stereospecific nature, for which ample evidence has been provided in a large number of functionally different organophosphorus compounds, including phosphines,[120,121] aminophosphines,[122] pholenes,[123] and phosphorinanes.[124] The characteristic pattern found in such systems is a smaller value of $^2J_{P,C}$ in cases where the phosphorus lone pair is *trans* oriented relative to the coupling carbon, as exemplified by the data for structures **54**–**7**.

	54	**55**
$^2J_{P,C}$ (Hz)	7.5[124]	0[124]

	56	**57**
$^2J_{P,C}$ (Hz)	0[123]	30[123]

The stereospecificity of $^3J_{P,C}$ is probably of more general applicability. It has been used very widely in conformational and configurational analysis. $^3J_{P,C}$ is usually positive and its magnitude covers a typical range of 5–45 Hz. Examples illustrating the angular dependence of vicinal P, C coupling are provided for the cyclic phosphites **58a** and **58b**:[125]

	58a	**58b**
$^3J_{P,C}$ (Hz)	4.2	13.5

It should be noted that, although the orientation of C-5 is the same in both stereoisomers, the phosphorus lone pair in **58a** is *trans–trans* disposed as opposed to *gauche–trans* in **58b**, obviously leading to a much larger coupling constant in the latter.

A more common Karplus relationship appears to hold for $^3J_{P,C}$ in phosphonates,[126] as demonstrated for systems **59a** and **59b**:[127]

	59a	**59b**
$^3J_{P,C}$ (Hz)	16.2	0.6

The data indicate nearly zero coupling for the *gauche* coupling path (59b), but a large coupling of 16 Hz in 59a, where the phosphonate group is *trans*-oriented.

Because of its biological ramifications, the dihedral angular dependence of $^3J_{POCC}$ in phosphate has been investigated particularly thoroughly. From studies in conformationally rigid cyclic sugar phosphates[128] it could be inferred that $^3J_{P,C}$ (*gauche*) ~ 2 Hz and $^3J_{P,C}$ (*trans*) ~ 8 Hz.

The typical ranges of P, C coupling constants for a number of phosphorus functional groups are listed in Table 2.23.

There are a number of examples that serve to demonstrate the general utility of $^3J_{P,C}$ couplings in conformational and configurational analysis. However, there is mounting evidence that the existence (or non-existence) of angular dependence of vicinal ^{31}P—^{13}C coupling constants depends in part on the oxidation state (i.e. hybridization) of the ^{31}P nucleus in question.[136] Thus, while a Karplus-type dihedral angle dependence for $^3J_{P,C}$ has been demonstrated for a variety of phosphine oxides,[137–9] phosphonates,[126,140–42] phosphine sulphides,[143] and phosphonium salts,[143] this does *not* appear to be the case for corresponding compounds that contain *trivalent* phosphorus (for example, dihalophosphines and dialkylphosphines[143]). Similar conclusions have been reached in the case of ^3J(POCH) couplings, for which separate correlations with dihedral angle are required for trivalent and pentavalent phosphorous systems.[144]

TABLE 2.23 Typical P, C spin–spin coupling constants (Hz)[a]

P Hybridization	System		$^1J_{P,C}$	$^2J_{P,C}$	$^3J_{P,C}$	Ref.
	Phosphines PR^{1}R^{2}R^3	R = Alkyl	$-5 \div -20$	$0 \div +20$	~ 10	129, 130
		R = Aryl, Alkenyl, Alkynyl	$-10 \div -15$	$-10 \div +20$	$-5 \div +10$	118, 119
P^{III}		Phospholes	$-4 \div -16$	$+2 \div +23$	$+5 \div +10$	121, 131
		R = N—Alkyl	—	$-6 \div +50$	$20 \div 25$	122
	Phosphites	R = O—Alkyl	—	$8 \div 10$	$3 \div 5$	132
	Ylides	$R^1R^2R^3\overset{+}{P}$—$\bar{C}R^{2'}$	$+80 \div 130$	$10 \div 20$		133
	Phosphonium salts	$R^1R^2R^3\overset{+}{P}X^-$				
P^{IV}		R = Alkyl	$+40 \div +50$	-5	~ 15	117, 129
		R = Aryl	$85 \div 90$	~ 10	~ 13	129, 133
	Phosphine oxides P(O)R^{1}R^{2}R^3	R = Alkyl	$+50 \div +60$	$2 \div 12$	$O \div 47$	134, 135
		R = Aryl	$+100$	10	5	133
P^V		R = Alkynyl	$+150 \div +300$	$+30 \div +60$	$+3 \div +5$	118
	Phosphonates RP(O)(OR')$_2$		$140 \div 160$	$0 \div 10$	$0 \div 12$	127
	Phosphates $P(O){\overset{OR^1}{\underset{OR^3}{\diagdown OR^2}}}$		—	$-4 \div -7$	$+2 \div +6$	128

[a]Signs have been indicated where they were either determined or where they follow by analogy.

Carbon, fluorine coupling

Fluorine-19 exhibits a less intricate coupling behaviour than that found for phosphorus-31. $^1J_{F,C}$ appears to be always negative and $^2J_{F,C}$ and $^3J_{F,C}$ are probably positive in all coupling situations.[145] Assignment is usually straightforward as F,C couplings progressively decrease with the number of intervening bonds, i.e. $|^1J_{F,C}| \gg |^2J_{F,C}| > |^3J_{F,C}| \gtrsim |^4J_{F,C}|$. A stereospecific behaviour has been reported for $^3J_{F,C}$,[146,147] for which there is evidence that $^3J(gauche) \ll {}^3J(trans)$, with the former being nearly zero, as, for example, demonstrated by the low-temperature data reported for 1,1-difluorocyclohexane (**60**)[147]

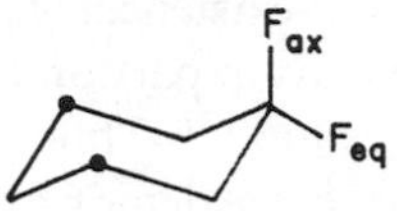

$$^3J_{F_{eq},C} = 9.5 \text{ Hz}$$
$$^3J_{F_{ax},C} = 0$$

60

Whereas at room temperature, at which rapid chair–chair interconversion is operative, $^3J_{F,C} = 4.7$ Hz is observed, (both fluorines coupling equally), twice this value is obtained at $-90°$, whereas the signal displays doublet multiplicity,

TABLE 2.24   F, C spin–spin coupling constants (Hz)[147-7]

	$^1J_{CF}$	$^2J_{CCF}$	$^3J_{CCCF}$	$^4J_{CCCCF}$
$F_3C{-}C(sp^3)$	$-270 \div -285$	$+38 \div +45$		
$F_3C{-}X$	$-260 \div -350$			
$F_3C{-}C(sp^2)$	~ -270	$+32 \div +40$	~ 4	≈ 1
$F_3C{-}C\overset{O}{\diagdown}$	$-280 \div -290$	$\sim +45$		
$F_3C{-}C(sp)$	$-250 \div +260$	$\sim +58$		
$F_2C\overset{X}{\underset{X}{\diagup}}$	$-280 \div -360$			
$F_2C\overset{C}{\underset{C}{\diagup}}$ (sp^3)	$-235 \div -260$	$+19 \div +25$	$0 \div 14$	
$FCH_2{-}X$	$-158 \div -180$			
$F_2C{=}C\diagdown$	~ -287			
$F{-}C(sp^2arom)$	$-230 \div -262$	$+16 \div +21$	$6 \div 8$	≈ 94

thus suggesting that only one of the fluorines (F_{eq}) couples. This is in analogy with the findings in 2, 2-difluoronorbornane (**61**),[146] where only the *endo* fluorine couples:

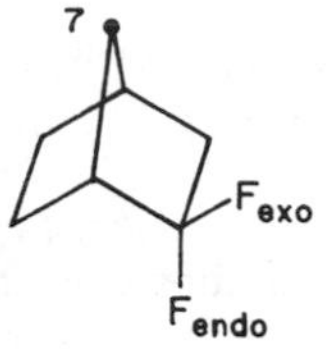

7
F exo
F endo

$^{3}J_{F(endo),C(7)} = 5$ Hz

$^{3}J_{F(exo),C(7)} = 0$

61

Some representative F, C coupling data are summarized in Table 2.24.[145–7]

Although relatively few $^{3}J_{F,C}$ and $^{4}J_{F,C}$ couplings have been measured, it appears that the magnitude of each of these coupling constants may display Karplus-type dihedral angle dependence.[146,148] Interestingly, the results of INDO–MO calculations for both $^{3}J_{F,C}$ and $^{4}J_{F,C}$ couplings (1) confirm their orientational dependence and (2) satisfactorily reproduce data obtained for the respective couplings in anomeric deoxyfluoro-D-glucopyranoses, in 2-deoxy-2-fluoro-D-mannopyranose, and in 4-deoxy-4-fluoro-D-galactose.[148]

Four-bond and more distant couplings involving carbon

Relatively few long-range couplings involving carbon (i.e. $^{n}J_{X,C}$, $n \geq 4$) have been reported. Some examples that involve long-range couplings to carbon in rigid bicyclic systems are shown below:

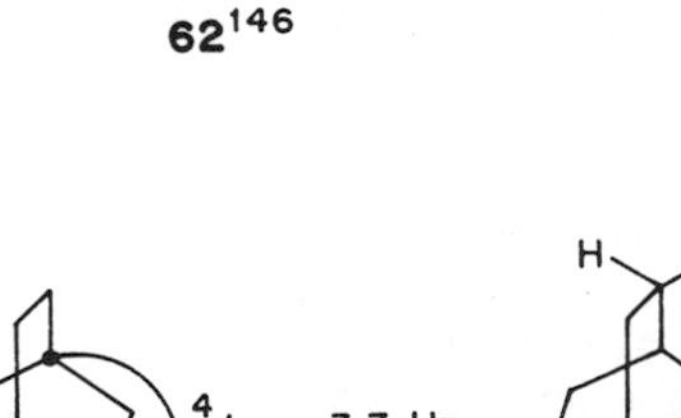

⁵J F,C = 1.0 ± 0.5 Hz
H3C
H
F exo
F endo

62[146]

H
CH3
⁴J F,C = 4.5 ± 0.5 Hz
⁴J F,C
< 1.0 ± 0.5 Hz
F exo
F endo

63[146]

⁴J F,C = 3.3 Hz
F

64[149]

S
H
PPh2
⁵J P,C = 3 Hz
H
Me2HC

65[150]

⁴J C,C =
0.29 ± 0.07 Hz
13COH
O

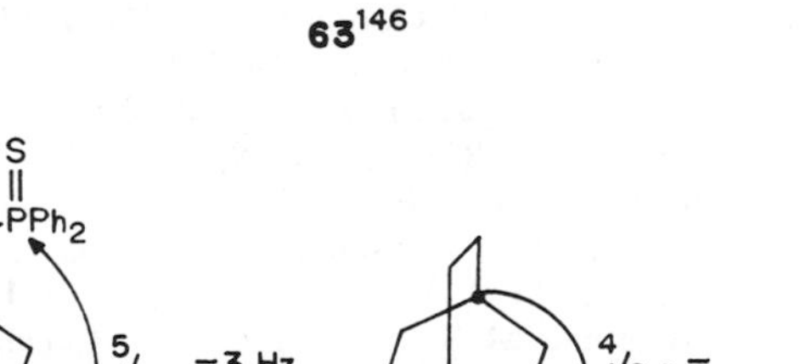

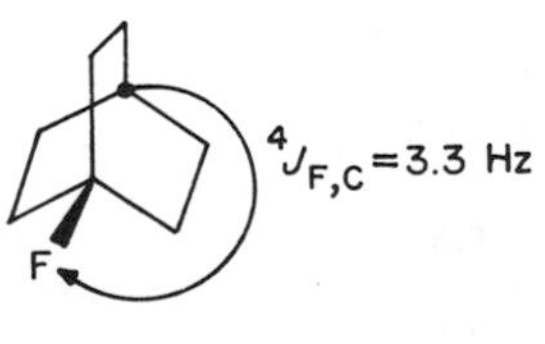

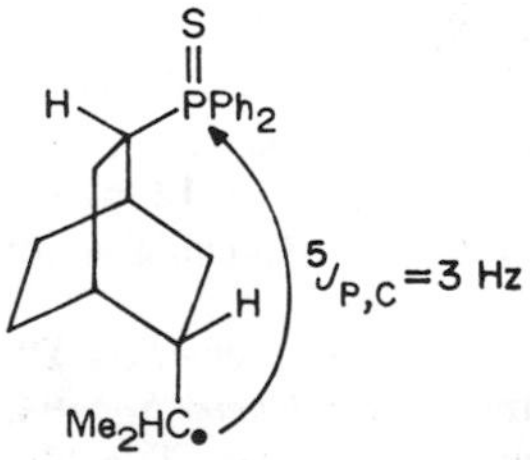

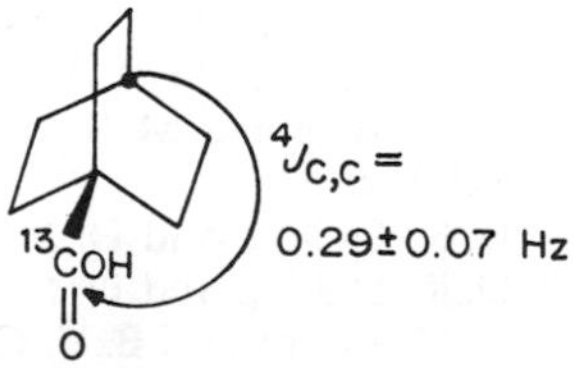

66[151]

References

1 W. T. Raynes, *Nuclear Shielding in Nuclear Magnetic Resonance. A Specialist Periodical Report*, Vol. 2 (1973), Vol. 3 (1974), Vol. 4 (1975). The Chemical Society. London.

2 R. Ditchfield and P. D. Ellis, *Theory of ^{13}C Chemical Shifts in Topics in Carbon-13 Nuclear Magnetic Resonance Spectroscopy* (Ed. G. C. Levy), Vol. 1 (1974), Vol. 2 (1976), Wiley-Interscience, New York.

3 G. E. Maciel, *Substituent Effects on ^{13}C Chemical Shifts. In Topics in Carbon-13 Nuclear Magnetic Resonance Spectroscopy*, cited above, ref. 2.

4 G. C. Levy, R. L. Lichter, and G. Nelson, *Carbon-13 Nuclear Magnetic Resonance Spectroscopy*, John Wiley, New York, 2nd Edition 1980.

5 E. Breitmaier and W. Voelter, ^{13}C *NMR Spectroscopy*, 2nd edn, Verlag Chemie, New York, 1978.

6 J. B. Stothers, *Carbon-13 Nuclear Magnetic Resonance Spectroscopy*, Academic Press, New York, 1972.

7 P. E. Hansen, *Org. Magn. Res.* **12**, 109 (1979).

8 B. Wrackmeyer, ^{13}C NMR spectroscopy of boron compounds. In *Progress in NMR Spectroscopy* (Eds J. W. Emsley, J. Feeney, and L. H. Sutcliffe), Vol. 12, Pergamon Press, New York, 1978.

9 E. L. Eliel and K. M. Pietrusiewicz, ^{13}C NMR of nonaromatic heterocyclic compounds. In *Topics in ^{13}C NMR* (Ed. G. C. Levy), Vol. 3, Chap. 3, Wiley-Interscience, New York, 1979.

10 T. A. Crabb, Nuclear magnetic resonance of alkaloids. In *Annual Reports on NMR Spectroscopy* (Ed. G. A. Webb), Vol. 8, Academic Press, London, 1978.

11 W. B. Smith, ^{13}C NMR spectroscopy of steroids. In Webb (ed.) *op. cit.*; J. W. Blunt and J. B. Stothers, *Org. Magn. Res.* **9**, 439 (1977).

12 H. Sugiyama, *Heterocycles* **11**, 615 (1978).

13 F. W. Wehrli and T. Nishida, The use of Carbon-13 NMR spectroscopy in natural products chemistry. In *Progress in the Chemistry of Organic Natural Products* (Eds W. Herz, H. Grisebach, and G. W. Kirby), Vol. 36, Springer, Vienna, 1979.

14 *Deuterated NMR Solvents—Handy Reference Data*, Merck and Co., Inc. 1979.

15 A. Saika and C. P. Slichter, *J. Chem. Phys.* **22**, 26 (1954).

16 W. E. Lamb, *Phys. Rev.* **60**, 817 (1941).

17 Ref. 5, p.104.

18 J. A. Pople, W. G. Schneider, and H. J. Bernstein, *High-Resolution Nuclear Magnetic Resonance*, McGraw-Hill, New York, 1959, p.166.

19 J. Mason, *J. Chem. Soc.* **A1**, 1038 (1971).

20 W. H. Flygare and J. Goodisman, *J. Chem. Phys.* **49**, 3122 (1968).

21 D. Zeroka, *Chem. Phys. Lett.* **14**, 471 (1972).

22 H. M. McConnell, *J. Chem. Phys.* **27**, 226 (1957).

23 J. A. Pople, *Proc. Roy. Soc.* **A239**, 550 (1957).

24 M. Karplus and J. A. Pople, *J. Chem. Phys.* **38**, 2803 (1963).

25 G. A. Olah and G. D. Matescu, *J. Am. Chem. Soc.* **92**, 1430 (1970).

26 C. Collier and G. A. Webb, *Org. Magn. Res.* **12**, 659 (1979).

27 J. A. Pople and M. S. Gordon, *J. Amer. Chem. Soc.* **89**, 4253 (1967).

28 Ref. 3, p. 64.

29 D. K. Dalling and D. M. Grant, *J. Am. Chem. Soc.* **89**, 6612 (1967).

30 D. K. Dalling and D. M. Grant, *J. Am. Chem. Soc.* **94**, 5318 (1972).

31 D. M. Grant and B. V. Cheney, *J. Am. Chem. Soc.* **89**, 5315 (1967).

32 A. D. Buckingham, *Can. J. Chem.* **38**, 300 (1960).

33 J. G. Batchelor, J. H. Prestgard, R. J. Cushley, and S. R. Lipsky, *J. Am. Chem. Soc.* **95**, 6358 (1973).

34 J. G. Batchelor, *J. Am. Chem. Soc.* **97**, 3410 (1975).
35 J. G. Batchelor, J. Feeney, and G. C. K. Roberts, *J. Magn. Res.* **20**, 19 (1975).
36 H. J. Schneider and W. Freitag, *J. Am. Chem. Soc.* **99**, 8363 (1977).
37 H. J. Schneider, W. Gschwendter, and U. Buchheit, *J. Magn. Res.* **26**, 175 (1977).
38 E. L. Eliel, W. F. Bailey, L. D. Kopp, R. L. Willer, D. M. Grant, R. Bertrand, K. A. Christensen, D. K. Dalling, M. W. Duch, E. Wenkert, F. M. Schell, and D. W. Cochran, *J. Am. Chem. Soc.* **97**, 322 (1975).
39 Ref. 9, p. 204.
40 A. R. Katritzky and R. D. Topsom, *J. Chem. Educ.* **48**, 427 (1971).
41 W. F. Reynolds, I. R. Peat, M. H. Freedman, and J. R. Lyerla, *Con. J. Chem.* **51**, 1857 (1973).
42 G. L. Nelson, G. C. Levy, and J. D. Cargioli, *J. Am. Chem. Soc.* **94**, 3089 (1972).
43 E. M. Schulman, K. A. Christensen, D. M. Grant, and C. Walling, *J. Org. Chem.* **39**, 2686 (1974).
44 C. G. Swain and E. C. Lupton, *J. Am. Chem. Soc.* **90**, 4328 (1968).
45 R. H. Levin and J. D. Roberts, *Tetrahedron Lett.* **135** (1973).
46 T. Pehk and E. Lippmaa, *Org. Magn. Res.* **3**, 679 (1971).
47 B. V. Cheney and D. M. Grant, *J. Am. Chem. Soc.* **89**, 5319 (1967).
48 S. H. Grover, J. P. Guthrie, J. B. Stothers, and C. T. Tan, *J. Magn. Res.* **10**, 227 (1973).
49 S. H. Grover and J. B. Stothers, *Can. J. Chem.*, **52**, 870 (1974).
50 K. R. Stephens, J. B. Stothers and C. T. Tan, Mass spectrometry and nuclear magnetic resonance spectroscopy. In *Pesticide Chemistry* (Eds R. Haque and F. J. Biros), Plenum Press, New York, 1974.
51 J. G. Batchelor, *J. Magn. Res.* **18**, 212 (1975).
52 D. M. Grant and E. G. Paul, *J. Am. Chem. Soc.* **86**, 2984 (1964).
53 L. P. Lindeman and J. Q. Adams, *Anal. Chem.* **43**, 1245 (1971).
54 D. E. Derman, M. Jautelat and J. D. Roberts, *J. Org. Chem.* **36**, 2757 (1971).
55 D. K. Dalling and D. M. Grant, *J. Am. Chem. Soc.* **89**, 6612 (1967).
56 D. K. Dalling, D. M. Grant, and E. G. Paul, *J. Am. Chem. Soc.* **95**, 3718 (1973).
57 D. K. Dalling and D. M. Grant, *J. Am. Chem. Soc.* **96**, 1827 (1974).
58 H. J. Schneider and V. Hoppen, *Tetrahedron Lett.* **579** (1974).
59 O. A. Subbotin and N. M. Sergeyev, *J. Am. Chem. Soc.* **97**, 1080 (1975).
60 F. Naf, B. L. Buckwalter, I. R. Burfitt, A. A. Nagel, and E. Wenkert, *Heiv. Chim. Acta* **58**, 1967 (1975).
61 I. Wahlberg, S. O. Almquist, T. Nishida, and C. Enzell, *Acta Chem. Scand.* **29B**, 1047 (1975).
62 J. T. Clerc, E. Pretsch, and S. Sternhell, *^{13}C-Kernresonanzspektroskopie*, Akademische Verlagsgesellschaft, Frankfurt/M, 1973.
63 U. Vögeli and W. von Philipsborn, *Org. Magn. Res.* **7**, 617 (1975).
64 U. Vögeli, D. Herz, and W. von Philipsborn, *Org. Magn. Res.* **13**, 200 (1980).
65 J. L. Marshall, Carbon–Carbon and carbon–proton couplings—applications to organic stereochemistry. In *Methods in Stereochemical Analysis* (Ed. A. P. Marchand), Vol. 2, Verlag Chemie International, Deerfield Beach, Florida, 1983.
66 J. L. Marshall, D. E. Miiller, S. A. Conn, R. Seinwell, and A. M. Ihrig, *Acc. Chem. Res.* **7**, 333 (1974).
67 G. E. Maciel, *$^{13}C-^{13}C$ Coupling Constants; in Nuclear Magnetic Resonance of Nuclei Other than Protons* (Eds T. Axenrod and G. A. Webb), Wiley-Interscience, New York, 1974.
68 N. K. Wilson, Stereochemical aspects of ^{13}C nuclear magnetic resonance spectroscopy. In *Topics in Stereochemistry* (Eds E. L. Eliel and N. L. Allinger), Vol. 8, Wiley-Interscience, New York, 1974.
69 P. E. Hansen, Long-range $^{13}C-^{13}C$ coupling constants. A review. *Org. Magn. Res.* **11**, 215 (1978).

70 For a review of the theory of spin–spin coupling and calculations of C, H and C, X spin–spin coupling constants see P. D. Ellis and R. Ditchfield, in *Topics in* ^{13}C *NMR* (Ed. G. C. Levy), Vol. 2, Chap. 8, Wiley-Interscience, New York, 1976.

71 J. A. Pople and D. P. Santry. *Mol. Phys.* **8**, 1 (1964).

72 N. Muller and D. E. Pritchard, *J. Chem. Phys.* **31**, 768, 1471 (1959).

73 M. D. Newton, J. M. Schulman, and M. M. Manus, *J. Am. Chem. Soc.* **96**, 17 (1974).

74 N. H. Werstiuk, R. Taillefer, R. A. Bell, and B. A. Sayer, *Can. J. Chem.* **51**, 3010 (1973).

75 N. J. Hoboken and E. R. Malinowski, *J. Am. Chem. Soc.* **83**, 1479 (1961).

76a P. E. Hansen Carbon–hydrogen spin–spin coupling constants. In *Progress in NMR spectroscopy* (Eds J. W. Ensley, J. Feeney, and L. H. Sutcliffe), Vol. 14, Pergamon, New York, 1981.

76b W. von Philipsborn, Review paper, Proceedings of the Third Conference of the Groupe d'Etudes Resonance Magnetique (GERM III), Vichy, 1979.

77 T. Spoormaker and M. J. A. deBie, *Recl. Trav. Chim. Pays-Bas* **99**, 194 (1980).

78 J. A. Schwarcz, N. Cyr, and A. S. Perlin, *Can. J. Chem.* **53**, 1872 (1975).

79 L. Ernst and V. Wray, *J. Magn. Res.* **25**, 123 (1977).

80 A. R. Tarpley and J. H. Goldstein, *J. Phys. Chem.* **76**, 515 (1972).

81 M. Hansen and H. J. Jakobsen, *J. Magn. Res.* **10**, 74 (1973).

82 F. J. Weigert and J. D. Roberts, *J. Am. Chem. Soc.* **90**, 3543 (1968).

83 T. N. Huckerby, *J. Molec. Struct.* **31**, 161 (1976).

84 R. E. Wasylishen and H. M. Hutton, *Can. J. Chem.* **55**, 619 (1977).

85 J. Uzawa and M. Uramoto, *Org. Magn. Res.* **12**, 612 (1979).

86 See, for example, Ref. 65 and references cited therein.

87 J. L. Marshall and R. Seiwell, *Org. Magn. Res.* **8**, 419 (1976).

88 J. A. Schwarcz and A. S. Perlin, *Can. J. Chem.* **50**, 3667 (1972).

89 R. Wasylishen and J. Schaefer, *Can. J. Chem.* **50**, 2710 (1972).

90 R. U. Lemieux, T. L. Nagabushan, and B. Paul, *Can. J. Chem.* **50**, 773 (1972).

91 L. T. J. Delbaere, M. N. G. James, and R. U. Lemieux, *J. Am. Chem. Soc.* **95**, 7866 (1973).

92 P. E. Hansen, J. Feeney, and G. C. K. Roberts, *J. Magn. Res.*, **17**, 249 (1975).

93 T. Spoormaker and M. J. A. deBie, *Recl. J. Roy. Neth. Chem. Soc.* **98**, 380 (1979).

94 F. J. Weigert and J. D. Roberts, *J. Am. Chem. Soc.* **94**, 6021 (1972).

95 H. Egli and W. von Philipsborn, *Tetrahedron Lett.* 4265 (1979).

96 J. A. Pople and M. Gordon, *J. Am. Chem. Soc.* **89**, 4253 (1967).

97 R. Richarz, W. Ammann, and T. Wirthlin, *J. Magn. Res.* **45**, 270 (1981).

98 See, for example, S. A. Linde and H. J. Jakobsen, *J. Am. Chem. Soc.* **98**, 1041 (1976).

99 Ref. 69, Table 1.

100 T. E. Walker, R. E. London, T. W. Whaley, R. Barter, and N. A. Matwiyoff, *J. Am. Chem. Soc.* **98**, 5807 (1976).

101 P. A. Chaloner, *J. C. S. Perkin Trans. II*, 1028 (1980).

102 J. L. Marshall and D. E. Miiller, *J. Am. Chem. Soc.* **95**, 8305 (1973).

103 M. Barfield, I. Burfitt and D. Doddrell, *J. Am. Chem. Soc.* **97**, 2631 (1975).

104 P. E. Hansen, O. K. Poulsen, and A. Berg, *Org. Magn. Res.* **12**, 43 (1979).

105 For a condensed review of some relevant literature see, for example, D. F. Erwing, Applications of Spin–spin couplings in nuclear magnetic resonance. In *Specialist Periodical Report of the Royal Society of Chemistry* (Senior Reporter: G. A. Webb), Vol. 10, Alden Press, London, 1981.

106 J. A. Pople, W. G. Schneider, and H. J. Bernstein, *High Resolution Nuclear Magnetic Resonance*, McGraw-Hill, New York, 1959, pp. 102, 216.

107 M. Witanowski and G. A. Webb, *Annual Reports on NMR Spectroscopy* (Ed. G. A. Webb), Vol. 5a, Academic Press, London, 1972, pp. 395–404.

108 M. J. O. Anteunis, F. A. M. Borremans, J. Gelan, A. P. Marchand, and R. W. Allen, *J. Am. Chem. Soc.* **100**, 4050 (1978).

109 A. P. Marchand, *Stereochemical Applications of NMR Studies in Rigid Bicyclic Systems*, Verlag Chemie International, Deerfield Beach, Florida, 1982, pp. 152–3, 164.

110 Y. Terui, K. Aono, and K. Tori, *J. Am. Chem. Soc.* **90**, 1069 (1968).

111 K. Tori, T. Iwata, K. Aono, M. Ohtsuru, and T. Nakagawa, *Chem. Pharm. Bull.* **15**, 329 (1967).

112 P. G. Gassman and D. C. Heckert, *J. Org. Chem.* **30**, 2859 (1965).

113 R. A. Ogg and J. D. Ray, *J. Chem. Phys.* **26**, 1339, 1340 (1957).

114 L. A. Valckx, F. A. M. Borremans, C. E. Becu, R. H. K. DeWaele, and M. J. O. Anteunis, *Org. Magn. Res.* **12**, 302 (1979).

115 V. N. Solkan and V. F. Bystrov, *Izv. Akad. Nauk SSSR Ser. Khim.* **1**, 102 (1974); see *Chem. Abstr.* **80**, 121315s (1974).

116 B. E. Mann, *Advanc. Organomet. Chem.* **12**, 135 (1974).

117 W. McFarlane, *Proc. Roy. Soc.* **A306**, 185 (1968).

118 R. -M. Lequan, M. -J. Pouet, and M. -P. Simonnin, *Org. Magn. Res.* **7**, 392 (1975).

119 T. Bundegaard and H. J. Jakobsen, *Acta Chem. Scand.* **26**, 2548 (1972).

120 S. Sørensen, R. S. Hansen, and H. J. Jakobsen, *J. Am. Chem. Soc.* **94**, 5900 (1972).

121 G. Gray and H. Nelson, *Org. Magn. Res.* **14**, 14 (1980).

122 See G. A. Gray and J. H. Nelson, *Org. Magn. Res.* **14**, 8 (1980) and references cited therein.

123 J. J. Breen, S. I. Featherman, L. D. Quin and R. C. Stocks, *Chem. Commun.* 657 (1972).

124 S. I. Featherman and L. D. Quin, *Tetrahedron Lett.* 1955 (1973).

125 M. Haemers, R. Ottinger, D. Zimmermann and J. Reisse, *Tetrahedron* **29**, 3539 (1973).

126 G. W. Buchanan and C. Benezra, *Can. J. Chem.* **54**, 231 (1976).

127 G. W. Buchanan and F. G. Morin, *Can. J. Chem.* **58**, 530 (1980).

128 R. D. Lapper, H. H. Mantsch, and I. C. P. Smith, *J. Am. Chem. Soc.* **94**, 6243 (1972).

129 F. J. Weigert and J. D. Roberts, *Inorg. Chem.* **12**, 313 (1973).

130 B. E. Mann, *J. C. S., Perkin Trans. II*, 30 (1972).

131 T. Bundegaard and H. J. Jakobsen, *Tetrahedron Lett.* 3353 (1972).

132 H. W. Tan and W. G. Bentrude, 619 (1975); W. G. Bentrude and W. H. Tan, *J. Am. Chem. Soc.* **98**, 1850 (1976).

133 G. A. Gray, *J. Am. Chem. Soc.* **95**, 7736 (1973).

134 G. A. Gray and S. E. Cremer, *J. Org. Chem.* **37**, 3458, 3470 (1972).

135 R. B. Wetzel and G. L. Kenyon, *Chem. Commun.* 287 (1973).

136 A. P. Marchand, Ref. 109, pp. 115, 164–6.

137 R. B. Wetzel and G. L. Kenyon, *J. Am. Chem. Soc.* **96**, 5189 (1974).

138 C. A. Kingsbury and D. Thoennes, *Tetrahedron Lett.* 3037 (1976).

139 J. R. Wiseman and H. O. Krabbenhoft, *J. Org. Chem.* **41**, 589 (1970).

140 G. W. Buchanan and F. G. Morin, *Can. J. Chem.* **55**, 2885 (1977).

141 G. W. Buchanan and J. H. Bowen, *Can. J. Chem.* **55**, 604 (1977).

142 L. Ernst, *Org. Magn. Res.* **9**, 35 (1977).

143 L. D. Quin and L. B. Littlefield, *J. Org. Chem.* **43**, 3508 (1978).

144 D. W. White and J. G. Verkade, *J. Magn. Res.* **3**, 111 (1970).

145 F. J. Weigert and J. D. Roberts, *J. Am. Chem. Soc.* **93**, 2361 (1971).

146 J. B. Grutzner, M. Jautelat, J. B. Dence, R. A. Smith, and J. D. Roberts, *J. Am. Chem. Soc.* **92**, 7107 (1970).

147 D. Doddrell, C. Charrier, and J. D. Roberts, *Proc. Nat. Acad. Sci.* **67**, 1649 (1970).

148 V. Wray, *J. C. S., Perkin Trans. II*, 1598 (1976).

149 G. E. Maciel and H. C. Dorn, *J. Am. Chem. Soc.* **93**, 1268 (1971).

150 M. Lauer, O. Samuel, and H. B. Kagan, *J. Organometal. Chem.* **177**, 309 (1979).

151 M. Barfield, S. E. Brown, E. D. Canada, Jr, N. D. Ledford, J. L. Marshall, and E. Yakali, *J. Am. Chem. Soc.* **102**, 3355 (1980).

152 R. L. Lichter and R. E. Wasylishen, *J. Am. Chem. Soc.* **97**, 1808 (1975).

3

Experimental Techniques for Spectral Assignment

3.1 SUMMARY OF METHODS

This chapter is intended to give the user of ^{13}C n.m.r. a detailed outline of the experimental techniques available for correlating spectral features with constitution and stereochemistry. Although ^{13}C spectral parameters are closely analogous to those in proton n.m.r. there are distinct differences between the two types of spectroscopy. Whereas the normal recording mode in proton n.m.r. is single resonance, the overwhelming majority of ^{13}C n.m.r. spectra are obtained under proton noise-decoupling conditions (double resonance). The purpose of the second irradiation frequency is to simultaneously remove all carbon–proton scalar couplings with a twofold benefit; (1) an increase of S/N by one to two orders of magnitude and (2) a simplification of otherwise very complex spectra. This is exemplified with the spectra in Figs 3.1(a) and (b), both obtained under identical experimental conditions (acquisition time, pulse flip angle, pulse repetition time, number of pulses, etc.) except that for the spectrum in Fig. 3.1(b) proton noise-decoupling had been employed, versus no decoupling at all in trace (a). Clearly, the enhanced S/N achieved by decoupling is at the expense of information, reducing this latter to the chemical shift. Although broadbanded proton decoupling will remain the standard mode of operation, advances in instrumentation over the past ten years have rendered possible recording of proton-coupled spectra without intolerable sacrifice in spectrometer time. Increased intrinsic detection sensitivity (by a factor of 5–10) has been complemented more recently by the advent of new pulse techniques such as INEPT and DEPT,[1-4] further enhancing S/N. Moreover, the average field strength at which ^{13}C spectra are being recorded is considerably higher than a decade ago, thus giving an additional impetus to proton-coupled spectra. Mention should also be made of new experiments such as selective excitation[5-7] and two-dimensional J-resolved n.m.r.[8,9] which permit extraction of single-isotopomer subspectra, therefore facilitating analysis. Last but not least, however, it is the proven stereochemical

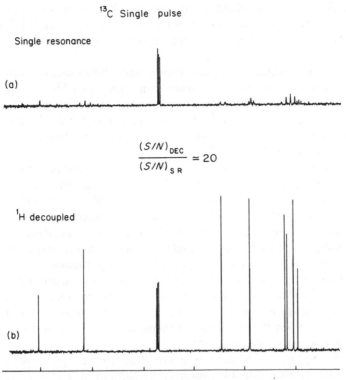

Fig. 3.1 ^{13}C n.m.r. spectra of 30% (v/v) β-pinene in deuterochloroform, both recorded under the same experimental conditions except that in (b) proton noise-decoupling was employed, versus no decoupling at all in experiment (a), demonstrating a roughly twenty-fold S/N increase, brought about by decoupling.

potential of carbon–proton coupling constants[10] and the large database which exists today that have further spurred the chemist's interest in proton-coupled spectra.

Single-frequency off-resonance decoupling (SFORD),[11-13] a double-resonance mode which, in its effect, is intermediate between full decoupling and full retention of spin coupling, provides spectra in which all ^{13}C—^{1}H couplings are reduced, ideally showing splittings from one-bond couplings only. The technique, used complementarily to broadband decoupling,[11,12] therefore allows the user to retrieve part of the lost coupling information without sacrificing the general benefits of proton decoupling (NOE) and being considerably more sensitive than fully coupled spectra. Off-resonance decoupling appears to have lost some of its appeal as a multiplicity assignment aid[13] because of competition from new methods such as INEPT with decoupling[3,4] or APT,[14] which provide multiplicity information while maintaining the sensitivity advantages of noise-decoupling. However, an in-depth analysis of the off-resonance data, we will see, can provide additional structural information beyond just multiplicity, and SFORD spectra will therefore, in the authors' view, remain a powerful assignment tool.

Another new technique, subsequently discussed, allows selective observation at natural abundance of ^{13}C isotopomers containing two ^{13}C isotopes in the same molecule.[15-19] From the carbon–carbon spin–spin coupling constants the connectivity among the various carbon atoms in the molecule can be derived. The method therefore bears the potential to map out the carbon skeleton of completely unknown molecules.[17-19]

Besides the new n.m.r. techniques which usually require state-of-the-art instrumentation, the methodology of signal identification will remain to rely on some of the more traditional methods such as chemical shift correlation, whose principle is to compare the shieldings within closely related molecular frameworks and with those in model compounds.[20] Incorporation of deuterium labels is often synthetically straightforward, and the resulting secondary isotope effects on nuclear shielding are well documented.[21,22] Lanthanide shift reagents, routinely used in proton n.m.r., have proven their usefulness also in ^{13}C n.m.r.[23,24] albeit there remains some doubt as to the reliability of a treatment of the incremental shifts in terms of a dipolar model.[25] Another benefit of the technique results from the concomitant increased dispersion of proton shifts, therefore enabling selective irradiation.

Apart from coupling of ^{13}C to protons, coupling to other spin-$\frac{1}{2}$ nuclei such as ^{19}F or ^{31}P may provide valuable assignment information. The stereospecificity has in many instances been applied to deduce stereochemistry and conformation of molecules.[26]

A further powerful structural parameter is the spin–lattice relaxation time, T_1.[27-31] Since interpretation of T_1 data requires a more detailed knowledge of the underlying mechanisms, this topic will be treated separately in Chapter 4.

3.2 PROTON-DECOUPLING TECHNIQUES

The initial ^{13}C experiment, conducted to obtain a survey spectrum is, as a rule, always recorded under proton broadband (noise) decoupling. As we have seen, this is the most efficient recording mode, providing the chemical shift and, ideally, the carbon number of the molecule. The term *broadband* or *noise decoupling* implies a modulation scheme designed to distribute the rf power so as to afford uniform decoupling across the band of proton frequencies.[11,32] We will not elaborate on the technicalities of noise decoupling, which has become a routine experiment (cf. also Chapter 1).

In this section we shall treat the various simple double-resonance methods leading to spectra in which carbon–proton spin–spin couplings are partially or fully retained. Hence what such experiments share in common is that during the acquisition period there is either some form of coherent irradiation or no irradiation at all. We shall exclude the more complex multiple-pulse experiments which are addressed in the subsequent sections of this chapter.

Single-frequency off-resonance decoupling (SFORD)

Coherent or single-frequency proton decoupling off-resonance[11-13] has the effect of reducing the one-bond ^{13}C—^1H coupling to a fraction of its actual value while in most cases removing the much smaller long-range couplings. The resulting reduced splitting J^r is related to the frequency offset, Δv, of the decoupler from resonance and the decoupler field amplitude, H_2, usually expressed in Hertz as $\gamma_x H_2/2\pi$, where γ_x is the magnetogyric ratio of the proton.

The experiment is best visualized in terms of a vector diagram, based on the rotating frame of reference.[33] For the sake of simplicity, we choose a two-spin system AX, where A and X denote ^{13}C and ^1H, respectively. This situation is displayed in Fig. 3.2. We further assume that the decoupler field H_2 is applied off-resonance by an amount $\Delta v = v_X - v_2$, and the x- and y-axes of the Cartesian co-ordinate system rotate in synchronism with the double-resonance frequency v_2 about the z-axis, therefore reducing the field in this direction to Δv Hz. Hence the effective field to which the nucleus is subjected is given by the vector sum of Δv and $\gamma_x H_2/2\pi$. However, the energy of X also depends on the polarization of the A-spin, i.e. on the magnetic quantum number m_A. In practice, this means that the z

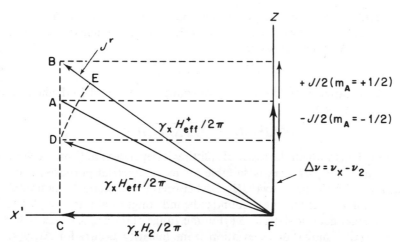

Fig. 3.2 Vector diagram representing the effective field H_{eff} acting on nucleus X of an AX spin system, in which X is irradiated Δv Hz away from its resonance with a field of amplitude H_2. The reference frame rotates with a frequency v_2 (decoupler frequency) about the z axis. The effective fields denoted H_{eff}^+ and H_{eff}^- correspond to different spin polarizations of the A spin ($m_A = +\frac{1}{2}$ and $-\frac{1}{2}$, respectively). The residual splitting J^r in the A doublet is obtained as the difference between $\gamma_x H_{eff}^+/2\pi$ and $\gamma H_{eff}^-/2\pi$.

component of the effective field varies with $m_A(= \pm\frac{1}{2})$ through spin–spin coupling, and it can therefore assume the values $\Delta v + J/2$ or $\Delta v - J/2$. This leads to the resultant effective fields

and

$$\gamma_X H_{eff}^+/2\pi = [(\Delta v + J/2)^2 + (\gamma_X H_2/2\pi)^2]^{1/2} \qquad (3.1)$$

$$\gamma_X H_{eff}^-/2 = [(\Delta v - J/2)^2 + (\gamma_X H_2/2\pi)^2]^{1/2} \qquad (3.2)$$

In order to calculate the total magnetic energy of the system we have further to specify the spin state of X ($m_X = \pm\frac{1}{2}$). The total energy can then be written as

$$E(m_A, m_X) = m_A v_A + (\gamma_X H_{eff}/2\pi)m_X \qquad (3.3)$$

In Eqn. (3.3) H_{eff}^+ or H_{eff}^- has to be inserted, depending on whether $m_A = \frac{1}{2}$ or $-\frac{1}{2}$. $-\frac{1}{2}$. For strong irradiating fields where H_{eff}^+ and H_{eff}^- become nearly parallel, it can be shown that transitions involving a change in m_X become forbidden,[33] and two lines will be obtained in the ^{13}C spectrum, corresponding to frequencies

$$v_A(m_X) = v_A + m_X[(\gamma_X H_{eff}^+/2\pi) - (\gamma_X H_{eff}^-/2\pi)] \qquad (3.4)$$

where $m_X = \pm\frac{1}{2}$.

Hence a doublet of reduced splitting J^r is predicted whose magnitude corresponds to the difference of the line frequencies, i.e.

$$J^r = \gamma_X H_{eff}^+/2\pi - \gamma_X H_{eff}^-/2\pi \qquad (3.5)$$

or explicitly:

$$J^r = [(\Delta v + J/2)^2 + (\gamma_X H_2/2\pi)^2]^{1/2} - [(\Delta v - J/2)^2 + (\gamma_X H_2/2\pi)^2]^{1/2} \qquad (3.6)$$

Equation (3.5) simplifies on condition that $\gamma_X H_2/2\pi \gg \Delta v, J$. This becomes evident from Fig. 3.2 if we replace the arc DE by a straight line intersecting AB at right angles. We then can write:

$$DE \simeq [J^2 - (J^r)^2]^{1/2} \qquad (3.7)$$

and from the approximate similarity of triangles AFC and BDE the following proportionality follows:[34]

$$J^r/[J^2 - (J^r)^2]^{1/2} \simeq \Delta v/(\gamma_X H_2/2\pi) \qquad (3.8)$$

Equation (3.8), in which the residual splitting is dependent only on the value of J and the ratio $\Delta v/(\gamma_X H_2/2\pi)$, is suitable for most practical purposes. Figure 3.3 shows a plot of J^r versus $\Delta v/(\gamma_X H_2/2\pi)$ between limits 0 and 1, computed from Eqn. (3.8) for $J = 125$ Hz. It was found to be indistinguishable from a plot derived from the exact equation (Eqn. (3.6)) in the range 500 Hz $\leqslant \gamma_X H_2/2\pi \leqslant 4000$ Hz. Figure 3.3 also shows that deviation from linearity occurs for $\Delta v/(\gamma_X H_2/2\pi) > 0.25$. Below this critical value Eqn. (3.8) simplifies further, since under these conditions $J^r \ll J$; leading to

$$J^r \simeq J \cdot \Delta v \cdot (\gamma_X H_2/2\pi)^{-1} \qquad (3.9)$$

i.e. J^r is simply proportional to the offset parameter $\Delta v/(\gamma_X H_2/2\pi)$.

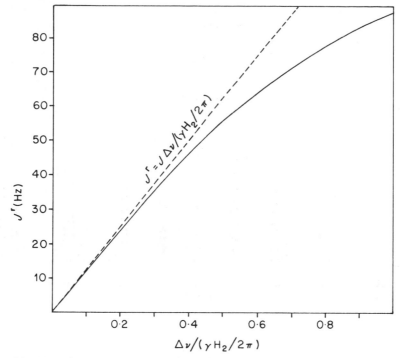

Fig. 3.3   Residual splitting J^r, calculated from Eqn. (3.8) for a coupling constant $J_{CH} = 125\,Hz$ as a function of the ratio $\Delta\nu/(\gamma H_2/2\pi)$. The straight line corresponds to the approximation implicit to Eqn. (3.9).

So far, we have dealt with the simplest case of a methine carbon (AX). Extension to the methylene (AX$_2$) and methyl (AX$_3$) spin system is straightforward. For the general AX$_n$ case the total spin quantum number for the X spins becomes $I_X = n/2$ and the total magnetic quantum number M_X consequently can assume $2n + 1$ values, ranging from $+ n/2$ to $- n/2$. Equation (3.4) then assumes the more general from

$$\nu_A(M_X) = \nu_A + M_X[\gamma_X H_{eff}^+/2\pi - \gamma_X H_{eff}^-/2\pi] \tag{3.10}$$

Multiplets will thus be observed as in the absence of decoupling, but with a splitting, given by Eqn. (3.5).

Figure 3.4(a) shows a series of SFORD spectra for methanol, obtained by incrementally varying the decoupler frequency offset, $\Delta\nu$, at constant power level ($\gamma H_2/2\pi = 3.7\,kHz$). It is seen that the residual splitting increases with increasing values of $|\Delta\nu|$, as predicted by Eqn. (3.9). A plot of J^r against $\Delta\nu$ exhibits the expected linearity within the range shown ($0 < |\Delta\nu|/(\gamma_X H_2/2\pi) < 0.24$). Conversely, this experiment may serve to calibrate the decoupling field amplitude from a measurement of J^r and known values of $\Delta\nu$ and J_{CH}.

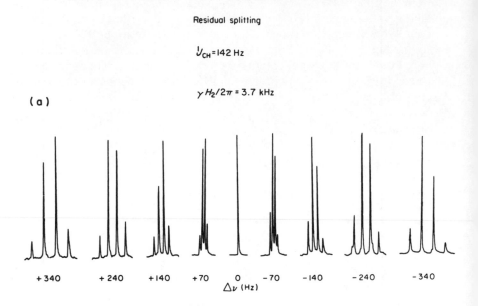

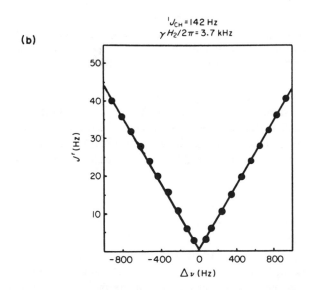

Fig. 3.4 (a) ^{13}C SFORD spectra, showing the methanol quartet, recorded at constant decoupler field amplitude ($\gamma H_2/2\pi = 3.7\,\text{kHz}$) and various settings of the decoupler frequency offset. (b) Experimental residual splittings, partly shown in (a), plotted against the decoupler offset $\Delta\nu$, demonstrate linearity within the range of offset values used.

From the obtained signal multiplicities in the off-resonance spectrum it is now possible to discriminate between methyl, methylene, methine, and quaternary carbons. However, further conclusions can be drawn from SFORD spectra, by exploiting not only multiplicities but also the magnitudes of the splittings. It is obvious from the foregoing that, provided both the one-bond coupling constant $^1J_{CH}$ and the proton chemical shifts are known, specific assignments can be made. Although $^1J_{CH}$ is seldom known in advance, reliable estimates can be made, based on Malinowski's additivity rule[35] (see Chapter 2).

The practical use of the technique is demonstrated in the following with a typical organic molecule. The proton noise-decoupled spectrum of brucine (1)[36] at 2.3 T reveals twenty-three separate resonance lines corresponding to the twenty-three carbons in the molecule. On chemical shift grounds alone, only two carbons (C-13 and C-15) can unambiguously be assigned:

Whereas the assignment of the quaternary carbons, which are found in the low-field region of the spectrum, requires additional experimental data,[36] the majority of aliphatic carbons can be assigned with the combined aid of SFORD and shielding arguments. An expansion of the high-field portion of the proton noise-decoupled together with the SFORD spectrum is displayed in Figs 3.5(a) and (b), respectively.

Out of a total of fourteen lines, two are identified as quartets, six as triplets, five as doublets, whilst only one has singlet multiplicity, in accordance with the number of methyls, methylenes, methines, and quaternary carbons anticipated to resonate in the region between 0 and 100 ppm from TMS.

The decoupler frequency was centred at 4 ppm towards lower frequency from TMS. This implies that the reduced splittings increase with decreased shielding of the protons attached to the relevant carbon for a given value of $^1J_{CH}$. However, electronegative substituents raise the s-character of a C—H bond and hence the value of $^1J_{CH}$. The proton chemical shifts follow the same trend, i.e. increased substituent electronegativity progressively deshields protons. One recognizes that by irradiating at high field (low frequency) with respect to all proton resonances the two effects governing residual splitting are mutually supportive.

Following these arguments one concludes that the highest-frequency triplet must be due to C-22 in α position to the ether oxygen and the olefinic double bond. This carbon signal exhibits the largest residual splitting among the

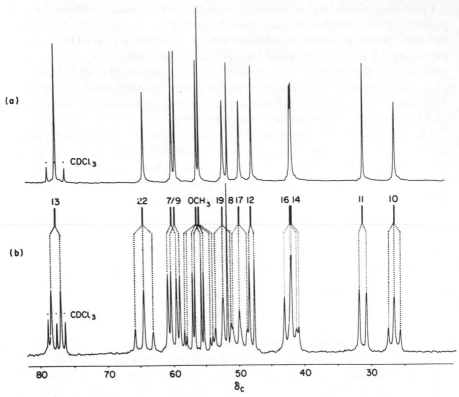

Fig. 3.5   25.2 MHz (2.3 T) ^{13}C spectrum, showing the aliphatic carbon region of brucine (1). (a) Proton noise-decoupled; (b) off-resonance decoupled, $\gamma H_2/2\pi = 3.6$ kHz, irradiation frequency centred at 4 ppm to low frequency from TMS.

methylene resonances. Since the protons at C-19 are deshielded by both the amine nitrogen and the olefinic double bond, its residual splitting should be larger than that of any remaining triplet signal. This allows the line at 52.6 ppm to be assigned to C-19 and the one at 50.0 ppm to C-17 or C-14, for which a larger residual splitting is expected than for C-10 or C-16. At this point we tentatively assign the line at 42.2 ppm to C-14 and the one at 50.0 ppm to C-17. It may be noted that, although the two lines around 42 ppm are closely spaced, they exhibit markedly different residual splittings. Shielding arguments finally permit the designation of the resonances at 26.7 and 42.4 ppm to C-10 and C-16, respectively. Among the methine carbons, the lines originating from carbons 7, 9, and 11 should have larger residual splittings than C-12. It is indeed found that the doublet 48.2 ppm exhibits by far the smallest splitting among the methine resonances. For chemical shift reasons the line at 31.5 ppm is assigned to C-11. No discrimination, however, is possible between the two lines belonging to C-7 and C-9. The only singlet, of course, pertains to the quaternary C-8, while the two quartets are due to the two methoxyl carbons.

These assignments are in almost complete agreement with those reported later.[37]

Second-order effect in SFORD spectra

The appearance of well-resolved sharp multiplets, as shown in the previously discussed example, is not mandatory by any means. Figure 3.6(a) shows the SFORD spectrum of thiamine hydrochloride (**2**) (low-frequency region). It is

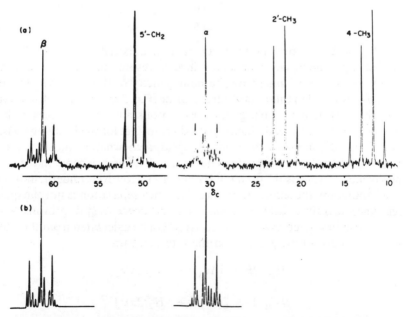

2

interesting to note that, among the three types of methylene carbons, only one give rise to a distinct three-line triplet, whereas the two remaining multiplets exhibit additional fine structure. An explanation of this anomalous behaviour had

Fig. 3.6   Aliphatic region of the 20 MHz (1.9 T) ^{13}C spectrum of thiamine hydrochloride (**2**) in D_2O, (a) SFORD, $H_2/2 = 5.1$ kHz, decoupler frequency centred at 15 ppm from TMS; (b) computed spectra for methylene carbon α and β, when treated as an A_2B_2X system. For further details see text.

been given[38] in terms of virtual coupling, a second-order effect, well known from proton n.m.r.[39] Virtual coupling implies a situation where two sets of weakly coupled nuclei A and X are present with one, for example A, being further strongly coupled to a species B. Even though X may not be coupled to B, its resonance can still show second-order characteristics.

A similar situation applies in the case of off-resonance decoupling, where the irradiated protons are part of a tightly coupled set of protons.[38] This may be found in the molecular framework —$^{13}CH_2$—$^{12}CH_2$—, where the two sets of methylene protons are isochronous or near-isochronous. It is even conceivable that the single-resonance spectrum is first order to a good approximation. A simple example illustrating this is ethylene glycol, whose proton-coupled spectrum is shown in Fig. 3.7(a). The first-order character of this spectrum is maintained because $^1J_{CH} \gg {}^2J_{CCH}, {}^3J_{HH}$. This inequality is no longer fulfilled when the decoupling field is applied, as this reduces $^1J_{CH}$ to $^1J_{CH}^r$ and the latter may now be of the same order of magnitude as $^3J_{HH}$. The SFORD spectrum in Fig. 3.7(b) illustrates this effect. The observed pattern is characteristic of the X-part of an $A_2A_2'X$ system as verified with the computed spectrum (Fig. 3.7(c)) which is based on the following parameter set:

$$^1J^r = 22.1 \, \text{Hz}$$

$$^2J^r = 0$$

$$^3J = 5.7 \, \text{Hz}$$

Since $^1J_{CH}^r = 22.1$ Hz corresponds to a more than a sixfold reduction, $^2J_{CCH}^r = 0$ is a justified approximation, in particular since the resolution in such experiments is typically 1–2 Hz (it is limited by the homogeneity of the H_2 field). The effect, previously observed in thiamine hydrochloride (Fig. 3.6(a)) is entirely analogous, and results from strong coupling of α and β protons, affording complex multiplets for the two contiguous methylenes, but a sharp triplet for the bridging CH_2. The result obtained for the former may appear somewhat astonishing at first, considering the fact that the proton chemical shift difference between the two sets of methylene protons is 0.7 ppm (56 Hz at 1.9 T). However, under conditions of strong irradiation, the single-resonance frequency separation is not the quantity of relevance. It is the effective chemical shift difference Δv_{eff}^{AB} of protons A and B which determines spectral appearance. Δv_{eff}^{AB} can be calculated from the effective fields the two protons experience in the rotating frame:

$$\gamma H_{eff}^A / 2\pi = [(\Delta v_A)^2 + (\gamma H_2/2\pi)^2]^{1/2}$$

and

$$\gamma H_{eff}^B / 2\pi = [(\Delta v_B)^2 + (\gamma H_2/2\pi)^2]^{1/2}$$

Insertion of the appropriate values for the frequency offsets Δv_A and Δv_B of the decoupler from resonance and for $\gamma H_2/2\pi$ yields an effective frequency difference $\Delta v_{eff}^{AB} = 10$ Hz, from which the computed spectra in Fig. 3.6(c) were obtained. It is

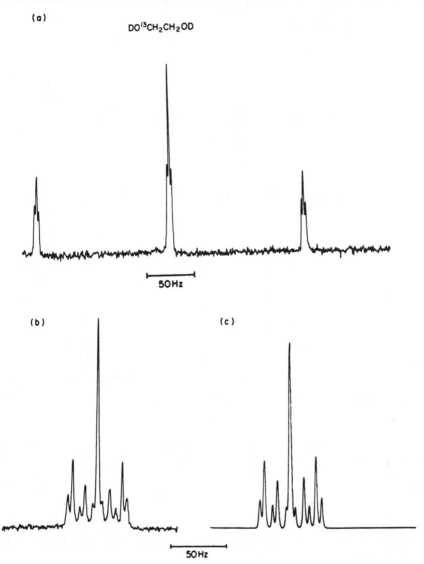

Fig. 3.7 20 MHz spectra of ethylene glycol in D_2O, (a) Proton coupled; (b) same experimental condition as in (a), but with coherent proton irradiation at -5 ppm from TMS and $\gamma H_2/2\pi = 5.1$ kHz; (c) computed SFORD spectrum obtained from the parameter set given in the text.

therefore not a necessary requirement that the adjacent methylene protons be tightly coupled in order to lead to a complex off-resonance spectrum. Rather it is requisite that the system be strongly coupled under the influence of the applied off-resonance decoupling field. It should be noted that the previously described

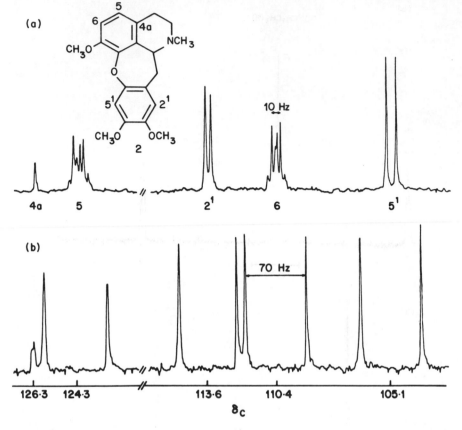

Fig. 3.8   Aromatic region of the 25.2 MHz (2.3 T) ^{13}C SFORD spectrum of the alkaloid cularine (3), obtained with different decoupler power settings, corresponding to residual splittings of 10 Hz (a) and 70 Hz (b).[42] Reproduced by permission of John Wiley & Sons.

approach is only approximately correct. A more rigorous treatment requires calculation of the transition energies and intensities in terms of the double-resonance Hamiltonian.[40,41]

The diagnostic value of second-order effects in SFORD spectra is apparent in the example of Fig. 3.6. There is another aspect, however, that deserves mention: whereas in ^{1}H n.m.r. second-order effects can normally not be influenced except by altering the field, in SFORD experiments they are largely under the experimenter's control.[42–6] Whether or not a spin system, consisting of, say, one ^{13}C spin and two protons, such as formed by the moiety

$$H_A - {}^{13}C_X - {}^{12}C - H_B$$

is second order depends primarily on the ratio $^1J^r_{CH}/^3J_{HH}$ which, within the limits of validity of Eqn. (3.9), is proportional to the decoupler offset parameter $\Delta v/(\gamma H_2/2\pi)$. In addition, spectral appearance depends on the effective chemical shift difference $\Delta v_{\text{eff}}^{AB}$. An illustrative example of the sensitivity of the coupling pattern to the decoupling conditions is shown in Fig. 3.8, displaying the aromatic portion of the SFORD spectrum of the alkaloid cularine (3),[42] obtained with

3

different decoupler power settings. The signals pertaining to the two adjacent carbons 5 and 6 give rise to a six-line pattern under strong irradiation conditions, characteristic of the X part of an ABX system (Fig. 3.8(a)). At lower H_2 amplitude, however, simple doublets are observed, since now $^1J_{AX}^r \gg {}^3J_{HH}$. By contrast, carbons 2' and 5' retain their doublet structure irrespective of decoupler rf amplitude, since $^3J_{HH} \sim 0$. In this case the two inner lines degenerate into one each and the two combination transitions have vanishing intensity. The evolution of a theoretical off-resonance spectrum for a $-{}^{13}CH-{}^{12}CH-$ fragment, calculated as a function of the off-resonance parameter $\Delta v/(\gamma H_2/2\pi)$, is reproduced in Fig. 3.9. The values chosen are those typically found in aryl systems ($^1J_{CH} = 160$ Hz, $^2J_{CCH} = 1$ Hz, $^3J_{HCCH} = 6$ Hz), while for the proton chemical shift difference $\Delta v_{AB} = 60$ Hz had arbitrarily been chosen. The approximation inherent to this approach was further improved by calculating the reduced coupling constants according to Eqn. (3.6). It should further be noted that the small splitting arising from the finite value of $^2J_{CCH}^r$ is masked by the use of a 2 Hz line broadening, as may typically result from H_2 rf inhomogeneity. Clearly, the effects are also field-dependent through Δv_{AB}. High-field spectra are therefore less likely to exhibit second-order splittings if similar H_2 amplitudes are used.

A second source of deviations from 'normal' behaviour occurs for methylene carbon SFORD signals whenever the two protons are magnetically non-

4

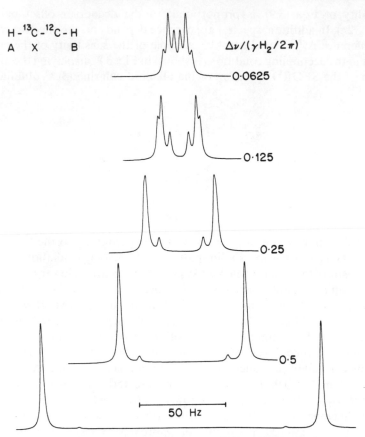

Fig. 3.9   Simulated SFORD spectra for a $—^{13}\text{CH}—^{12}\text{CH}—$ moiety, calculated for various values of the off-resonance parameter $\Delta v/(\gamma H_2/2\pi)$ under
the following assumptions: $^1J_{CH} = 160\,\text{Hz}$, $^2J_{CCH} = 1\,\text{Hz}$, $^3J_{HH} = 6\,\text{Hz}$, line
broadening 2 Hz. Reduced coupling constants were calculated according to
Eqn. (3.6) and effective chemical shifts as discussed in the text.

equivalent.[42] In the tribenzocyclononane derivative (4) the methylene protons
are highly anisochronous even at 1.9 T, thus giving rise to a pair of doublets[47]
(Fig. 3.10). It has been shown[42] that the inner lines of such four-line patterns
remain invariant, irrespective of the off-resonance parameter, therefore being
dependent only on the frequency separation of the two proton resonances. It is
clear that this effect will become much more pronounced at high magnetic field, in
contrast to the first-discussed second-order effects. It should further be noted that
second-order effects caused by magnetic non-equivalence of methylene protons
are unlikely to occur[45] because the difference in the two reduced one-bond
coupling constants is not large enough to afford significant intensities for the
combination lines.

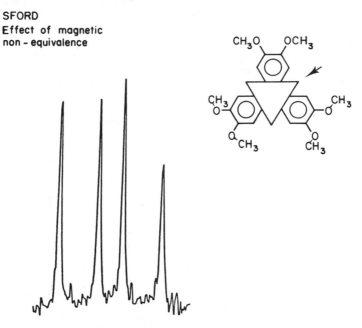

SFORD
Effect of magnetic
non - equivalence

Fig. 3.10   20 MHz (1.9 T) SFORD for the methylene carbon in the tribenzononane derivative (4), showing effect of magnetic non-equivalence of methylene protons.[47]

Evaluation of off-resonance data with the aid of graphical methods

We have seen in Fig. 3.4(b) that the point of intersection of the straight line, defined by the function $J^r = f(\Delta v/(\gamma H_2/2\pi))$ with the abscissa, essentially corresponds to the proton chemical shift. This is the basic idea underlying the graphical evaluation of SFORD spectra,[48,49] which allows correlation of carbon chemical shifts with their proton counterparts. In practice, one proceeds as follows. A series of SFORD spectra is recorded by stepping the decoupler frequency in increments of, say, 1 ppm, and the peak frequencies are subsequently plotted against the decoupler frequency. This is illustrated in Fig. 3.11 for a series of 9.4 T SFORD spectra displaying the aromatic region in 6-methylcoumarine (5), obtained in the way described previously. Since the intersection of the straight

lines occurs at the exact resonance frequency of each of the protons, these points can now be correlated with the proton spectrum plotted on the ordinate. It is

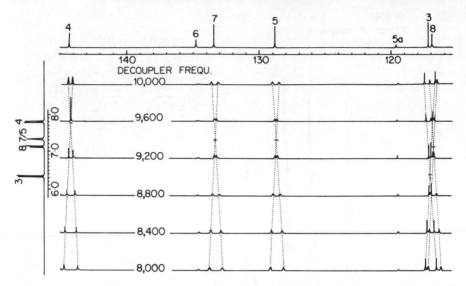

Fig. 3.11 Series of ^{13}C SFORD spectra, recorded at 9.4 T, showing the aromatic region in 6-methylcoumarine (**5**). Each trace corresponds to a different decoupler frequency offset, as indicated in the figure. The straight lines connecting the line frequencies in successive experiments intersect at the respective proton resonance frequencies, as indicated by the proton spectrum, which is plotted on the ordinate.

interesting to note that this experiment unambiguously assigns the two closely spaced resonances pertaining to C-3 and C-8 ($\Delta\delta = 0.25$ ppm).

As is evident from Fig. 3.11, the determination of the points of intersection becomes problematic when their slopes are small (relative to the decoupler frequency axis as abscissa). Since the slope b of the two straight lines, in the case of a methine carbon, for example, is given by

$$b = \tfrac{1}{2}{}^{1}J_{CH}/(\gamma H_2/2\pi) \qquad (3.11)$$

the choice of weak decoupler fields would favourably influence the precision of a graphical determination of the proton resonance positions. However, straight lines are only obtained on the premise that $\gamma H_2/2\pi \gg J$ and Δv (condition for the validity of Eqn. (3.9)). It is therefore mandatory to use sufficiently large decoupler power, but instead to determine the points of intersection analytically rather than graphically.

We realize that this method does not only permit correlation of carbon chemical shifts with known proton shieldings but it may in fact be a method for identifying not directly observable protons in complex molecules where individual protons are masked by a broad signal envelope.

Selective decoupling

Instead of decoupling off-resonance with respect to all protons, the decoupler frequency may be chosen so as to be coincident with a particular proton resonance. If the condition $\gamma H_2/2\pi > {}^1J_{CH}$ is met, the affected resonance collapses into a singlet.[50,51] The general purpose of the selective decoupling experiment is to establish a one-to-one correspondence between proton and carbon shieldings. Hence the objective is the same as in the previously discussed multiple-irradiation SFORD experiment. Since, except at very high fields, the typical frequency separation of two protons is less than or comparable to the magnitude of the one-bond coupling constant, the spectra obtained from selective decoupling exhibit the general appearance of SFORD spectra with one signal showing complete decoupling. In order to ensure a high degree of selectivity the decoupler power is reduced to typically 500–1000 Hz. This experiment is greatly facilitated by the availability of a dual-observation ${}^{13}C/{}^1H$ probe, permitting measurement of both carbon and proton spectrum on the same sample and with the same experimental set-up. After recording of a survey proton spectrum the frequencies of the protons to be irradiated are then entered into a list of decoupler frequencies and an array experiment is set up in such a way that the decoupler frequency is switched to the next value in the list after a predetermined number of scans were taken, affording n spectra for the n irradiated protons. Figure 3.12 shows such an array of selectively decoupled spectra obtained on the sample of 6-methyl-coumarine (5). A notable feature in this high-field (9.4 T) experiment is retention of some of the long-range couplings, in spite of the relatively large decoupler field amplitude used ($\gamma H_2/2\pi \simeq 1.9$ kHz). Irradiation of H-5/7, for example, converts the carbon signals pertaining to C-5 and C-7 into residual quartets, obviously originating from a long-range coupling of these carbons with the methyl protons (cf. inset). This phenomenon was first reported in conjunction with the accurate assignment of the *ortho* and *meta* carbon resonances in toluene.[52] Coherent irradiation at the resonance frequency of the *ortho* protons was found to remove the one-bond coupling but partially retain ${}^3J_{CCCH_3}$, resulting in a quartet for the *ortho* carbon but a singlet for the *meta* carbon resonance. Likewise, a residual quartet was observed for the *ipso* carbon resonance, caused by ${}^2J_{CCH_3}$. The two coupling constants are known to have values of -7 Hz (^2J) and 5.8 Hz (^3J),[53] and are probably of very similar magnitude in formula (5). A calculation of the respective reduced splitting, based on $\gamma H_2/2\pi = 1.9$ kHz and $\Delta v = 1900$ Hz ($\delta_{H-7} - \delta_{CH_3} = 4.8$ ppm) readily shows that ${}^3J_{CCCH_3}$ is almost completely retained under these conditions. Similarly, irradiation of the methyl protons produces a triplet (cf. inset) for the methyl carbon signal, caused by unremoved long-range coupling involving protons H-5 and H-7. It should be pointed out that retention of long-range coupling constants in selective decoupling experiments will generally be of much more frequent occurrence at high magnetic field.

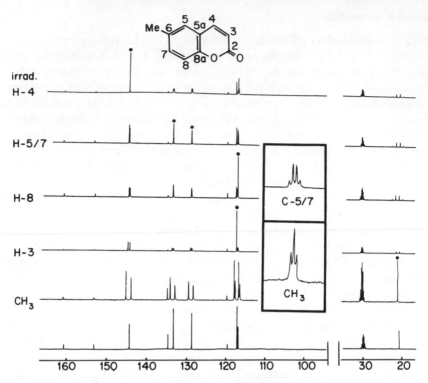

Fig. 3.12   Array of selectively decoupled 9.4T ^{13}C spectra of 6-methylcoumarine (**5**), using $\gamma H_2/2\pi = 1.9\,kHz$. Signals of fully decoupled carbons in each trace are labelled with a dot (●). The insets are expansions of the C-5/7 and methyl carbon signals, observed while their directly bonded protons are irradiated, showing effect of retained long-range couplings.

The observation of singlets in selective decoupling experiments is therefore not obligatory.

Selective decoupling has lost some of its popularity because of the excessive time requirements, as each trace, in order to be useful, needs at least the S/N of a noise-decoupled (though not necessarily the SFORD) spectrum. Moreover, the technique has received competition from two-dimensional correlated n.m.r., which will be discussed later in this chapter.

NOE-enhanced proton-coupled carbon spectra

The previously described experiments are designed to reduce the great information density attendant on the single-resonance spectrum and further to increase efficiency in terms of S/N. We have seen in Chapter 1 that saturation of protons is accompanied by an up to 200% enhancement of the carbon signals. The NOE,

(a)

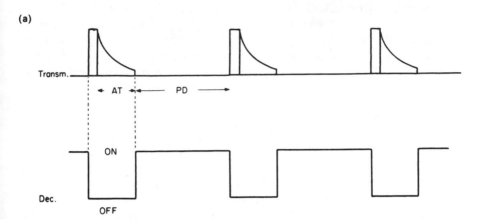

(b)

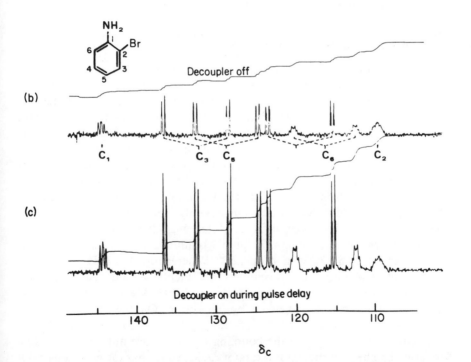

Fig. 3.13   (a) Timing diagram in a gated decoupling experiment, permitting retention of heteronuclear spin–spin coupling and NOE. The decoupler is gated on during a pulse delay (PD), but off during data acquisition (AT). (b) 20 MHz proton-coupled spectrum of 2-bromoaniline in acetone-d_6, recorded in the single-resonance mode; (c) same experimental conditions as in (b), but with the decoupler gated as described under (a).

when the decoupler is initially turned on, builds up with a time constant $T = T_1$ (^{13}C) and decays with a time constant $T' > T_1$ (^{13}C),[54,55] when irradiation ends. By contrast, the coherent motion of the proton spins, driven by the H_2 field, vanishes almost instantly upon discontinuing decoupling. Hence by gating the decoupler in such a way that it is on during a polarization period (pulse delay) but off during the acquisition period,[56,57] the NOE can be retained without otherwise modifying the single-resonance spectrum. The fraction of the NOE retained depends on the duty cycle of the decoupler, i.e. the delay (on period) should be long relative to the acquisition period. In order to maintain optimum conditions for the experiment the pulse flip angle has to be adjusted as dictated by Eqn. (1.21) by setting $t_p = AT + PD$, where PD designates the pulse delay. Figure 3.13(a) displays the timing digram for the experiment. The two traces represented by Figs 3.13(b) and (c) were obtained with no proton irradiation and pulse-modulated irradiation, respectively. It is interesting to note that all signals, including the quaternary carbon 1, show substantial enhancement, except carbon 2, which is bonded to bromine. In this case the mechanism of relaxation is not dipolar and consequently no NOE can build up.[58]

Analysis of proton-coupled carbon spectra

One of the principal problems associated with the interpretation of proton-coupled spectra is their often exorbitant complexity, compounded with second-order splitting characteristics. Even though the proton spectrum of the ^{12}C isotopomers may be first order, this is not a sufficient criterion for the ^{13}C spectra to be first-order as well. As a consequence of the large one-bond couplings, some energy levels may accidentally approach each other, so that mixing between them occurs. A conspicuous symptom for deviations from first order is the asymmetry of spin multiplets. The fine structure on the low-frequency half of a CH doublet differs in appearance from its high-frequency counterpart.[59] The importance of high magnetic fields has often mistakenly been regarded as less critical in ^{13}C n.m.r. However, it is clear that high field offers basically the same advantages as in proton n.m.r. Figure 3.14 illustrates the effect of magnetic field on spectral complexity in propene.[60] Whereas at 25.2 MHz (2.3 T) the spectra of the C-1 and C-2 isotopomers are highly second order, all three isotopomers give rise to first-order spectra at 100.6 MHz (9.4 T) and most coupling constants can be extracted by inspection. On the other hand, this example also demonstrates that the spectral information content at the higher field is reduced as the signs of the coupling constants are not amenable. Sign information can be of particular interest for $^2J_{CCH}$, which has the notorious habit of being either positive or negative (cf. Chapter 2). It may therefore be advantageous to have access to spectrometers operating at different magnetic field and to use the data from the high-field spectrum as a starting parameter set for simulation, complemented by proton chemical shifts and coupling

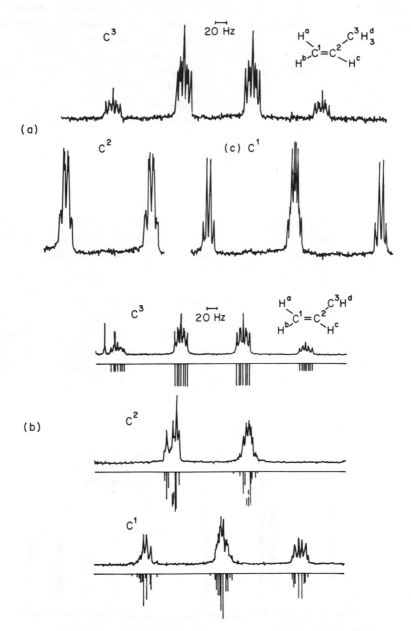

Fig. 3.14   Proton-coupled ^{13}C spectra of propene at $-53\,^\circ$C, recorded at 9.4 T (a) and 2.3 T (b), showing effect of spin–spin splitting characteristics on field strength.[60] Reproduced by permission of John Wiley & Sons.

constants.[61] In this particular example the assignment of the three carbons is trivial; it directly follows from the multiplicities. The designation of the various spin–spin coupling constants, on the other hand, is not straightforward.

We will address this particular aspect of assignment later and now turn to the task of signal assignment with the aid of long-range coupling. A precondition clearly is the knowledge of the typical order of magnitude of the spin–spin coupling constants for the various paths. An illustrative example is 2,2′-dipyridylamine (**6**), for which we shall use as a model pyridine, whose coupling

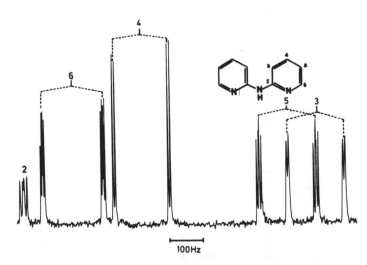

constants have been derived from a complete simulation of the experimental spectra.[59] (Compare Table 2.21). In aromatic systems the three-bond coupling $^3J_{CCCH}$ is usually largest (7–12 Hz), whereas the four-bond coupling is always very small (1–2 Hz). $^2J_{CCH}$, on the other hand, exhibits a great deal of variance[53] and it can be positive or negative. As we will confine ourselves to first-order analysis, we will, for the time being, ignore the sign. In addition, we should be cognisant of the fact that in heterocyclic aromatic systems such as pyridine, pyrrole, furan, etc., $^2J_{C_\beta H_\alpha}$ is comparable in magnitude to $^3J_{CCCH}$.[62-4] Following this general pattern it is straightforward to assign the resonances in the proton-coupled spectrum of **6** (Fig. 3.15). Chemical shift criteria suggest $\delta_{C(2)} > \delta_{C(6)} > \delta_{C(4)} > \delta_{C(3)}$. The ambiguity in the assignment of C-3 and C-5 is readily

Fig. 3.15   20 MHz proton-coupled ^{13}C spectrum of 2,2′-dipyridylamine (**6**) in acetone-d_6.

eliminated from an inspection of the long-range couplings by considering that C-5 should have two large (8–10 Hz) such couplings ($^2J_{C(5)H(3)}$, $^3J_{C(5)H(3)}$), but C-3 only one ($^3J_{C(3)H(5)}$). All remaining couplings are too small to be resolved. From this one infers that the highest-field signal belongs to C-3. Analogous arguments may be used to confirm the designation of C-4 and C-6.

We will now turn to the more complex spin system of 6-methylcoumarine (5), a molecule we had previously dealt with in connection with coherent on- and off-resonance decoupling, and we shall attempt to corroborate some of the previously made assignments. Figure 3.16 displays the high-frequency region of the 9.4 T spectrum between 115 and 162 ppm, together with an expansion of the section between 128 and 135 ppm. The four quaternary carbons have unique chemical shifts and can therefore immediately be designated. C-6, resonating at 134.9 ppm, has a double-quartet structure accountable to coupling with the methyl protons ($^2J_{CCH_3} = 6.0$ Hz) and vicinal coupling with H-8 ($^3J_{CCCH} = 7.5$ Hz). In addition, there is a further small doublet splitting (~ 0.5 Hz) arising from two-bond coupling with either H-5 or H-7. The assignment of the two lowest-frequency doublets had previously been secured by correlating their shieldings with those of the appended protons. A conspicuous feature

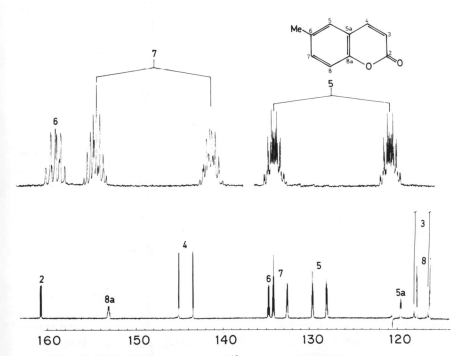

Fig. 3.16   100.6 MHz (9.4 T) proton-coupled ^{13}C spectrum of the high-frequency region of 6-methylcoumarine (5). The expansion shows the effect of long-range couplings to methyl protons on the multiplet structure of carbons 5, 6, and 7.

of the signal of the less shielded of the two methines is the absence of any measurable long-range coupling. It is well known[65] (cf. also Chapter 2) that $^2J_{CCH}$ is very small in olefinic systems of type X—*CH=CH with X=COOR and transoid configuration of the X—C—C—H bonds. This further confirms assignment of this resonance to C-3. The expanded middle region encompassing the two methine multiplets belonging to C-5 and C-7 shows the highest degree of complexity. This is, of course, caused by additional coupling involving the methyl protons. It should further be noted that the more shielded of the two methine carbons exhibits a completely symmetric first-order pattern, in marked contrast to its neighbour. Ignoring long-range couplings, C-7 is predicted to constitute the X part of a tightly coupled ABX system with A = H-7 and B = H-8. Since $\Delta v_{AB} = 0.2$ ppm (80 Hz at 400 MHz proton frequency), corresponding to just half the one-bond coupling constant ($^1J_{C(7)H} \sim 160$ Hz), the conditions for strong coupling are fulfilled.[65] This warrants assignment of the less shielded of the two CH carbons to C-7, which could further be confirmed by simulation.

Selective low-power irradiation

The spectra discussed in the previous section were all obtained without any irradiation during the data acquisition period, i.e. with full retention of all CH spin–spin couplings. Selective low-power irradiation aims at selectively removing one long-range coupling with the objective to (1) reduce the number of transitions for facilitating analysis; (2) assign a resonance by locating a coupling partner; (3) assign a coupling constant to a particular coupling path; and (4) determine the relative signs of two coupling constants. An illustration of (1) and (2) is provided by Fig. 3.17, showing once more the same spectral region of 5 as in Fig. 3.16, except that the data were obtained by low-power irradiation of the methyl protons ($\gamma H_2/2\pi \simeq 65$ Hz). Most affected are, as expected, the two carbons undergoing three-bond coupling to the methyl protons (C-5, 7; cf. inset). The quaternary C-6, whose fully coupled spectrum consists of a doublet of quartets, now simplifies to a doublet, whose splitting is consistent with a three-bond ring coupling ($^3J_{C(6)H(8)} = 7.5$ Hz). The more highly shielded methine carbon, previously assigned to C-5, converts to a first-order triple doublet, whose splittings are attributed as follows: $^3J_{C(5)H(7)} = 8.5$ Hz, $^3J_{C(5)H(4)} = 4.0$ Hz, and a coupling constant of 1 Hz involving either H-3 or H-8. Since the reduction factor J^r/J for all one-bond couplings does not measurably deviate from unity, the thus-obtained coupling constants do not need correction. As previously noted, the C-7 subspectrum is of higher order and coupling constants cannot be extracted by inspection.

In pterine (7)[66] and its derivatives, quaternary carbons 1a and 4a exhibit vicinal transoid coupling to the pyrazine protons H-7 and H-6, respectively. With the aid of selective low-power irradiation at the resonance frequency of the two protons the chemical shifts of carbons 1a and 4a could be established unequivocally as

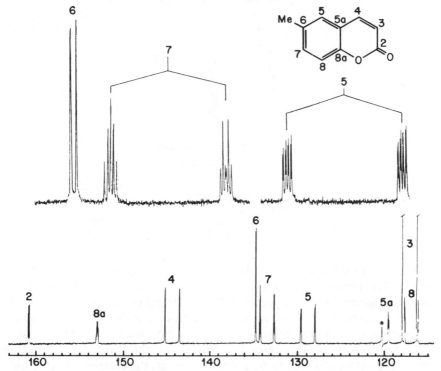

Fig. 3.17   100.6 MHz (9.4 T) proton-coupled spectrum of **5**, obtained under conditions of Fig. 3.16, except that during data acquisition the methyl protons were selectively irradiated at $\gamma H_2/2\pi = 65$ Hz, simplifying the resonances of carbons 5, 6, and 7.

$^3J_{C(1a)H(7)} = 12$ Hz

$^3J_{C(4a)H(6)} = 10$ Hz

7

well as their respective coupling constants.

Considerable effort has been put into the assignment of carbonyl carbon resonances in peptides,[67,68] for which it could be shown that proton-coupled spectra provided the clue for discerning some of the closely spaced resonances. A particularly instructive example is N-acetylaspartic acid (**8**), where the three

$$HOOC - \overset{\beta}{C}H_2 - \overset{\alpha}{C}H - NHCOCH_3$$
$$|$$
$$COOH$$

8

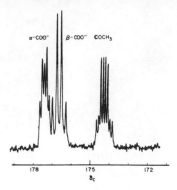

Fig. 3.18   25.2 MHz (2.3 T) proton-coupled ^{13}C spectrum of the carbonyl region of N-acetyl aspartic acid in D_2O at pD = 4.4.[68]

carbonyl carbons resonate within a 4 ppm window. An expansion of the carbonyl region of the proton-coupled spectrum is shown in Fig. 3.18.[68] Since the multiplets originate from geminal and vicinal coupling to the α and β protons as well as those of the N-acetyl group, and the magnitude of the coupling constants is not known *a priori*, interpretation is again supported by double-resonance experiments. Irradiation at the resonance frequency of the α proton converted the doublet of quartets at 174 ppm to a simple quartet, the centre quartet (which originates from two overlapping triplets) to a triplet, and the highest-frequency multiplet from a double to a single triplet. On the assumption that the vicinal is smaller than the geminal coupling, the assignments given in Fig. 3.18 follow. The magnitudes of the long-range coupling constants deduced from the spectrum in Fig. 3.1 are listed in Table 3.1.

The third possible goal of a selective low-power decoupling experiment is to identify the origin of a long-range coupling constant. An elegant method for tackling this problem relies on the differential reduction factor of two coupling constants $^nJ_{CH}$ and $^mJ_{CH}$ upon low-power irradiation off-resonance with respect to the two protons involved.[69] The site of irradiation may (but need not) be chosen such as to simplify the spectrum by eliminating coupling to a third group of protons, say a methyl group. In *trans*-crotonaldehyde (**9**), for example, the spectrum of the carbonyl carbon represents the X part of a weakly coupled

TABLE 3.1   Long-range ^{13}C—H coupling constants in N-acetyl aspartic acid (**8**)[68]

COCH₃	α—COO⁻	β—COO⁻
$^2J_{C-CH_3} = 6.1\,Hz$	$^2J_{C-C_\alpha H} = 6.1\,Hz$	$^2J_{C-C_\beta H} = 6.5\,Hz$
$^3J_{C-C_\alpha H} = 3.1\,Hz$	$^3J_{C-C_\beta H} = 3.2\,Hz$	$^3J_{C-C_\alpha H} = 5.6\,Hz$

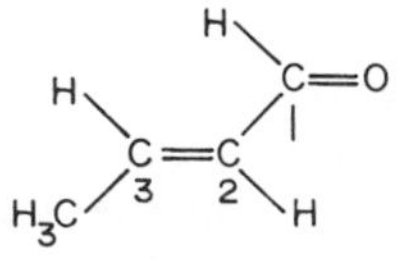

9

A_3KLMX spin system that can easily be analysed following first-order splitting rules. Selective irradiation of the methyl protons reduces the spectrum to an eight-line pattern. However, an ambiguity arises regarding the assignment of the coupling constants $^2J_{CCH}$ and $^3J_{CCCH}$. Selective irradiation of the methyl protons reduces the magnitude of all splittings. Since irradiation occurs at low frequency, and $\delta H_3 > \delta H_2$, $^3J_{CCCH}$ should be reduced by a smaller fraction than $^2J_{CCH}$. The residual splittings for the three long-range couplings are listed in Table 3.2. The larger fractional reduction of the smaller of the two coupling constants clearly relates it to proton C(2)—H.

A rather exceptional opportunity for ascertaining a coupling path offers itself if the spin–spin coupling involves exchangeable protons. In order to be able to observe coupling at all, the residence time of the proton in the X—H (X = O, S, N) moiety has to be larger than the inverse of the coupling constant, i.e. $\tau > 1/J$. This condition is often satisfied for hydroxyl protons which are stabilized by an intramolecular hydrogen bond. In naringenin (**10**), for example, spin–spin

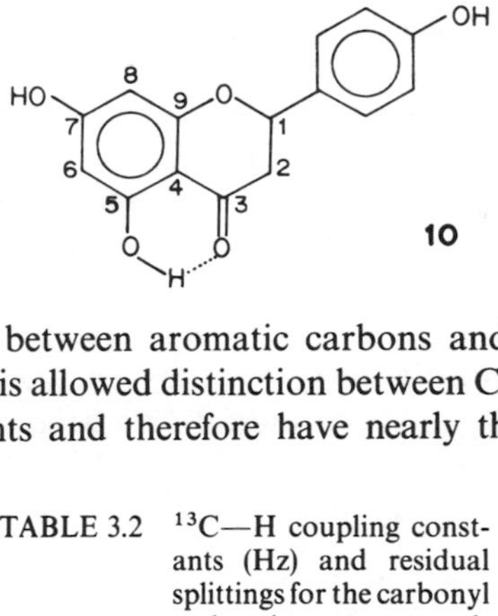

10

coupling was observed between aromatic carbons and the chelated hydroxy proton C(5)—OH.[70] This allowed distinction between C-6 and C-8, which are in very similar environments and therefore have nearly the same chemical shifts.

TABLE 3.2 ^{13}C—H coupling constants (Hz) and residual splittings for the carbonyl carbon in transcrotonaldehyde (**9**)[69]

	Single resonance	CH_3 protons irrad.
$^1J_{CH}$	171.3	165.5 (97%)
$^2J_{CCH}$	2.3	2.0 (87%)
$^3J_{CCCH}$	8.9	8.2 (92%)

In the proton-coupled spectrum, however, one of the two signals exhibits a more complex splitting pattern. While both show vicinal coupling (to H-6 and H-8, respectively), C-6 is further coupled to C(5)—OH ($^3J_{C(6)COH} = 4.5\,Hz$), which could easily be verified by exchanging the hydroxyl protons by deuterons. Likewise, deuterium exchange was found to convert the quartet due to C-4 (equally coupled to H-6, H-8 and C(5)—OH) to a triplet, thus distinguishing it from C-9.

Stereospecificity of the three-bond coupling constant $^3J_{CH}$

So far, we have almost exclusively dealt with chemical shift assignment. However, proton-coupled ^{13}C spectra have aroused the chemist's interest primarily because of their potential ability to provide stereochemical information through the dependence of the magnitude and sign of a coupling constant on the geometry of the coupling path.[10] Although it is mainly ^3J, which shows the characteristic dependence of its magnitude on the dihedral angle (Karplus relationship),[71,72] similar dependencies were also found for ^2J, ^4J, and even ^5J.[10]

In the original description of the Karplus relationship[73] for the coupling path H_1—C_1—C_2—H_2 the torsion or dihedral angle ϕ has been defined as the angle between the two planes formed by H_1—C_1—C_2 and C_1—C_2—H_2, respectively. In the analogous three-bond carbon–proton coupling network hydrogen 2 is substituted by a carbon (cf. Chapter 2). There exists a vast literature on the use of carbon–proton coupling constants for stereochemical analysis and the interested reader is referred to Chapter 2 for details or one of the specialized reviews such as Reference 10. We shall therefore confine ourselves to briefly discuss two examples from conformational analysis with the aid of $^3J_{CCCH}$. The prime motivation for resorting to carbon–proton coupling rather than using their still-better documented and more easily amenable ^1H—^1H analogues is the frequent non-existence of a suitable coupling path. This, however, is not an exceptional occurrence in heavily substituted carbon networks.

In a conformational study of the symmetric dicarboxylic acid (11)[74] a value of 2 Hz was found for $^3J_{COOH,H(3)}$, thus indicating a *gauche* disposition for the two

bonds in question. More intriguing was the finding for $^3J_{CN,H(3)}$ (6 Hz), a value intermediate between *gauche* and *trans*. Because of the substituent dependence of ^3J the limiting figure for $\phi = 180°$ was first ascertained in the cyclohexane derivative (12) of established stereochemistry, for which $^3J_{CN,H(3)} = 9\,Hz$ was

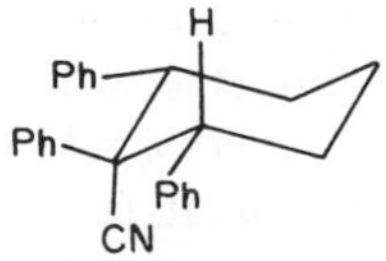

12

found. It is readily seen from the Newman projections **11a** and **11b** that H-3 is *gauche* whereas the magnetically equivalent H-5 in *trans*. The experimental value

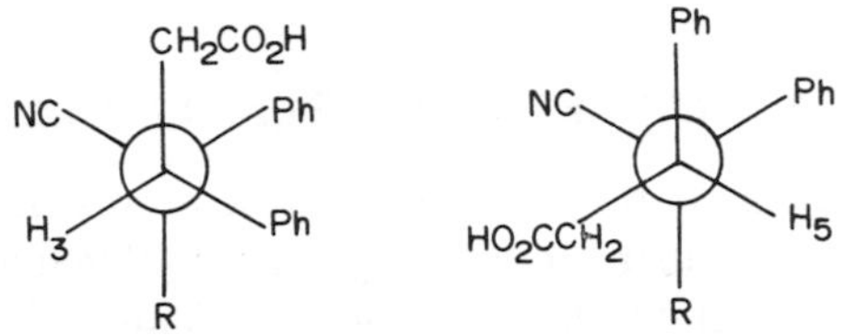

11a **11b**

of 6 Hz is an average of a *gauche* and a *trans* contribution, thus establishing the conformation shown by formula **11**.

In the recent past considerable effort has been expended on the characterization of the conformation of biomolecules in solution. In the nucleoside cytidine (**13**) the glycosidic conformation, determined by the torsional angle, defined by the $C_2-N_1-C_1'-H_1'$ has been of interest.[75] From the value of $^3J_{C(2)H(1')} = 3 \pm 1$ Hz, a dihedral angle $\phi = \pm 35°$ or $\pm 130°$ could be determined. The latter possibility was ruled out on the basis of severe steric repulsion that would result from the interaction of the keto group with H-2′. Hence the anti-conformation

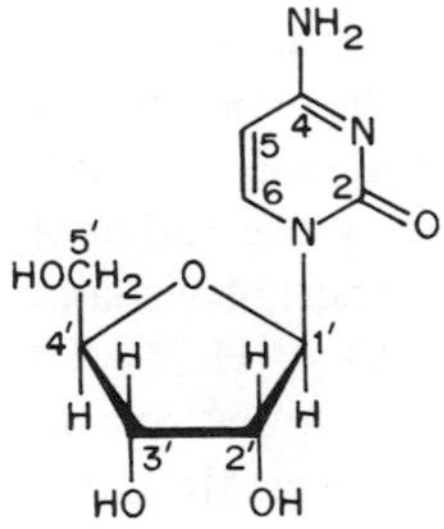

13 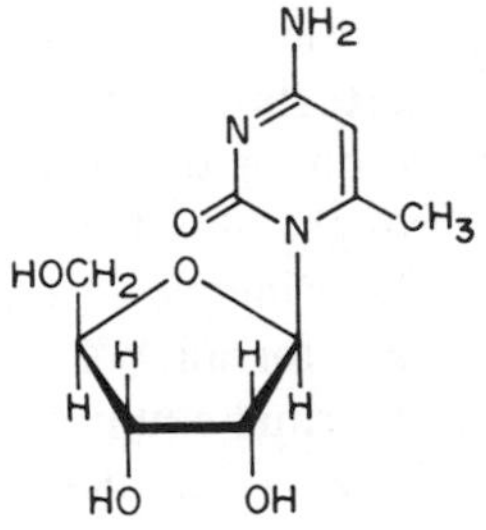 **14**

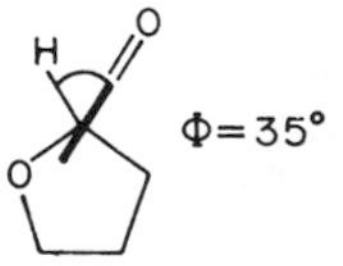

$\Phi = 35°$ $\Phi = 165°$

13a 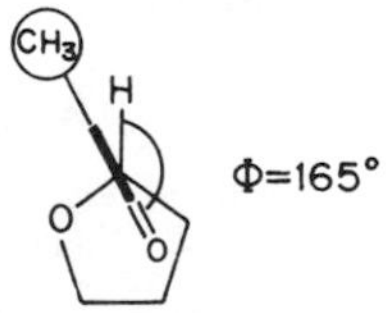 **14a**

13a was assigned to cytidine. In contrast, for 6-methylcytidine (**14**) a value of 6 ± 1 Hz was obtained for $^3J_{C(2)H(1')}$, consistent with a dihedral angle $\phi = \pm 165°$, or alternatively $\pm 15°$. The latter, however, would place the 6-methyl group in close contact with the $5'$-CH$_2$OH substituent, favouring therefore conformation **14a**. The data clearly show that an ambiguity remains and that complementary information is required. An example of joint use of proton–proton and proton–carbon couplings is discussed in Chapter 5. In the present case, however, no vicinal proton–proton coupling path exists.

3.3 POLARIZATION TRANSFER AND RELATED EXPERIMENTS

The physical commonality of the group of experiments discussed in this section is a ^{13}C polarization enhancement, induced by transfer of magnetization via scalar coupling from the proton to the ^{13}C spins.[1-4] This can result in intensity enhancements of up to a factor γ_H/γ_C, irrespective of the mechanism governing relaxation. The major benefit of the experiment is therefore increased sensitivity. We will see, however, that multiplicity information can be obtained as a by-product, therefore providing an alternative to off-resonance decoupling. Moreover, polarization transfer constitutes the basis for heteronuclear two-dimensional correlated spectroscopy,[8,76-82] discussed in the subsequent paragraph.

Selective population inversion[83-5]

The idea of this experiment is to selectively invert the spin populations on the energy levels defining a proton transition. The resulting population change will then affect the intensities of those carbon transitions that share an energy level with those of the perturbed proton transition. The experiment may therefore be regarded an extension of the generalized NOE experiment,[86] except that rf perturbation is brought about by an inversion pulse rather than continuous-wave saturation. The principle is simple, and can best be visualized in terms of an energy level diagram for an AX spin system (A = ^{1}H, X = ^{13}C), as shown in Fig. 3.19. Since in thermal equilibrium the spin populations are approximately proportional to the Zeeman energies, we can designate them in terms of the quantities $\Delta \propto \frac{1}{2}\nu_H$ and $\delta \propto \frac{1}{2}\nu_C$, providing the relative populations given in Table 3.3. The transition intensities for the two proton and two carbon transitions are then proportional to the respective population differences, i.e.

$$I_{H(1)} = I_{H(2)} \propto 2\Delta$$

and

$$I_{C(1)} = I_{C(2)} \propto 2\delta$$

Let us now assume that the populations between levels labelled $\alpha\beta$ and $\beta\beta$ are inverted by means of a selective 180° proton (decoupler) pulse, resulting in the

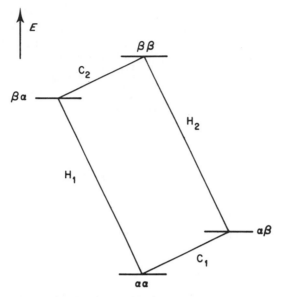

Fig. 3.19   Schematic energy level diagram for an AX spin system ($A = {}^1H, X = {}^{13}C$). Allowed proton and ^{13}C transition are labelled H_1, H_2, and C_1, C_2, respectively.

TABLE 3.3   Energy states and relative spin population for the AX spin system described in the text

Energy	Spin state	Spin population a	b
$\frac{1}{2}\nu_C + \frac{1}{2}\nu_H$	$\alpha\alpha$	$\Delta + \delta$	$\Delta + \delta$
$-\frac{1}{2}\nu_C + \frac{1}{2}\nu_H$	$\alpha\beta$	$\Delta - \delta$	$-\Delta - \delta$
$\frac{1}{2}\nu_C - \frac{1}{2}\nu_H$	$\beta\alpha$	$-\Delta + \delta$	$-\Delta + \delta$
$-\frac{1}{2}\nu_C - \frac{1}{2}\nu_H$	$\beta\beta$	$-\Delta - \delta$	$\Delta - \delta$

[a] At thermal equilibrium.
[b] Following a 180° inverting pulse of H(2).

relative populations given in the last column of Table 3.3. It is readily verified that the new populations now become

$$I_{H(1)} \propto 2\Delta \qquad\qquad I_{C(1)} \propto 2\Delta + 2\delta$$
$$I_{H(2)} \propto (-2\Delta) \qquad\qquad I_{C(2)} \propto (-2\Delta + 2\delta)$$

Of only concern to us are the carbon transitions, which experience a fourfold increase and a fourfold decrease, respectively, as seen if one takes into account the fact that $\gamma_H/\gamma_C \simeq 4$, consistent with $\Delta = 4\delta$.

The method requires accurate calibration of the decoupler field amplitude,

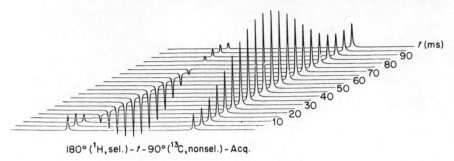

180° (^{1}H, sel.) - t - 90° (^{13}C, nonsel.) - Acq.

Fig. 3.20   Series of selective population inversion ^{13}C spectra, obtained by the pulse sequence set out in the figure. The time delay t between selective and non-selective pulse is indicated on the right of the plot. The sample was chloroform and the high-frequency proton transition was inverted.

which can, for example, be achieved by centering the decoupler frequency on one of the ^{13}C satellites in chloroform, a convenient reference sample, and using the pulse sequence

$$\alpha(^1H) - \pi/2(^{13}C) - \text{Acquisition}$$

Whereas the initial decoupler pulse is selective, the following ^{13}C transmitter pulse is, of course, non-selective, i.e. equally affecting all ^{13}C transitions. When varying the duration of the former pulse, maximum positive and negative excursion of the ^{13}C doublet components is obtained for the condition

$$\gamma_H H_2 \tau(^1H) = \pi \qquad (3.12)$$

as exemplified by the traces shown in Fig. 3.20, obtained by incrementally increasing the width of the decoupler pulse.

High selectivity inevitably demands long (weak) pulses. Selectively inverting one of two transitions, which are, say, 5 Hz apart, obviously requires $\gamma H_2/2\pi \ll 5$ Hz, i.e. $\tau(^1H) \gg 0.1$ s.

From an assignment point of view the method lends itself for identifying quaternary carbons by making use of coupling to geminal and vicinal protons. An illustrative example is given in Fig. 3.21, depicting the high-frequency portion of the proton-coupled spectrum (Fig. 3.21(a)) of the alkaloid oxaline (15).[87] The

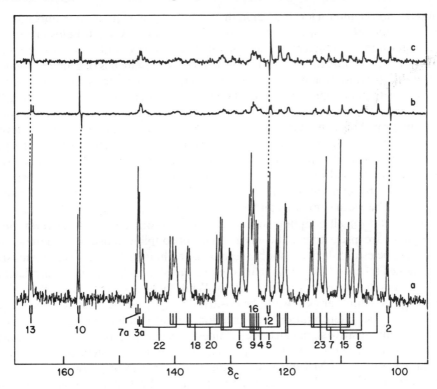

Fig. 3.21 High-frequency region of the 25.2 MHz (2.3 T) ^{13}C spectra of oxaline (15), (a) Proton-coupled; (b) SPI experiment, low-frequency transition of C-8 proton selectively inverted using $\tau_\pi = 0.08$ s; (c) same as (b) but π pulse acting on high-frequency transition of C-15 proton.[87]

selective population inversion (SPI) experiments represented by traces (b) and (c) were obtained by selectively inverting one of the proton doublet components of the C-8 proton (Fig. 3.2(b)) and C-15 proton (Fig. 3.21(c)). In the first case the doublet splittings pertaining to C-10 and C-2 are affected, both carbons being coupled to H-8 by vicinal coupling. In the second case enhancement is observed on the signals identifiable as C-12 and C-13. It should be noted that the proton transitions inverted by a single SPI experiment belong to different isotopomeric species, which cannot directly be observed in the proton spectrum but whose frequencies can be predicted on the basis of coupling constants in model compounds.

Common to all SPI experiments is an intensity change of connected transitions.[88,89] These are defined as transitions sharing an energy level with that of the irradiated line. From Fig. 3.19 we recognize that two different types of transition connectivity can be distinguished. Let us again assume perturbation of H(2) causing an intensity change for C(1) and C(2). With respect to C(1) we note that the non-common levels differ in their total magnetic quantum number M by ± 2 ($\alpha\alpha$ and $\beta\beta$ states) in contrast to C(2), for which $\Delta M = 0$ holds. The two distinguishable configurations are denoted *progressive* and *regressive*, respectively. From the foregoing we can now deduce a rule, which is of general validity: progressive connection invokes an intensity enhancement, regressive connection an intensity reduction. SPI and related experiments such as spin tickling,[88] generalized NOE experiments,[86] etc. can furnish energy level connectivity information. Since the assignment of transitions to a particular energy level depends on the relative signs of the coupling constants, the problem of sign determination is translated to the designation of energy levels in the energy level diagram. In second-order spectra the signs are, of course, more conveniently determined by spin simulation, as an exact match of transition frequencies and intensities between simulated and experimental spectrum requires a unique sign combination for all coupling constants.[39] In first-order spectra, however, line intensities are insensitive to the relative signs of coupling constants and the signs can only be established via double-resonance techniques.[88] We have previously mentioned (cf. also Chapter 2) that coupling constants characterize a particular coupling path and electron distribution. While one- and three-bond $^{13}C-^{1}H$ coupling constants are known to be positive,[10] $^{2}J_{CCH}$ can be positive or negative and its correct interpretation may therefore require the knowledge of its sign.

Since only relative signs can be determined by n.m.r., at least three magnetically non-equivalent spins have to be present. The simplest system meeting these requirements is the AMX spin system, consisting, for example, of two protons and one carbon. One such case is the spectrum of the C-3 isotopomer of 2,3-dibromothiophene (**16**)[84,90] where the two protons, denoted A and B, together

(**16**)

with C-3, form a weakly coupled three-spin system. The twelve transitions are schematically shown in Fig. 3.22(a). The relative magnitudes of the three coupling constants are $|^{3}J_{C(3)H(5)}| > |^{3}J_{HH}| > |^{2}J_{C(3)H(4)}|$. Assuming positive signs for all three quantities the spin polarizations for the various transitions are as indicated in Fig. 3.22(a). The energy-level diagram, which has the topology of a cube, can now be constructed (Fig. 3.22(b)). From this we recognize that A_1, the highest-frequency H-4 transition, is connected to B_1, X_1 and B_2, X_2, respectively. A population inversion on the $\alpha\alpha\alpha$ and $\alpha\beta\alpha$ levels should therefore affect the intensities of the above-mentioned lines. Provided that relaxation is negligible during the 180° decoupler pulse, the X_1 transition should result in an inverted negative and X_2 in an enhanced positive signal. The condition of negligible

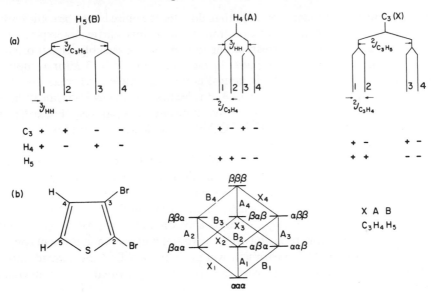

Fig. 3.22 (a) Schematic spectrum of a weakly coupled unsymmetric three-spin system assuming $|J_{BX}| > |J_{AB}| > |J_{AX}|$. The assignment of spin states given below each line assumes positive signs for all three coupling constants. (b) Energy-level diagram for the three-spin system defined in (a).

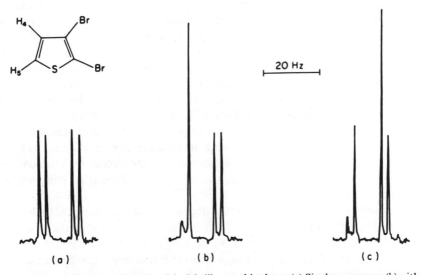

Fig. 3.23 Spectrum of carbon-3 in 2,3-dibromothiophene. (a) Single resonance; (b) with population inversion on the highest-frequency H-4 transition (line A1 in Fig. 3.22). In order to ensure that only one transition was perturbed, a small H_2 amplitude was used for the experiment ($\gamma H_2/2\pi = 0.2$ Hz). Therefore the time required for selective inversion became 2 s, long enough to allow for relaxation during the pulse. For this reason no inverted (negative) signals were obtained.[84]

relaxation during the pulse is difficult to realize, for the pulse has to be sufficiently selective so as to perturb only one transition. In this particular experiment $\gamma H_2/2\pi = 0.2\,\text{Hz}$ was used, corresponding to $\tau_\pi = 2\,\text{s}$. Appreciable relaxation can therefore take place during the pulse. In the spectrum of Fig. 3.23(b) we notice that line 1 is not inverted but is almost completely nulled, whereas line 2 is enhanced. Likewise, a perturbation of B_1, the highest-frequency H-5 transition, is accompanied by alterations in the relative intensities of X_1 and X_3 (Fig. 3.23(c)), in accordance with the scheme of Fig. 3.22(b), hence clearly establishing like signs for all three coupling constants.

An interesting extension of the simple SPI experiment has recently been reported in conjunction with the determination of the relative signs in proton-coupled satellite spectra[91] of isotopomers containing two rare spins (for example, two ^{13}C's). The technique, which was given the acronym DSPT (double selective population transfer), is based on the idea of driving two proton transitions simultaneously, therefore resulting in (negative and positive) enhancements on all rare-spin transitions. The experiment will probably not find widespread application, since it requires hardware not normally provided on commercial spectrometers.

INEPT (insensitive nucleus enhancement by polarization transfer)[1-4]

The utilization of selective pulses imposes a severe limitation on the applicability of the SPI technique as a general-purpose sensitivity enhancement method. The inversion pulse, inducing population changes, by virtue of its design affects only one particular transition. A somewhat related experiment overcoming this drawback by employing non-selective pulses has been proposed recently.[1] In order to outline its principle let us again return to the energy-level diagram of Fig. 3.19. The objective of the experiment essentially remains the same, i.e. to selectively invert the populations on one of two proton transitions, except that this is to be accomplished for all protons, irrespective of their chemical shift. The experiment is best understood by following the proton magnetization in the now-familiar rotating frame of reference, assuming that the decoupler frequency is applied off-resonance relative to the Larmor frequency of the precessing proton spins.

The starting point is a non-selective $90°$ decoupler pulse, nutating both doublet components into the transverse plane (Figs 3.24(a) and (b)). During the following period τ_1 the two vectors precess at a rate $\nu_H \pm J/2$, with J denoting the CH coupling constant, thus accumulating a relative phase angle $\phi = 2\pi J \tau_1$. If τ_1 is chosen such as to meet the condition

$$\tau_1 = 1/(4J) \tag{3.13}$$

it is readily seen that the two proton magnetization components make up an angle of $90°$, as illustrated in Fig. 3.24(c). A $180°$ proton pulse acting at time $t = \tau_1$

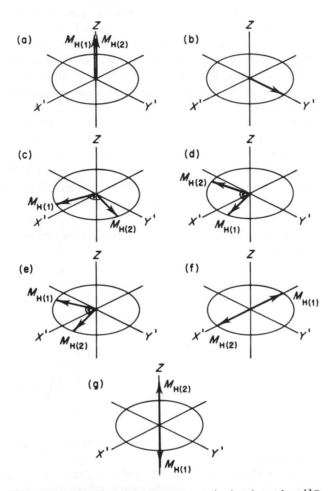

Fig. 3.24   Behaviour of the proton magnetization shown for a ^{13}C, ^{1}H (AX) spin system during an INEPT pulse sequence in the rotating frame. (a) At thermal equilibrium; (b) following a $90^\circ_x(^1$H) pulse the proton magnetization is aligned along y'; (c) $(4J)^{-1}$ s after the initial 90° proton pulse the two proton magnetization components have built up a relative phase angle of 90°; (d) a $180^\circ_x(^1$H) refocusing pulse converts the two components into their mirror images (relative to $x'z$ plane); (e) the $180^\circ_x(^{13}$C) pulse applied in synchronism with $180^\circ_x(^1$H) exchanges the labels (H(1) and H(2)); (f) at time $t = (2J)^{-1}$ the two components $M_{H(1)}$ and $M_{H(2)}$ align along $-x'$ and x', respectively. (g) A$90^\circ_y(^1$H) pulse converts $M_{H(1)}$ and $M_{H(2)}$ into longitudinal magnetization corresponding to an inversion of the populations on the $\alpha\alpha$ and $\beta\alpha$ levels (cf. Fig. 3.19).

flips the two components $M_{H(1)}$ and $M_{H(2)}$ into their mirror images (Fig. 3.24(d)), resulting at a time $t = 2\tau_1$, without any other provisions made to the experiment, in complete refocusing, i.e. realignment of the combined proton magnetization, along the $-y'$ axis. If, on the other hand, a 180° carbon pulse is applied in synchronism with the refocusing proton pulse, this causes reversal of the spin states[92] (Fig. 3.24(e)), corresponding to an exchange of fast and slow components. At time $t = 2\tau_1$, $M_{H(1)}$ and $M_{H(2)}$ will therefore end up in an antiphase situation with $M_{H(1)}$ pointing along $-x'$ and $M_{H(2)}$ along $+x'$ (Fig. 3.24(f)). Establishment of the desired non-equilibrium configuration with $M_{H(2)}$ being inverted is effected by means of a proton pulse along y' (Fig. 3.24(g)). It should be noted that the 90° phase shift of this pulse relative to the two preceding pulses is essential, as a $90°_x$ pulse would not nutate the magnetization. A 90° ^{13}C detection pulse will now, following Fourier transformation, create a spectrum in which C(1) and C(2) have relative intensities of 5 and -3, respectively, compared with the single-resonance spectrum.

In its simplest version the INEPT experiment can be summarized by the following sequence of events:

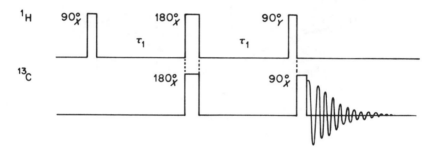

The relative intensities found for a CH_2 and CH_3 moiety are $-1, 0, +1$ and -1, $-1, +1, +1$, respectively.

From Fig. 3.24(f) we further see that a $90°_{-y}$ pulse at time $t = 2\tau_1$ establishes the opposite polarization of $M_{H(1)}$ and $M_{H(2)}$, therefore reversing the intensities of the ^{13}C doublet. By inverting the transmitter phase on alternate pulses and alternately adding and substracting the resulting FID, symmetric patterns can be obtained.

The delay τ_1 is determined by the magnitude of $^1J_{CH}$ according to Eqn. (3.13). Because of the range of one-bond coupling constants (given essentially by carbon hybridization, cf. Chapter 2), a compromise setting has to be chosen. In practice, it turns out that this choice is not critical and deviations from the condition $\tau_1 = 1/(4J)$ will simply cause lower than optimum polarization transfer.

The gain in sensitivity achieved by the INEPT experiment is twofold. It results, on the one hand, from the redistribution of spin populations on the ^{13}C levels, leading to a maximum enhancement $\varepsilon_{max} = \gamma_H/\gamma_C \simeq 4$ in the AX case. On the other hand, the experiment tolerates shorter recycle times. This is because the latter are dictated by the proton rather than the normally longer ^{13}C spin–lattice relaxation time.

The power of the technique is illustrated with the proton-coupled INEPT spectra in Fig. 3.25. The polarization transfer-enhanced spectrum (traces (a) and (b)), obtained from a 20 millimolar solution of naphthacene quinone at 90.5 MHz

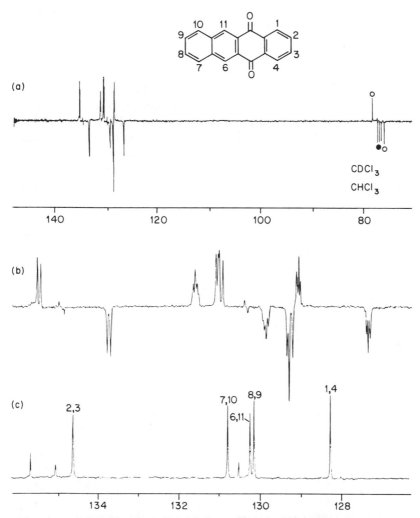

Fig. 3.25   (a), Proton-coupled 90.5 MHz (8.4 T) ^{13}C INEPT spectra of the aromatic region of naphthacene quinone, 20-millimolar in deuterochloroform. The solvent doublet (trace a) arises from polarization of residual non-deuterated chloroform (*ca* 0.1%). (b) expanded. Spectrum (c), which was recorded under standard proton noise-decoupling conditions from the same sample in the same total accumulation time, is shown for comparison.

(8.4 T), required about the same spectrometer time as the broadband-decoupled spectrum (trace (c)) and exhibits a S/N which is only a factor of 2 lower than the latter. A notable point and manifestation of the extraordinary sensitivity of the experiment is the solvent doublet arising from the residual non-deuterated chloroform (0.1%). This resonance cannot be detected in the classical proton noise-decoupled spectrum.

A limitation of the simple polarization transfer scheme is the differential transfer of coherence, in that multiplet lines become positively and negatively polarized. This clearly impedes broadband decoupling during the acquisition period, as this would lead to mutual cancellation of symmetric multiplet components. The simplest remedy to this problem is to introduce a delay between the 90° read pulse and the beginning of acquisition.[2] The purpose of this delay is to allow for spontaneous rephasing of antiphase multiplet components. It is readily seen that for a ^{13}C—^1H doublet spontaneous refocusing occurs at time $\Delta = (2J)^{-1}$. However, the outer components of a $^{13}CH_2$ triplet, which differ in frequency by $2J$, change their phase angle twice as fast, therefore demanding $\Delta = (4J)^{-1}$. The situation for a methyl signal is somewhat more complex, since the two pairs of multiplet components rotate at frequencies which differ by J and $3J$, respectively. It is obvious that a choice of $\Delta = (2J)^{-1}$, for example, affords a positive signal for a CH moiety while nulling that for CH_2.

More effective refocusing than by free precession is achieved by applying a 180° spin echo pulse at the midpoint of the Δ period ($\Delta/2 = \tau_2$) to both ^{13}C and ^1H spins,[3] as shown in the timing diagram below:

INEPT with refocusing and decoupling

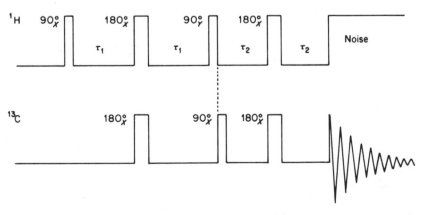

This modification of the basic polarization transfer experiment is sometimes referred to as INEPT with refocusing and decoupling. The extent of polarization transfer clearly is related to the two time periods τ_1 and τ_2 and, of course, to the multiplicity of the signal, as the magnetization components in the transverse

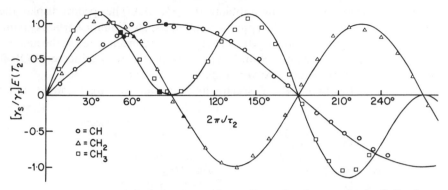

Fig. 3.26 Dependence of the enhancement factor E on the phase angle $\phi = 2\pi J\tau_2$ for a CH circles), CH_2 (triangles), and CH_3 (squares) signal, calculated according to Eqns (3.14)–(3.16)[3] for an INEPT polarization transfer experiment, assuming $\tau_1 = (4J)^{-1}$. Reproduced by permission of the author.

plane evolve with relative frequencies J (methine), $2J$ (methylene) and J and $3J$ (methyl). Calculations for the enhancement factor E yield for the three coupling situations:

$$E_{CH} = (\gamma_H/\gamma_C) \sin(2\pi J\tau_1) \sin(2\pi J\tau_2) \tag{3.14}$$

$$E_{CH_2} = (\gamma_H/\gamma_C) \sin(2\pi J\tau_1) \sin(4\pi J\tau_2) \tag{3.15}$$

$$E_{CH_3} = (3\gamma_H/4\gamma_C) \sin(2\pi J\tau_1)[(\sin(2\pi J\tau_2) + \sin(6\pi J\tau_2)] \tag{3.16}$$

For all three spin systems the optimum value for τ_1 is $\tau_1 = (4J)^{-1}$. The dependence of the enhancement factor E on the refocusing period τ_2 is plotted in Fig. 3.26, assuming $\tau_1 = (4J)^{-1}$. It is evident that there is no value of τ_2 which optimizes the signal enhancement for all three spin configurations. Optimum average enhancement, however, is predicted for the condition $2\pi J\tau_2 = \pi/3$ (60°) or $\tau_2 = (6J)^{-1}$. Apart from enhancing the sensitivity in broadband-decoupled ^{13}C spectra this experiment lends itself for multiplicity sorting and therefore represents a powerful alternative to single-frequency off-resonance decoupling. It is, for example, recognized from Fig. 3.26 that for the condition $2\pi J\tau_2 = \pi/2$ ($\tau_2 = (4J)^{-1}$) both methyl and methylene signals are suppressed, whereas $2\pi J\tau_2 = 3/4\pi$ ($\tau_2 = (2.7J)^{-1}$) affords positive signals for CH and CH_3 but inverted ones for CH_2. This behaviour is illustrated with the spectra of lasalocid (**17**) (Fig. 3.27), a

17

ionophorous, transport-inducing antibiotic ($C_{34}H_{54}O_8$). The spectra display the aliphatic region between 0 and 90 ppm with all of the 26 aliphatic carbons giving rise to separate signals. However, 10 and 8 signals are concentrated in the narrow bands between 7 and 20 and 29 and 38 ppm, respectively, producing SFORD spectra (Fig. 3.2.7(b)), which are very difficult to unravel.

By contrast, the two INEPT spectra in traces (c) and (d), recorded under

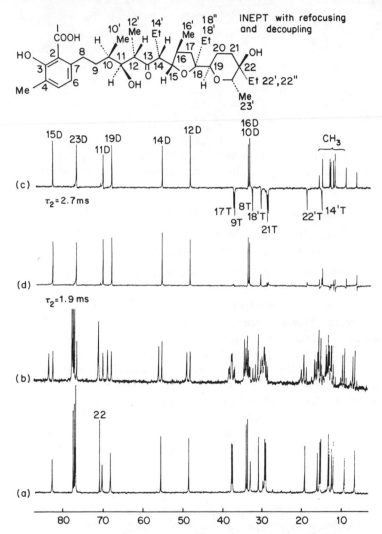

Fig. 3.27   90.5 MHz (8.4 T) ^{13}C spectra of the aliphatic region in lasalocid (**17**). (a) (Proton noise-decoupled; (b) SFORD; (c), (d) INEPT with refocusing and decoupling, (c) $\tau_2 = 2.7$ ms ($\sim 1/(2.7J)$): CH, CH$_3 > 0$, CH$_2 < 0$; (d) $\tau_2 = 2.0$ ms ($\sim 1/(4J)$): CH$_3$, CH$_2 = 0$, CH > 0.

refocusing conditions with broadband decoupling during the acquisition period (as previously described) establish the multiplicities immediately. Both spectra were obtained under identical experimental conditions, except for a different choice of the refocusing period τ_2. Assuming an average one-bond coupling constant of 140 Hz, $\tau_2 = 2.7$ ms should be the condition for furnishing positive signals for methyl and methine resonances and negative ones for methylenes. Analogously, setting $\tau_2 = 2.0$ ms affords positive signals for CH carbons while nulling both CH_3's and CH_2's. The somewhat arbitrary choice of $^1J_{CH}$ appears to best match the real values for the centre carbons while being too large for carbons without oxygen substituents and too small for some of the ether carbons. This explains the unequal intensities for the methine carbon resonances. The experiments corroborate the presence of eight methyls, eight methylenes, and eight (aliphatic) methines, as suggested by the formula. It should further be noted that the quaternary carbons are almost completely suppressed, irrespective of the choice of τ_2.

For the sake of completeness we should mention another polarization transfer experiment, which is analogous to cross-polarization in solids.[93] The experiment requires to spin-lock the protons by means of a coherent rf field for a period of the order of milliseconds, immediately following a 90° proton pulse and 90° rf phase shift. During the spin-locking period the ^{13}C transmitter is activated so that the condition $\gamma_C H_{1C} = \gamma_H H_{2H}$ (Hartmann–Hahn condition) is fulfilled. Proton polarization is then transferred to the ^{13}C spins via dipolar coupling (solids) or J-coupling (liquids).[94] The experiment, however, is critical to implement on high-resolution spectrometers and has therefore not found broad acceptance as a method for polarization transfer in liquids, whereas it is an established technique in solid-state ^{13}C n.m.r. (see Section 3.10).

Attached proton test[14,95,96]

A method which bears some resemblance to the previously discussed version of the INEPT experiment with its multiplicity sorting capabilities also exploits the idea of introducing an amplitude modulation into the signals of proton noise-decoupled ^{13}C spectra. However, the technique strictly does not belong to the category of polarization transfer experiments because the attached proton test (APT) pulse sequence does not, in fact, entail any transfer of polarization from protons to ^{13}C. On the other hand, it does, as suggested by its designation, represent another, particularly simple, method for determining multiplicity in the fully decoupled recording mode.

The principle of the experiment is to let the spins precess during a delay period following an initial non-selective 90° observation pulse. During this free precession period the magnetization components, characterized by the various spin states of the attached protons, build up a relative phase angle, determined by the magnitude of $^1J_{CH}$ and multiplicity. Figure 3.28 shows the evolution of the

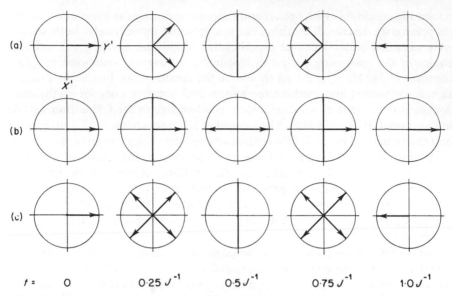

Fig. 3.28 Evolution of the transverse ^{13}C magnetization, following a 90° pulse, shown schematically in the $x'y'$ plane of the rotating frame at resonance, for the three spin systems ^{13}CH (a), ^{13}CH$_2$ (b), and ^{13}CH$_3$ (c).

transverse magnetization in time increments $t = 0.25 J^{-1}$ for a methine, methylene, and methyl spin system. It is evident that at time $t = J^{-1}$ the magnetization components are realigned for all three coupling situations, therefore permitting spin decoupling. However, the phase of the methine and methyl signals is opposite to that of the methylene signals. Not shown in the figure is the behaviour of quaternary carbons, which, of course, are unmodulated and give rise to the same relative phase as the methylenes.

A disadvantage of this simple version of the experiment are the large, frequency-dependent phase shifts, caused by the acquisition delay. This problem can be circumvented by introducing a refocusing pulse at the end of the τ period, thereby reversing the dispersal of the magnetization components caused by chemical shift. The relative phases of the signal components, however, can be maintained if at this point in time the proton noise-decoupler is turned on while data acquisition commences at time $t = 2\tau$ only, as shown in the timing diagram below.

ATP is definitely not quite as powerful a method as INEPT for multiplicity selection since it does not permit distinction between CH and CH$_3$ in a straightforward way, and, of course, it lacks the benefit of sensitivity enchancement. On the other hand, it is simple to implement and does not require sophisticated instrumentation.

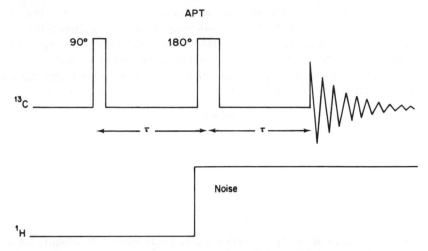

It has recently been pointed out that the efficiency of the experiment can be improved by using, instead of the initial 90° pulse, a flip angle $\alpha < 90°$ (Ernst angle, cf. Eqn (1.21)). This obviously means that the residual z magnetization following the dephasing period τ is inverted by the 180° refocusing pulse.[96] It has therefore been suggested that an additional 180° pulse following the second

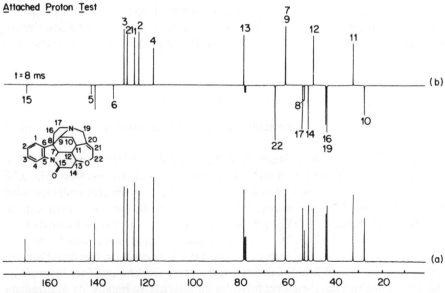

Fig. 3.29 Proton noise-decoupled ^{13}C spectra of strychnine (**18**), (a) Standard spectrum; (b) spectrum obtained by the refocused APT technique described in the text, using a value of 8 ms for the dephasing period. Assignments are those of Ref. 37, except for C-14, C-17, and C-19, which were revised in the light of more recent findings (cf. p. 106).

period be applied.[95] The sole purpose of this pulse is to re-invert the $-z$ magnetization in analogy to the driven equilibrium Fourier transform (DEFT)[97] method, which was designed to recover transverse magnetization by spin echo techniques in order to improve efficiency in situations where $T_2 \gg T_2^*$.[97]

Figure 3.29 illustrates the application of the refocused ATP technique for multiplicity determination in strychnine (**18**), showing inverted signals for

18

carbons with even multiplicity (methylenes and quaternaries). It should also be noted that, in contrast to INEPT, the quaternary signals are not attenuated and appear with the same intensity they would in normal FT n.m.r.

DEPT

Another polarization transfer method suited for spectral editing has been dubbed DEPT (a, b) (distortionless enhancement by polarization transfer).[98-100] Unfortunately, a full explanation of this sequence in terms of a classical vector model for the magnetization is not possible. We therefore confine ourselves to a description of the sequence and the spectral phenomena. The sequence can be summarized as follows:

^{1}H: $90° - \tau - 180° - \tau - \Theta$

^{13}C: $90° - \tau - 180° - \tau -$ Acquisition

τ is set to $(2J)^{-1}$, where J represents the ^{13}C—^{1}H one-band spin–spin coupling constant.

The signal intensity as a function of flip angle Θ is plotted in Fig. 3.30 for typical methyl, methylene, and methine resonances. It is obvious from Fig. 3.30 that by taking intensity-weighted linear combinations of spectra collected with $\Theta = 90°, 45°$, and $135°$, subspectra pertaining to a particular multiplicity can be obtained. For example, S(90°) provides CHs only, S(45°) – S(135°) affords CH_2 subspectra. Analogously, CH_3 subspectra are obtained from S(45°) + S(135°) $- 0.707$ S(90°). One of the virtues of the DEPT spectra is the fact that the multiplet components are not distorted, therefore lending itself to decoupling. The DEPT technique, therefore, provides an alternative means for multiplicity determination. Further, the nulling condition for CH_2 and CH_3 is not very critical to the magnitude of $^1J_{CH}$. However, it is sensitive to the setting of the proton flip angle Θ.

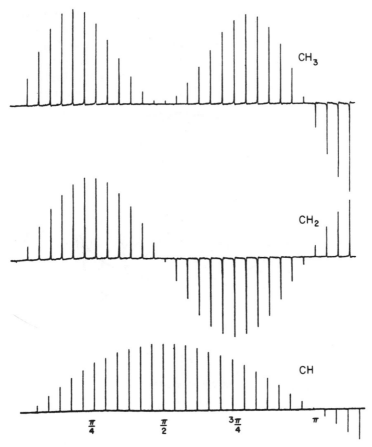

Fig. 3.30   Proton-decoupled ^{13}C DEPT signals obtained with $\tau = (2J)^{-1}$ and various flip angles.[100] Reproduced by permission of Electrospin AG.

3.4 SELECTIVE EXCITATION[6,7]

We have earlier seen that unravelling the various signal components in complex proton-coupled spectra, even at high magnetic field, poses a considerable problem. In such instances it would therefore be desirable to selectively excite the subspectrum pertaining to a particular ^{13}C isotopomer. Several approaches have been put forward to tackle this task. The most straightforward one consists of narrowing the excitation spectrum of the pulse train. We have seen in Chapter 1 that the excitation breadth is governed by the duration of the rf pulses. Hence it suffices to lower the amplitude of the H_1 field and lengthen the pulses accordingly so that the width of the sin x/x curve just covers the bandwidth of the spectral pattern to be observed. This approach has been applied with success to blank out

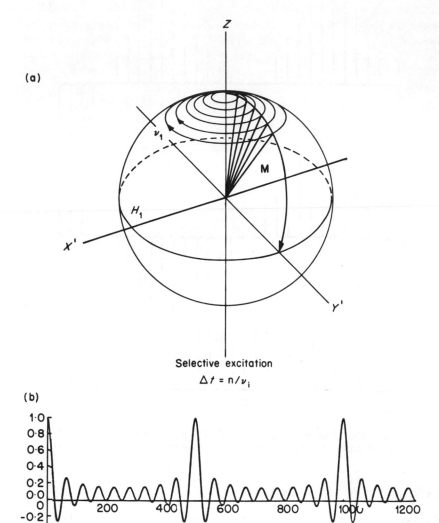

Fig. 3.31   (a) Principle of selective excitation, shown in a reference frame rotating at the carrier frequency. A train of short pulses of flip angle $\Delta\alpha$ consecutively tips the magnetization following each full precession, therefore leading to a coherent nutation of angle $\alpha = m \cdot \Delta\alpha$. (b) Magnetization M_y at the end of a train of 10 pulses spaced 2 ms apart, each of flip angle $\pi/20$ radians, calculated from the Bloch equations as a functions of carrier offset Δv.[5] Reproduced with permission.

undesired solvent resonances in proton n.m.r.[101,102] The method, however, suffers from several drawbacks. It is not convenient to use and requires additional equipment such as precision attenuators. The very versatile tailored excitation scheme[103] consists of letting the computer generate a pulse sequence whose Fourier transform has a prespecified shape. The disadvantage of this experiment are the requirement of very large computer memories for retaining the data necessary to define the very complex pulse trains, as well as the stringent demands put on the hardware.

A simplified version of tailored excitation, designed to excite only one particular frequency in the spectrum, consists of applying a train of strong equidistant pulses of interval Δt and duration τ in such a way that the sum of these 'minipulses' corresponds to a 90° flip.[6,7] By choosing the pulse interval so that the magnetization carries out an integral number n of precessions between successive minipulses, the magnetization of the desired signal is cumulatively tipped as outlined in Fig. 3.31(a). The condition for frequency selectivity is therefore defined as follows:

$$\Delta v \Delta t = n \qquad (3.17)$$

In order for relaxation to be negligible during the duration of the full pulse train $n = 1$ is chosen (first sideband excitation). It is easy to realize that magnetization vectors corresponding to a different offset Δv will have rotated through a precession angle $\theta \neq 2\pi n$, therefore experiencing less nutation. A detailed analysis based on the Bloch equations shows that the transverse component of these magnetizations follows a circular path, moving them away from the y' axis of the rotating frame.

Figure 3.31(b) shows the dependence of M_y, the y component of the magnetization on the frequency offset from the carrier. The number of the weak maxima between the main excitation peaks of the various sidebands is determined by the number m of the minipulses. It can further be shown that the selectivity of excitation increases with the number of pulses in the train. Typical values for m are between 20 and 100. At high transmitter power it may be necessary to lengthen $\tau_{90°}$ by putting an attenuator into the transmitter line, so that the duration of the individual pulses can be controlled with sufficient accuracy.

By combining selective excitation with gated coupling it now becomes possible to observe individual multiplets which may otherwise be obscured by spectral overlap. For this purpose, the decoupler is turned on during the pulse train (a necessary condition) and during an optional pulse delay, but turned off during the acquisition period. By determining first the frequency offsets for each carbon from the proton noise-decoupled spectrum, a series of experiments can be programed, each affording the spectrum of a particular isotopomer, as illustrated in Fig. 3.32 for the methylene and methyl region in β-pinene (**19**). The remarkable selectivity, achievable by this experiment becomes evident when one considers,

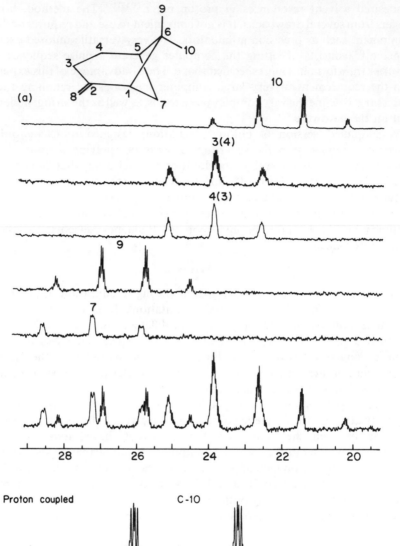

(a)

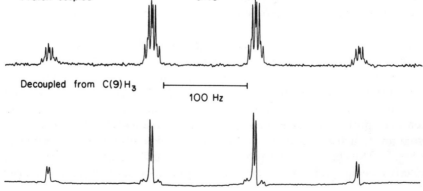

(b)

Proton coupled   C-10

Decoupled from C(9)H₃   ├─────────────┤
                              100 Hz

19

for example, the two methylene resonances at 23.875 and 23.816 ppm ($\Delta\delta =$ 0.059 ppm or 5.8 Hz at 9.4 T). It is interesting to note that the more shielded of the two resonances exhibits considerable fine structure and, to conclude from the width of the signal pattern, involves a larger number of vicinal couplings. In particular, three-bond coupling to the exocyclic olefinic methylene protons is expected to be comparatively large for C-3 (8–12 Hz[10]), thus favouring assignment of the line at 23.816 ppm to C-3. This assignment could, of course, be verified by selective irradiation of the olefinic protons.

Instead of turning the decoupler off during the acquisition period, affording a proton-coupled spectrum, one may as well selectively irradiate a proton which is a suspected coupling partner of the carbon, whose subspectrum is recorded. This is demonstrated with the selective excitation spectrum in Fig. 3.32(b), for C-10, obtained by irradiating the C-9 methyl protons during acquisition. This removes the three-bond methyl–methyl coupling and simplifies the fine structure on each of the C-10 methyl components to doublets, thus showing that only one of the bridge protons is coupled.

Another useful application of selective excitation, which shall not be discussed at great length here, concerns experiments requiring selective inversion of individual lines, as, for example, in saturation transfer studies[104] in slow-exchange kinetics. If the exchange rate is slow enough to yield separate lines for each of the exchanging sites, but comparable in magnitude to $1/T_1$, saturation is carried over to the exchanging site, when the resonance at one of the sites is saturated. Instead of saturating a line, as proposed in the original experiment, an enhanced effect is observed by inverting the signal by a selective excitation pulse train, whose total nutation is 180°.[6] This is then followed by a non-selective detection pulse τ later. The evolution of the intensities of the affected signals is a

---

Fig. 3.32 (a) Selective excitation spectra, recorded at 100.6 MHz (9.4 T), showing the methylene and methyl region in β-pinene (**19**). For effective separation of the methylene resonances C-5 and C-6 ($\Delta\delta = 0.059$ ppm), a train of 300 pulses ($\Delta\alpha = 0.3°$) was required. The decoupler was gated on during the pulse train and a delay period, but off during data acquisition. The bottom trace represents the conventional proton-coupled spectrum. (b) Top: expansion of the lowest-frequency methyl (C-10) carbon subspectrum (top trace in (a)); bottom: same experiment with low-power irradiation of the C-9 methyl protons during the acquisition period. Assignments are those of Bohlmann et al. (*Org. Magn. Res.* **7**, 426 (1975)).

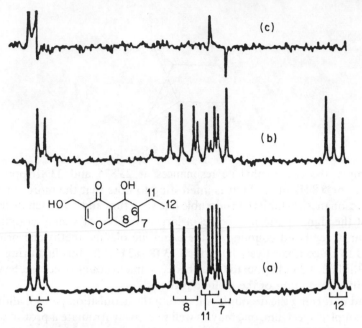

Fig. 3.33   Proton noise-decoupled ^{13}C spectra of diplosporin (**20**), derived from [1,2-^{13}C$_2$] acetate. (a) Conventional spectrum; (b) spectrum obtained by selective inversion of highest-frequency line of C-6, followed by a non-selective detection pulse; (c) difference spectrum ((a) minus (b)).[105] Reproduced by permission of the author.

function of the interval τ, the exchange rates, and the longitudinal relaxation times of the exchanging nuclei. In favourable cases both relaxation times and exchange rates can be extracted.

Another useful application of the selective excitation sequence has been reported in connection with the assignment of carbon–carbon couplings in biosynthetic studies[105] (cf. also subsequent sections). The spectra of molecules, derived from biosynthetic incorporation of [1, 2-^{13}C$_2$] acetate display ^{13}C—^{13}C spin–spin couplings. Because of the similarity of the coupling constants, assignment in spectrally congested areas is not always straightforward, and the previously described version of selective excitation, starting with a selective 180° pulse followed immediately by a non-selective observe pulse, affects the two

20

connected transitions of the AB system. This is essentially a selective population inversion experiment, entirely analogous to the one discussed at the beginning of this section. The induced population changes become particularly conspicuous in the difference spectrum, obtained by subtracting the SPI from the unperturbed proton noise-decoupled spectrum, as shown in Fig. 3.33 for $[1, 2\text{-}^{13}C_2]$ acetate-derived diplosporin (20):

3.5 TWO-DIMENSIONAL EXPERIMENTS[8,9,106]

Principle

Two-dimensional n.m.r. comprises a class of experiments which have in common an array of spectral data, plotted as a function of two frequency variables. In this sense of the definition a series of SFORD spectra, plotted against observe frequency (x axis) and decoupler frequency (y-axis), can be regarded as a two-dimensional spectrum. The group of two-dimensional experiments that we shall discuss here differ from the latter in one aspect, in that the data are initially collected as a function of *two time variables* t_1 and t_2. For the sake of clarity, it is convenient to divide the time axis into three periods as depicted in Fig. 3.34.[107]

During the *preparation period* the spin system is prepared in a suitable way, for example by having an Overhauser polarization built up. During the *evolution period*, characterized by a time variable t_1, the spin system evolves in a way specified by the purpose of the experiment. During the *detection period* t_2, finally, the data are acquired by the usual method, affording a FID, in which every point is a function of both, t_1 and t_2. By collecting m FIDs, each differing in the value of t_1, starting with $t_1 = \Delta t_1$ and ending with $t_1 = m.\Delta t_1$, a time domain data matrix $s(t_1, t_2)$ is obtained. Assuming each FID to consist of n data points, $s(t_1, t_2)$ has a

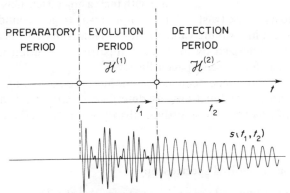

Fig. 3.34   Schematic representation of the three time periods in two-dimensional experiments. t_1 and t_2 are the time variables defining the signal $s(t_1, t_2)$.[107] Reproduced by permission of the author and the American Institute of Physics.

total of $m \times n$ points. Fourier transformation of each of the m FIDs yields m spectra $S(t_1, v_2)$, with v_2 designating the conventional chemical shift co-ordinate. The essence of the experiment now is that at a given frequency v_2, signals may have a periodicity in t_2, thus creating a motivation to carry out a second Fourier transformation for corresponding points within each column of the data matrix, i.e. according to t_2. This second FT leads to a frequency-domain data matrix $S(v_1, v_2)$. It should be pointed out that such a two-dimensional data array constitutes a surface of intensities. We are therefore confronted with the problem of projecting this data surface which can, for example, be achieved by stacking of consecutive traces. This presentation mode probably best provides the three-dimensional aspect of the data surface, in particular, when those sections of each trace, that appear hidden, are blanked ('white-washing'). An alternative presentation is the *contour plot*, which gives a top view of the data surface. This is obtained by mutually connecting equal-intensity points, resulting in a series of concentric circles for each peak. Often it is desirable to reduce the abundance of data by plotting only specific segments, i.e. by cutting profiles at operator-selected values of v_1 or v_2 parallel to the abscissa or ordinate. Finally, there is the possibility to project the surface onto the xz or yz plane (the z co-ordinate is assumed to define intensity). The specific benefits of the various presentation modes will become more lucid in the examples discussed below.

A peculiarity of two-dimensional n.m.r. is the unusual phase behaviour. In one-dimensional spectroscopy we have seen that the cosine transform ideally yields the absorption, the sine transform the dispersion spectrum. In two-dimensional n.m.r. the situation is more complex.[108,109] Although it turns out that cross sections through the centre of the peak, parallel to either one of the frequency axes, can be phased for pure absorption, this is no longer the case when the sections are offset from the centre. Such sections are found to be in dispersion, with the degree of dispersion increasing with increasing offset. This behaviour has been referred to as 'phase twist'.[108] The simplest way to get around this problem, of course, is to plot in absolute value mode.[110] Unfortunately the absolute value line shape is characterized by long tails, impairing resolution and leading to interference effects for closely spaced lines.

The former inconvenience can be overcome by convoluting the FID with resolution enhancement functions, as described in Chapter 1, although more recently phasing techniques affording absorption-mode 2DFT spectra have become available.

Two-dimensional *J*-resolved spectroscopy

In Section 3.4 we addressed the problem of reducing the data density in complex proton-coupled spectra with the aid of selective excitation.[5-7] The objective of two-dimensional *J*-resolved spectroscopy is similar. Its principle is to separate the two interfering parameters, chemical shift and spin–spin coupling, without sacrificing information, as we do in selective excitation (because of its reduced excitation bandwidth).

The simplest scheme, which leads to the desired separation of the two parameters, is that originally proposed,[111] which in fact is a two-dimensional extension of the APT experiments[14,95,96] in its simplest form. In this case the time interval between the 90° pulse and the beginning of data acquisition becomes the variable evolution period.

A more elaborate version of two-dimensional J-resolved n.m.r. makes use of a phenomenon denoted J *modulation*.[119] Let us once again consider the simplest possible case of J coupling, that occurring in an AX spin system, and follow the evolution of the magnetization in the transverse plane after a 90° pulse. We further assume that one of the spins, say A, is ^{13}C and X is a proton and, for the sake of sufficient generality, that excitation takes place slightly off resonance. In the rotating frame the A magnetization then carries out a slow precession about z, while at the same time splitting up into a slow and a fast component, differing in their frequencies by J. This situation is depicted in Figs 3.35(a) and (b). A 180° refocusing pulse around y at time $t = \tau$ flips the two components labelled F (fast) and slow (S) into mirror-image positions, as shown in Fig. 3.35(c). Since the direction of precession remains unaltered, this would result in complete

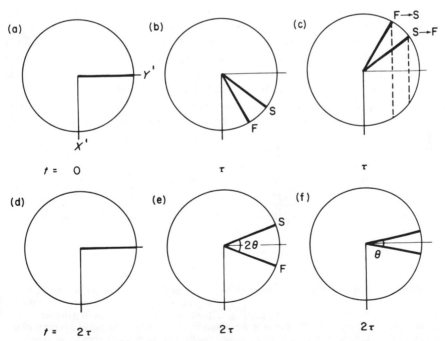

Fig. 3.35 Principle of J-modulation shown schematically for a two-spin system AX in the rotating frame as described in the text.

refocusing at time $t = 2\tau$, resulting in a spin echo with both components aligned along y' (Fig. 3.35(d)). If, however, a 180° pulse is applied simultaneously to the protons, the spin states of the protons are interchanged and the fast component becomes slow, and conversely. Instead of forming an echo, the two components continue to diverge, i.e. the relative phase angle between the two components increases, while the dispersing effect of chemical shift is reversed[5] (Fig. 3.35(e)).

We realize at this point a certain analogy to the INEPT experiment, the principal difference being in the latter that $\tau = (4J)^{-1}$ is a fixed delay and a mandatory setting, whereas in the two-dimensional J experiment $t_1 = 2\tau$ becomes the variable evolution period.

An even simpler method of introducing J modulation consists of turning on the proton noise-decoupler at time $t = \tau$ instead of applying a 180° decoupler pulse. The two magnetization components in Fig. 3.35(d) will then maintain their relative phase, leading to the situation shown in Fig. 3.35(f), which differs from (e) only in the phase angle between fast and slow component, which is reduced by half. The pulse timing diagrams in Fig. 3.36 summarize the two versions of the experiment.

It should be noted that at the end of the evolution period the effects of chemical shift and field inhomogeneity have been refocused. The relative phase angle between the multiplet components is determined by t_1 and the magnitude of the spin–spin coupling constant. By stepping t_1 in small increments, a series of FIDs is obtained which, following Fourier transformation according to t_2, afford spectra in which the multiplet lines change their phases in a cyclic fashion. Under proton noise-decoupling conditions during the acquisition period an amplitude modulation occurs, as shown in Fig. 3.37(a). This amplitude modulation reflects spin–spin coupling and there is no chemical shift superimposed on this frequency, for reasons stated earlier. With progressing t_1 the amplitude decays, but it does so

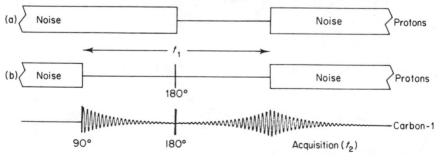

Fig. 3.36   Pulse sequence used for heteronuclear two-dimensional J-spectroscopy. In both modes of operation a spin echo is generated by a 180° refocusing pulse at the midpoint of the t_1 period. In the gated decoupler mode (a) phase modulation is maintained by turning the decoupler on during the refocusing period, whereas in the proton flip method (b) a 180° pulse in synchronism with the 180° ^{13}C pulse is applied, having the effect of interchanging proton spin states, therefore preventing reconvergence of magnetization components. For further details see text and Fig. 3.34.[8] Reproduced by permission of the International Society of Magnetic Resonance.

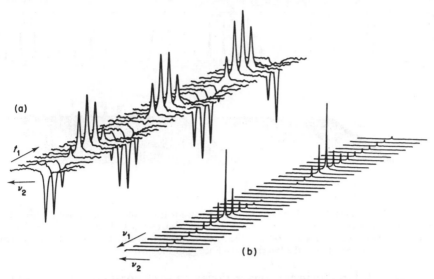

Fig. 3.37 ^{13}C two-dimensional-J spectra for chloroform, obtained by the sequence outlined in Fig. 3.36 (a). (a) After transformation in the t_2 domain, illustrating that the phase modulation caused by heteronuclear spin–spin coupling is converted to amplitude modulation by decoupling. (b) Same data after second Fourier transformation along the t_1 co-ordinate, plotted in the magnitude mode.

with a time constant T_2 rather than T_2^*, therefore resulting in spectra which, ideally, are free of inhomogeneity broadening. A second transformation in the t_1 domain therefore reproduces the spin–spin coupling multiplets, in this simple case just the chloroform doublet. Since the v_1 dimension is devoid of chemical shift, the width of the spectrum in this dimension is quite narrow (a few hundred Hertz, at most), therefore allowing very high resolution to be obtained.

Figure 3.38(a) shows a multiple-trace plot of the two-dimensional J-resolved spectrum of β-pinene (19), obtained by double Fourier transformation of a 256 $(t_1) \times 2048(t_2)$ time domain data array. Since the experiment was conducted in the gated decoupler version (cf. Fig. 3.36(a)), the splittings in the v_1 dimension represent half the actual coupling constants. With $t_{1\,\mathrm{max}} = 0.512\,\mathrm{s}$ ($256 \times 0.002\,\mathrm{s}$) the digital resolution in the J dimension is limited to 2 Hz, which is certainly insufficient for most practical purposes. However, if the resolution to be improved to, say, 0.5 Hz, this would increase the data size from 0.5 to 2 M words. On the other hand, one could, of course, sacrifice resolution in the δ dimension. In either event, however, this will lead to a prolongation of the total accumulation time.

The traces in Fig. 3.38(b) represent cross sections taken at the centre of the chemical shift of each carbon, thus providing the proton-coupled spectra of the individual isotopomers.

In connection with a discussion of the various techniques for assignment of

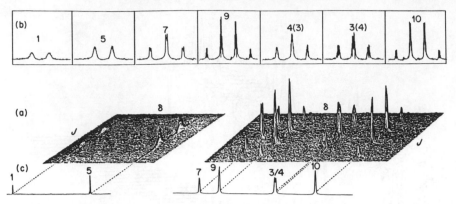

Fig. 3.38   75.4 MHz (7.0 T) ^{13}C heteronuclear two-dimensional J-resolved spectrum of β-pinene (**19**), obtained by the gated decoupler mode as described in the text. (a) Stacked plot; (b) cross sections at the chemical shift of each carbon, parallel to the J axis; (c) projection onto the δ axis. Reproduced by permission of the International Society of Magnetic Resonance.

proton-coupled spectra in previous sections we also elaborated on the question of assigning long-range ^{13}C—^{1}H couplings to specific protons. Complementing SPI and selective low-power decoupling, an alternative method, which is a modification of the proton flip version (cf. Fig. 3.36(b)) of the heteronuclear two-dimensional J-resolved experiment, has recently been suggested.[120] Instead of using a non-selective 180° proton decoupler pulse for reversing the proton spin states, this pulse is made frequency-selective, thus affecting only long-range couplings originating from one particular proton in the molecule. Since only coupling across two and more bonds are of interest, the frequency range required in the v_1 dimension is very small (10–20 Hz), enabling very accurate digitization and therefore high digital resolution. This is exemplified with the spectra in Fig. 3.39, obtained as cross sections parallel to the v_1 axis and showing the doublets arising from long-range coupling between proton h in carvone (**21**). The splittings observed are equal to the respective long-range couplings as long as the perturbed proton is only weakly coupled to other protons in each ^{13}C isotopomer.

21

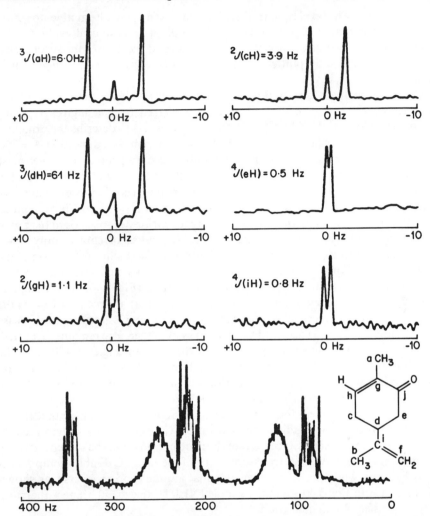

Fig. 3.39   Long-range ^{13}C—^1H couplings observed between the various carbons in carvone (21) and proton h. Each trace represents a cross section at the respective carbon chemical shift in a frequency-selective two-dimensional-J experiment, obtained by the proton h with a 'soft' 180° decoupler pulse.[120] Reprinted with permission from Bax and Freeman, *J. Am. Chem. Soc.* **104**, 1099. Copyright (1982) American Chemical Society.

Chemical shift correlation[76–82]

Probably the most powerful application of two-dimensional n.m.r. to date is the carbon–proton chemical shift correlation experiment, which provides both carbon and proton chemical shifts at the same time, but, more importantly, establishes a one-to-one correspondence for each pair of directly bonded carbon and hydrogen. The experiment therefore represents an alternative to more traditional techniques such as selective decoupling or graphical evaluation of SFORD data. A two-dimensional $^{13}C/^{1}H$ correlated spectrum consists of a map of carbon chemical shifts plotted in an xy co-ordinate system against their proton counterparts.

The experiment is based on polarization transfer and bears some resemblance to INEPT, where magnetization is also transferred from a high-γ (typically protons) to a low-$\gamma(^{13}C, ^{15}N, ^{29}Si,$ etc.) nucleus. In order to describe the principle, we first resort once again to the simple AX spin system, assuming that A $= ^{13}C$ and X $= ^{1}H$. The experiment begins with a non-selective proton decoupler pulse to create transverse magnetization. This is followed t_1 s later by a second 90_x° proton pulse. During the interval t_1 the proton magnetization precesses at a rate determined by the proton transmitter offset. The effect of the second proton pulse, whose purpose is to rotate the transverse magnetization M_{xy} back onto the z axis, depends on the relative phase angle ϕ of M_{xy}, since, of course, only the y component is affected by the pulse. The phase angle ϕ is a linear function of the time interval t_1. For $t_1 = 0$ the combined effect of the two pulses obviously is that of a 180° pulse, inverting the proton spin populations (Fig. 3.40). If we designate the decoupler frequency offset Δv, it is readily seen that at time $t_1 = 1/(4\Delta v)$ the magnetization has precessed by an angle of 90°, aligning it along x'. In this case the second pulse has no effect and the proton system therefore remains saturated. If $t_1 = 1/(2\Delta v)$ the magnetization precesses through an angle of 180° and the second pulse will therefore simply re-establish the Boltzmann populations, if we ignore relaxation during the t_1 period. If we increase t_1 to $3/(4\Delta v)$ the magnetization precesses 270°, ending up along $-x'$ and the subsequent proton pulse will again not affect the spin system, and so forth. We can see that in this fashion the proton longitudinal magnetization varies cyclically, determined by the proton chemical shift. Since each of the proton transitions has an energy level in common with the two ^{13}C transitions, one of the two ^{13}C doublet components gains intensity while the other loses an equal amount and vice versa, as shown schematically in Fig. 3.40. As in the simple INEPT experiment, there is therefore no net transfer of magnetization.

Immediately following the second proton pulse a 90° ^{13}C detection pulse generates an FID, carrying with it the proton chemical shift as an amplitude modulation. The experiment is repeated m times by incrementally varying t_1, the evolution period of the experiment. A second Fourier transformation in the t_1 dimension then provides a two-dimensional array in which the projection onto the v_1 axis represents the proton chemical shift.

One readily realizes that the experiment is impractical in this form, as both proton and carbon spectra are complicated by heteronuclear coupling. Because of the antiphase characteristics of symmetric multiplet components, proton noise-decoupling during the acquisition period would cause complete annihilation of the spectrum. The generally adopted solution to this problem[77] consists

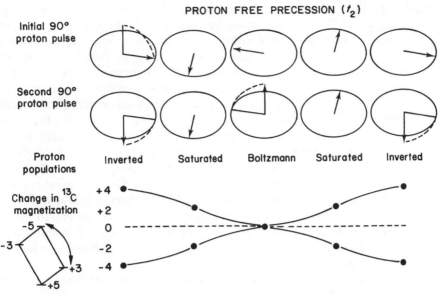

Fig. 3.40 Principle of the magnetization transfer experiment used in ^{13}C/^{1}H two-dimensional chemical shift correlation experiments, shown for the AX (A = ^{1}H, X = ^{13}C) spin system. The initial 90° proton pulse creates transverse magnetization. Following a precession period a second 90° proton pulse converts the y component of M_{xy} back to longitudinal magnetization. Depending on the phase of the magnetization at the time of the second pulse, the proton spin populations are inverted, saturated, or the equilibrium is restored. This affects the relative populations on the energy levels of the connected ^{13}C transitions. The longitudinal ^{13}C magnetizations are therefore modulated as shown on the bottom of the figure.[8] Reproduced by permission of the International Society of Magnetic Resonance.

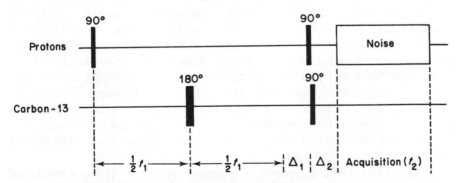

Fig. 3.41 Pulse timing diagram for the ^{13}C/^{1}H chemical shift experiment as described in the text. t_1 is the variable evolution time and the fixed delays Δ_1 and Δ_2 are related to the magnitude of $^1J_{CH}$ and signal multiplicity. Their purpose is to prevent cancellation of antiphase signals. 'Spin decoupling' is accomplished by the ^{13}C 180° pulse (^{1}H\{^{13}C\}) and proton noise-decoupling (^{13}C\{^{1}H\}).[8]

of applying a 180° pulse on ^{13}C, at the midpoint of the t_1 period, as shown in the timing diagram of Fig. 3.41. This pulse has the effect of exchanging the ^{13}C spin states, thus invoking convergence with respect to spin–spin coupling (though not with respect to chemical shift) at time t_1. A fixed delay Δ_1 then follows, whose duration is chosen such that the proton doublet components carry out a 180° relative phase shift, i.e. $\Delta_1 = 1/(2J)$. Proton noise-decoupling and acquisition start simultaneously Δ_2 s later. This period clearly depends on multiplicity, since it determines the relative phases of the ^{13}C signal components. For a CH carbon $\Delta_2 = 1/(2J)$; for a CH_2 carbon, however, the optimum choice would be $\Delta_2 = 1/(4J)$. In practice, it turns out that a compromise setting of $\Delta_2 \gtrsim 1/(3J)$ gives the best results.[82] Signals transmitted through long-range couplings do not appear since their relative phases change more slowly, thus cancelling one another upon decoupling. In summary, the 180° carbon pulse has the effect of decoupling in the ν_1 dimension, whereas proton noise decoupling provides decoupling in the ν_2 domension. Homonuclear (^1H—^1H) spin–spin couplings are still retained and may be useful for assignment. However, they are often not visible due to lack of digital resolution.

Two important aspects relating to the sensitivity of this experiment deserve mention. Because the detected carbon magnetization originates entirely from transfer of proton magnetization, the experiment has essentially the sensitivity of protons (though at the 1.1% abundance level, as only ^{13}C-containing iso-topomers are detectable). Further, the recycle time is determined by the usually shorter proton spin–lattice relaxation time, as in the INEPT experiment. It has been shown[81] that the efficiency of the experiment can further be enhanced if provisions are made to allow placement of the decoupler frequency at the centre of the proton spectrum. This is not possible with the previously discussed sequence, as no discrimination can be made between positive and negative frequencies produced by amplitude modulation. The modified sequence, which we will not discuss in detail, converts amplitude modulation into phase modulation, therefore allowing this distinction to be made. We have seen in Chapter 1 that centering the observe carrier reduces the digitization rate by half. The same holds true for sampling in the t_1 dimension of this two-dimensional experiment. Since we halve the t_1 sampling rate we also halve the number of t_1 samples at a given digital resolution in the ν_1 dimension, therefore increasing the efficiency of the experiment by a factor of two. Moreover, the phase cycling scheme proposed in Reference 81 eliminates the unwanted signals otherwise arising from the residual ^{13}C signals not originating from polarization transfer.

We would like now to demonstrate the application of $^{13}C/^1H$ two-dimensional chemical shift correlation for the complete assignment of the carbon chemical shifts in the alkaloid strychnine (**22**), a molecule, which differs from the earlier assigned brucine (**1**) only in the substitution of its aromatic ring. This will therefore give us an opportunity to cross-check the earlier assignments made. A precondition for a cross-assignment via proton chemical shifts, of course, is an assigned proton spectrum. Although the 400 MHz proton spectrum is largely first-order, its complexity would require extensive homonuclear double resonance for a complete assignment. We have alternatively chosen homonuclear two-dimensional correlated spectroscopy,[26] a technique which establishes the J-coupling connectivities between the various protons in a molecule. We will ignore here the physics and technicalities of the experiment. For the present purpose it suffices to briefly point out the phenomenology of a two-dimensional correlation (COSY) plot, which consists of a quadratic data matrix with both axes carrying chemical shift information. Each proton appears at its appropriate position on the diagonal. In addition, however, there are off-diagonal (cross) peaks, caused by spin–spin coupling between two protons. To identify the chemical shifts of two coupling protons it suffices to draw a horizontal and a vertical line through a cross peak. The intersections with the diagonal locate the chemical shifts of the coupling protons. The COSY contour plot for formula **22** is reproduced in Fig. 3.42(a). The starting point for the analysis is a proton resonance whose identity follows from its unique chemical shift or coupling characteristics. Such a conspicuous proton is the olefinic H-21. In the aromatic part it is H-4, which exhibits this property, as this proton is deshielded by the amide carbonyl group. Heteroatoms and quaternary carbons interrupt the coupling chain (for example, N-18, O-24). The only link between H-21 and H-11, for example, is a weak cross peak, originating from the small allylic coupling between the two protons. The location of H-11 could be confirmed via H-12, which is coupled to H-13. The latter, however, follows from the $^{13}C/^1H$ correlated spectrum in Fig. 3.42(b), because C-13 has a unique chemical shift. It is by far the least shielded CH carbon in the aliphatic region. The three cross peaks on the vertical line passing through the diagonal peak pertaining to H-13 are caused by vicinal coupling of this proton to H-12 and H-14a and b. The ambiguities in the assignment of the latter three protons can be resolved with the aid of the $^{13}C/^1H$ correlated spectrum. Since H-14a and b are geminal they belong to a single carbon resonance. Moreover, there should be a strong cross peak due to geminal coupling in the COSY plot, which indeed exists.

We will not carry this any further, and leave it to the reader to verify the remaining assignments. It turns out that the assignments given in Reference 36 for brucine are correct, whereas those for C-14 and C-19 in Reference 37 have to be interchanged.

A final point to note in the correlation chart of Fig. 3.42(b) is the lack of parallelity in the shieldings of C/H pairs. This is particularly striking in the high-frequency region. Whereas H-4 is by far the least shielded proton, its appended carbon is the most shielded within this group. Similar deviations from linearity are also found in the aliphatic region. This is certainly a manifestation of the importance of susceptibility anisotropy contributions to proton shieldings, which play an insignificant role in carbon shieldings.

(a)

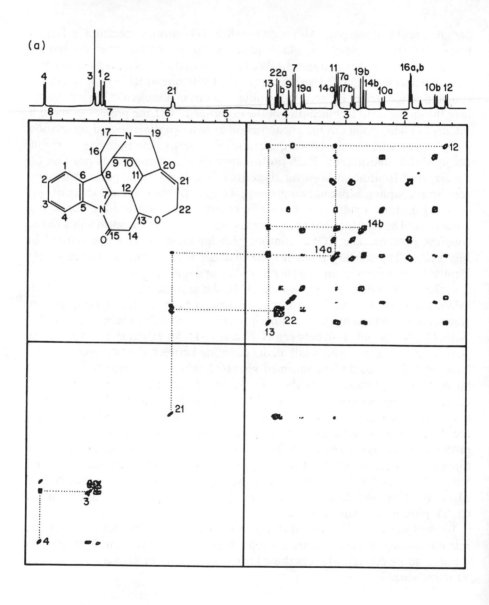

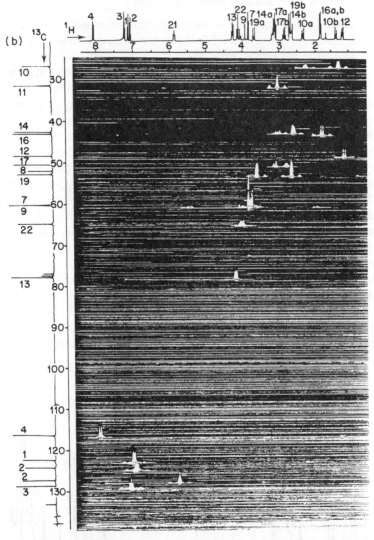

Fig. 3.42   (a) 400 MHz ^{1}H COSY contour plot of strychnine (**22**). Off-axis peaks are indicative of homonuclear spin–spin coupling. Coupling partners can be located by tracing horizontal and vertical straight lines through the cross peaks and by determining the intersections with the diagonal. (b)75.5 MHz (7.0T) ^{13}C/^{1}H two-dimensional correlated spectrum of **22**, obtained by the scheme of Fig. 3.41. Together with some unique chemical shifts as a starting point, the two experiments provided the couplet proton and ^{13}C spectral assignments.

3.6 CARBON–CARBON CONNECTIVITY EXPERIMENTS

^{13}C—^{13}C spin–spin coupling at natural abundance

Until recently, carbon–carbon coupling has not been used frequently as an assignment criterion, mainly because of the instrumental difficulties associated with obtaining the data in natural abundance. In effect, the vast majority of ^{13}C—^{13}C spin–spin coupling parameters appeared so far in the literature have resulted from measurements on singly and even doubly labelled species.[10,121-3] Observation of such couplings in natural abundance[122] requires the study of isotopomers containing two ^{13}C isotopes, i.e. molecules whose relative abundance is one in 10^4. Under proton noise-decoupling conditions, the signals resulting from such molecules appear in the form of weak satellite lines, being part of AB systems. Since the relative intensity of each of the satellites is of the order of 0.5% with respect to that of the centre peak, the experiment requires exceptionally high S/N ratios. Apart from high sensitivity, good line shape is a precondition for a successful experiment, as otherwise the weak satellite lines are obscured by the base of the intense parent signal. For this reason, it is difficult to detect other than on-bond couplings, as long-range ^{13}C—^{13}C couplings are substantially smaller than $^1J_{CC}$ (cf. Chapter 2).[121] A further difficulty may arise from impurities and interference of the satellite signals with spinning sidebands.

The analytical potential of one-bond carbon–carbon couplings is obvious. It allows identification of a carbon adjacent to one whose identity is already established. Such assignments are based on the dual appearance of the coupling constant. This is usually further aided by an analysis of the exact line frequencies and intensities when second-order coupling is involved.

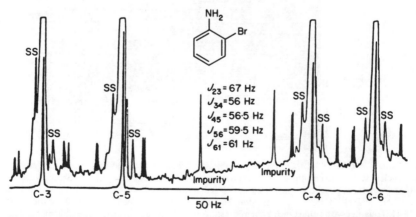

Fig. 3.43   25.2 MHz (2.3 T) proton noise-decoupled spectrum of 2-bromoaniline, showing region of proton-bearing carbons. In the upper trace the gain was increased, displaying ^{13}C satellites due to molecules containing two adjacent ^{13}C isotopes at natural abundance. Peaks denoted SS are spinning side bands.[125]

An early example demonstrating the usefulness of ^{13}C—^{13}C couplings deals with the unambiguous assignment of the proton-bearing carbons in 2-bromoaniline,[125] whose partial, proton noise-decoupled ^{13}C spectrum is displayed in Fig. 3.43. Although chemical shift criteria suggest that the two lowest-frequency resonances should be assigned to C-4 and C-6, and likewise the high-frequency lines to C-3 and C-5, an unequivocal assignment on an individual basis is not straightforward. The 25 MHz proton-coupled spectrum was not found suitable either because of its second-order character.

Inspection of the upper trace in Fig. 3.43, obtained by increasing the gain, reveals pairs of doublets flanking either side of the intense main peak. Two pairs of satellite lines appear, of course, because each carbon is bonded to two other carbons. For example, the four satellite lines associated with C-5 result from formulae **23a** and **23b**.

23a   **23b**

Since the satellite signals are clearly separated for the highest-frequency resonance, this lends itself best as a starting point. The doublets are found to be consistent with coupling constants of 67 and 56 Hz. As it can easily be verified that a coupling constant of 67 Hz does not match any of the remaining splittings, this carbon is identified as C-3 (C-6 can be ruled out for shielding reasons). The coupling constants are therefore assigned as follows: $J_{23} = 67$ Hz; $J_{34} = 56$ Hz. A coupling constant of 56 Hz is further found for the second most shielded carbon, which is now assigned to C-4. The couplings found for the remainder of signals corroborate these assignments.

Centre-band suppression techniques

It is clear that, without attenuation of the intense parent signal, the measurement of carbon–carbon couplings at natural abundance would remain limited to small molecules of little chemical interest. Recently, however, a new technique has been proposed, which effectively suppresses the signals from the major ^{13}C iso-topomers.[15] The method is based on the generation of double-quantum coherence.[126] There is, unfortunately, no simple pictorial description of the physics of this experiment, in terms of the macroscopic magnetization. For our purposes, it suffices to recall that normal n.m.r. absorption entails a transition between two energy levels differing in their magnetic quantum number by ± 1. In a double-quantum transition two quanta of rf energy are absorbed. It has been shown[127] that two successive 90° pulses, applied to a homonuclear scalar coupled spin system, generate multiple quantum coherence. Since in molecules containing two ^{13}C spins the spin systems are of type AB or AX, provided broadband proton decoupling is operative, double-quantum are the highest order transitions that can take place (transitions between the $\alpha\alpha$ and $\beta\beta$ levels).

Multiple-quantum transitions cannot directly be observed.[127] It is possible, however, to convert induced double or multiple quantum coherence back into measurable transverse magnetization by means of a third pulse.

The pulse sequence suggested[15] consists of two 90°_{x} ^{13}C pulses with a 180°_{y} refocusing pulse at the mid-point, all while protons are noise-decoupled. Two signal components are obtained in this way, one resulting from the (unwanted) major isotopomers, containing only one ^{13}C. Its magnetization at the end of the second 90° pulse is all longitudinal and directed along $-z$. The second component, which at this point is not a true magnetization, carries information about $[^{13}C_2]$ isotopomers. Maximum double quantum coherence is achieved by complying with the condition

$$\tau = (2n + 1)/(4J_{CC}) \qquad (3.18)$$

which is strictly valid only for weakly coupled systems. The period τ in Eqn. (3.18) represents the time interval between the initial 90° and the 180° refocusing pulses. A further 90° pulse generates transverse magnetization containing the signals from both isotopomers. The full pulse sequence is given below.

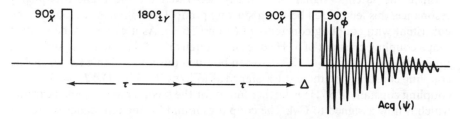

Δ is a short delay introduced for purely technical reasons, required for switching the rf phase. The crux of the experiment lies in the characteristic response of the two signal components to a change in rf phase. If the phase of the detection pulse is cycled in successive steps of 90° $(x, y, -x, -y)$ the transverse signal from the $[^{13}C_1]$ component follows in step $(-y, x, y, -x)$, whereas the $[^{13}C_2]$ signal generated from double-quantum coherence cycles in -90° steps for every $+90^{\circ}$ increment $(x, -y, -x, y)$ of ϕ. If the receiver reference phase follows this latter signal it is readily seen that the undesired centre-band signal is cancelled, since the two components are always 90° out of phase relative to one another.

The use of a four-phase cycle sequence, in principle, is not mandatory. Its sole purpose is to eliminate effects of pulse imperfections such as resonance offset effects (deviations from the condition $\gamma H_1/2\pi \gg \Delta F$; cf. Chapter 1). However, an analysis shows that even minute pulse imperfections can lead to sizable spurious

TABLE 3.4   Phase of the read pulse (ϕ) and receiver reference phase (ψ) compared with the three principal signal components $S_0([^{13}C_1]$ isotopomers), S_1(spurious response), and $S_2([^{13}C_2]$ isotopomers)[15]

ϕ	S_0	S_1	S_2	ψ
$+x$	$-y$	$+x$	$+x$	$+x$
$+y$	$+x$	$+y$	$-y$	$-y$
$-x$	$+y$	$+x$	$-x$	$-x$
$-y$	$-x$	$+y$	$+y$	$+y$

signals which were found to alter their phase as x, y, x, y. It is readily seen that the four-step cycling of the read pulse (phase ϕ) and receiver (phase ψ) cancel out these unwanted signals (Table 3.4).

A better degree of suppression is further reached by alternating the phase of the refocusing pulse every four steps. If, in addition, one increments the phase of every pulse by 90° at the beginning of every eight-step cycle, imperfections resulting from inaccurate phase settings are corrected. It should be noted, however, that such a 32-pulse sequence only corrects for coherent pulse deficiencies and that random effects, such as phase jitter, are incorrigible.

Provided that the basic instrumental requirements are met, excellent centre-band suppression can be achieved (up to a factor of 1000). From Eqn. (3.18) it becomes clear that the condition for the creation of double quantum coherence cannot be met for all coupling constants at once, as these cover a fairly large range (0–100 Hz), although one can, in practice, observe all one-bond couplings in a single experiment. It is further possible that a single value of τ satisfies Eqn. (3.18) for two different values of J. Let us assume that we arbitrarily choose $n = 0$ for the smaller (J') of two coupling constants J' and J''. We then can write

$$1/(4J') = \tau = (2n + 1)/(4J'') \tag{3.19a}$$

or

$$J''/J' = 2n + 1 \tag{3.19b}$$

Hence if the ratio of the two types of coupling constants to be evaluated (for example, $^1J_{CC}$ and $^2J_{CC}$) is near 3, 5, 7, etc., the conditions for the generation of double-quantum coherence are met for both. This is shown, for example, in the spectra for piperidine[15] in Fig. 3.44, showing ^1J, ^2J, and ^3J, all obtained from a single experiment.

In this spectrum we note that the doublet components are in antiphase (up–down or down–up). Which of the two modes appears depends on whether n in Eqn. (3.18) is odd or even.

The coupling constant of most analytical interest undoubtedly is $^1J_{CC}$, because it bears the potential to map out the carbon bonding network of unknown molecules.[16–19] However, since a carbon can have as many as four

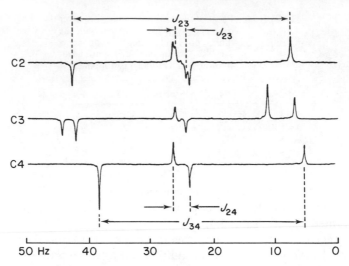

Fig. 3.44 Sections of the proton noise-decoupled ^{13}C spectrum of piperidine, obtained by double-quantum coherence transfer as described in the text. The spectrum exhibits signals due to the various naturally abundant $[^{13}C_2]$ isotopomers, whereas those of the major $[^{13}C_1]$ isotopomers are almost completely suppressed.[15] Reproduced with permission from Bax, Freeman and Kemsell, *J. Am. Chem. Soc.* **102**, 4849. Copyright 1980 American Chemical Society.

neighbours, up to eight often closely spaced satellite lines are grouped around that particular resonance site. Sorting out the various matching pairs of satellites pertaining to a particular C—C bond therefore can become a formidable problem. Calculations show that for an 18-carbon molecule, for example, approximately 2500 satellite combinations have to be analysed.[19] One possible approach to the solution of this problem is recourse to a computer program, as recently suggested.[19] The 'COSMIC' program assembles a list of all possible satellite pairs and checks all possible combinations of carbons, using two major matching criteria; (1) minimal difference between two coupling constants and (2) minimal difference between chemical shifts calculated from AB analysis and those observed in the spectrum of the major isotopomers (conventional ^{13}C spectrum). The applicability of the technique was examined on molecules as complex as codeine (C_{18}) and, together with multiplicity arguments and the magnitudes of the one-bond coupling constants, it was shown that the correct structure could be reproduced.

An alternative approach to the evaluation of satellite spectra for establishing carbon connectivities is based on an extension of the simple double-quantum coherence experiment to a second dimension.[17,18] The only difference to the sequence previously discussed is the delay Δ, which now becomes the variable evolution period t_1. Double Fourier transformation of the data array thus obtained then results in a two-dimensional spectrum in which the satellites appear in the v_2 dimension together with the chemical shift, whereas the v_1 dimension provides the double-quantum frequencies. These are given as the sum of the single-quantum frequencies relative to the rf carrier. Directly bonded carbons can be identified by the fact that they belong to the same double-quantum frequency as schematically outlined in Fig. 3.45(a) for a hypothetical three-carbon chain.[128] Figure 3.45(b) shows a two-dimensional carbon–carbon connectivity plot in contour representation for linalool (24).[128] Spin–spin

24

coupling doublets corresponding to bonded carbons have the same double-quantum frequency (ordinate) and have been connected by horizontal lines. It is to be noted that there are as many double quantum frequencies as there are carbon–carbon bonds in the molecule; in the present case, nine. The satellite doublets appear at the co-ordinates determined by the chemical shift and the double-quantum frequency. Three pairs of doublets are therefore predicted for a triply carbon-substituted site such as C-2, which is easily verified.

At present the time requirements for such experiments are still formidable, even on state-of-the-art instrumentation. The connectivity plot in Fig. 3.45(b), for example, which was recorded from a nearly neat liquid, required 22 h of accumulation. However, the potential of the method is enormous, considering the fact that, in principle, it allows one to deduce a structure (carbon skeleton) without resorting to any prior hypotheses.

We will now proceed to the determination of long-range coupling constants as they occur in isotopomers in which the two carbon sites are separated by more than one bond. Excitation of double-quantum coherence depends on how well the condition of Eqn. (3.18) is fulfilled. Since the body of reference data for long-range coupling constants is still relatively small, an arbitrary choice for the pulse interval τ has to be made. It has therefore been proposed[16] to make τ a variable in the experiment and to use $t_1 = 2\tau$ as the evolution period in a two-dimensional experiment. Following double Fourier transformation one then obtains a two-dimensional data matrix in which the v_2 dimension represents the conventional satellite spectrum, whereas the v_1 dimension carries the satellites centred around zero frequency. Sorting of the various doublets is facilitated by cutting cross sections perpendicular to the v_2 frequency axis. Figure 3.46(a) shows a small section of such a two-dimensional spectrum, centred on the chemical shift of the methyl carbon in tetramethyladamantane (25).[16] The spectral traces in

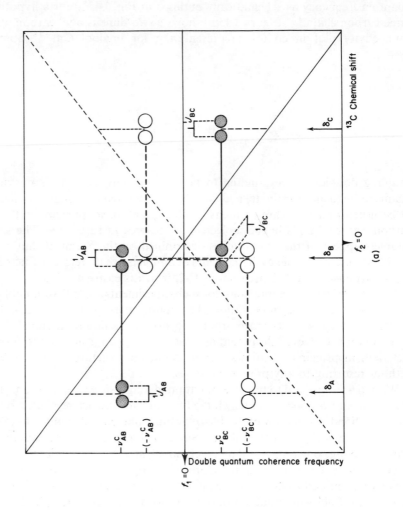

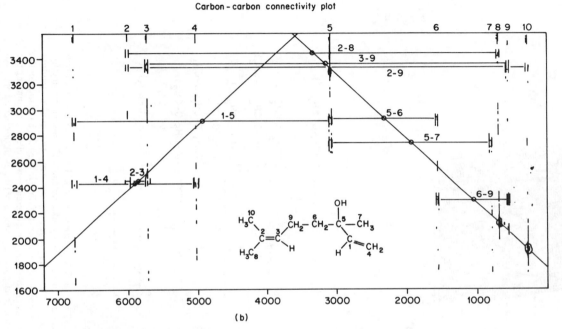

Carbon - carbon connectivity plot

(b)

Fig. 3.45 (a) Schematic $^{13}C—^{13}C$ two-dimensional connectivity plot for a three-carbon chain, showing carbon–carbon spin–spin coupling constants on the v_2 axis and double-quantum frequencies on the v_1 axis. Bonded carbons are identified by their common double quantum frequency. (b) 4.7 T two-dimensional connectivity plot for linalool (**24**), obtained in 21 h from a neat sample. The spectrum permits the delineation of the complete carbon skeleton without further assumptions to be made.[128] Reproduced by permission of Varian AG.

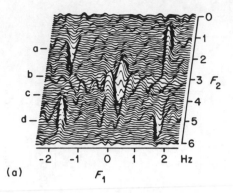

(a)

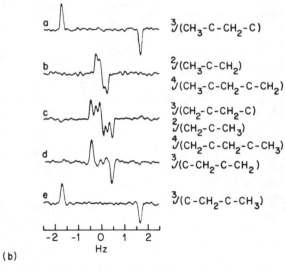

(b)

Fig. 3.46   (a) Small section of the 4.7T two-dimensional carbon spectrum of tetramethyladamantane (**25**), displaying methyl region. The data were obtained by the double-quantum coherence pulse sequence described in the text. Excitation of the various double-quantum coherences was effected by making the period 2τ (cf. pulse sequence, p. 66) the evolution period of a two-dimensional experiment. (b) Cross sections from the same data set at each of the three carbon sites, showing long-range couplings.[16] Reproduced with permission.

Fig. 3.46(b) were obtained by plotting cross sections in the way described.

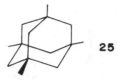

25

Biosynthetic experiments

[13]C labelling of putative precursors and study of the biosynthetic product by [13]C n.m.r. undoubtedly added a new dimension to the methodology of biogenetic research.[129-32] Contrary to the utilization of radioactive isotopes such as ^{14}C or tritium, label location can be effected on the intact metabolites without the need of degradation.

We will briefly elaborate on this subject, because the analysis of spectra resulting from singly or multiply labelled precursors poses a particular challenge and requires new experimental techniques and concepts. While early work was mainly concerned with singly labelled precursors, such as [^{13}C] acetate, this method was soon recognized to suffer from a deficiency; it requires relatively high specific incorporation, affording n.m.r. signal enhancements of at least 30–40% to allow unambiguous conclusions. Whereas such levels of incorporation are not unusual in natural products of microbial origin,[126-8] they have been found in exceptional cases only in the biosynthesis of higher plants,[133] where generally much higher label dilution takes place. These limitations led to the idea of using doubly labelled precursors.[134] Since the natural abundance of isotopomers with two contiguous ^{13}C sites is only 0.0123%($(1.11)^2$), a much greater level of dilution can be tolerated.

We have previously seen that, on condition that the two adjacent sites are magnetically non-equivalent, homonuclear spin–spin coupling is observed. One readily realizes that the spectra of molecules, derived from multiply labelled substrates, can become quite complex. What we observe are the superpositions of several different multiply labelled isotopomers. In order to lower spectral complexity it was therefore suggested that the labelled precursor be diluted with natural-abundance material for the case of high specific incorporation.[134] In this way the probability for two adjacent subunits (for example acetate) to be labelled in the same molecule is lowered.

In those situations where the analysis is encumbered by overlap of signals due to different isotopomers, homonuclear decoupling[135] has been shown to facilitate assignment or, in the case of strong coupling, selective population inversion, as previously shown for biosynthetic diplosporin (**20**).[105]

It has recently been shown that there are certain advantages to the

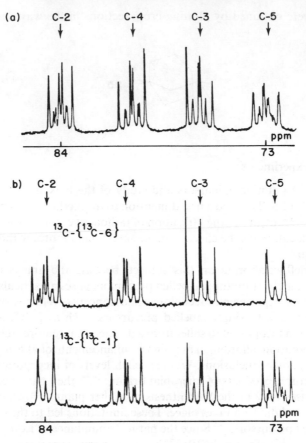

Fig. 3.47 (a) 62.9 MHz (5.9 T) proton noise-decoupled ^{13}C spectra displaying section between 70 and 85 ppm in 3-O-acetyl-1, 2:5, 6-di-O-isopropylidene-α-D-[U-^{13}C] gluco-furanose (**25**). The spectra were found to be analysable as superpositions of the various mono and multiply labelled isotopomers occurring as a result of random labelling. (b) Same region with homonuclear decoupling (triple resonance) from C-6 and C-1, respectively.[136] Reproduced by permission of John Wiley & Sons.

administration of statistically multiply labelled precursors.[136,137] We will not address the biosynthetic ramifications of this experiment but rather point to the phenomenology of the resulting spectra, be it that of the precursor or of the biosynthetic product. The complexity of such spectra is exemplified by Fig. 3.47, displaying the high-field portion of the 5.9 T proton noise-decoupled spectrum of 3-O-acetyl-1, 2:5, 6-diO-isopropylidene-α-D[U-^{13}C] glucofuranose (**25**).[136] The designation '[U-^{13}C]' indicates uniform labelling.

25

The eight-line pattern for C-2 in trace (a), for example, arises from the superposition of the spectra of four isotopomeric species:

(1) A large double doublet ($^1J(1,2) = 33.6$ Hz and $^1J(2,3) = 44.1$ Hz) from $^{13}C(1)$—$^{13}C(2)$—$^{13}C(3)$;
(2) A single doublet from $^{13}C(1)$—$^{13}C(2)$—$^{12}C(3)$;
(3) A single doublet from $^{12}C(1)$—$^{13}C(2)$—$^{13}C(3)$;
(4) A singlet from $^{12}C(1)$—$^{13}C(2)$—$^{12}C(3)$.

Traces (b) and (c) show the effect of homonuclear decoupling, which, considering the fact that protons are broadband decoupled, is a triple-resonance experiment $^{13}C\{^{13}C, ^{1}H\}$. We notice, for example, that irradiation of C-1 collapses two of the C-2 doublets.

3.7 LANTHANIDE SHIFT REAGENTS

Dipolar (pseudocontact) shifts

Paramagnetic rare earth complexes have been widely used in proton n.m.r. as chemical shift dispersing agents.[24] The induced shifts, caused by secondary internal fields originating from unpaired electron spins, usually result in greatly simplified spectra. Requisite for the applicability of the technique is the presence of polar groups in the substrate molecules, allowing binding to the lanthanide ion (for example, OH, COOH, NR_2, $= O$, etc.).

The chemical shift spreading effect, often the prime motivation for the routine use of these auxiliary reagents in proton n.m.r., is undoubtedly of less importance in ^{13}C n.m.r., and most applications exploit the differential chemical shifts as an assignment aid or for the elucidation of stereostructure.[138-40]

It has been pointed out[141] that the use of lanthanide shift reagents for the derivation of geometric parameters is subject to the fulfilment of the following conditions:

(1) The shifts are of purely dipolar origin;
(2) Only a single stoichiometric complex exists in solution;
(3) There is only a single geometric isomer of the complex;
(4) The complex has axial magnetic symmetry;

(5) The principal magnetic axis has known orientation with respect to the substrate ligand.

Whereas requirements (2) to (5) are usually met to a sufficient degree of approximation, condition (1), concerning the mechanism of the paramagnetic shift, is often not satisfied, as we will see later.

The origin of the dipolar shift lies in the anisotropy of the electronic **g**-tensor, a characteristic property of many of the $4f$-electron ions. The resulting anisotropy of the paramagnetic susceptibility produces an isotropic shift. Because the electron relaxation time is short compared with the molecular reorientational correlation time, i.e. $T_{1e} \ll \tau_R$ (cf. Chapter 4), electron relaxation does usually not significantly affect n.m.r. linewidths, in contrast to the contact interaction caused, for example, by transition metals. Assuming axial symmetry for the susceptibility tensor, the dipolar shift can be expressed by the McConnell–Robertson equation:[142]

$$(\Delta H_i/H) = K r_i^{-3}(3\cos^2\chi_i - 1) \qquad (3.20)$$

In Eqn. (3.20) K depends in a complex manner on the components of the electronic g-tensor, electron spin, etc., and can be regarded constant at a given temperature and for a given lanthanide ion; r_i represents the distance of the ith nucleus relative to the site of the lanthanide ion; and χ_i is the angle between the principal magnetic axis of the complex and the distance vector r_i. The incremental shifts are thus predicted to be proportional to the inverse cube of the internuclear distance. The relative sign, on the other hand, is given by χ_i and the sign of K. We see from Eqn. (3.20) that a critical angle is defined ($\chi_i = 54°\,44'$), at which sign reversal occurs. This has been confirmed experimentally.[143] Quantitative information is difficult to obtain from Eqn. (3.20), mainly because of uncertainties regarding the geometry of the complex. Moreover, the magnetic axis does not necessarily coincide with the metal–ligand bond axis.[144]

Several attempts have been made to computer-fit experimental shifts to geometrical parameters.[144-6] The starting point of such an evaluation of lanthanide-induced shift (LIS) data is the determination of molecular co-ordinates, which are derived either from molecular models or X-ray data. The location of the lanthanide ion is then taken as an adjustable parameter, which is varied until agreement with the experimental shifts is obtained. For this purpose, the direction of the magnetic axis is chosen coincident with the lanthanide–ligand bond. An optimum is sought by moving the Ln atom over the surface of a sphere, whose radius is given by the lanthanide–substrate bond. This procedure is repeated by incrementally varying the bond length until optimum agreement is obtained.

When applied to conformational analysis a critical question is whether or not lanthanide complexation perturbs the conformational equilibrium. One also realizes that only if the lanthanide ion complexes equally with all possible interconverting conformers does the conformation derived represent that in the uncomplexed molecule

Another problem in evaluating LIS data relates to the labile nature of the complexes formed, i.e. exchange is usually rapid on the n.m.r. time scale, resulting in averaged shifts, involving those of the free and bound substrate. Assuming a single equilibrium

$$L + S \rightleftharpoons LS$$

the observed LIS, $\Delta\delta$, can be written as

$$\Delta\delta = \frac{[LS]}{[S_0]}.\Delta \tag{3.21}$$

In Eqn. (3.21) $[LS]$ and $[S_0]$ denote the complex and total substrate concentrations, respectively, and Δ is the induced shift in the bound state.[147] For a sufficiently large equilibrium constant K, $\Delta\delta = f([L_0])$ ($[L_0]$ = total lanthanide concentration) is linear over a wide range of $[L_0]$ and the bound shifts $\Delta\delta$ are obtained by extrapolation of $\Delta\delta$ to $[L_0]/[S_0] = 1$.

The most commonly applied shift reagents involve chelate complexes of Eu[III], Pr[III], and Yb[III]. Among these three elements, Eu and Yb deshield ($\chi < 54°44'$) while Pr shields. To give an idea of the magnitudes of the induced paramagnetic shifts, LIS are listed in Table 3.5[148] for some simple organic molecules with oxygen and nitrogen functions. The shifts were corrected for complex formation (diamagnetic effect) and extrapolated for equimolar [Ln]/[substrate] ratio. The shift reagents used were Eu(DPM)$_3$, Pr(DPM)$_3$, and Yb(DPM)$_3$, with 'DPM' standing for 'dipivalomethanato', a bidentate organic ligand, most commonly employed. The data indicate that, among the three lanthanides, ytterbium produces the largest shifts, followed by prasaeodynium and europium. Further, the shifts are dependent upon substrate functionality and steric effects.

26   27   28   29

30   31   32 X=OH   33 X=NH₂

TABLE 3.5   Lanthanide-induced (bound) shifts for compounds **26–37** determined as described in the text[148]

Compound	Shift reagent	Shifts (ppm) induced at carbon atoms					
		1	2	3	4	5	6
26	Eu(DPM)$_3$	48.9	17.2	11.5	7.5		
	Pr(DPM)$_3$	− 65.9	− 27.9	− 14.9	− 10.5		
	Yb(DPM)$_3$	146.2	61.6	28.6	21.2		
27	Eu(DPM)$_3$	48.4	14.9	17.7	8.5	10.7	18.6
	Pr(DPM)$_3$	− 66.1	− 26.2	− 14.1	− 8.5	− 13.5	− 28.9
28	Eu(DPM)$_3$	47.9	16.0	9.5	5.3	2.5	0.9
	Pr(DPM)$_3$	− 60.3	− 23.9	− 13.3	− 8.7	− 2.9	− 1.8
	Yb(DPM)$_3$	149.7	61.7	28.4	20.7	9.0	5.5
29	Eu(DPM)$_3$	25.9	42.4	11.4	16.0		
	Pr(DPM)$_3$	− 28.8	− 58.3	− 21.6	− 18.8		
30	Eu(DPM)$_3$	15.7	20.1	29.4	21.9	17.8	
	Pr(DPM)$_3$	− 16.0	− 22.1	− 40.1	− 22.1	− 20.1	
31	Eu(DPM)$_3$	23.5	60.1	15.3	14.1		
	Pr(DPM)$_3$	− 42.3	− 92.1	− 36.0	− 23.6		
32	Eu(DPM)$_3$	13.7	19.2	51.5	16.6	14.5	12.5
	Pr(DPM)$_3$	− 22.4	− 35.1	− 79.3	− 34.6	− 22.9	− 21.4
33	Eu(DPM)$_3$	24.0	33.6	82.3	− 11.8	17.6	18.3
	Pr(DPM)$_3$	− 31.1	− 53.8	− 114.1	− 31.7	− 28.5	− 31.3
34	Eu(DPM)$_3$	52.9	116.2	− 30.7	18.8	19.6	
	Pr(DPM)$_3$	− 72.1	− 143.3	− 31.5	− 31.9	− 28.4	
35	Eu(DPM)$_3$	53.7	22.6	9.6	8.4		
	Pr(DPM)$_3$	− 72.3	− 32.6	− 13.1	− 11.1		
	Yb(DPM)$_3$	127.7	63.7	27.5	21.7		
36	Eu(DPM)$_3$	44.3	12.7	7.4	7.3		
	Pr(DPM)$_3$	71.2	− 28.9	− 12.8	− 10.3		
	Yb(DPM)$_3$	110.0	57.0	24.4	19.1		
37	Eu(DPM)$_3$	142.8	− 8.2				
	Pr(DPM)$_3$	− 149.5	− 40.0				

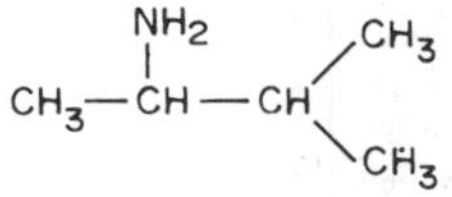

34

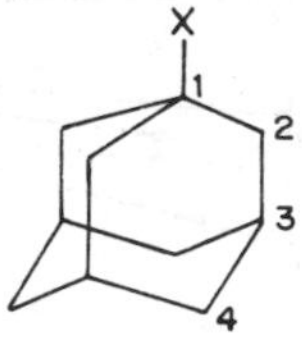

35 X=OH
36 X=NH$_2$

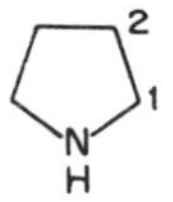

37

Spectral assignment

Although in principle the angular term in Eqn. (3.20) cannot be neglected, in conformationally rigid systems a near-proportionality between the observed shifts and $1/r^3$ is usually found,[149] since for most sites in the molecule $\chi < 54°$ holds. Even if used in this purely qualitative way, the method turns out to be an extremely potent assignment tool, because it enables one to differentiate between carbons located in different spatial positions of the molecule. This may be illustrated with an early assignment study, concerned with isoborneol (**38**), whose

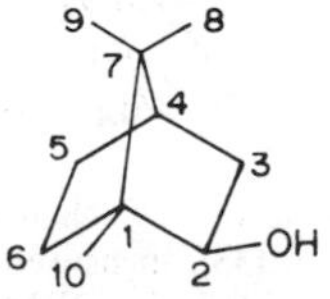

38

^{13}C chemical shifts had been reported earlier,[150] except for an assignment of methyl carbons 8 and 9, which have nearly the same chemical shifts. Figure 3.48(a)[149] shows a plot of chemical shifts as a function of the concentration of Eu(DPM)$_3$. Figure 3.48(b) gives the bound shifts, obtained by extrapolation to 1:1 Eu/substrate mole ratio. LIS found for the two methyl resonances were 10 and 4.2 ppm, thus assigning the former to C-8, which is closer to the binding site.

Since shift reagents affect both proton and ^{13}C spectra, use can be made of the enhanced chemical shift dispersion in a different way. We have seen earlier that selective irradiation is often hampered by insufficient separation of proton resonances. Thus if one succeeds in adequately spreading the proton spectrum, correlation between proton and ^{13}C shieldings can be achieved by selective decoupling, as discussed earlier. An example is provided in Fig. 3.49, showing the europium chemical shift-enhanced proton spectrum of geraniol (**39**) (Fig. 3.49(a)), along with the partial ^{13}C spectra, obtained by selective low-power irradiation ($\gamma H_2/2\pi \simeq 1.5\,\text{kHz}$) of protons C(7)-H and C(3)-H (Figs 3.49(b) and (c), respectively). The data prove the more highly shielded of the two olefinic carbon resonances to belong to C-3. It should be noted that this assignment refers to the

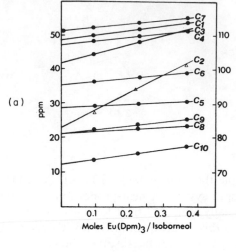

Fig. 3.48   (a) Plot of ^{13}C chemical shifts of 1-molar isoborneol (**38**), measured at several concentrations of added Eu(DPM)$_3$; ● refer to 0–60 ppm scale; Δ to 60–120 ppm scale. (b) Europium-induced chemical shifts in **38**, observed at 1:1 europium: substrate ratio.[149]

europium-shifted spectrum only. In this case it could, in fact, be shown that addition of shift reagent changes the relative positions of the two carbons. In the spectrum of the diamagnetic sample these assignments have therefore to be reversed, which is in accordance with literature data.[151]

Fermi contact shifts

The ubiquitous question confronting the user of lanthanide shift reagents is whether or not the observed shifts are dipolar. While it has been widely

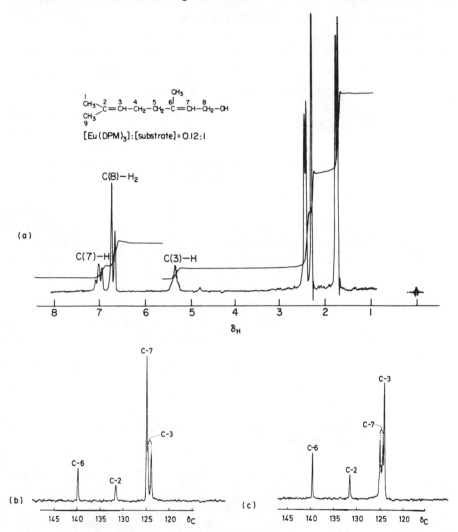

Fig. 3.49 Europium chemical shift-enhanced 90 MHz proton spectrum of geraniol (**39**). (b) Olefinic region of the corresponding 25.2 MHz (2.3 T) ^{13}C spectrum, obtained by selective irradiation of proton C(7)-H. (c) Same as (b) but with irradiation of C(3)-H. In both decoupling experiments $\gamma H_2/2\pi$ = 1.5 kHz was used.

recognized that proton LIS are essentially free from contact contributions,[23,24] there is as much evidence that this is not the case for carbon-13.[52]

The contact shift arises from a scalar interaction between nuclear and electronic spins via hyperfine coupling. Because of very fast electron spin relaxation, however, this does not produce observable splittings as in the case of scalar coupling between nuclear spins. A displacement of the nuclear resonance occurs because the two electron spin states are unequally populated.

Although not normally fulfilled for lanthanide ions, we shall, for the sake of simplicity, assume an isotropic g-tensor. Under this condition the contact contribution to the chemical shift can be formulated as follows:[25]

$$(\Delta H_i/H)^c = a_i g_e^2 \beta_e^2 S(S + 1)/(\gamma_N 3kT)$$ (3.22)

In Eqn. (3.22) a_i is the hyperfine coupling constant for nucleus i, g_e and β_e are the electronic g-factor and Bohr magneton, respectively, and S is the electron spin. The other symbols have their usual significance. Since the hyperfine interaction is proportional to the unpaired electron spin density at the nucleus, $|\psi(0)|^2$, it is clear that this interaction must be larger for heavier atoms such as carbon, nitrogen, and phosphorus.

Whereas excellent agreement is usually obtained when proton LIS are matched to the McConnell–Robertson equation, the ^{13}C data could not always be reproduced satisfactorily by this dipolar model.[153,154] Experimental evidence for other mechanisms contributing to the observed effects had first been established for amines.[154,155] In pyridine bases,[154] for example, a sign alternation was found for the observed shifts, which could not be rationalized in terms of geometrical factors. A separation of the dipolar contribution from the experimental shifts was achieved in the following way. A least-squares regressional analysis of the proton data yielded the spatial co-ordinates of the lanthanide ion. These, in turn, were used to compute the dipolar shifts for ^{13}C. The difference between experimental and computed values could then be considered to arise from Fermi contact interaction. Contact shifts obtained in this way are listed in Table 3.6.

TABLE 3.6   Europium-induced ^{13}C chemical shifts for some pyridine bases, separated into contact and dipolar contributions[154]

	Distance (Eu–N)		$^1H_{obs}$	$^1H_{calc}$	$^{13}C_{obs}$	$^{13}C_{calc}$	$^{13}C_{diff}$
Pyridine	2.7	α	-31.02	-31.11	-90.00	-67.78	-22.22
		β	-10.68	-10.59	$+\ 0.88$	-24.33	$+25.21$
		γ	$-\ 9.71$	$-\ 9.33$	-30.22	-18.73	-11.49
3,5-Lutidine	2.8	α	26.7	26.66	67.80	51.66	16.14
		β			$-\ 3.10$	19.44	-22.54
		γ	7.50	7.55	22.70	14.95	7.75
		β-CH$_3$	4.91	5.08	6.00	6.24	0.24
Isoquinoline	3.2	1	23.7	23.6	70.2	34.2	36.0
		3	25.7	25.7	64.0	37.5	26.5
		4	9.11	9.3	0.17	16.3	-16.0
		5	6.06	5.3	8.9	6.9	2.0
		6	3.44	3.4	2.8	5.0	$-\ 2.2$
		7	3.44	3.5	2.1	5.2	$-\ 3.2$
		8	5.77	6.6	10.1	8.4	1.8
		9			3.64	16.1	-12.5
		10			21.6	13.0	8.6

The data show that experimental LIS can be misleading if solely interpreted in terms of a dipolar interaction. In contrast to pseudocontact shifts, contact shifts are not geometry-related. Their magnitude rather depends on the ability of the system to delocalize electrons. Hence the effect is expected to become more severe in conjugated π-electron systems. In saturated molecules contact shifts are usually confined to α and β, and possibly γ, carbons.

A second approach toward a mechanistic separation of experimental LIS makes use of the unique temperature dependencies of the two contributions. Such data have been reported for the lanthanide complexes of 2, 6-pyridinedicarboxylic acid (DPA) and 4-methyl-2, 6-pyridinedicarboxylic acid (MDPA).[156] These complexes have a well-defined axial ligand symmetry and are kinetically stable, therefore affording separate resonances for complexed and free substrate. The theoretical temperature dependence for contact and dipolar shifts are:[157]

$$(\Delta H/H)^c \propto T^{-1} \tag{3.23}$$

$$(\Delta H/H)^d \propto T^{-2} \tag{3.24}$$

$(\Delta H/H)^{exp}$ can therefore be expressed as

$$(\Delta H/H)^{exp} = AT^{-1} + BT^{-2} \tag{3.25}$$

By plotting $T(\Delta H/H)^{exp}$ against T^{-1} a straight line of intercept A and slope B should therefore be obtained. In this particular example proton LIS were found to be linear with T^{-2}, thus confirming their dipolar nature, whereas the corresponding carbon shifts did not exhibit this behaviour. On the other hand, a plot according to Eqn (3.25) showed high linearity and permitted a separation of the two quantities.

A third approach uses the fact that the total paramagnetic shift can be expressed as the sum of two products,[158] representing the two terms:

$$(\Delta H/H)^t_{ij} = F_i \langle S_z \rangle_j + G_i C^D_j \tag{3.26}$$

$\langle S_z \rangle_j$, the expectation value of the z component of the electron spin and the quantity C^D_j, relating to the paramagnetic susceptibility, are both specific to a particular lanthanide ion j and have been tabulated.[159] F_i and G_i, on the other hand, are related to physical properties of nucleus i, or to its spatial co-ordinates, respectively. From shift measurements on a particular nucleus i in a series of isostructural lanthanide complexes a number of simultaneous equations are obtained, from which terms F_i and G_i can be computed. The method, which is strictly valid only for effective axial symmetry, has been tested for a variety of Ln (DPA)$_3$ complexes[158] and some representative data are reproduced in Table 3.7. It is interesting to note that, except for ytterbium, which is least prone to contact interactions, the magnitude of the contact shifts dominate the relative shielding of the β carbon, although it is four bonds removed from the paramagnetic centre.

TABLE 3.7   Experimental LIS[a] for Ln(DPA)$_3$ complexes and computed values for contact and dipolar contributions[158]

Carbon		Pr	Tb	Ho	Yb	F_i	G_i
C_α	tot	−10.9	−133.0	−75.0	+39.0	0.09	−0.022
	cont	0.0	0.7	0.4	0.1		
	dip	−10.9	−133.7	−74.4	+38.9		
C_β	tot	−19.9	+ 41.3	+27.4	+17.5	−2.623	0.554
	cont	−11.2	+ 86.9	+51.8	+ 6.3		
	dip	− 8.7	− 45.8	−24.4	+11.2		

[a]ppm rel. to diamagnetic La(DPA)$_3$.

3.8   CHEMICAL SHIFT COMPARISON

A total assignment of complex natural products such as terpenes, alkaloids, and steroids may still require methods beyond those already discussed. In almost all of the examples we have dealt with, the chemical shift has been a key element. The purpose of this section is to outline a more systematic approach to the exploitation of chemical shift criteria for spectral assignment. Chemical shifts have been discussed extensively in Chapter 2. We have seen that this quantity, unfortunately, cannot be predicted with sufficient accuracy to allow distinction of two carbons, situated in very similar chemical environments. Each functionality is characterized by a chemical shift range and sometimes the bandwidth is quite large (cf. chemical shift table inside the back cover of this book). Depending on substitution of nearest and next-nearest neighbouring carbons, a methylene carbon in an acyclic aliphatic hydrocarbon, for example, may resonate between 10 and 55 ppm. A detailed analysis of the chemical shift in linear and branched alkanes revealed a correlation with the number of α, β, and γ substituents.[162]

In this way, chemical shifts could be systematized and empirical rules established (for example, Lindeman–Adams rules[161]). Similar rules have been devised for polycyclic condensed aliphatics and a host of other systems (cf. Chapter 2). These rules typically permit prediction of the chemical shift with a precision of ± 2 ppm or better. It is clear that this approach breaks down in larger molecules. This may be exemplified with the spectrum of the hydrocarbon steroid cholestane (**40**) in Figure 3.50. None of the methods previously discussed probably allow us to distinguish between the resonances of the angular methyl C-18 and C-19, which differ in their chemical shifts by less than 0.5 ppm.

In their early work the research groups of J. D. Roberts and D. M. Grant pointed the way to tackle the problem.[162] Their strategy consisted of (1) approximating the shieldings in a substructure of the molecule, (2) analysing the effect of specific substitution, and (3) comparing the chemical shifts in a closely related series of very similar compounds (for example, structural and stereoiso-

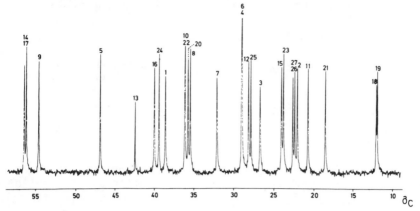

Fig. 3.50 25.2 MHz (2.3 T) proton noise-decoupled spectrum of cholestane (**40**), 1-molar in chloroform-*d*. Assignments shown are those of Ref. 20.

mers). Although this concept is now widely used by natural products chemists,[163] we wish to review part of the early fundamental work because of its particular didactic value.

Returning to the spectrum in Fig. 3.50 we find that, at 2.3 T field strength, twenty-five out of the twenty-seven carbons of the molecule give rise to separate resonances, but the chemical shift differences for a number of lines are as little as 0.2 ppm. Sorting out the signals pertaining to carbons of the C-17 sidechain is comparatively straightforward. In a series of cholestane-type derivatives (cholestan-3-one, cholestan-3-β-ol and its acetate, cholestan-3-α-ol and its acetate) seven resonances were found to exhibit very small variations within the series.[20] The argument in this case is that the sidechain carbons least sense a perturbation at the other end of the molecule. Moreover, these lines are absent in androstane (**41**), which lacks this sidechain. Individual assignment of sidechain

carbons was accomplished by chemical shift comparison with 2,6-dimethyl-octane (**42**) as a model compound in which chemical shifts can be predicted on the basis of empirical additivity rules[160] (cf. Chapter 2).

42

43

Three of the backbone signals were found to be significantly affected by introduction of a ketone function at C-3. These could thus be assigned to C-2, C-3, and C-4 (22.6, 27.3, and 29.6 ppm). Moreover, C-3 is anticipated to have a chemical shift deviating little from that found for cyclohexane (27.3 ppm) and for C-4 in trans-1, 2-dimethylcyclohexane (27.0 ppm). The ambiguity in the assignment of C-2 and C-4 could be eliminated by considering that C-4 should be less shielded because of the larger degree of α-substitution. Hence the line at 29.6 ppm was assigned to C-4. Introduction of the axial acetoxy group in cholestan-3α-yl acetate induces steric compression shifts at C-1 and C-5 through interaction with the syndiaxial protons at these carbons (cf. formula **44**). Together with different multiplicities for C-1 and C-5, the lines at 39.2 and 47.6 ppm were immediately

44

assigned. The quaternary carbon resonances belonging to C-10 and C-13 were assigned to the lines at 36.8 and 43.1 ppm, respectively. Distinction was made based on the larger variation of the former value within the series of 3-substituted cholestanes. This is predicted for C-10, which is closer to the site of perturbation. The same reasoning applies to the assignment of the high-frequency methine resonances, where the signal at 55.5 ppm was found to undergo larger changes as a function of the substituent at position 3, which favours identification of this line with C-9. In a similar way, distinction could be made between the two most shielded methyl resonances, originating from the angular methyl carbons C-18 and C-19. The most shielded methylene signals, among those not yet assigned, were expected to be those belonging to C-11 and C-15, because both these carbons experience considerable shielding due to steric perturbation[163] of their attached axial hydrogens by axial methyls C-18 and C-19, and C-18 and C-21, respectively, as shown in formula **45**. However, whereas the resonance at

24.6 ppm remains almost constant throughout the series, that at 21.4 ppm shows a larger standard deviation and is therefore assigned to C-11.

It is evident from the foregoing that chemical shift correlation can afford internal consistency even though some of the assignments may be incorrect and errors are therefore easily propagated. A later, thorough examination of the ^{13}C spectrum of cholestane and related steroids,[164] making extensive use of deuteration (see above), showed, for example, that the original assignments for C-12 and C-16 had to be reversed. C-12, bound to a quaternary and a methylene carbon, has more deshielding β interactions than C-16, which is adjacent to a methylene and a methine carbon. More importantly, however, C-16 experiences steric upfield shifts due to 1, 3-diaxial interactions of its axial hydrogens with C-18 and C-21.

The model study of the isomeric androstanone series[50] is a particularly vivid example of the application of chemical shift correlation techniques, as in this case all possible isomers with respect to the location of the keto function were available. It is interesting to note that those isomers bearing the keto group in either ring A or ring D are particularly suited for correlation. Since a shielding effect invoked by a substituent is attenuated with distance it is expected for these systems that a structural modification at one end of the molecule will only minimally affect the shieldings at the other extreme. By contrast, one predicts a perturbation closer to the centre of the molecule (i.e. in ring B or ring C) to significantly alter the shieldings of nearly all carbons. These predictions are verified by the experimental data. Whereas the spectra of ring D-substituted androstanes bear a close resemblance, there is very little correlation among, for example, 6- and 7-ketoandrostane.

A correlation diagram is obtained by vertically lining up the spectra (or stick plots) of the related compounds and by connecting signals of corresponding carbons. This is illustrated in Fig. 3.51 for the three androstanones bearing the keto function in ring D. The multiplicities and, where relevant, residual splittings were previously determined from SFORD experiments.

Inspection of the low-frequency quartet resonances shows that the signal of the most highly shielded carbon remains unaltered within the series. Hence this line is assigned to C-19, whose chemical shift should not vary appreciably upon altering the substitution site in ring D. On the other hand, the second quartet (C-18) resonance experiences shielding in 17-one as a consequence of steric proximity of this methyl carbon to the carbonyl group. The quaternary carbons C-10 and C-13

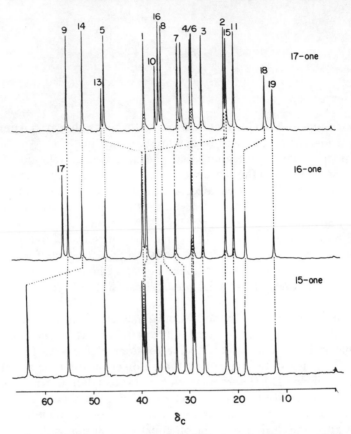

Fig. 3.51 Chemical shift correlation within the series 15-, 16-, and 17-
ketoandrostane. Assignments were made with the aid of residual splittings
and compared chemical shifts within the series and with cholestane.[2,50]

can be assigned on the same arguments. In 17-one C-13 is deshielded by 8 ppm
due to the α carbonyl. Among the high-frequency methine resonances, C-14 is
readily identified because of its 11 ppm deshielding in 15-one. C-8 can be assigned
on the basis of its characteristically high shielding, resulting from syn 1, 3
interaction of C(8)-H with methyls C-18 and C-19. Moreover, its resonance shifts
to lower frequency in 15-one, on account of steric interaction with the carbonyl
group. According to the findings in cholestane and its derivatives, C-2, C-11, and
C-15 exhibit unusually large shieldings, resonating in the band between 20 and
25 ppm. Whereas in 17-one three lines appear in this region, this reduces to two in
16- and 15-one. The remaining lines show little variance in the three isomers. The
displaced resonance is therefore identified as C-15. The very small standard

deviation found for the less shielded of the two remaining resonances suggests that it belongs to C-2.

We have seen in the previous example that the steric effect is a particularly valuable assignment criterion. In the following we would like to highlight its use as a configurational probe in connection with spectral correlation.

In the tricarbocyclic ring-C aromatic diterpenoids (46)–(48)[165] the chemical shifts of the aliphatic ring-A and ring-B carbons are all within *ca* 1 ppm from each other, with the exception of the carbinyl carbon C-18 (C-19 in 47 and 48), methyl

46   **47**   **48**

C-19 (C-18 in 47 and 48), and C-5. It is readily seen that with CH_2OH equatorially disposed (46a) C-5 experiences a γ-*gauche* effect, which is not

46 a   **47 a , 48 a**

operative when the configuration at C-4 is inverted. Hence one predicts C-5 to be more highly shielded in the former stereoisomer. Conversely, the axial carbinyl in 47 and 48 suffers a relative shielding, caused by 1, 3-diaxial interaction with $C(2)\text{-}H_{ax}$. Both effects are clearly recognizable from the chemical shifts listed in Table 3.8.[165]

A precondition for the applicability of chemical shift correlation, i.e. the availability of a sufficient number of structurally related molecules, is not always given. An alternative approach of spectral correlation relies on a comparison of

TABLE 3.8   Chemical shifts of aliphatic carbons in diterpenoids 46–8[165]

Compound	1	2	3	4	5	6	7	10	15	16	17	18	19	20
46	38.5	18.7	35.1	37.9	44.0	18.9	30.1	37.4	33.4	24.0	24.0	72.2	17.4	25.3
47	38.9	19.0	35.1	38.7	51.2	19.2	30.0	37.9				26.8	65.1	25.6
48	39.0	19.0	35.2	38.8	51.4	19.4	30.4	37.8	26.4	22.7	22.9	26.8	65.4	25.7

chemical shifts with those calculated for possible isomers. This, of course, presupposes that additivity rules exist for the compound class of interest. An example illustrating this deals with the determination of the stereochemistry of the hydrogenation products of synthetic fragrants belonging to the ambergis family.[167] The decalin derivative **49** has four diastereomers (**49a–d**):

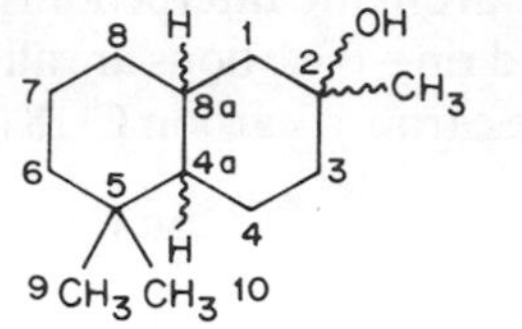

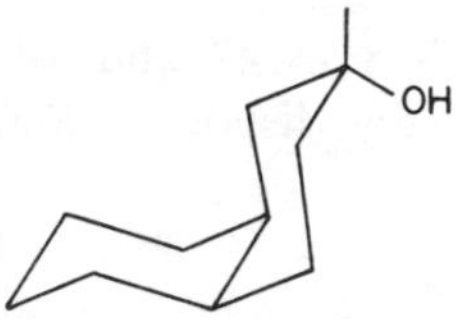

With the aid of the shielding increments derived for methyl decalins[166] the chemical shifts were calculated for stereoisomers **49a–d** and compared with the two synthetic products. It was further assumed that a hydroxyl group acts like a methyl substituent except, of course, that it invokes much larger deshielding of the substituted carbon than the former (cf. substituted methylcyclohexanes, Chapter 2). The striking feature of the data collated in Table 3.9 is the close resemblance of the experimental shieldings of corresponding carbons, which differ all by less than 2 ppm except C-8a and 2-Me ($\Delta\delta = 2.4$ ppm and 4.5 ppm, respectively). From a comparison of calculated values it becomes evident that this behaviour is only consistent with a *trans* junction geometry. Hence the two compounds are identical with **49a** and **49c**. Since the relative shielding of 2-Me is sensitive to the configuration of C-2 it can further be inferred that the compound exhibiting a larger shielding for 2-Me has to be assigned to **49a**. This results from the *γ-gauche* effect exerted by C(8a)-H.

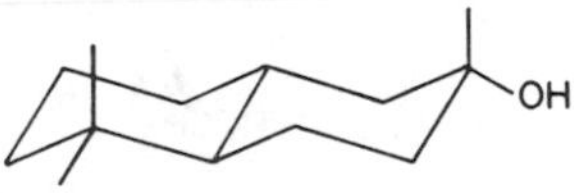

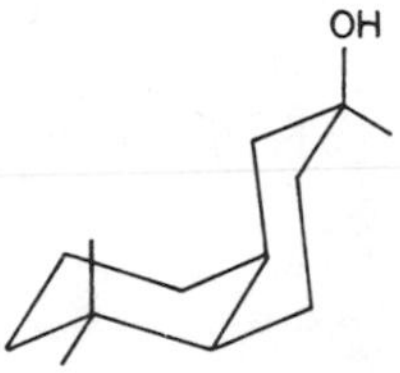

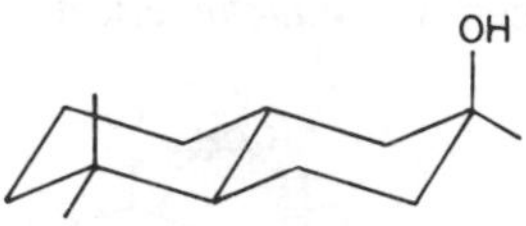

3.9 ISOTOPE EFFECTS

Isotopic labelling as a means to induce spectral effects has been one of the classic techniques of signal assigment in ^{13}C n.m.r. The most inexpensive, synthetically

TABLE 3.9   Experimental and calculated ^{13}C shieldings for the four stereoisomers of 2-hydroxy-2, 5, 5-trimethyldecalin[167]

	Compound					
	49a		49b	49c		49d
Carbon	Calc.	Exptl	Calc.	Calc.	Exptl	Calc.
1	48.8	49.3	45.3	48.8	47.8	45.3
2	a	71.1	a	a	70.0	a
3	40.3	41.3	40.3	40.3	39.8	40.3
4	26.1	24.4	21.5	26.1	22.4	21.5
4a	50.3	51.9	46.8	50.3	51.6	46.8
5	34.0	33.0	34.0	34.0	33.1	34.0
6	40.3	42.8	35.7	40.3	42.9	35.7
7	22.2	22.4	22.2	22.2	22.1	22.2
8	35.3	35.3	30.7	35.3	35.1	30.7
8a	34.6	35.3	34.1	34.6	32.9	34.1
2-Me	28.4	26.3	27.8	32.4	30.8	32.4
5-Me$_{ax}$	23.8	20.2	28.4	23.8	20.5	28.8
5-Me$_{eq}$	32.4	30.8	32.4	32.4	31.7	32.4

^{a}Not amenable to calculation.

least demanding, and probably spectroscopically most useful label is deuterium. Weakly acidic protons, such as those α to a carbonyl function, can easily be exchanged by deuterons. This procedure has been used extensively in early assignment work,[20] for example for the identification of the α carbon resonances in keto steroids. In 3, 17-diketoandrostane, base-catalysed exchange gives the 2, 2, 4, 4, 16, 16-hexadeutero derivative (50), whose spectrum has three fewer

visible resonances than the protio compound. These signals are usually greatly attenuated due to lengthened relaxation times (comparable to quaternary carbons), spin–spin coupling to deuterium, and, possibly, a reduction of the NOE. A further effect impairing detection is broadening of the multiplet components due to modulation of spin–spin coupling by deuterium relaxation (scalar transverse relaxation, cf. Chapter 4), an effect which becomes noticeable in larger molecules. In fact, if $(J_{CD})^{-1} \ll T_1(D)$, coalescene into a single line occurs, in which event detection becomes easier again. Whereas the one-bond coupling constant is usually observable if the S/N is sufficient, $^2J_{CCD}$ and $^3J_{CCCD}$ are often not resolved, for the reasons discussed previously. Moreover, it should be borne in mind that ^{13}C—D coupling constants are scaled by a factor 6.5 relative to their ^{13}C—^1H counterparts.

From an assignment point of view more useful than deuterium spin coupling is the isotope effect exerted on ^{13}C shieldings.[22] The physical origin of this secondary isotope effect is a perturbation of the intramolecular dynamics on the level of the vibrational and rotational states.[21,168] In order to distinguish it from indirect perturbations of magnetic shielding, caused by isotopic substitution, the effect has also been denoted the *intrinsic isotope effect*. It obeys the following qualitative rules:

(1) The nucleus in the heavier isotopomer is usually more shielded. This may be rationalized in terms of a shortening of the carbon–hydrogen bond (ca 0.003–0.005 Å) for replacement of C—H by C—D.[21]

(2) The magnitude of the isotope effect decreases with increasing distance from the site of isotopic substitution.

(3) The magnitude of the relative shielding is roughly proportional to $(m_1^{-1/2} - m_h^{-1/2})$,[169] where m_1 and m_h denote the masses of the light and heavy isotopes, respectively. It should therefore be largest for the isotopic pairs $^3H/^1H$, followed by $^2H/^1H$, etc.

(4) The isotope shift is usually additive and proportional to the number of isotopically substituted atoms. Experimentally, it was shown[22,68,164,170-74] that the intrinsic deuterium-induced isotope effect on carbon shieldings is of the order of 0.25–0.40 ppm/deuterium for a directly bonded carbon and ca 0.1 ppm for a geminal carbon.

An example of the deuterium-induced isotope effect as an assignment aid and, conversely, for the determination of the deuterium labelling site, is shown by the spectra of cyclooctanone in Fig. 3.52.[174] Although mass spectrometry provided proof for the presence of a monodeuterated species, it failed to unequivocally establish the deuteration site. The spectrum of Fig. 3.52(b), which was obtained from a mixture of protio and deutero isotopomers, indicates the label to be at C-5. This assignment follows from the fact that C-5 has a unique chemical shift among the two non-degenerate carbons. The spectrum afforded, besides the single line at 24.797 ppm, pertaining to the protio isotopomer, a triplet centred at 24.405 ppm ($\Delta\delta_C = -0.392$ ppm). From an assignment point of view more intriguing was the differentiation of the closely spaced degenerate lines belonging to C-3 and C-4. Based on the isotope shift of -0.100 ppm, the signal at 27.239 ppm (^1H isotopomer) could unambiguously be designated to C-4. This could further be confirmed by C-3 (^2H isotopomer) exhibiting a triplet (splitting ($^3J_{CCCD} = 0.75$ Hz) and a vicinal isotope effect (-0.033 ppm).

An interesting application of the two-bond deuterium isotope effect has been reported in connection with the assignment of carbonyl carbons in peptides.[68] Replacement of a proton on the nitrogen atom in a peptide bond was found to lead to small low-frequency shifts of the resonances of the α carbon and the

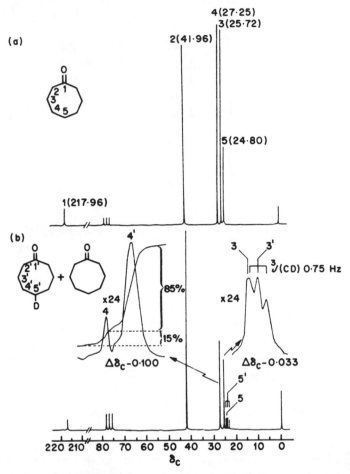

Fig. 3.52   25.2 MHz (2.3 T) proton noise-decoupled ^{13}C spectra of cyclooctanone in chloroform-d. (a) Pure protio compound; (b) mixture of protio and [5-^{2}H] isotopomer, showing isotope effect on shielding and ^{13}C—^{2}H spin–spin coupling.[174] Reproduced by permission of John Wiley & Sons.

carbonyl. This was used for distinguishing peptide carbonyl signals from amide and carboxylic carbonyls. Figure 3.53 shows the spectrum of the carbonyl carbons of reduced glutathione (51) in a 1:1 mixture of H_2O and D_2O as solvent.

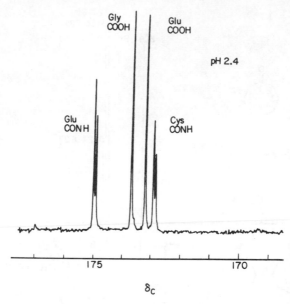

Fig. 3.53 Proton noise-decoupled ^{13}C spectrum of the carbonyl region in reduced glutathione (**51**), recorded at pH 2.4 in 1:1 (v/v) $H_2O:D_2O$.[68]

The two peptide carbonyl carbon signals can immediately be recognized because they appear twice each (belonging to CONH and COND, respectively). On the other hand, the carboxylate carbons give rise to singlets because the protons of the carboxylate and α-amine groups are in rapid exchange with the solvent. Requisite for the detectability of such isotope shifts is that the lifetime of the proton at a particular site is longer than $(\Delta v_{HD})^{-1}$, the inverse of the isotope shift in Hertz.

So far, we have dealt with the intrinsic isotope effect only and we will see that the general rules stated earlier are only valid for this type of isotope effect. In particular, the myth that isotope effects are necessarily shielding has to be abandoned. In effect, deshielding isotope effects have been observed in many situations.[175-81] Besides the intrinsic isotope effect there are at least three causally distinctly different isotope effects which can, but need not, cause deshielding: (1) the *hyperconjugative isotope effect*,[180-82] (2) the *equilibrium isotope effect*,[183-7] and (3) the *through-space* or *steric isotope effect*.[181]

The hyperconjugative isotope effect rests on the fact that a CD_3 group, for example, has a slightly lower ability for electron release than a CH_3 group. Such an effect has been reported for the phenyldimethyl carbenium ion (**52**).[180] The

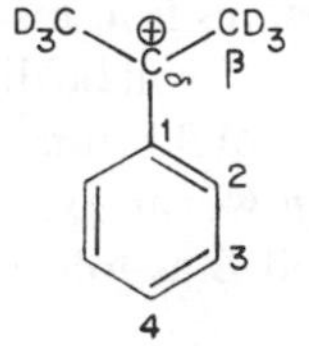

52

deshielding effects on the *ortho* and *para* carbons ($+0.16$ and $+0.22$ ppm, respectively) are far too large to be accounted to an intrinsic isotope effect across four and six bonds. An effect of the same origin has also been observed in neutral molecules such as coniferyl alcohol (**53**).[188]

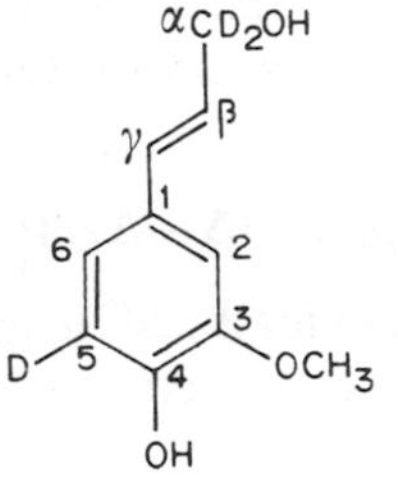

53

The deshielding isotope effect of 0.15 ppm for the γ carbon clearly has to be attributed to hyperconjugation of the α-CD$_2$ group. By contrast, C$_\beta$ exhibited the expected shielding ($\Delta\delta = -0.14$ ppm). Hyperconjugative isotope effects appear to be a general phenomenon, expected to occur in unsaturated systems whenever deuterium-substituted alkyl groups are present α to an unsaturated carbon. There is no doubt, however, that the effect has its origin ultimately also in the different vibrational amplitudes, caused by the heavier isotope.

The equilibrium isotope effect occurs in rapidly equilibrating molecules, be it fast conformational interconversion[185-7] or degenerate intramolecular rearrangements as they are often found for carbonium ions.[183] A further situation giving rise to this effect may be hydrogen bonding, caused by the slightly different hydrogen bonding abilities of OD relative to OH groups.[177,189]

An illustrative example of an isotope perturbation of an otherwise degenerate equilibrium is 1,1,3,3-tetramethylcyclohexane,[186] which is known to inter-convert fast on the n.m.r. time scale, therefore showing only a single resonance in its ^{13}C spectrum. However, substitution of one CH$_3$ by CD$_3$ breaks the symmetry of the molecule, therefore lifting the potential energy degeneracy for the two conformational states **54a** and **54b**.

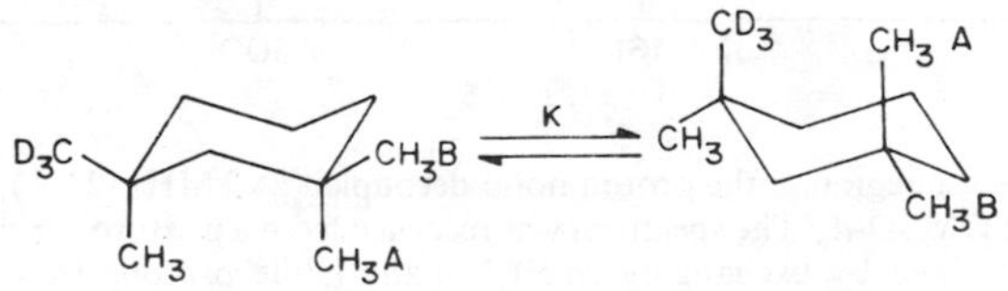

54a   **54b**

Even though interconversion remains fast, resulting in chemical shift averaging, the two methyl groups in position 3 will be distinguishable and their relative chemical shifts are related to the equilibrium constant K. By denoting the populations of **54a** and **54b** p and $1-p$, we can express the averaged chemical shifts δ_A and δ_B for methyl carbons A and B as follows:

$$\delta_A = p\delta_{ax} + (1-p)\delta_{eq} \tag{3.27}$$

$$\delta_B = p\delta_{eq} + (1-p)\delta_{ax} \tag{3.28}$$

or

$$\Delta\delta \equiv \delta_A - \delta_B = p(\delta_{ax} - \delta_{eq}) + (1-p)(\delta_{eq} - \delta_{ax}) \tag{3.29}$$

$\delta_{ax} - \delta_{eq} \equiv \Delta$ is the chemical shift difference for the two methyl carbons in the static conformer (obtainable from low-temperature spectra). We can thus write for the equilibrium constant

$$K = (1-p)/p = (\Delta - \Delta\delta)/(\Delta + \Delta\delta) \tag{3.30}$$

Insertion of the experimental values for Δ and $\Delta\delta$ gave $K = 1.042$; hence there is a slight preference of the molecule for axial orientation of the CD_3 group (conformer **54b**).

These data demonstrate that isotope shifts, when used for assignment purposes, have to be used with caution whenever conformationally labile molecules are involved, as introduction of the isotopic label inevitably displaces the equilibrium. This, in turn, may produce long-range isotope shielding effects,

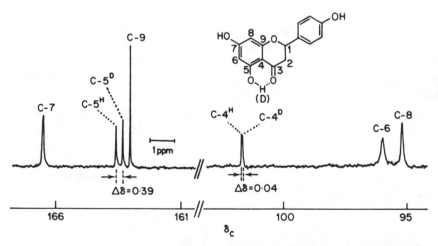

Fig. 3.54 Aromatic region of the proton noise-decoupled 25.2 MHz (2.3 T) ^{13}C spectrum of naringenin (9) in DMSO-d_6. The spectrum was recorded from a mixture of protio and deutero isotopomers, obtained by exchanging ca 50% of the labile protons by deuterons. Peaks labelled 'H' and 'D' originate from different isotopomeric species.[190] Reproduced by permission of Springer, New York.

which are totally unrelated to the number of bonds separating the carbon under study from the labelling site.

Isotope effects caused by replacement of an intramolecular hydrogen-bonded proton by a deuteron in naringenin (9, p. 125) has been shown to afford an unusually large isotope effect on the shielding of C-5 ($\Delta\delta = -0.39$ ppm), almost four times the value one would expect for an intrinsic geminal isotope effect.[190] In this case the splittings, observed in dimethylsulphoxide solution, have been measured following a partial exchange of the labile protons by addition of D_2O (Fig. 3.54). The effect served to distinguish the resonances caused by C-5 and C-9, which have nearly the same chemical shifts. Simultaneous observation of the signals of both isotopomers obviously requires that the exchange rate be sufficiently low so that no averaging ensues, as mentioned earlier.

Finally, the least-understood isotope effect so far, which had first been reported for proton chemical shifts,[191] has meanwhile also been observed in [13]C n.m.r.[181] in some methyl-substituted paracyclophanes such as 55.

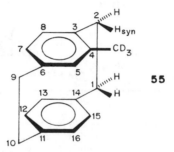

Whereas deshielding relative to the protio isotopomer, found for C-3 and C-7 (+ 26 and + 7 parts per billion (ppb), respectively), undoubtedly is to be accounted to the hyperconjugation effect discussed earlier, the deshielding of C-15 ($\Delta\delta = +35$ ppb) can only be rationalized in terms of a through-space effect, as seven bonds separate this carbon from the deuterium site. Conformational equilibrium perturbation could be ruled out as an explanation as the same order of magnitude effects could be observed in related totally rigid structures.

The more widespread availability of very high field instruments ($\geqslant 8.4$ T) has certainly contributed to spurring chemists' interest in n.m.r. isotope effects, and this field is therefore anticipated to expand rapidly over the next few years. Researchers have already begun hunting for even subtler effects, such as those caused by substituting [16]O by the (non-magnetic) [18]O.[192-7] Because of the smaller fractional mass difference between the two isotopes, the expected shifts should be smaller than for the ^{2}H/^{1}H isotope pair. Typically one-bond isotope shifts of 10-50 ppb have been reported,[196] depending on oxygen functionality and structure. The effects detected so far are definitely of the intrinsic type, as they qualitatively comply with the rules stated earlier. This is, for example, reflected by

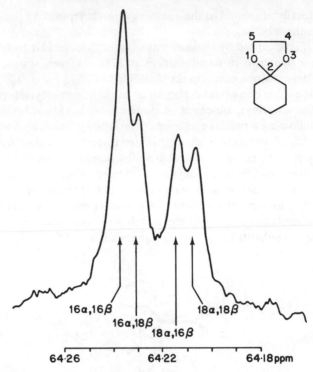

Fig. 3.55   Proton noise-decoupled 100.6 MHz (9.4 T) ^{13}C spectrum displaying region of C-4 and C-5 carbon resonance in the
cyclic ketal **56**, which is a mixture of three isotopomers: [^{16}O$_2$],
[^{16}O, ^{18}O], and [^{18}O$_2$]. The labelling of peaks refer to the isotopes
of oxygen in α and β position.[197] Reproduced by permission of the
Royal Society of Chemistry.

the additivity of the shielding, found for successive substitution of ^{16}O by ^{18}O in
ortho esters.[195] Additivity and long-range shielding effects are illustrated by the
high-field (9.4 T) ^{13}C spectrum in Fig. 3.55,[197] displaying the expanded region of
the ethylene carbons in the spirocyclic ketal (**56**). The spectrum was recorded
from a mixture of the [^{16}O$_2$], [^{18}O^{16}O] and [^{18}O$_2$] isotopomers. It should be
noted that in the unsymmetrically labelled isotopomer the symmetry is lifted,

thus rendering the two methylene carbons C-4 and C-5 non-equivalent and
giving rise to separate signals (labelled 16α, 18β and 18α, 16β in the figure). The

smaller of the two splittings is therefore attributed to a vicinal isotope effect ($\Delta\delta = -7$ ppb).

The structural dependence of the ^{18}O isotope effect is already well documented and the following qualitative trends have been found: ketones $\geqslant$ aldehydes $\geqslant$ esters $\geqslant$ amides $\gg$ tertiary alcohols $\geqslant$ secondary alcohols $\geqslant$ primary alcohols $\geqslant$ phenols.[196] Electron donation appears to lower the isotope shift, as, for example, evidenced by the data reported for averufin (57), biosynthetically derived from [1-^{13}C, ^{18}O$_2$] acetate and ^{18}O$_2$ gas.[198] C-9 clearly experiences electron donation

from hydroxyls in positions 1, 6, and 8, accounting for the lower value of $\Delta\delta_{16/18}$ (29 ppb), compared with C-10 (42 ppb). This experiment also demonstrated that ^{18}O-labelled biosynthetic precursors lend promise as a new method of establishing the biogenetic origin of oxygen. In this particular case it was shown that all but the C-10 carbonyl oxygen are acetate-derived.

The spin–zero nature of ^{18}O in some instances gives it an edge over deuterium, as the carbon spectra are unencumbered by spin–spin coupling and relaxation effects. On the other hand, the relatively high cost of ^{18}O-labelled compounds will probably confine the use of ^{18}O as a tracer to specialized experiments such as biosynthesis and kinetic and mechanistic studies.[192]

3.10 SOLID-STATE ^{13}C N.M.R.

Until about 1975, high-resolution n.m.r. studies were performed almost exclusively on samples in the liquid state or on liquid solutions of samples. There are a number of instances when it would be highly desirable to obtain ^{13}C n.m.r. spectra of solid samples; such is the case when studying, for example, high-molecular-weight and/or highly crosslinked polymers, which are often insoluble or only very slightly soluble in common organic solvents. However, early attempts to obtain ^{13}C n.m.r. spectra of solid samples by using standard high-resolution n.m.r. techniques afforded generally poor-quality spectra or failed to provide spectra at all.

Characteristically, ^{13}C n.m.r. spectra of solids reflect the cumulative line-broadening effects of ^{1}H—^{13}C dipole–dipole interactions and ^{13}C chemical shift anisotropy. As a result, only broad, rather featureless resonances that contain very little useful structural information are obtained under these conditions. An

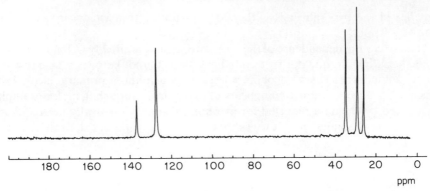

Fig. 3.56   62.7 MHz ^{13}C n.m.r. spectrum of a sample of a rubber suction cap with proton noise-decoupling, showing resonances characteristic of polyisoprene, obtained in less than 1 min total acquisition time.

exception in this regard are materials that are not truly solids; for example, it is possible to obtain a useful ^{13}C n.m.r. spectrum of a piece of rubber tubing (Fig. 3.56) under standard high-resolution conditions. In such materials, the interactions that are responsible for line broadening in the solid state are partially averaged due to internal and segmental motion.

Most instrument manufacturers now offer either (1) spectrometers that are capable of obtaining high-resolution n.m.r. spectra directly on solid samples or (2) accessories that are compatible with existing, routine high-resolution n.m.r. spectrometers. Although the main thrust of this book is devoted to discussion of solution-state n.m.r., it would seem appropriate, nevertheless, to review briefly the factors responsible for line broadening in solid samples and to discuss the methods that are employed to overcome this problem. For more detailed discussions of the intricacies of high-resolution solid-state n.m.r. spectroscopy, the reader is referred to specialized reviews.[199]

Characteristically, ^{13}C n.m.r. spectra of non-crystalline solids reflect the cumulative line-broadening effects that arise from the operation of two major phenomena: (1) ^{1}H—^{13}C dipole–dipole interactions and (2) ^{13}C chemical shift anisotropy. The primary factor that is responsible for the very broad (typically unobservable) resonance patterns in ^{13}C n.m.r. spectra of solids is dipole–dipole coupling between ^{13}C nuclei and neighbouring protons. Consider the case of two neighbouring spins that possess magnetic moments μ_C and μ_H, respectively, and that are located r_{CH} Å apart such that the vector r_{CH} joining these two spins makes an angle θ relative to the orientation of the static magnetic field. In such a situation, each nucleus generates a field at the other's location; the local field, h_1, at the site of nucleus C can shown[199b,200] to be given by:

$$h_1 = \pm \mu_H (3 \cos^2 \theta - 1) r_{CH}^{-3} \qquad (3.31)$$

The sign of h_1 in Eqn. (3.31) depends upon the spin state of nucleus H. Hence, the carbon resonance is split into a doublet (frequencies $\omega_C \pm \mu_H . h_C$) except when $\theta = 54.7°$, i.e. for the so-called 'magic angle' (see below).[199b] Incidentally, this same interaction gives rise to relaxation phenomena in solution, provided that there exists a suitable mechanism for modulation (cf. discussion in Chapter 4). Of course, all orientations are possible in a rigid polycrystalline sample; this results in a Gaussian distribution of the resonance frequency. In solution, rapid molecular tumbling averages $< 3\cos^2 \theta - 1 >$, and thus h_1 vanishes. All line-narrowing techniques, therefore, concentrate upon averaging the dipolar interaction energy.

A major advance in solid-state ^{13}C n.m.r. spectroscopy was provided by the application of high-power proton decoupling ($\gamma B_2/2\pi = ca$ 20 kHz), which effectively neutralizes the effects of ^{1}H—^{13}C dipolar broadening.[201] This is analogous to the corresponding rf decoupling of protons that was discussed in Section 3.2 for the removal of scalar coupling. However, the interaction energies involved in dipolar coupling are two to three orders of magnitude larger than those encountered in scalar coupling; hence, dipolar decoupling requires the application of significantly increased decoupling power levels.

Figures 3.57(a)-(c), respectively, depict 15 MHz ^{13}C n.m.r. spectra of the bisacetonide of 4, 4'-bis[(2, 3-dihydroxypropyl)oxy]benzil that have been obtained first under solution conditions (a), with conventional, low-power decoupling (b), and with high-power decoupling (c).[199c] Examination of Fig 3.57 (c) reveals that spectra obtained with high-power decoupling still exhibit very little structural detail with almost no distinction among chemically shifted resonances. The relatively poor quality of these spectra is due to the operation of a second line-broading mechanism which reflects the orientational dependence of nuclear shielding (i.e. chemical shift anisotropy). This effect, discussed in Section 4.3 in the context of relaxation phenomena, is averaged in solution due to rapid molecular reorientation (as is the case with dipole–dipole interactions).

In particular, sp^2- and sp-hybridized carbon atoms have large shielding anisotropies. For example, in benzene, carbon nuclei are more highly shielded by ca 150 ppm when the molecule is oriented with its plane perpendicular to rather than parallel with the applied magnetic field.[202] In polycrystalline samples all orientations of the shielding tensor relative to the direction of the magnetic field are equally probable. This results in a broad envelope that encompasses the shieldings in their various orientations. Linder and coworkers[202] discuss the low-temperature (14 K) proton-decoupled ^{13}C n.m.r. spectrum of polycrystalline benzene. At this temperature, in-plane rotation of the benzene ring ceases, with the result that an asymmetric shielding tensor is generated. This tensor possesses two principal elements (σ_{11}, σ_{22}) that define the two in-plane orientations and a third, σ_{33}, that represents the element perpendicular to the molecular plane, as shown below.[201-3]

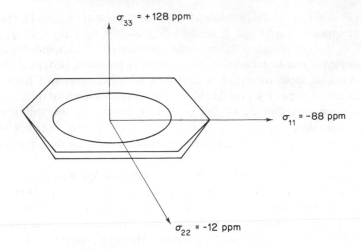

The isotropic (solution-state) chemical shift is equal to one-third the trace of the principal elements,[199b] i.e.

$$\langle \sigma \rangle = 1/3(\sigma_{11} + \sigma_{22} + \sigma_{33})$$

It is obvious that, in more complicated molecules, overlap of the shielding anisotropy powder patterns will impede both routine spectral interpretation and analysis of orientational effects in n.m.r. spectra of solid samples. In order to obtain an isotropic spectrum it will be necessary to average the various orientation-dependent shieldings. This desirable end can be achieved by spinning the sample about an axis oriented at the 'magic angle' (i.e. 54.7°) with respect to the external magnetic field.[199b,204] The effect of magic-angle spinning (MAS) on resolution is dramatic, as evidenced by the quality of the n.m.r. spectrum in Fig. 3.57(d). In this example, the quality of the n.m.r. spectrum obtained for a solid sample compares favourably with the corresponding solution-state spectrum (Fig. 3.57(e)).

It is particularly important to achieve appropriate spinning rates when performing MAS on solid samples. Ideally, the sample must be spun at a rate given by the expression $(\gamma H_0/2\pi)|\sigma_{33} - \sigma_{11}|$; the difference $|\sigma_{33} - \sigma_{11}|$ defines the width of the powder pattern, in Hertz.[199b] Hence, at field strengths of 2.3 and 4.7 T and a shielding anisotropy of 100 ppm, spinning rates in excess of 2500 and 5000 Hz, respectively, are required. If spinning is performed at rates that are below these lower limits, modulation sidebands will appear in the n.m.r. spectrum at frequencies corresponding to $\nu_{iso} \pm n\nu_s$, where ν_{iso} is the isotropic resonance frequency and ν_s is the rate of sample spinning. The multiple sidebands thereby encountered are reminiscent of the sidebands that occur in solution-state n.m.r. spectroscopy when the experiment is performed under conditions of poor magnetic field homogeneity.

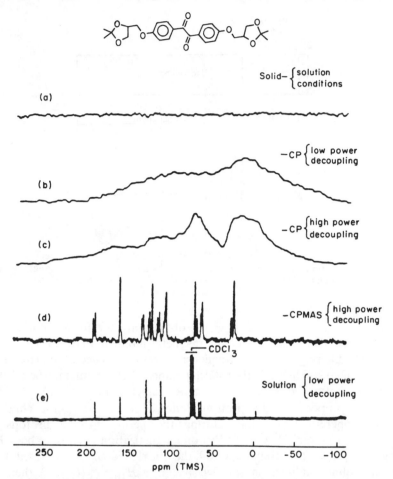

Fig. 3.57 (a)–(d): Solid-state ^{13}C n.m.r. spectra obtained at 15 MHz: conventional data acquisition (60° pulses, 10 s pulse decay and 6 kHz proton decoupling) from a 680 mg sample; (b) same as (a) with proton–carbon cross polarization using a 1-s pulse decay; (c) same as (b) but with 43 kHz proton decoupling; (d) same as (c) with addition of magic angle spinning from a 34 mg sample of the same compound; (e) solution spectrum (34 mg). Total acquisition times: (a) and (e): 12 h; (b)–(d): 1.2 h.[199c] Reprinted with permission from *Acc. Chem. Res.*, **15**, 201. Copyright (1982) American Chemical Society.

An additional problem that is encountered when attempting to obtain high-quality ^{13}C n.m.r. spectra of solids is low signal-to-noise ratio concomitant with (1) the low natural isotopic abundance of the ^{13}C nucleus, (2) its small magnetic moment, and (3) its typically long relaxation times (relative to proton T_1 values). Sensitivity can be enhanced by using the cross-polarization (CP) technique first described by Pines and coworkers,[201] which employs transfer of magnetization

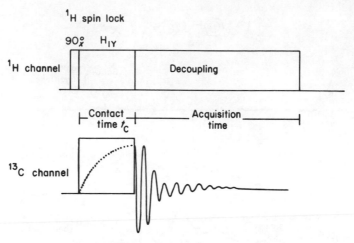

Fig. 3.58 Pulse sequence for ^{13}C—^{1}H cross-polarization experiment. Protons are excited by a 90° pulse, followed by a spin lock period during which polarization of the ^{13}C nuclei occurs due to Hartmann–Hahn contact.

from ^{1}H nuclei to nearby ^{13}C nuclei. Not only are the former nuclei more abundant, they also possess a larger magnetic moment than do ^{13}C nuclei. Despite their differing resonance frequencies, contact between the ^{1}H and ^{13}C reservoirs can be established in the rotating frame.[206] This result may be achieved by using a pulse sequence of the type depicted in Fig. 3.58. A $(90°)_x$ pulse first is applied to the protons, followed by a 90° phase shift and continued coherent irradiation. During the period t_C, the protons are said to be 'spin-locked'.[199c,201,205] Proton magnetization precesses about H_{1H}. During this spin-lock or contact period, the ^{13}C rf field H_{1C} is activated. If the Hartman–Hahn condition is satisfied (i.e. if $\gamma_H H_{1H} = \gamma_C H_{1C}$),[205] then ^{13}C transverse magnetization will build up resulting in a FID which can be sampled in the usual way. The S/N gain from cross-polarization is twofold: (1) the pulse recycle time is determined by the proton T_1 rather than by the much longer ^{13}C T_1 value, and (2) the available magnetization is governed by the (much larger) ^{1}H magnetic moment. The combination of cross-polarization with magic angle spinning (CP/MAS) techniques continues to be widely employed for the study of high-resolution n.m.r. in solids.[199,206]

References

1 A comprehensive review of 1-dimensional multipulse experiments, including polarization transfer-type experiments is C. Turner, Multipulse NMR in liquids, In *Progress in NMR Spectroscopy*, Vol. 16, (Eds. J. W. Emsley, J. Feeney, and L. H. Sutcliffe), Pergamon, New York, 1984.

2a G. A. Morris and R. Freeman, *J. Am. Chem. Soc.* **101**, 760 (1979).

2b G. A. Morris, *J. Am. Chem. Soc.* **102**, 428 (1980).

3 D. P. Burum and R. R. Ernst, *J. Magn. Res.* **39**, 163 (1980).

4 For a comprehensive review on multiple pulse techniques, see, for example G. A. Morris, *Magn. Res. Chem.* **24**, 371 (1986).

5 G. Bodenhausen, R. Freeman, and G. A. Morris, *J. Magn. Res.* **23**, 171 (1976).

6 G. A. Morris and R. Freeman, *J. Magn. Res.* **29**, 433 (1978).

7 R. Freeman, G. A. Morris, and M. J. T. Robinson, *J. C. S. Chem. Commun.* 754 (1976).

8 R. Freeman and G. A. Morris, *Bull. Magn. Res.* **1**, 5 (1979).

9. D. Turner, Basic two-dimensional NMR. In *Progress in NMR Spectroscopy*, Vol. 17 (Eds. J. W. Emsley, J. Feeney, and L. S. Sutcliffe), Pergamon, New York, 1985.

10 See, for example, J. L. Marshall, *Methods in Stereochemical Analysis* (Ed. A. P. Marchand), Vol. 2, Chemie International, Deerfield Beach, Florida, 1982.

11 R. R. Ernst, *J. Chem. Phys.* **45**, 3845 (1966).

12 F. J. Weigert, M. Jautelat, and J. D. Roberts, *Proc. Nat. Acad. Sci. US* **60**, 1152 (1968).

13 L. F. Johnson, *Topics in ^{13}C Nuclear Magnetic Resonance* (Ed. G. C. Levy), Vol. 3, Chap. 1, Wiley-Interscience, New York, 1979.

14 C. LeCocq and J. -Y. Lallemand, *J. C. S. Chem. Commun.* 150 (1981).

15 A. Bax, R. Freeman, and S. P. Kemsell, *J. Am. Chem. Soc.* **102**, 4849 (1980).

16 A. Bax., R. Freeman, and S. P. Kemsell, *J. Magn. Res.* **41**, 349 (1980).

17 A. Bax, R. Freeman, and T. A. Frenkiel, *J. Am. Chem. Soc.* **103**, 2102 (1981).

18 A. Bax, R. Freeman, T. A. Frenkiel, and M. J. Levitt, *J. Magn. Res.* **43**, 478 (1981).

19 R. Richarz, W. Ammann, and T. Wirthlin, *J. Magn. Res.* **45**, 270 (1981).

20 H. Reich, M. Jautelat, M. T. Messe, F. J. Weigert, and J. D. Roberts, *J. Am. Chem. Soc.* **91**, 7445 (1969).

21 C. J. Jameson, *Bull. Magn. Res.* **3**, 3 (1980).

22 J. B. Stothers, *Topics in ^{13}C Nuclear Magnetic Resonance* (Ed. G. C. Levy), Vol. 1, p. 229, Wiley-Interscience, New York, 1974.

23 J. Reuben, *Progress in Nuclear Magnetic Resonance Spectroscopy* (Eds J. Emsley, J. Feeney, and L. S. Sutcliffe), Vol. 9, Pergamon Press, Oxford, 1973.

24 R. E. Sievers, *Nuclear Magnetic Resonance Shift Reagents*, Academic Press, London and New York, 1973.

25 See, for example, K. G. Orell, *Annual Reports on NMR Spectroscopy*, (Ed., G. A. Webb), Vol. 9, Academic Press, London, 1979.

26 See, for example, A. Bax and R. Freeman, *J. Magn. Res.* **44**, 542 (1981), and references cited.

27 J. R. Lyerla, Jr and D. M. Grant, *Int. Rev. Science Phys. Chem. Series* (Ed. C. A. McDowell), Vol. 4, Chap. 5, Medical and Technical Publishing Company, Chicago, 1972.

28 J. R. Lyerla, Jr and G. C. Levy, *Topics in ^{13}C Nuclear Magnetic Resonance* (Ed. G. C. Levy), Vol. 1, Wiley-Interscience, New York, 1974.

29 F. W. Wehrli, *Topics in ^{13}C Nuclear Magnetic Resonance* (Ed. G. C. Levy), Vol. 2, Wiley-Interscience, New York, 1975.

30 S. Berger, *Advances in Physical Organic Chemistry* (Eds V. Gold & D. Bethel), Vol. 16, 1978.

31 D. A. Wright, D. E. Axelson, and G. C. Levy, *Topics in ^{13}C Nuclear Magnetic Resonance* (Ed. G. C. Levy), Vol. 3, Wiley-Interscience, New York, 1979.

32 L. F. Johnson, Spin Decoupling Methods in ^{13}C NMR Studies, pp. 2–16 in *Topics in Carbon-13 NMR Spectroscopy* (Ed. G. C. Lovy), Vol. 3, Wiley–Interscience, New York, 1979.

33 R. A. Hoffmann and S. Forsén, *High-Resolution Nuclear Magnetic Double and*

Multiple Resonance (Eds J. Emsley, J. Feeney, and L. H. Sutcliffe), Pergamon Press, Oxford, 1966.

34 K. G. R. Pachler, *J. Magn. Res.* **7**, 442 (1972).

35 E. R. Malinowski, *J. Am. Chem. Soc.* **83**, 4479 (1961).

36 F. W. Wehrli, *Advances Mol. Relax. Proc.* **6**, 139 (1974).

37 P. R. Srinivasan and R. L. Lichter, *Org. Magn. Res.* **8**, 198 (1976).

38 R. A. Newmark and J. R. Hill, *J. Am. Chem. Soc.* **95**, 4435 (1973).

39 E. D. Becker, *High-Resolution NMR, Theory and Chemical Applications*, Chap. 7, Academic Press, New York, 1980.

40 G. Jikeli, W. Herrig, and H. Günther, *J. Am. Chem. Soc.* **96**, 323 (1974).

41 Ref. 39, Chap. 9.

42 E. W. Hagaman, *Org. Magn. Res.* **8**, 389 (1976).

43 H. Fritz and H. Sauter, *J. Magn. Res.* **15**, 177 (1974); **18**, 527 (1975).

44 R. Radeglia, *Org. Magn. Res.* **9**, 164 (1977).

45 R. Radeglia, *Z. Phys. Chem.* **260**, 990 (1979).

46 R. Radeglia, *Z. Phys. Chem.* **261**, 617 (1980).

47 S. Altenburger–Wehrli, private communication.

48 R. Freeman and H. D. W. Hill, *J. Chem. Phys.* **54**, 3367 (1971).

49 B. Birdsall, N. J. M. Birdsall, and J. Feeney, *Chem. Commun.* 316 (1972).

50 F. W. Wehrli, *Nuclear Magnetic Resonance Spectroscopy of Nuclei Other than Protons* (Eds T. Axenrod and G. A. Webb), Wiley-Interscience, New York, 1974.

51 N. S. Bhacca, F. W. Wehrli, and N. H. Fischer, *J. Org. Chem.* **38**, 3618 (1973).

52 S. Sørensen, M. Hansen, and H. J. Jakobsen, *J. Magn. Res.* **12**, 340 (1973).

53 L. Ernst and V. Wray, *J. Magn. Res.* **25**, 123 (1977).

54 R. K. Harris and R. H. Newman, *J. Magn. Res.* **24**, 449 (1976).

55 D. Canet, *J. Magn. Res.* **23**, 361 (1976).

56 O. Gansow and W. Schittenhelm, *J. Am. Chem. Soc.* **93**, 4294 (1971).

57 R. Freeman and H. D. W. Hill, *J. Magn. Res.* **5**, 278 (1971).

58 R. Freeman and H. D. W. Hill, *Molecular Spectroscopy 1971*, Inst. of Petroleum, London, 1971; J. R. Lyerla, Jr, D. M. Grant, and R. D. Bertrand, *J. Phys. Chem.* **75**, 3967 (1971).

59 M. Hansen and H. J. Jakobsen, *J. Magn. Res.* **10**, 74 (1973).

60 R. V. Dubs and W. von Philipsborn, *Org. Magn. Res.* **12**, 326 (1979).

61 L. F. Williams and A. A. Bothner-By, *J. Magn. Res.* **11**, 314 (1973).

62 M. Hansen, R. S. Hansen, and H. J. Jakobsen, *J. Magn. Res.* **13**, 386 (1974).

63 F. J. Weigert and J. D. Roberts, *J. Am. Chem. Soc.* **89**, 2967 (1967).

64 F. J. Weigert and J. D. Roberts, *J. Am. Chem. Soc.* **90**, 3543 (1968).

65 U. Vögeli, D. Herz, and W. von Philipsborn, *Org. Magn. Res.* **13**, 200 (1980).

66 G. Müller and W. von Philipsborn, *Helv. Chim. Acta* **56**, 2680 (1973).

67 Ch. Grathwohl, R. Schwyzer, A. Tun-kyi, and K. Wüthrich, *FEBS Lett.* **29**, 271 (1973).

68 J. Feeney, P. Partington, and G. C. K. Roberts, *J. Magn. Res.* **13**, 268 (1974).

69 W. von Philipsborn, *Pure Appl. Chem.* **40**, 159 (1974).

70 F. W. Wehrli, *J. C. S. Chem. Commun.* 663 (1975).

71 R. U. Lemieux, T. L. Nagabushan, and B. Paul, *Can. J. Chem.* **50**, 113 (1972).

72 R. Wasylishen and T. Schaefer, **50**, 2710 (1972).

73 M. Karplus, *J. Chem. Phys.* **30**, 11 (1959); *J. Am. Chem. Soc.* **85**, 2870 (1963).

74 C. A. Kingsbury and M. E. Jordon, *J. C. S. Perkin II* 364 (1977).

75 M. P. Schweizer and G. P. Kreishman, *J. Magn. Res.* **9**, 334 (1973).

76 A. A. Maudsley and R. R. Ernst, *Chem. Phys. Lett.* **50**, 368 (1977).

77 A. A. Maudsley, L. Müller, and R. R. Ernst, *J. Magn. Res.* **28**, 463 (1977).

78 R. Bodenhausen and R. Freeman, *J. Magn. Res.* **28**, 471 (1977).

79 R. Freeman and G. A. Morris, *J. C. S. Chem. Commun.* 684 (1978).
80 L. Müller and R. R. Ernst, *Mol. Phys.* **38**, 909 (1979).
81 A. Bax and G. A. Morris, *J. Magn. Res.* **42**, 501 (1981).
82 G. A. Morris and L. D. Hall, *J. Am. Chem. Soc.* **103**, 4703 (1981).
83 K. G. R. Pachler and P. L. Wessels, *J. Magn. Res.* **13**, 337 (1973).
84 S. Sørensen, R. S. Hansen, and H. J. Jakobsen, *J. Magn. Res.* **14**, 243 (1974).
85 H. J. Jakobsen, S. A. Linde, and S. Sørensen, *J. Magn. Res.* **15**, 385 (1974).
86 R. Kaiser, *J. Chem. Phys.* **39**, 2435 (1963).
87 D. W. Nagel, K. G. R. Pachler, P. S. Steyn, R. Vleggaar, and P. L. Wessels, *Tetrahedron* **32**, 2625 (1976).
88 W. von Philipsborn, *Angew. Chem.* **83**, 470 (1971).
89 R. Freeman and W. A. Anderson, *J. Chem. Phys.* **37**, 2053 (1962).
90 A. A. Chalmers, K. G. R. Pachler, and P. L. Wessels, *J. Magn. Res.* **15**, 415 (1974).
91 O. W. Sørensen, H. Bildsøe, and H. J. Jakobsen, *J. Magn. Res.* **45**, 325 (1981).
92 See, for example, R. Freeman and H. D. W. Hill, *Dynamic Nuclear Magnetic Resonance* (Eds F. A. Cotton and L. M. Jackman), Academic Press, New York, 1975.
93 A. Pines, M. G. Gibby, and J. S. Waugh, *J. Chem. Phys.* **59**, 569 (1973).
94 R. D. Bertrand, W. B. Moniz, A. N. Garroway, and G. C. Chingas, *J. Am. Chem. Soc.* **100**, 5227 (1978); *J. Magn. Res.* **32**, 465 (1978).
95 D. L. Rabenstein and T. T. Nakashima, *Anal. Chem.* **51**, 1465 (1979).
96 S. Patt and J. N. Shoolery, *J. Magn. Res.* **46**, 535 (1982).
97 E. D. Becker, J. A. Ferretti, and T. C. Farrar, *J. Am. Chem. Soc.* **91**, 7784 (1969).
98 D. M. Doddrell, D. T. Pegg, and M. R. Bendall, *J. Magn. Res.* **48**, 323 (1982).
99 M. R. Bendall, D. M. Doddrell, and D. T. Pegg, *J. Am. Chem. Soc.* **103**, 4603 (1981).
100 M. R. Bendall, D. M. Doddrell, D. T. Pegg, and W. E. Hull, *High-Resolution Multipulse NMR Spectrum Editing and DEPT*, Bruker Analytische Messtechnik, Karlsruhe, 1982.
101 A. G. Redfield and R. K. Gupta, *J. Chem. Phys.* **54**, 1418 (1971).
102 J. D. Stoesz, A. G. Redfield, and D. Malinowski, *FEBS Lett.* **91**, 320 (1978).
103 B. L. Tomlinson and H. D. W. Hill, *J. Chem. Phys.* **59**, 1775 (1973).
104 See, for example, Ref. 33, p. 15.
105 A. A. Chalmers, *J. Magn. Res.* **38**, 565 (1980).
106 For a comprehensive treatment of the subject, see A. D. Bax, *Two-dimensional Nuclear Magnetic Resonance in Liquids*, Delft University Press, R. D. Reidel, Dordrecht, Holland, 1982.
107 W. P. Aue, E. Bartholdi, and R. R. Ernst, *J. Chem. Phys.* **64**, 2229 (1976).
108 G. Bodenhausen, R. Freeman, R. Niedermeyer, and D. L. Turner, *J. Magn. Res.* **26**, 133 (1977).
109 P. Bachmann, W. P. Aue, L. Müller, and R. R. Ernst, *J. Magn. Res.* **28**, 29 (1977).
110 See, for example, E. D. Becker and T. C. Farrar, *Pulse and Fourier Transform NMR*, Chap. 5, Academic Press, New York, 1971.
111 L. Müller, A. Kumar, and R. R. Ernst, *J. Chem. Phys.* **63**, 5490 (1975).
112 G. Bodenhausen, R. Freeman, and D. L. Turner, *J. Chem. Phys.* **65**, 839 (1976).
113 G. Bodenhausen, R. Freeman, R, Niedermeyer, and D. L. Turner, *J. Magn. Res.* **24**, 291 (1976).
114 L. Müller, A. Kumar, and R. R. Ernst, *J. Magn. Res.* **25**, 383 (1977).
115 R. Freeman, G. A. Morris, and D. L. Turner, *J. Magn. Res.* **26**, 373 (1977).
116 G. Bodenhausen, R. Freeman, and D. L. Turner, *J. Magn. Res.* **27**, 511 (1977).
117 G. Bodenhausen, R. Freeman, G. A. Morris, and D. L. Turner, *J. Magn. Res.* **28**, 17 (1977).
118 L. D. Hall and G. A. Morris, *Carbohydr. Res.* **82**, 175 (1980).

119 R. Freeman and H. D. W. Hill, *J. Chem. Phys.* **54**, 301 (1971).

120 A. Bax and R. Freeman, *J. Am. Chem. Soc.* **104**, 1099 (1982).

121 P. E. Hansen, *Org. Magn. Res.* **11**, 215 (1978).

122 G. Maciel, *Nuclear Magnetic Resonance Spectroscopy of Nuclei Other than Protons* (Eds T. Axenrod and G. A. Webb), Chap. 13, John Wiley, New York, 1974.

123 J. -R. Llinas, E. J. Vincent, and G. Pfeiffer, *Bull. Soc. Chim. France* **11**, 3209 (1973).

124 See, for example, Ref. 119 and references cited.

125 F. W. Wehrli, unpublished data.

126 See, for example, Ref. 33.

127 A. Wokaun and R. R. Ernst, *J. Chem. Phys.* **67**, 1752 (1977).

128 W. Ammann, R. Richarz, and T. Wirthlin, *Varian Instruments at Work*, Z-12, 1981.

129 M. Tanabe, *Biosynthesis* **2**, 241 (1973); **3**, 247 (1975); **4**, 204 (1976).

130 U. Seguin and A. I. Scott, *Science* **186**, 101 (1974).

131 A. G. McInnes, J. A. Walter, J. L. C. Wright, and L. C. Vining, *Topics in ^{13}C Nuclear Magnetic Resonance* (Ed. G. C. Levy), Vol. 2, Chap. 3, Wiley-Interscience, New York, 1976.

132 E. Leete, *Rev. Latinoamer. Quim.* **11**, 8 (1980).

133 See, for example, C. R. Hutchinson, A. H. Heckendorf, P. E. Daddona, E. Hagaman, and E. Wenkert, *J. Am. Chem. Soc.* **97**, 1988 (1975).

134 H. Seto, W. Cary, and M. Tanabe, *J. C. S. Chem. Commun.* 867 (1973); H. Seto, T. Sato, and H. Yonehara, *J. Am. Chem. Soc.* **95**, 8461 (1973).

135 See, for example, H. C. Dorn and G. E. Maciel, *J. Phys. Chem.* **76**, 2972 (1972); A. G. McInnes, D. G. Smith, J. A. Walter, L. C. Vining, and J. L. C. Wright, *J. C. S. Chem. Commun.* 66 (1975).

136 D. Gagnaire, H. Reutenauer, and F. Taravel, *Org. Magn. Res.* **12**, 679 (1979).

137 D. Gagnaire and F. R. Taravel, *J. Am. Chem. Soc.* **101**, 1625 (1979).

138 R. J. Abraham, J. M. Bovill, D. J. Chadwick, L. Griffith, and F. Sancassan, *Tetrahedron* **36**, 279 (1980).

139 K. L. Servis and F. F. Shue, *J. Magn. Res.* **40**, 293 (1980).

140 J. Mossoyan, M. Asso and D. Benlian, *Org. Magn. Res.* **13**, 287 (1980).

141 G. N. LaMar, W. DeW. Horrocks, and R. H. Holm, *NMR of Paramagnetic Molecules: Principles and Applications*, Chap. 12, Academic Press, New York, 1973.

142 H. M. McConnell and R. E. Robertson, *J. Chem. Phys.* **29**, 1361 (1958).

143 B. L. Shapiro, J. R. Hlubucek, G. R. Sullivan, and L. F. Johnson, *J. Am. Chem. Soc.* **93**, 3281 (1971).

144 G. E. Hawkes, D. Leibfritz, D. W. Roberts, and J. D. Roberts, *J. Am. Chem. Soc.* **95**, 1659 (1973).

145 J. Reuben, *J. Am. Chem. Soc.* **98**, 3726 (1976).

146 R. H. Newman, *Tetrahedron* **30**, 969 (1974).

147 M. L. Martin, J. J. Delpuech, and G. J. Martin, *Practical NMR Spectroscopy*, Chap. 10, Heyden, London, 1980.

148 D. J. Chadwick and D. W. Williams, *J. C. S. Perkin II*, 1202 (1974).

149 O. A. Gansow, M. R. Willcott, and R. E. Lenkinski, *J. Am. Chem. Soc.* **93**, 4295 (1971).

150 E. Lippmaa, T. Pehk, J. Paassivirta, N. Belikova, and A. Plate, *Org. Magn. Res.* **2**, 1581 (1970).

151 F. Bohlmann, R. Zeisberg, and E. Klein, *Org. Magn. Res.* **7**, 426 (1975).

152 See, for example, C. M. Dobson and B. A. Levine, *New Techniques in Biophysics and Cell Biology*, Vol. 3, Chap. 2, Wiley-Interscience, New York, 1976.

153 G. E. Hawkes, C. Marzin, S. R. Johns, and J. D. Roberts, *J. Am. Chem. Soc.* **95**, 1661 (1973).

154 O. A. Gansow, P. A. Loeffler, R. E. Davis, M. R. Willcott, and R. E. Lenkinski, *J. Am. Chem. Soc.* **95**, 3390 (1973).

155 A. K. Bose and P. R. Srinivasan, *J. Magn. Res.* **15**, 592 (1974).
156 J. F. Desreux and C. N. Reilley, *J. Am. Chem. Soc.* **98**, 2105 (1976).
157 See references cited in Ref. 153.
158 C. N. Reilly, B. W. Good, and J. F. Desreux, *Anal. Chem.* **47**, 2110 (1975).
159 R. M. Golding and M. P. Halton, *Aust. J. Chem.* **25**, 2577 (1972).
160 D. M. Grant and E. G. Paul, *J. Am. Chem. Soc.* **86**, 2984 (1964).
161 L. P. Lindeman and J. Q. Adams, *Anal. Chem.* **43**, 1245 (1971).
162 See, for example, J. B. Stothers, *Carbon-13 Nuclear Magnetic Resonance Spectroscopy*, Academic Press, New York, 1972.
163 See, for example, F. W. Wehrli and T. Nishida, *Progress in the Chemistry of Organic Natural Products* (Eds W. Herz, H. Grisebach and G. W. Kirby), Vol. 36, Springer, Vienna, 1979.
164 H. Eggert and C. Djerassi, *J. Org. Chem.* **38**, 3788 (1973).
165 E. Wenkert, B. L. Buckwalter, I. R. Burfitt, M. J. Gasić, H. E. Gottlieb, E. W. Hagaman, F. M. Schell and P. M. Wovkulich, *Topics in* ^{13}C *Nuclear Magnetic Resonance* (Ed. G. C. Levy), Vol. 2, Chap. 2, Wiley-Interscience, New York, 1976; T. Nishida, I. Wahlberg and C. R. Enzell, *Org. Magn. Res.* **9**, 203 (1977).
166 D. K. Dalling, D. M. Grant, and E. G. Paul, *J. Am. Chem. Soc.* **95**, 3718 (1973).
167 P. A. Christensen, B. J. Willis, F. W. Wehrli, and S. Wehrli, *J. Org. Chem.* **47**, (1982).
168 H. Batiz-Hernandez and R. A. Bernheim, *Progress in Nuclear Magnetic Resonance Spectroscopy* (Eds J. Emsley, J. Feeney and L. H. Sutcliffe), Vol. 3, Pergamon Press, Oxford, 1967.
169 O. Lutz, A. Nolle and P. Kroneck, *Z. Phys.* **A282**, 157 (1977).
170 R. A. Bell, C. A. Chan, and B. G. Sayer, *Chem. Commun.* 67 (1972).
171 D. Doddrell and I. Burfitt, *Aust. J. Chem.* **25**, 2239 (1972).
172 J. B. Stothers, C. T. Tan, A. Nickon, F. Huang, R. Sridhar, and R. Weglein, *J. Am. Chem. Soc.* **94**, 8581 (1972).
173 G. L. Lebel, J. D. Laposa, G. G. Sayer, and R. A. Bell, *Anal. Chem.* **43**, 1500 (1971).
174 S. Milosavljević, D. Jeremić, M. Mihailović, and F. W. Wehrli, *Org. Magn. Res.* **13**, 299 (1981).
175 G. E. Maciel, P. D. Ellis, and D. C. Hofer, *J. Phys. Chem.* **71**, 2160 (1967).
176 D. G. Morris and A. M. Murray, *J. C. S. Perkin II*, 1579 (1976).
177 D. H. O'Brien and R. D. Stipanovic, *J. Org. Chem.* **43**, 1105 (1978).
178 F. W. Wehrli, D. Jeremić, M. L. Mihailović, and S. Milosavljević, *J.C.S. Chem. Commun.* 302 (1978).
179 R. A. Newmark and J. R. Hill, *Org. Magn. Res.* **13**, 40 (1980); P. E. Hansen & J. J. Led, *Org. Magn. Res.* **15**, 288 (1981).
180 D. A. Forsyth, P. Incas, and R. M. Burk, *J. Am. Chem. Soc.* **104**, 240 (1982).
181 L. Ernst, S. Eltamany, and H. Hopf, *Org. Magn. Res.* **104**, 299 (1982).
182 K. L. Servis and F. F. Shue, *Org. Magn. Res.* **102**, 7233 (1980).
183 For an excellent description of the equilibrium isotope effect see, for example, M. Saunders, *Stereodynamics of Molecular Systems* (Ed. R. H. Sarma), p. 171–84, Pergamon Press, New York, 1979.
184 M. Saunders, L. Telkowski, and M. R. Kates, *J. Am. Chem. Soc.* **99**, 8070 (1977).
185 K. W. Baldry and M. J. T. Robinson, *J. Chem. Res.* (S) 86 (1977).
186 F. A. L. Anet, V. J. Basus, A. P. W. Hewett, and M. Saunders, *J. Am. Chem. Soc.* **102**, 3945 (1980).
187 H. Booth and J. Ramsey Everett, *Can. J. Chem.* **58**, 2714 (1980).
188 D. Gagnaire, R. Nardin, and D. Robert, *Org. Magn. Res.* **12**, 405 (1980).
189 I. Kurobane, L. C. Vinings, and A. G. McInnes, *Tetrahedron Lett.* 4633 (1978).

190 F. W. Wehrli and T. Nishida, 'The use of carbon-13 nuclear magnetic resonance spectroscopy in natural products chemistry.' In *Progress in the Chemistry of Organic Natural Product's* Vol. 36 (Eds W. Herz, H. Grisebach and G. W. Kirby), Springer, New York, 1979.

191 F. A. L. Anet and A. H. Dekmezian, *J. Am. Chem. Soc.* **101**, 5449 (1979).

192 J. M. Risley and R. L. VanEtten, *J. Am. Chem. Soc.* **101**, 252 (1979).

193 J. C. Vederas, *J. Am. Chem. Soc.* **102**, 374 (1980).

194 J. M. Risley and R. L. VanEtten, *J. Am. Chem. Soc.* **102**, 4609 (1980).

195 J. M. Risley and R. L. VanEtten, *J. Am. Chem. Soc.* **102**, 6699 (1980).

196 J. Diakur, T. T. Nakashima, and J. C. Vederas, *Can. J. Chem.* **58**, 1311 (1980).

197 R. N. Moore, J. Diakur, T. T. Nakashima, S. L. McLaren and J. C. Vederas, *J.C.S., Chem. Commun.* 501 (1981).

198 J. C. Vederas and T. T. Nakashima, *J.C.S., Chem. Commun.* 183 (1980).

199 (a) M. Mehring, *Principles of High Resolution NMR in Solids*, Springer-Verlag, Berlin, 1983; (b) J. R. Lyerla in *Contemporary Topics in Polymer Science* (Ed. M. Shen), Vol. 3, Plenum, New York, 1979, pp. 143–213; (c) C. S. Yannoni, *Acc. Chem. Res.* **15**, 201 (1982); (d) D. E. Axelson, *Solid State Nuclear Magnetic Resonance of Fossil Fuels: An Experimental Approach*, Multiscience Publications, Montreal, Canada, 1985; (e) T. Terao and F. Imashiro in *Applications of NMR Spectroscopy to Problems in Stereochemistry and Conformational Analysis* (Eds Y. Takeochi and A. P. Marchand), VCH Publishers, Deerfield Beach, Florida 1986, pp. 125–154.

200 J. S. Waugh, *Ann. N. Y. Acad. Sci.* **70**, 900 (1958).

201 (a) A. G. Pines, M. G. Gibby, and J. S. Waugh, *J. Chem. Phys.* **56**, 1776 (1972); (b) A. G. Pines, M. G. Gibby and J. S. Waugh, *J. Chem. Phys.* **57**, 569 (1973).

202 M. Linder, A. Hoehener, and R. R. Ernst, *J. Magn. Res.* **35**, 379 (1979).

203 (a) S. Pausak, J. Tegenfeldt, and J. S. Waugh, *J. Chem. Phys.* **61**, 1338 (1974); (b) J. van Dongen Torman and W. S. Veeman, *J. Chem. Phys.* **68**, 3233 (1978).

204 J. Schaefer and E. O. Stejskal in *Topics in Carbon-13 NMR Spectroscopy* (Ed. G. C. Levy), Vol. 3, Wiley-Interscience, New York, 1979, pp. 283ff.

205 S. R. Hartmann and E. L. Hahn, *Phys. Rev.* **128**, 2042 (1962).

206 For reviews of ^{13}C CP/MAS theory and applications, see: (a) R. E. Wasylishen and C. A. Fyfe, *Ann. Rep. NMR Spectr.* **12**, 1 (1982); (b) J. R. Lyerla, C. S. Yannoni, and C. A. Fyfe, *Acc. Chem. Res.* **15**, 208 (1982).

207 (a) C. S. Yannoni and R. D. Kendrick, *J. Chem. Phys.* **74**, 747 (1981); (b) D. Horne, R. D. Kendrick, and C. S. Yannoni, *J. Magn. Res.* **52**, 299 (1983); (c) T. C. Clarke, C. S. Yannoni, and T. J. Katz, *J. Am. Chem. Soc.* **105**, 7787 (1983).

208 T. J. Katz, S. M. Hacker, R. D. Kendrick, and C. S. Yannoni, *J. Am. Chem. Soc.* **107**, 2182 (1985).

209 See footnote 11 in Reference 207.

4

Nuclear Spin Relaxation

In this chapter we attempt to acquaint the reader with a further important aspect of ^{13}C n.m.r., nuclear spin relaxation. While high-resolution n.m.r., as characterized by nuclear shielding and spin–spin coupling constants, has rapidly attracted the interest of chemists, because of its obvious links to chemical structure, relaxation has only recently more broadly been utilized. The reason for this late discovery of relaxation as a complementary aid to the solution of chemical problems is the difficulty measuring relaxation rates in multi-line spectra. Moreover, relaxation data are less straightforwardly analysable than shielding and coupling parameters. In fact, the correct interpretation of the phenomenon requires some knowledge of the dynamics of the molecular processes that underlie relaxation. However, contrary to relaxation of other nuclei such as protons, the physics of ^{13}C relaxation is relatively simple. This is related to the fact that (1) ^{13}C is a rare spin and (2) the measurements are carried out under simultaneous proton noise decoupling. Along with the wide chemical shift range of ^{13}C nuclei, this offers a number of advantages over T_1 measurements on protons; for example:

(1) The relaxation behaviour is not complicated by interactions between like spins.
(2) ^{13}C signals appear as single lines and the relaxation of individual nuclei can usually be approximated by a single exponential time constant.[1]
(3) Since carbon atoms are located at the interior (rather than at the periphery) of a molecule, intermolecular interactions, which are known to play an important role in proton relaxation, are usually insignificant in ^{13}C relaxation.

Because of the limited scope of this book, most of the discussion will be centred around dipolar relaxation, which is the dominant mechanism in diamagnetic molecules of chemical interest. This will be followed by a short outline of other mechanisms, notably electron–nuclear relaxation. Other topics touched upon include anisotropic rotational diffusion and internal motion and a brief discussion of the experiment. For a more profound treatment of the physicochemical aspects of ^{13}C relaxation the reader is referred to the appropriate reviews.[2-4]

Since the measurement of spin–spin relaxation rates puts more severe demands on a spectrometer's performance,[5] most relaxation data appearing in the literature are those derived from spin–lattice relaxation measurements.[6] Moreover, it can be shown that in the absence of chemical exchange, in most organic molecules of practical interest, spin–lattice and spin–spin relaxation rates are equal. For these reasons, the present chapter will almost exclusively be devoted to T_1 relaxation.

4.1 CORRELATION AND SPECTRAL DENSITY FUNCTION

As outlined in Chapter 1, spin–lattice relaxation involves energy exchange between the spin system and its environment, the lattice. Eventually, a nuclear spin in its excited state will return to the ground state by dissipating its excess energy to the lattice. Since spontaneous emission is insignificant in nuclear magnetic resonance, such transitions occur through interaction with the lattice. This requires the presence of an oscillating magnetic field at the site of the nucleus. A further requirement is that the frequency of the local magnetic field, which we denote H_{loc}, be equal to the Larmor frequency of the nucleus in question. A transition is induced only if these conditions are fulfilled. We may now inquire about the nature of these fields which obviously must have their origin in lattice fluctuations caused by Brownian motion. In order to gain a better understanding of the processes leading to relaxation, we must learn more about the microdynamic behaviour of molecules in the liquid phase. For this purpose let us consider an ensemble of moving particles, each rotating at a specific angular frequency ω_i. The particles, which are represented by vectors (A, B, C, D, ...), are initially assumed to be in phase relative to one another. Then after a time τ_1, which is short compared with $2\pi/\omega_i$, they have rotated through angles $\omega_A \tau_1$, $\omega_B \tau_1$, $\omega_C \tau_1$, ..., but the relative phases are still very similar.

At a time $\tau_2 > \tau_1$ the phase relationship is less obvious and, as time proceeds, phase correlation progressively decreases. Mathematically, such a process is described by the correlation function. The autocorrelation function of H_{loc} is given by[7]

$$G(\tau) = \langle H_{loc}(0) \cdot H_{loc}(\tau) \rangle_0 \qquad (4.1)$$

The bracket in Eqn. (4.1) indicates an ensemble average. $G(\tau)$ could be defined as describing the persistence of motional order of an ensemble of particles. From the above it is further recognized that $G(\tau)$ decays to zero. Physically more relevant than $G(\tau)$ is its Fourier transform, the power spectral density $J(\omega)$ which expresses the frequency distribution of H_{loc}. $J(\omega)$ therefore provides the amplitude of the local magnetic field at the Larmor frequency of the relaxing nucleus. The relationship between the correlation function $G(\tau)$ and the spectral density

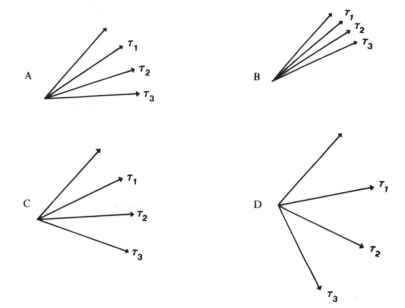

function $J(\omega)$ can thus be formulated as

$$J(\omega) = \int_{-\infty}^{\infty} G(\tau)\, e^{i\omega\tau}\, d\tau \qquad (4.2)$$

Assuming an exponential decay of the motional order, $G(\tau)$ can be expressed as

$$G(\tau) = \langle H_{loc}^2(0)\rangle e^{-\tau/\tau_c} \qquad (4.3)$$

In Eqn (4.3), τ_c designates the rotational correlation time, the time constant for the loss of phase coherence of a system of rotating particles.* Substitution of Eqn. (4.3) into Eqn. (4.2) yields[7,8]

$$J(\omega) = \langle H_{loc}^2(0)\rangle_0 \int_{-\infty}^{\infty} e^{(-1/\tau_c + i\omega)\tau}\, d\tau = \langle H_{loc}^2(0)\rangle_0 \frac{2\tau_c}{1 + \omega^2\tau_c^2} \qquad (4.4)$$

It is recognized from Eqn. (4.4) that the spectral density function is Lorentzian. There is thus a close analogy to the free induction decay of a single line and its frequency spectrum, except for the very different time scale (rotational correlation times in non-viscous liquids are typically of the order of 10^{-12}–10^{-3} s.

Figure 4.1 shows $J(\omega)$ plotted against ω for fast $(1/\tau_c \gg \omega_0)$, intermediate $(1/\tau_c \sim \omega_0)$ and slow motion $(1/\tau_c \ll \omega_0)$, where ω_0 denotes the Larmor resonance frequency. Since $J(\omega)$ expresses the power of the local fluctuating field, it can be

*Sometimes τ_c is defined as the average time required for a molecule to rotate through an angle of one radian.

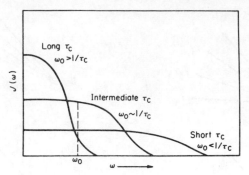

Fig. 4.1 Spectral density $J(\omega)$ as a function of the frequency of molecular reorientation for different values of the correlation time τ_c in the case of rotational correlation.

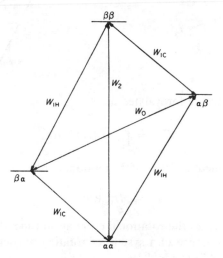

Fig. 4.2 Energy-level diagram for a two-spin system ^{13}C—^{1}H. The subscript i of the transition probabilities W_i refers to the change in total magnetic spin quantum number M.

inferred that relaxation should be most efficient in the case where the frequency of the molecular motions is comparable to ω_0, whereas it is less efficient for the two extrema. Hence relaxation rates governed by rotational correlation (dipole–dipole, chemical shift anisotropy) are expected to pass through a maximum. Within the motional narrowing limit, i.e. for the condition[8] $\omega_0\tau_c \ll 1$ which applies to all small (fast-moving) molecules, $J(\omega)$ becomes frequency independent over a wide range of resonance frequencies as is readily seen from Eqn. (4.4) and Fig. 4.1.

4.2 DIPOLE–DIPOLE RELAXATION

After we have learned about the frequency characteristics of the local magnetic fields inducing transitions in a spin–lattice relaxation process we may now further inquire about the origin of H_{loc}. Most frequently, these fields are generated by magnetic dipoles of nearby magnetic nuclei such as a proton attached to the carbon in question. The field produced at the site of a nucleus j by a nucleus i is given by[2]

$$H_{loc} = \pm \mu_i \frac{3\cos^2\theta_{ij} - 1}{r_{ij}^3} \qquad (4.5)$$

In Eqn. (4.5), μ_i represents the magnetic dipole moment of nucleus i, r_{ij} is the internuclear distance, and θ_{ij} is the angle between r_{ij} and the external field H_0. From the inverse third-power relationship of r_{ij} it is seen that H_{loc} rapidly decreases with increasing separation of the two interacting nuclei.

Let us now derive the general equation for dipolar relaxation in a simple two-spin system defined by two nuclei ^{13}C and ^1H. The energy-level diagram for such a spin system is given in Fig. 4.2.[1] The transition probabilities W_{1H} and W_{1C} refer to transitions of a proton and ^{13}C spin, respectively. In addition, there exists a finite probability for a simultaneous flip of carbon and proton spins. The latter can occur in two different ways depending on the change in the total magnetic quantum number $M = m_C + m_H$, where m_C and m_H represent the individual magnetic quantum numbers for ^{13}C and ^1H, which, of course, can assume values of $+\frac{1}{2}$ or $-\frac{1}{2}$. Accordingly, we denote a $\alpha\alpha \to \beta\beta(+\frac{1}{2}+\frac{1}{2} \to -\frac{1}{2}-\frac{1}{2})$ transition a double-quantum transition, involving two photons of energy and corresponding to $|\Delta M| = 2$. By contrast, a transition $\beta\alpha \to \alpha\beta(-\frac{1}{2}+\frac{1}{2} \to +\frac{1}{2}-\frac{1}{2})$ does not entail any net energy change and is therefore designated zero quantum transition. The probabilities for the two types of transitions are W_2 and W_0, respectively.

The quantum-mechanical treatment of H_{loc} shows that a close relationship exists between the spectral density function $J(\omega_{ij})$ and the transition probability W_{ij}, where i and j denote the energy levels between which the transition ensues:[1]

$$W_{ij} = F_{ij}^2 J_m(\omega_{ij}) \qquad (4.6)$$

Equation (4.6) implies that the transition probabilities are proportional to the spectral density at the frequency ω_{ij}. From a calculation of the matrix elements F_{ij} accounting for the dipolar interaction involving spin states i and j, one obtains for the four transition probabilities:[1]

$$W_0 = (\tfrac{1}{20})K^2 J_0(\omega_H - \omega_C) = (\tfrac{1}{10})K^2 \frac{\tau_c}{1 + (\omega_H - \omega_C)^2 \tau_c^2} \qquad (4.7)$$

$$W_{1C} = (\tfrac{3}{40})K^2 J_1(\omega_C) = (\tfrac{3}{20})K^2 \frac{\tau_c}{1 + \omega_C^2 \tau_c^2} \qquad (4.8)$$

$$W_{1H} = (\tfrac{3}{40})K^2 J_1(\omega_H) = (\tfrac{3}{20})K^2 \frac{\tau_c}{1 + \omega_H^2 \tau_c^2} \qquad (4.9)$$

$$W_2 = (\tfrac{3}{10})K^2 J_2(\omega_H + \omega_C) = (\tfrac{3}{5})K^2 \frac{\tau_c}{1 + (\omega_H + \omega_C)^2 \tau_c^2} \qquad (4.10)$$

The subscript in J indicates the change in the total magnetic quantum number upon transition, and $K \equiv \gamma_H \gamma_C \hbar r_{CH}^{-3}$.

It can be shown that in the absence of an irradiating field at the resonance frequency of the proton, the rate of change of the longitudinal magnetization, $M_z(t)$, is governed by two different time constants:[2,3]

$$T_{1C} = (W_0 + 2W_{1C} + W_2)^{-1} \qquad (4.11)$$

and

$$T_{1CH} = (W_2 - W_0)^{-1} \qquad (4.12)$$

Relaxation becomes independent of the coupling term T_{1CH} only at saturation of the proton transitions. It can now be described by a single exponential time constant $T_1^{DD} = T_{1C}$. The theoretical prediction of different recovery rates in the proton-coupled and decoupled case has been experimentally corroborated.[9] The rate for dipolar relaxation, $1/T_1^{DD}$ (defined as the inverse of the relaxation time T_1^{DD}), of a ^{13}C nucleus by a single proton can now be derived by combining Eqns (4.7)–(4.10) with Eqn. (4.11):

$$1/T_1^{DD} = \frac{\hbar^2 \gamma_H^2 \gamma_C^2}{20 r_{CH}^6} \{ J_0(\omega_H - \omega_C) + 3J_1(\omega_C) + 6J_2(\omega_H + \omega_C) \} \qquad (4.13)$$

Figure 4.3 shows a doubly logarithmic plot of T_1 as a function of τ_C, computed from Eqn. (4.13) for five different values of the magnetic field strength (1.4, 2.3, 4.7,

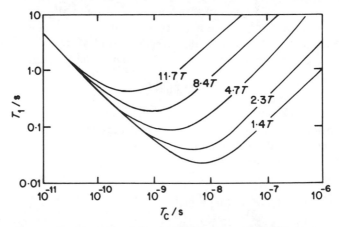

Fig. 4.3 Log–log plot of the dipolar relaxation time, T_1^{DD}, versus rotational correlation time, τ_c.[10] Reproduced by permission of John Wiley & Sons Inc.

8.4, and 11.7 T).[10] This illustrates that the dipolar relaxation time is frequency-dependent in the general case, with minima near the condition $(\omega_H + \omega_C)\tau_c = 1$. The frequency dependence vanishes if the extreme narrowing limit conditions apply, i.e. for $(\omega_H + \omega_C)\tau_c \ll 1$, so that the denominator in the spectral density terms (Eqns (4.7)–(4.10)) becomes unity, and Eqn. (4.13) simplifies to

$$1/T_1^{DD} = \hbar^2 \gamma_C^2 \gamma_H^2 r_{CH}^{-6} \tau_c \qquad (4.14)$$

or in the more general case of dipolar interaction with several protons i:

$$1/T_1^{DD} = \hbar^2 \gamma_C^2 \gamma_H^2 \sum_i r_{CH_i}^{-6} \tau_c \qquad (4.15)$$

For N directly bonded hydrogens, assuming equal C—H bond distances and neglecting non-bonded interactions, Eqn. (4.15) can be reformulated as

$$1/T_1^{DD} = N\hbar^2 \gamma_C^2 \gamma_H^2 r_{CH}^{-6} \tau_c \qquad (4.16)$$

In order to compare relaxation rates of proton-bearing carbons within the same molecule it is practical to divide the observed relaxation rates by N.

Because of the inverse sixth-power relationship between r_{CH} and the dipolar relaxation rate, it is readily recognized that proton-bearing carbons are bound to relax much more efficiently than their quaternary analogues. Quaternary carbon relaxation rates are a function of the number and distance of geminal and vicinal protons. Typically, such carbons relax one to two orders of magnitude more slowly than protonated ones.

Although we have so far ignored T_2, we will, for the sake of completeness, also give the corresponding expression for the dipolar spin–spin relaxation time. In contrast to T_1^{DD}, T_2^{DD} is also dependent upon the spectral densities at the proton frequency, $J_1(\omega_H)$, and at zero frequency, $J(0)$:

$$1/T_2^{DD} = \frac{\hbar^2 \gamma_H^2 \gamma_C^2}{20 r_{CH}^6} \{2J(0) + \tfrac{1}{2}J_0(\omega_H - \omega_C) + \tfrac{3}{2}J_1(\omega_C) + 3J_1(\omega_H) + 3J_2(\omega_H + \omega_C)\} \qquad (4.17)$$

It is readily verified from Eqn. (4.17) that, in the motional narrowing limit, the expression in the bracket becomes equal to $20\tau_c$. Hence under these conditions $T_2^{DD} = T_1^{DD}$ holds.

It has been shown[13] more recently that the simple model predicting exponential recovery with a single time constant as given by Eqns (4.13)–(4.16) is not entirely correct if several protons contribute to relaxation of a carbon spin. For the latter more general case a cross-relaxation term has to be taken into account. The complication arises because the two (methylene carbon) or three (methyl carbon) C—H bond vectors have different orientations at a given point in time. It turns out, however, that for most practical purposes cross-correlation effects are negligible and relaxation is exponential with the limits of experimental error.

It should be further pointed out that the simple description of dipolar relaxation (Eqns (4.13)–(4.17)), in terms of a single correlation time τ_c for all carbons of the molecule, is only valid on condition that the molecule rotates isotropically, i.e. if there are no preferred axes for reorientation.[12] A molecule which meets these criteria is adamantane (1).[1] Because of its tetrahedral

$T_1(CH) = 20.5\,s$

$T_1(CH_2) = 11.4\,s$

1

symmetry, adamantane is an isotropic tumbler and therefore the correlation time is the same for each C—H bond vector of the two non-equivalent carbons. Taking only directly bonded protons into consideration and assuming dipolar relaxation as the sole significant relaxation mechanism, then Eqn (4.16) is applicable, which predicts a 2:1 ratio for the methine and methylene relaxation times, respectively. Considering also next-nearest protons, this ratio becomes 1.82:1. This reduction is a consequence of the unequal number of geminal hydrogens for the methine and methylene carbons (six versus two). The ratio $T_1(CH)/T_1(CH_2)$ of the experimentally observed relaxation times is 1.80, in excellent agreement with theory.

If the dipolar coupling between ^{13}C and ^1H is dominant, irradiation of protons manifests itself in non-Boltzmann populations. Saturation of the proton transitions in Fig. 4.2 results in a population equalization of the $\alpha\alpha$ and $\alpha\beta$, and $\beta\alpha$ and $\beta\beta$ levels. However, the steady-state populations depend on the various transition probabilities W_{ij}. For a predominance of cross relaxation, i.e. for the condition $W_2 > W_0, W_{1C}$, it is readily seen that the net population difference of the ^{13}C levels increases and hence a signal enhancement results. It has been shown[14] that the nuclear Overhauser enhancement factor (NOEF), η, defined as the change in total ^{13}C intensity upon irradiation of the proton resonances, is given by

$$\eta = (\gamma_H/\gamma_C)\frac{W_2 - W_0}{W_0 + 2W_{1C} + W_2} \qquad (4.18)$$

We have previously seen that, in general, the transition probabilities are frequency-dependent through the spectral density terms (Eqns (4.7–4.10)). Hence the NOE is also a frequency-dependent quantity. In Fig. 4.4 plots of η against the rotational correlation time τ_c are given for five different field strengths. Note that, like the dipolar relaxation rate, η becomes frequency-invariant in the motional narrowing limit. From the spectral density terms given by Eqns (4.7)–(4.10), one obtains $W_0:W_1:W_2 = \frac{1}{6}:\frac{1}{4}:1$ for the relative transition probabilities. Inserting

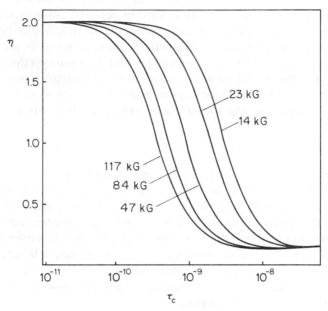

Fig. 4.4   Log plot of the nuclear Overhauser enhancement factor, η, versus the rotational correlation time, τ_c, for various field strengths.

these into Eqn (4.18) yields

$$\eta_0 = 0.5(\gamma_H/\gamma_C) = 1.988 \tag{4.19}$$

It is further noted from Eqn (4.18) that the NOE is independent of the number and distance of the protons which induce relaxation, because η is given by a ratio of spectral density terms. This, of course, is only valid as long as dipolar relaxation dominates.

The first experimental evidence of the existence of a heteronuclear Overhauser effect of the kind described had been provided with a sample of ^{13}C-enriched formic acid,[15] showing the full 200% enhancement predicted by Eqn (4.19). We will see later, however, that carbons in small molecules are not always relaxed by the intramolecular dipolar mechanism.

4.3   OTHER RELAXATION MECHANISMS

Spin rotation (SR)

In addition to dipolar relaxation, where H_{loc} has its origin in the magnetic dipole of neighbouring nuclei with $I \neq 0$, the molecule itself may have a magnetic moment originating from a modulation of the magnitude and direction of the angular momentum vector associated with the rotation of the molecule.[8] This may lead to molecular magnetism, independent of the symmetry of the electronic charge distribution. The induced magnetic moment increases as the molecular motion becomes faster. Hence it is anticipated that relaxation via this mechanism becomes more efficient for small, rapidly tumbling molecules and at elevated temperatures, in particular in the vapour phase. For a carbon in a cylindrically symmetric environment, the spin–rotation (SR) relaxation rate $1/T_1^{SR}$ is given by[16]

$$1/T_1^{SR} = (2kT/3\hbar^2)I_m(C_\parallel^2 + 2C_\perp^2)\tau_{SR} \qquad (4.20)$$

In Eqn (4.20), I_m designates the moment of inertia relative to axis m; $C_\parallel$ and $C_\perp$ are the spin rotation constants parallel and perpendicular to the symmetry axis, respectively, and τ_{SR} is the angular momentum correlation time. The remaining symbols have their usual meaning. τ_{SR} is a measure for the persistance of a given angular momentum. Spin rotation is best identifed as a dominating mechanism through the temperature dependence of relaxation rates. Whereas dipolar relaxation rates decrease with increasing temperature due to the concomitant shortening of τ_c, under motional narrowing conditions, the opposite behaviour is observed in the case of spin rotation.

For the condition $\tau_{SR} \ll \tau_c$, the two correlation times can be converted into one another through the Hubbard relation[17]

$$\tau_c\tau_{SR} = I_m/(6kT) \qquad (4.21)$$

from which it follows that τ_c and τ_{SR} are inversely proportional.

While exclusive relaxation by spin rotation is a rare occurrence, probably confined to small molecules where dipolar relaxation is precluded due to absence of protons, SR relaxation is often found in those instances where fast internal motion lowers the efficiency of the dipolar mechanism. Prime candidates for partial SR relaxation are methyl groups in small to intermediate-sized molecules. In cycloalkanes C_nH_{2n}, for example,[18] the percentage spin–rotation relaxation decreases from 50% $(n = 3)$ to 0% $(n = 6)$.

Due to the decreased efficiency of DD relaxation for quaternary carbons, the latter are often partially relaxed by spin–rotation, at least at moderate field strengths and sp^3 carbon hybridization.

Chemical shift anisotropy (CSA)

Basically, the chemical shift is an anisotropic quantity. If, for example, a C—X bond is placed with its axis parallel to the external field, the carbon shielding will be different from that experienced when the bond is perpendicular to the field. In liquids Brownian motion will, of course, lead to averaging over all molecular orientations so that only a single chemical shift is observed. However, the nucleus still experiences a fluctuating magnetic, field brought about by modulation of the chemical shift tensor by Brownian motion. If the local field thus produced has

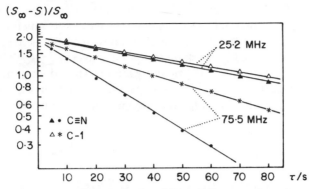

Fig. 4.5   Semilogarithmic plot of $(S_\infty - S)/S_\infty$ versus inversion delay τ for carbons C-1 and CN in benzonitrile, derived from inversion–recovery spin–lattice relaxation experiments performed at 25.2 and 75.5 MHz.[20] S and S_∞ denote signal intensities at $t = \tau$ and $t \gg T_1$ (see p. 230 for exact definitions).

frequency components near the Larmor frequency, relaxation can ensue. As for spin rotation, we will formulate the chemical shift anisotropy relaxation rate for an axially symmetric system:[8]

$$1/T_1^{CSA} = (\tfrac{2}{15})\gamma_C^2 H_0^2 (\sigma_\parallel - \sigma_\perp)^2 \tau_c \tag{4.22}$$

In Eqn (4.22), $\sigma_\parallel$ and $\sigma_\perp$ are the components of the chemical shift tensor parallel and perpendicular to the external magnetic field axis, respectively. A notable property of the chemical shift anisotropy relaxation rate is its dependence upon the square of the external field.

CSA relaxation is thus more likely to prevail at high magnetic field strength. Prime candidates are quaternary sp² and sp-hybridized carbons,[19] as exemplified by the C-1 and CN carbon relaxation behaviour in benzonitrile.[20] Figure 4.5 shows the ^{13}C signal recovery at 2.3 and 7.0 T field strength for the two carbons in question, indicating much faster relaxation at the higher field. The data demonstrate that at 7.0 T, ca 80% of the relaxation rate for the cyano carbon is CSA-induced.

A field dependence characteristic of CSA relaxation has also been reported for quaternary carbons in acetylenes and diacetylenes.[21] Moreover, CSA appears to be the dominating relaxation mechanism of quaternary sp² carbons in general at magnetic field strengths above 4.7 T.[19]

Scalar relaxation (SC)[2,3]

Scalar coupling of ^{13}C to another magnetic nucleus of spin S may give rise to a fluctuating field at the site of the carbon nucleus, provided the coupling spin undergoes rapid change of its spin state, or, in other words, if it relaxes rapidly. On condition that $1/T_{1S} \gg 2\pi J$, i.e. if the relaxation rate of nucleous S is fast compared to $2\pi J$, where J stands for the scalar spin–spin coupling constant, no spin coupling will be observed. However, the ^{13}C nucleus to which spin S is spin-coupled experiences a local field that fluctuates at a frequency $1/T_{1S}$. This condition is often satisfied for such quadrupolar nuclei as ^{14}N, ^{35}Cl, ^{37}Cl, ^{79}Br, ^{81}Br, ^{127}I, etc. Scalar relaxation is frequently found to be a dominant mechanism for transverse relaxation, to the extent where it manifests itself in appreciable line broadening. This scalar process for the 'second kind' may be formulated as

$$1/T_1^{SC} = \frac{8\pi^2 J^2}{3} S(S+1) \frac{\tau_{SC}}{1+(\omega_S - \omega_C)^2 \tau_{SC}^2} \qquad (4.23)$$

and

$$1/T_2^{SC} = 1/(2T_1^{SC}) + \frac{4\pi^2 J^2}{3} S(S+1)\tau_{SC} \qquad (4.24)$$

The symbols τ_{SC} and ω_S in Eqns (4.23) and (4.24) designate the relaxation time of the interacting nucleus S and its Larmor frequency, respectively. It is obvious that the conditions for scalar spin–lattice relaxation of ^{13}C to become efficient are rather narrow. First, we note that the mechanism is field-dependent through the Larmor frequencies of the two interacting spins, whose difference occurs in the denominator of the equation. Normally, $(\omega_S - \omega_C)^2 \tau_{SC}^2 \gg 1$, so that $\tau_{SC}/[1+(\omega_S - \omega_C)^2 \tau_{SC}^2]$ becomes very small. Unless this is offset by a very large spin–spin coupling constant, scalar spin–lattice relaxation is negligible. If, on the other hand, $(\omega_S - \omega_C)^2 \tau_{SC}^2 \ll 1$, due to a very short relaxation time of the coupling nucleus S, the quotient $\tau_{SC}/[1+(\omega_S - \omega_C)^2 \tau_{SC}^2]$ is still very small and the mechanism is of no importance unless it is balanced by an exceptionally large coupling constant. Although this situation does not appear to have been investigated, a likely candidate is ^{127}I, which is known to give rise to very large coupling constants.[22]

A common case of scalar ^{13}C spin–lattice relaxation, however, is that involving ^{79}Br,[23-5] whose Larmor frequency at 2.3 T field strength differs from the ^{13}C resonance frequency by only 100 kHz. In p-bromobenzonitrile, for example,[23] T_1 of the quaternary carbon 1 was found to be 47 s, whereas C-4 (adjacent to bromine) exhibited a value of 8.8 s, close to that of the protonated carbons, for which 6 s had been determined. Since bromine–^{13}C dipolar interaction could be ruled out as a substantial contributor, spin–lattice relaxation for this nucleus was mainly attributed to the scalar mechanism. A complicating factor is the circumstance that bromine is composed of ^{79}Br and ^{81}Br isotopes. The measurements therefore yield only a mean value for the two isotopic species. This is also the reason for the non-exponential behaviour, which has been observed in such situations.[25] From an analysis of the recovery curve it is possible to retrieve the individual relaxation rates, $1/T_1^{SC}(79)$, and $1/T_1^{SC}(81)$ which, together with a knowledge of $\tau_{SC}(= T_1(^{79}Br)$ and $T_1(^{81}Br)$, respectively), yield the spin–spin coupling constants $J(^{79}Br, ^{13}C)$ and $J(^{81}Br, ^{13}C)$.[26]

For comparative purposes, some of the main characteristics of the four

TABLE 4.1 The mechanisms dominating ^{13}C spin–lattice relaxation

Mechanism	$1/T_1 \propto K \cdot f(\omega, \tau)^a$ K	Correlation time τ
Dipole–dipole (DD)	$\gamma_C^2 \gamma_H^2 \sum_i r_{CH_i}^{-6}$	Rotational
Spin rotation (SR)	$IkT(C_\parallel^2 + 2C_\perp^2)^b$	Angular momentum
Chemical shift anisotropy (CSA)	$H_0^2(\sigma_\parallel - \sigma_\perp)^{2b}$	Rotational
Scalar (SC)	$J^2 S(S+1)$	Relaxation time of nucleus S

^{a}The spin–lattice relaxation rate can be formulated as the product of a term K which relates to the origin of the local magnetic field H_{loc}, and a term $f(\omega, \tau)$ which, in the general case, is a function of both the correlation time τ and the Larmor resonance frequency ω of the relaxing nucleus.
^{b}Assumes cylindrical symmetry.

mechanisms relevant to ^{13}C relaxation in diamagnetic systems are compiled in Table 4.1.

It is probably fair to state that in organic molecules more than 95% of all carbons, including quaternaries, are relaxed by the magnetic dipoles of nearby protons. Within the motional narrowing limit (valid at intermediate field strengths for molecular masses up to 1000–1500 Daltons), the relaxation rate typically increases with molecular mass and size. Quaternary carbons relax at least one order of magnitude more slowly than proton-bearing carbons.

Electron–nuclear relaxation

Unpaired electrons can interact with nuclear spins by virtue of two principally different mechanisms: *dipole–dipole* or, alternatively, *scalar* interaction through *hyperfine coupling*. Both are closely analogous to mechanisms previously dealt with. In the former case it is the modulated dipole moment of the unpaired electron, which creates a fluctuating field at the site of the carbon nucleus. Provided it has components at the ^{13}C Larmor frequency, it may give rise to spin–lattice and spin–spin relaxation. By contrast, scalar electron–nuclear relaxation arises from a modulation of the hyperfine coupling constant, due to fast electron spin relaxation.

We shall first discuss dipolar electron–nuclear relaxation and assume that the paramagnetic species (for example, a transition metal ion or a complex thereof) binds to a diamagnetic substrate molecule. Supposing electron interaction in the form of a point dipole at distance r from the carbon in question, the electron–nuclear relaxation rate, $1/T_1^e$, can be formulated as follows:[27,28]

$$\frac{1}{T_1^e} = \frac{2S(S+1)\gamma_S^2\gamma_C^2}{15r_{CS}^6}\left[\frac{3\tau_{C1}}{1+\omega_C^2\tau_{C1}^2} + \frac{7\tau_{C2}}{1+\omega_e^2\tau_{C2}^2}\right] \tag{4.25}$$

In Eqn. (4.25) S represents the electron spin, γ_S the magnetogyric ratio of the electron, and ω_S its Larmor frequency. The equation, not surprisingly, bears a distinct similarity with Eqn. (4.13), except that the spectral density terms involving the sum and difference of the two frequencies merge into one, which is because at typical field strengths $\omega_e \gg \omega_C$. More significant, however, is the circumstance that there exist various ways in which the static dipolar interaction can be modulated. The two correlation times τ_{c1} and τ_{c2} contain the earlier introduced rotational correlation time, τ_c, but, additionally, contributions from the electron relaxation times τ_{e1} and τ_{e2}. Unlike nuclear spins, electron spins can relax very rapidly, often of the same order of magnitude or even faster than reorientational motion. Moreover, the paramagnetic complexes are often labile, therefore introducing a further dynamic element, characterized by the lifetime, τ_m, of the paramagnetic complex. Accordingly, the total correlation time can be written as

$$\tau_{c1,2}^{-1} = \tau_c^{-1} + \tau_{e,1,2}^{-1} + \tau_m^{-1} \tag{4.26}$$

Often τ_m^{-1}, $\tau_e^{-1} \ll \tau_c^{-1}$, i.e. rotational diffusion dominates electron–nuclear relaxation.

Experimental relaxation rates are normally composed of diamagnetic and paramagnetic contributions. An additional complication arises from outer-sphere contributions, i.e. non-bonded interactions between carbon spins and electron dipoles. It is readily recognized that if one succeeds in separating the three contributions and thus in determining T_1^e, distance information can be deduced from Eqn. (4.25), provided the correlation time can be ascertained.

Specific chemical bonding of the species accommodating the paramagnetic site to a substrate molecule is not a prerequisite for the efficacy of this mechanism. In fact it suffices that the electronic dipole is close enough to produce a magnetic field of adequate amplitude at the site of the nucleus and that there is some mechanism causing a time variance of this field. Such a modulation is induced by translational diffusion.

Paramagnetic relaxation with various degrees of interaction between para-magnetic and substrate have recently been discussed.[29] Paramagnetic molecules with no tendency to associate with solute molecules serve as so-called inert relaxation reagents, whose purpose is to equally enhance the relaxation rates of all nuclei within a molecule.

For the simple model of dipole–dipole interaction between two hard-shell spheres of equal diamater[30] the electron–nuclear relaxation rate can be approximated as:

$$1/T_1^e = \frac{16\pi^2}{15} N_S \hbar^2 S(S + 1) \gamma_S^2 \gamma_C^2 \chi / kT \tag{4.27}$$

In Eqn. (4.27) N_S represents the concentration of paramagnetic species per cubic centimetre and χ is the solution viscosity. The remaining symbols have their usual meaning. For a given paramagnetic relaxation reagent, the only variables of the system are its concentration, the solution viscosity, and the absolute temperature. The specific nature of the solute molecules, on the other hand, is of only secondary importance.

In order to become operative, the scalar mechanism requires chemical bonding, permitting delocalization of unpaired electron spin density. The analogy to scalar relaxation (cf. Eqns (4.23), (4.24)) is borne out by Eqns (4.28) and (4.29) for hyperfine exchange longitudinal and transverse relaxation:[31]

$$1/T_1^{hf} = \frac{8\pi^2 A^2}{3} S(S+1) \frac{\tau_{e2}}{1 + (\omega_C - \omega_e)^2 \tau_{e2}^2} \tag{4.28}$$

$$1/T_2^{hf} = \tfrac{1}{2} 1/T_1^{hf} + \frac{4\pi^2 A^2}{3} S(S+1)\tau_{1e} \tag{4.29}$$

A is the hyperfine coupling constant in Hz and the remaining symbols have previously been defined.

Scalar or hyperfine relaxation and contact shifts have the same underlying cause, i.e. hyperfine coupling. Whereas the relaxation rate is proportional to A^2, the contact shift is linearly related to the hyperfine coupling constant. Hyperfine relaxation is therefore usually insignificant in those instances where no paramagnetic shift is observed. It should be noted, however, that hyperfine T_2 relaxation can, in principle, still act as a line-broadening mechanism, without a contact shift being present. In analogy to scalar nuclear–nuclear relaxation, the nuclear–electron hyperfine mechanism is more often a T_2 than a T_1 mechanism.

One final point of interest is the possibility of dipole–dipole electron–nuclear relaxation arising from interactions of the dipoles produced by delocalization of unpaired electron spin density, as reported for some transition metal acetylacetonates.[32]

Besides specific metal–substrate binding studies, ideally affording thermodynamic and kinetic parameters for ligand exchange,[26] paramagnetics play a role as relaxation reagents (PARRs = paramagnetic relaxation reagents[33]). Most often the purpose of these additives is an indiscriminate shortening of the carbon spin–lattice relaxation times, usually aimed at enhancing sensitivity. Assuming a totally inert relaxation reagent and equal spatial access to all sites of the solute molecule, then according to Eqn. (4.27) the electron–nuclear relaxation rate, $1/T_1^e$, defined as the difference

$$1/T_1^e = 1/T_1^{exp} - 1/T_1^{dia} \tag{4.30}$$

should be equal for all carbon sites. $1/T_1^{exp}$ and $1/T_1^{dia}$ in Eqn. (4.30) are the experimental and diamagnetic relaxation rates, respectively.

Typical relaxation reagents for organic solvents are bis- and tris-β-diketonates of Ni^{2+}, Co^{2+}, Mn^{2+}, Fe^{3+}, Cr^{3+}, and Gd^{3+}, with the most common ligands being acetylacetonate (acac) and dipivaloylmethanate (dpm).[34] In aqueous solution ligands with polar groups are required, such as cryptands[35] or diethylenetriaminepentaacetic acid (dtpa).[36] The latter appear to be particularly suited, as the ion is well encapsulated and the complexes are stable over the pH range 1–12.

The question of specific interactions between the relaxation reagent and the solute has been extensively studied in various organic solvents.[37] In this context it was found that random relaxation is the exception rather than the rule. Polar molecules with OH and NH functions tend to undergo hydrogen bonding with the carbonyl oxygens of the diketonate ligands, therefore preferentially augmenting the relaxation rate of carbons close to the bonding centres. This may be exemplified with the paramagnetic relaxation times in borneol (2), measured

upon addition of 5×10^{-2} M Fe(acac)$_3$,[34] showing distinctly enhanced relaxation for carbons near the OH function.

Chelates of gadolinium-III, with its seven unpaired electrons, have been found to be particularly potent relaxation reagents.[38] The tris-chelates of the rare earth ions are co-ordinatively unsaturated, and binding to polar molecules occurs by extension of the metal ion's co-ordination sphere. The same holds true for the bis-chelates of Ni-II and Mn-II. These reagents are therefore particularly useful, as an alternative to shift reagents, for probing molecular geometry.

4.4   ANISOTROPIC ROTATIONAL DIFFUSION

In our discussion of dipolar relaxation we have implicitly assumed that the modulation of the dipole field occurs by random isotropic reorientation of the molecule. This assumption is strictly only correct for spherical molecules. In all other cases reorientation around axes of small momentum of inertia will be favoured. The simplest case of an anisotropic tumbler is that of a symmetric top, in which two of the three principal axes for rotational diffusion are equal for symmetry reasons. A molecule with threefold overall symmetry such as [2.2.2]-bicyclooctane (3) satisfies the criteria for a symmetric top. Diffusion around the two axes perpendicular to the symmetry axis is characterized by a single diffusion constant, $D_\perp$. The general relationship between a correlation time τ_{ci} for rotation around an axis i and its corresponding diffusion constant D_i is given as[39]

$$D_i = 1/(6\tau_{ci}) \tag{4.31}$$

According to Woessner[40] and Huntress[41] the dipolar relaxation rate, $1/T_1^{DD}$, is related to the two diffusion constants as follows:

$$1/T_1^{DD} = N\hbar^2\gamma_C^2\gamma_H^2 r_{CH}^{-6}\left[\frac{A}{6D_\perp} + \frac{B}{5D_\perp + D_\|} + \frac{C}{2D_\perp + 4D_\|}\right] \tag{4.32}$$

The constants A, B, and C are functions of the directional cosine $l = \cos\theta$, where θ represents the angle between the C—H bond vector and the C_3 symmetry axis:

$$A = \tfrac{1}{4}(3l^2 - 1)^2 \tag{4.32a}$$
$$B = 3l^2(1 - l^2) \tag{4.32b}$$
$$C = \tfrac{3}{4}(l^2 - 1)^2 \tag{4.32c}$$

Clearly, the methine carbon in **3** does not sense diffusion around the C_3 axis as its C—H vector is collinear with C_3. Since $\cos\theta = 1$, $A = 1$ and $B = C = 0$; $D_\perp$ can be extracted in a straightforward manner. $D_\|$ is then determined by solving Eqn. (4.32) using the experimental methylene carbon relaxation rate, and inserting the respective value for $\cos\theta$. In this way, $D_\| = 2.2 \times 10^{11}$ and $D_\perp = 1.1 \times 10^{11}\,\mathrm{rad\,s^{-1}}$ were obtained for [2.2.2]-bicyclooctane,[2] showing that diffusion around the molecule's symmetry axis is twice as fast as that perpendicular to it.

Most molecules are asymmetric, and a correct description of rotational diffusion requires the three principal axes of the rotational diffusion tensor. Determination of the three quantities in turn demands that at least three relaxation times be known, each resulting from a carbon whose C—H orientation is unique. For a detailed treatment of this case the reader is referred to the original literature.[40]

One of the problems encountered in such studies is the choice of the co-ordinate system. Typically, it is assumed that the principal diffusion axes coincide with those of the momentum of inertia tensor, and the direction cosines l_{xi}, l_{yi}, l_{zi} for each carbon i are then inserted into the equations $1/T_1^i = f(l_{xi}, l_{yi}, l_{zi}, D_x, D_y, D_z)$, from which the diffusion constants are deduced by computer analysis.[18,42] The assumption of coincidence of the two co-ordinate systems is warranted for apolar molecules and the absence of heavy atoms. An alternative approach therefore consists of making both diffusion constants and directional cosines adjustable parameters.

A further complication arises from internal rotation about carbon–carbon single bonds. In fact, relatively few molecules are totally rigid. The simplest case of intramolecular mobility is that of a methyl group attached to an isotropic rigid rotor. Since a methyl group is a spinning top, this case can be treated by the formalism of Eqn. (4.32), by substituting R_{int} for $D_{\parallel}$ and D for $D_{\perp}$. R_{int} is then the diffusion constant for internal methyl rotation, while D describes overall reorientation. In the absence of internal motion a methyl carbon should relax three times as fast as a methine carbon, i.e. $T_1(CH_3) = (1/3)T_1(CH)$. In practice, $T_1(CH_3) > T_1(CH)/3$ is always found,[43] which is a manifestation of fast internal methyl spinning. It is readily recognized from Eqn. (4.32) that in the extreme where $D_{\parallel} \gg D_{\perp}$ (i.e. $R_{int} \gg D$) the second and third terms in the bracket vanish. By inserting $\theta = 109.47°$ (tetrahedral angle) into Eqn. (4.32a), the relaxation rate predicted is a factor of 9 lower than that for carbons not subjected to internal motion. This behaviour may be exemplified with the relaxation data reported for 3-β-hydroxyandrostane (4).[44] The very nearly equal NT_1 values for skeletal

carbons (average 2.2 s) are indicative for isotropic overall reorientation. By contrast, C-18 and C-19 exhibit values of 9.0 and 13.2 s, respectively, showing the effect of methyl rotation. The difference results from different degrees of steric hindrance, an effect which will be discussed in more detail in Chapter 5.

The limiting factor, predicted for the ratio $NT_1^{int}/N'T_1^{isotr}$ in the case of fast internal motion is critically dependent on the angle θ. An increase to $112°$, for example, increases the free rotation limit from 9 to 12. Such an increase in the CCH bond angle has been suggested to occur as a result of steric stress[45,46] to alleviate non-bonded interactions.

4.5 EXPERIMENTAL TECHNIQUES FOR THE MEASUREMENT OF T_1 AND THE NUCLEAR OVERHAUSER EFFECT (NOE)

Unlike the chemical shift, which can be obtained from a single measurement, the determination of relaxation times requires at least two, for accurate data, however, a series of measurements. The latter permits to map out the evolution of the magnetization with time, as predicted by theory.

At present, most high-resolution T_1 studies are carried out by means of one of the common methods, referred to as (1) *inversion recovery*,[47] (2) *progressive saturation*,[48] or (3) *saturation recovery*.[50] In this section we will briefly outline the principle of each method and point to possible systematic[48,51,52] and random errors.[52,53]

Inversion recovery[47]

We have seen in Chapter 1 that the macroscopic spin magnetization in thermal equilibrium is aligned along the $+z$ axis, i.e. the direction of the magnetic field H_0. Application of a 180° pulse nutates it onto $-z$, as depicted in Fig. 4.6. In terms of spin population, this situation corresponds to an exact inversion of the equilibrium populations of the two Zeeman levels. Since return to thermal equilibrium usually is exponential with time, and the equilibrium value of $M_z(t)$ is M_∞, the process obeys the differential equation

$$\frac{dM_z}{dt} = -\frac{M_\infty - M_z}{T_1} \tag{4.33}$$

which, after integration, yields

$$M_\infty - M_z = A e^{-t/T_1} \tag{4.34}$$

The constant A in Eqn. (4.34) depends on the initial conditions. In the case of an inversion-recovery experiment, $M_z(0) = -M_\infty$, by definition, and thus $A = 2M_\infty$, which, upon insertion into Eqn. (4.34), gives

$$M_\infty - M_z = 2M_\infty e^{-t/T_1} \tag{4.35}$$

At a time $t(t \lesssim T_1)$ after the intial perturbing pulse has elapsed, the magnetization has partially recovered. If a 90° observing pulse is applied and the resulting FID is collected and Fourier transformed in the usual way, the relative intensities of the various lines reflect their individual relaxation times. The experiment is thus characterized by the following pulse sequence

$$(180° - t - 90° - T)_n$$

From Fig. 4.6 it is recognized that, for an absorption mode phase setting, the line of a particular nucleus may appear positive or negative, depending on

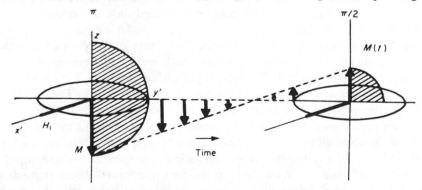

Fig. 4.6 Diagram illustrating the principle of the inversion-recovery experiment. On the left, the H_1 field has nutated the magnetization through 180°; on the right, the effect of the observing pulse on the partly recovered magnetization is shown.

whether the observing pulse is applied before or after the magnetization has passed through zero. In principle, two different measurements suffice to determine T_1: one each to obtain $M_z(t)$ and M_∞. However, more precise data are retrieved from a set of spectra obtained by varying t. A typical set of inversion-recovery spectra, three-dimensionally stacked in order of increasing t values, is given in Fig. 4.7(a). By plotting $\ln[M_\infty - M(t)]/M_\infty$ against t, the time interval between the inverting and observing pulse, ideally a straight line with slope $-1/T_1$ should be obtained.

This method is inaccurate and mathematically untenable for two reasons. First, the uncertainty in $\ln(M_\infty - M(t))$ increases with increasing t/T_1 ratio. Second, it can be shown that the slope of the straight line critically depends on the accuracy of the M_∞ value. Both problems can be circumvented by fitting the line intensities $S(t) \propto M(t)$ to the exponential

$$S(t) = S_\infty(1 - k\,e^{-t/T_1}) \tag{4.36}$$

by non-linear three-parameter regression, with the adjustable parameters being S_∞, k, and T_1.[52] It is essential that S_∞ and k be optimized as well since they are functions of pulse imperfections[51] such as resonance offset effects and pulse flip angle missettings. Only in the case of a perfect $180°$ pulse does $k = 2$ hold for the inversion parameter.

A calculated recovery curve, along with the experimental data points, is displayed in Fig. 4.7(b) for C-3 in 2-octanol.[54] Most modern spectrometers provide software routines for online calculation of T_1 by means of the spectrometer's data system.

Systematic errors may arise if the waiting time T in the sequence $180°-t-90°-T$ is not chosen sufficiently long between the pulses for the system to return to equilibrium. However, it has been shown[55] that considerably shorter values for $T(\lesssim T_{1\max})$ are tolerable on condition that the FID of the first cycle is ignored and that the transverse magnetization has decayed. This modified version of the inversion-recovery experiment has also been termed 'fast inversion-recovery Fourier transform' (FIRFT).[55]

A further source of systematic errors arises from pulse imperfections due to the finite strength of the H_1 field. This leads to transverse components of the magnetization following the $180°$ inverting pulse which, for short values of t, may interfere with the signal generated by the $90°$ observe pulse. The resulting spectrum then exhibits phase errors which are not corrigible by first-order phase-correction procedures. Such phase anomalies can be suppressed by alternately inverting the phase of either the inversion or detection pulse[51] and subtracting alternate FIDs. In quadrature phase alternation is substituted by stepping the transmitter and receiver phase through the four quadrants. Phase anomalies, corrected in this way, and the underlying resonance offset effects were shown not to falsify T_1,[51] as signal recovery in the invesion-recovery experiment remains exponential.

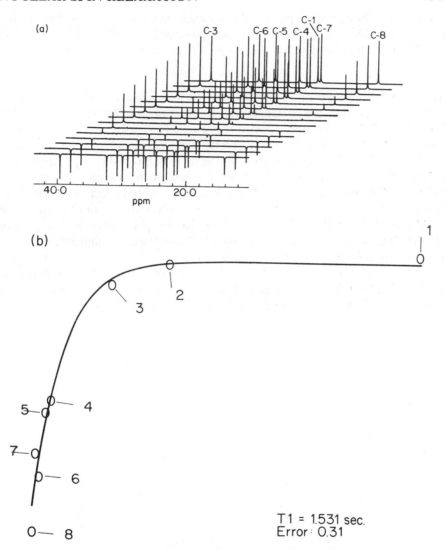

Fig. 4.7   (a) Stack of inversion-recovery spectra plotted for 15 values of the inversion delay τ (from 0.05 to 14 s). The compound was d-2-octanol.[54] (b) Computer-drawn plot of signal intensities for C-8 from the spectrum shown in (a). The solid line was computed as a best fit to Eqn.(4.36), affording $T_1 = 1.53 \pm 0.31$ s.[54]

Progressive saturation[48]

If the spin system is subjected to repetitive 90° pulses, a dynamic equilibrium will eventually be established in which the saturation effect of the rf pulses and that of the relaxation balance each other. Provided that the magnetization is all transverse at the end of the pulse and exclusively longitudinal at the moment the pulse is initiated, the spin dynamics are governed by Eqn. (4.36). The initial conditions in this case demand that $M_z(0) = 0$ and thus $A = M_\infty$, which leads to

$$M_\infty - M_z = M_\infty e^{-t/T_1}$$

The steady-state situation is usually reached after three to four pulses, which means that the first few FIDs should be disregarded. Apart from this, the experiment requires no other provisions, and the spectra are obtained by simply varying the pulse interval. A typical spectra set is shown in Fig. 4.8. The T_1 data are determined in the way described for the inversion-recovery experiment. The progressive saturation method lends itself in particular to situations where, for sensitivity reasons, a large number of transients are to be collected. The method, however, is restricted to T_1s that are not substantially shorter than the acquisition time, which sets a lower limit to the pulse interval. Progressive saturation experiments put more stringent demands on instrumentation[56] and are thus more susceptible to systematic errors.

They are, in particular, not very tolerant to offset-related pulse imperfections.[48] It is mainly for this reason that the progressive saturation method has largely been abandoned for high-resolution T_1 determinations.

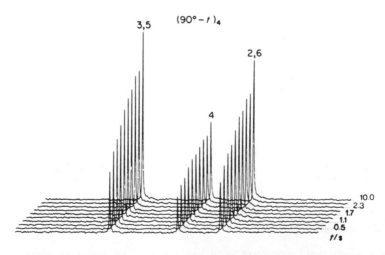

Fig. 4.8   A set of progressive saturation spectra, obtained from a sample of 80% phenol in hexadeuteroacetone, recorded at 30°C at 2.3 T field strength. The pulse interval t is indicated on the right of each spectral trace.

Saturation recovery[49,50]

The initial complete elimination of the magnetization, not only along the z-axis but also in the xy-plane, is peculiar to this technique. This is accomplished, for example, by a 90° pulse followed by a field gradient pulse along the z-axis,[50] which has the effect of dispersing M_{xy}, as illustrated in Fig. 4.9(a). After a time $t \lesssim T_1$ the magnetization has partially recovered and can be measured in an analogous way from a Fourier transform of the FID following a 90° pulse. The sequence required can thus be written as

$$(90° - \text{HSP} - t - 90°)_n$$

in which HSP stands for a homogeneity spoiling pulse. Figure 4.9(b) shows some traces obtained from a typical saturation recovery experiment.

Like inversion-recovery, saturation-recovery T_1 experiments are quite insensitive to pulse flip angle missettings and resonance offset effects. However, the need for a field gradient pulse for dispersal of the transverse magnetization is a

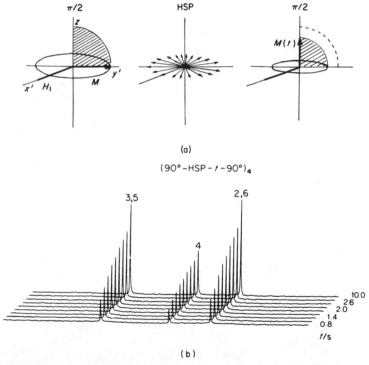

(a)

$(90° - \text{HSP} - t - 90°)_4$

(b)

Fig. 4.9 (a) Principle of the saturation-recovery experiment; (left) the 90° perturbing pulse rotates the magnetization onto xy; (centre) the field gradient pulse disperses M_{xy}; (right) the 90° observing pulse monitors the partly recovered magnetization. (b) Set of saturation-recovery spectra with the sample conditions being those used in Fig. 4.8.

handicap, as not all commercial spectrometers are normally fitted with the circuitry to apply a software-controlled field gradient pulse.

NOE measurements

Early NOE measurements[15] relied on a comparison of the integrated intensities with and without proton decoupling. This method is inconvenient, as integration of the proton-coupled spectra is complicated by signal overlaps. However, it has been shown that by suitably gating the decoupler[56,57] so that it is active only during the acquisition period, but not during a pulse delay, the NOE can be suppressed. In this way, data evaluation reduces to a simple peak-height determination, although integration is probably more accurate. The accuracy of the experiment is further enhanced by gating the decoupler in such a way that it is always on during acquisition but alternatively on and off during the pulse delay, as illustrated in Fig. 4.10. The NOE-retained and suppressed FIDs are then stored in different memory areas and subtracted from one another following completion of the experiment. In this way, effects of slow spectrometer drift or of sample heating are minimized.

A critical quantity in this experiment is the duration of the pulse delay. Unlike originally assumed, $5T_1$ is inadequate,[58,59] since during the off period of the decoupler the signal decays non-exponentially in a complex fashion as a function of T_{1C}, the carbon-13 relaxation time, T_{1CH}, the cross-relaxation time, and T_{1H}, the proton relaxation time. Figure 4.11 shows a plot of calculated and experimental values of the apparent NOEF, η^{app}, against the ratio T_d/T_{1C} for the β carbon in triptycene[58] (5), using experimental values for the three above-mentioned relaxation times. Since during the on period the NOEF builds up with a time constant T_{1C} (the dipolar ^{13}C relaxation time measured in the presence of ^1H decoupling), optium efficiency suggests a shorter delay for the decoupler on period ($\sim 5T_1$).

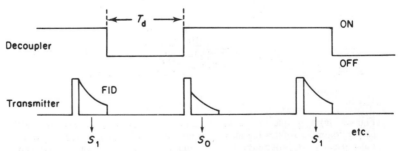

Fig. 4.10   Decoupler gating scheme used in proton-decoupled NOE experiment. The NOEF, η, is obtained as the ratio $(S_1 - S_0)/S_0$.

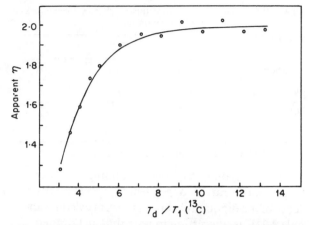

Fig. 4.11   Effect of pulse delay T_d (compare Fig. 4.10) on the apparent NOEF, η^{app}, for the β carbon in triptycene (5).[58] The solid line was calculated on the basis of the Bloch equations by inserting the following values for the experimentally determined relaxation times: $T_{1C} = 1.95$ s; $T_{1CH} = 1.42$ s; $T_{1H} = 5.3$ s.

   **5**

4.6   TEMPERATURE AND CONCENTRATION DEPENDENCE OF THE ROTATIONAL CORRELATION TIME

The rotational correlation time, which governs intramolecular dipolar relaxation, is temperature-dependent. Often this temperature dependence can be expressed by an Arrhenius-type relationship:[2]

$$\tau_c = \tau_c^0 \exp(-E_a/kT) \tag{4.37}$$

where E_a represents the activation energy for rotational diffusion, which has been shown to vary between about 1.5 and 5 kcal/mole.[51] Some caution should therefore be exercised when relaxation times are compared between different samples. According to Eqn. (4.37), the rotational correlation time and therefore the dipolar relaxation rate decrease exponentially with increasing temperature. The effects are particularly large if intermolecular association (for example, hydrogen bonding) is involved. Under these circumstances, Eqn. (4.37) is no longer valid, since E_a is different in the aggregate.

The rotational correlation time is also dependent on concentration. With increasing concentration, diffusion slows down, caused by enhanced friction among solute molecules. This effect is particularly pronounced for polar molecules for which electrostatic interaction acts as a solute-ordering mechan-

TABLE 4.2 Dipolar relaxation times T_1^{DD} in cyclohexanol as a function of concentration[a]

Carbon	T_1^{DD}/s		
	$1\,m$	$4\,m$	$12\,m$
1	17.6	10.7	3.4
2, 6	11.5	7.8	2.4
3, 5	11.1	7.8	2.4
4	8.7	5.1	1.3

[a]At 25.2 MHz and 30°C in hexadeuteroacetone.

ism.[21] Molecules undergoing hydrogen bonding are expected to show a particularly large concentration dependence.[21,60] The very marked concentration dependence of the dipolar relaxation times in cylcohexanol is illustrated by the data in Table 4.2. It is interesting to note that at 12-molar concentration the ratio $T_1(2,3,5,6)/T_1(4)$ is larger than for the 1-molar solution (ca 1.8 versus 1.3, respectively). This is because in the dimer, motion about the aggregate's symmetry axis is particularly favoured, thus rendering diffusion more anisotropic than in the monomeric state. Similar effects have been observed for the pyridine carbon relaxation times in pyridine/methanol complexes.[61]

4.7 RELATIVE CONTRIBUTIONS FROM INDIVIDUAL RELAXATION MECHANISMS

Often the experimentally determined relaxation rate $1/T_1$ is composed of contributions from different mechanisms. Since the relationship is additive with respect to rates, it can be formulated as[2]

$$1/T_1 = \sum_j 1/T_1^j \qquad (4.38)$$

Equation (4.38) conveys that the experimentally observed relaxation rate, $1/T_1$, is the sum of individual contributions $1/T_1^j$, with j indicating a particular mechanism. In order to examine whether contributions other than those of dipolar origin occur to a significant extent, the NOEF has to be determined. If $\eta < \eta_0$ ($= 1.988$ for the irradiation of protons), this is indicative that the relaxation of that particular carbon is not entirely dipolar.[1] Since non-dipolar mechanisms affect only the transition probability W_{1C} involving the frequency ω_C, Eqn. (4.18) for the NOEF may be modified by replacing W_{1C} by $W_{1C}^{DD} + W_{1C}^O$, where W_{1C}^O refers to other than dipolar mechanisms. Hence the equation for the fractional NOEF becomes

$$\eta = (\gamma_H/\gamma_C)\frac{W_2 - W_0}{W_0 + 2W_{1C}^{DD} + 2W_{1C}^O + W_2} \qquad (4.39)$$

By substituting $1/T_1 = W_0 + 2W_{1C}^{DD} + 2W_{1C}^O + W_2$ and $\eta_0/T_1^{DD} = (\gamma_H/\gamma_C)(W_2$

$- W_0$) (Eqns (4.11) and (4.18)), Eqn. (4.39) becomes

$$\eta = \eta_0 T_1/T_1^{DD} \qquad (4.40)$$

which relates the fractional (observed) NOEF η to the maximum value η_0.

If the (very rarely occurring) scalar mechanism is absent, the non-dipolar contributions to the relaxation rate are composed of $1/T_1^{CSA} + 1/T_1^{SR}$, from which the CSA contribution can be separated by experiments carried out at two different field strengths, say H_0' and H_0''. Then the relaxation rate for other than dipolar relaxation, $1/T_1^O$, can be expressed as

$$1/T_1^O(H_0') = 1/T_1^{SR} + (\tfrac{2}{15})\gamma_C^2(H_0')^2(\sigma_\parallel - \sigma_\perp)^2\tau_c \qquad (4.41a)$$

and

$$1/T_1^O(H_0'') = 1/T_1^{SR} + (\tfrac{2}{15})\gamma_C^2(H_0'')^2(\sigma_\parallel - \sigma_\perp)^2\tau_c \qquad (4.41b)$$

The CSA contribution can be determined by subtracting Eqns (4.41a) and (b) from each other, and since DD and CSA are both governed by the same correlation time, in addition $\sigma_\parallel - \sigma_\perp$, the chemical shift anisotropy is obtained in ideal cases.

Because of the different temperature dependencies of the various relaxation mechanisms, the relative contribution from each mechanism is itself a function of temperature.

Temperature-dependent relaxation data are usually plotted as the logarithm of the relaxation time or rate against inverse absolute temperature (Arrhenius plot). According to Eqn. (4.37), a straight line should be obtained in the event that relaxation is governed by a rotational correlation time. Depending on whether the relaxation rate or time is plotted, the slope of the straight line is positive or negative, respectively. This, for example, is the case for intramolecular dipolar or chemical shift anisotropy relaxation. Spin–rotation relaxation, by contrast, would yield a straight line of opposite slope, since $\tau_c \propto 1/\tau_{SR}$ (cf. Eqn. (4.21)). If, on the other hand, spin rotation competes with either DD or CSA, or both, the Arrhenius plot exhibits curvature. The situation of three competing mechanisms is illustrated in Fig. 4.12, showing an Arrhenius plot for carbon 1 in benzonitrile, recorded at 2.3 T.[62] T_1^{exp} represent the experimental spin–lattice relaxation times, T_1^{dd} the dipolar relaxation times. The latter separation was achieved from a measurement of the NOEF and Eqn. (4.40). T_1^{nd}, finally, in Fig. 4.12, are the relaxation times from non-dipolar mechanisms, calculated as $T_1^{nd} = (1/T_1^{exp} - 1/T_1^{dd})^{-1}$. It is interesting to note that the latter shows linearity over a wide range of temperatures with a slope characteristic of spin–rotation relaxation. The curvature seen at low temperatures is indicative of the onset of competition with CSA relaxation ($ca\ 10\%$ of the total relaxation rate at 2.3 T and room temperature[20]).

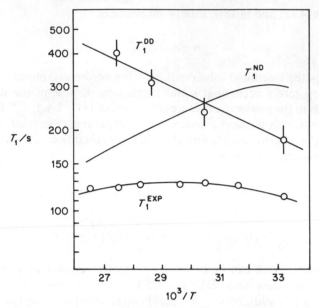

Fig. 4.12 Experimental (T_1^{exp}), dipolar (T_1^{dd}), and non-dipolar (T_1^{nd}) relaxation time for carbon-1 in benzonitrile,[62] plotted against inverse absolute temperature.

References

1 K. F. Kuhlmann, D. M. Grant and R. K. Harris, *J. Chem. Phys.* **52**, 3439 (1970).
2 J. R. Lyerla, Jr and D. M. Grant, *Int. Rev. Science, Phys. Chem. Series* (Ed. C. A. McDowell), Vol. 4, Chap. 5, Medical and Technical Publishing Co., Chicago, 1972.
3 J. R. Lyerla Jr and G. C. Levy, *Topics in Carbon-13 Nuclear Magnetic Resonance Spectroscopy* (Ed. G. C. Levy), Vol. 1, Chap. 3, Wiley-Interscience, New York, 1974.
4 D. A. Wright, D. E. Axelson, and G. C. Levy, *Topics in Carbon-13 Nuclear Magnetic Resonance Spectroscopy* (Ed. G. C. Levy), Vol. 3, Chap. 2, Wiley-Interscience, New York, 1980.
5 See, for example, R. Freeman and H. D. W. Hill, *Dynamic Nuclear Magnetic Resonance Spectroscopy* (Eds L. M. Jackman and F. A. Cotton), Academic Press, London and New York, 1975.
6 E. Breitmaier, K. H. Spohn, and S. Berger, *Angew. Chem.* **87**, 152 (1975).
7 C. P. Slichter, *Principles of Magnetic Resonance*, Chap. 5, Harper and Row, New York, 1963.
8 A. Carrington and A. D. McLachlan, *Introduction to Magnetic Resonance*, Chap. 11, Harper and Row, New York, 1967.
9 I. D. Campbell and R. Freeman, *J. Chem. Phys.* **58**, 266 (1973).
10 G. C. Levy, R. L. Lichter, and G. L. Nelson, *Carbon-13 Nuclear Magnetic Resonance Spectroscopy*, Chap. 8, 2nd edn, Wiley-Interscience, New York, 1980.
11 G. C. Levy, *Acc. Chem. Res.* **6**, 161 (1973); F. W. Wehrli, *Advanc. Molec. Relaxation Processes* **6**, 139 (1974).
12 An illustrative example of anisotropic overall motion is described in G. C. Levy, D. M. White, and F. A. L. Anet, *J. Magn. Res.* **6**, 453 (1972).

13 See, for example, L. G. Werbelow and D. M. Grant, *Advanc. Magn. Res.* **9**, 189 (1977).
14 I. Solomon, *Phys. Rev.* **99**, 559 (1955).
15 K. F. Kuhlmann and D. M. Grant, *J. Am. Chem. Soc.* **90**, 7355 (1968).
16 C. Deverell, *Mol. Phys.* **18**, 319 (1970).
17 P. S. Hubbard, *Phys. Rev.* **131**, 1155 (1963).
18 S. Berger, F. R. Kreissl, and J. D. Roberts, *J. Am. Chem. Soc.* **96**, 4348 (1974).
19 G. C. Levy and U. Edlund, *J. Am. Chem. Soc.* **97**, 5031 (1975).
20 F. W. Wehrli, *J. Magn. Res.* **32**, 45 (1979).
21 G. C. Levy, J. D. Cargioli, and F. A. L. Anet, *J. Am. Chem. Soc.* **95**, 1527 (1973).
22 See, for example, M. Brownstein and H. Selig, *Inorg. Chem.* **11**, 656 (1972).
23 R. Freeman and H. D. W. Hill, *Molecular Spectroscopy 1971*, Institute of Petroleum, London, 1971.
24 J. R. Lyerla, Jr D. M. Grant, and R. D. Bertrand, *J. Phys. Chem.* **75**, 3967 (1971).
25 G. C. Levy, *Chem. Commun.* 352 (1972).
26 See, for example, S. Hayashi, K. Hayamizu, and O. Yamamoto, *J. Magn. Res.* **37**, 17 (1980) and papers cited.
27 N. Bloembergen, *J. Chem. Phys.* **27**, 572 (1957).
28 I. Solomon, *Phys. Rev.* **99**, 559 (1955).
29 For a summary of the recent literature see, for example, Ref. 4, p. 112.
30 A. Abragam, *The Principles of Nuclear Magnetism*, Oxford University Press, Oxford, 1961.
31 See, for example, K. G. Orrell, *NMR Spectroscopy of Paramagnetic Species, Annual Reports on NMR Spectroscopy*, (Ed. G. A. Webb), Vol. 9, Academic Press, London, 1979.
32 D. M. Doddrell and A. K. Gregson, *Chem. Phys. Lett.* **29**, 512 (1974).
33 G. N. LaMar, *Chem. Phys. Lett.* **10**, 230 (1971); R. Freeman, K. G. R. Pachler, and G. N. LaMar, *J. Chem. Phys.* **55**, 4586 (1971); O. A. Gansow, A. R. Burke, and W. D. Vernon, *J. Am. Chem. Soc.* **94**, 2550 (1972).
34 G. C. Levy and R. Komoroski, *J. Am. Chem. Soc.* **96**, 678 (1974).
35 O. A. Gansow, A. R. Kausar, K. M. Triplett, M. J. Weaver, and E. L. Lee, *J. Am. Chem. Soc.* **99**, 7087 (1977).
36 T. J. Wenzel, M. E. Ashley, and R. Sievers, *Anal. Chem.* **54**, 615 (1982).
37 G. C. Levy, U. Edlund, and C. Holloway, *J. Magn. Res.* **24**, 375 (1976).
38 J. J. Dechter and G. C. Levy, *J. Magn. Res.* **39**, 207 (1980).
39 D. Wallach and W. T. Huntress, *J. Chem. Phys.* **50**, 1219 (1969).
40 D. E. Woessner, *J. Chem. Phys.* **37**, 647 (1962).
41 W. T. Huntress, *Advanc. Magn. Res.* **4**, 1 (1970).
42 S. Berger, F. R. Kreissl, D. M. Grant, and J. D. Roberts, *J. Am. Chem. Soc.* **97**, 1805 (1975).
43 A. Allerhand, D. Doddrell, and R. Komoroski, *J. Chem. Phys.* **55**, 189 (1971).
44 J. W. ApSimon, H. Beierbeck, and J. K. Saunders, *Can. J. Chem.* **53**, 338 (1975).
45 D. E. Axelson and C. E. Holloway, *Can J. Chem.* **54**, 2820 (1976).
46 J. W. Blunt and J. B. Stothers, *J. Magn. Res.* **27**, 515 (1976).
47 R. L. Vold, J. S. Waugh, M. P. Klein, and D. E. Phelps, *J. Chem. Phys.* **48**, 3831 (1968); R. Freeman and H. D. W. Hill, *J. Chem. Phys.* **51**, 3140 (1969).
48 R. Freeman and H. D. W. Hill, *J. Chem. Phys.* **54**, 3367 (1971).
49 J. L. Markley, W. J. Horsley, and M. P. Klein, *J. Chem. Phys.* **55**, 3604 (1971).
50 G. C. McDonald and J. S. Leigh, Jr, *J. Magn. Res.* **9**, 358 (1973).
51 For a detailed discussion of the various experimental techniques, including a discussion of error sources, see, for example, G. C. Levy and I. R. Peat, *J. Magn. Res.* **18**, 500 (1975).
52 M. Sass and D. Ziessow, *J. Magn. Res.* **25**, 263 (1977).

53 R. Gerhards and W. Dietrich, *J. Magn. Res.* **23**, 21 (1976).
54 G. C. Levy and C. Dumoulin, Unpublished data.
55 D. Canet, G. C. Levy, and I. R. Peat, *J. Magn. Res.* **18**, 199 (1975).
56 R. Freeman, H. D. W. Hill and R. Kaptein, *J. Magn. Res.* **7**, 82 (1972).
57 R. Freeman, H. D. W. Hill, and R. Kaptein, *J. Magn. Res.* **7**, 327 (1972).
58 R. K. Harris and R. H. Newman, *J. Magn. Res.* **24**, 449 (1976).
59 D. Canet, *J. Magn. Res.* **23**, 361 (1976).
60 T. D. Alger, D. M. Grant, and J. R. Lyerla, Jr, *J. Phys. Chem.* **75**, 2539 (1971).
61 I. D. Campbell, R. Freeman, and D. L. Turner, *J. Magn. Res.* **20**, 172 (1975).
62 F. W. Wehrli, unpublished data.

5

Applications

5.1 INTRODUCTION

Our objective in this chapter is to guide the reader through areas in chemistry and biology where ^{13}C n.m.r. spectroscopy has proved to be a particularly powerful problem-solving technique. Of particular interest in this regard are applications of ^{13}C n.m.r. to such diverse areas as structure elucidation, characterization of macromolecules (natural and synthetic), conformational analysis, molecular dynamics, kinetics, reaction mechanisms, and quantitative analysis of mixtures. Recently, this list has been complemented by two novel applications: ^{13}C n.m.r. studies of solid samples and of whole cells, even of intact living tissues. The last of these applications opens new vistas for n.m.r. as a non-invasive device for *in vivo* diagnostic and physiological studies.

In view of the enormous literature that has been generated during the past decade, any attempt at comprehensive coverage certainly would be doomed to failure. Rather than to attempt to review the subject, the authors instead have chosen to present some characteristic examples and to discuss these in significant detail. Since familiarity with an area cannot be acquired until one has applied it, ^{13}C n.m.r. applications are frequently presented in problem form, and the reader is encouraged to solve these problems by using the information contained in the previous four chapters. Solutions to all such problems, containing full particulars of the arguments that lead to relevant conclusions in each case, are presented in an appendix located at the end of this chapter.

Much of the material presented in this chapter results from the authors' work during the past ten years and from workshops and seminars.

5.2 STRUCTURE ELUCIDATION OF ORGANIC MOLECULES

The vast majority (perhaps as much as 95%) of ^{13}C n.m.r. spectra are obtained for the purpose of structural elucidation of an unknown (or partially unknown) sample. Carbon-13 n.m.r. spectroscopy, particularly in combination with two-dimensional proton n.m.r. spectroscopy, is capable of providing unique insights into structural, stereochemical, and conformational details of carbon-containing molecules. Therefore, it is both unrealistic and inefficient to regard ^{13}C n.m.r. spectra in an isolated way; ^{13}C and ^{1}H n.m.r. spectroscopy should be used jointly, complemented if necessary by additional spectral information. For this reason, we have included relevant ^{1}H n.m.r. spectral data in many of the interpretation problems that are discussed subsequently (unless they were either unavailable or considered to provide little useful supplementary information).

Prior to attempting n.m.r. spectral interpretation, users are advised to examine the quality of the spectrum and to ascertain whether the data were obtained under appropriate experimental conditions. The use of inadequate spectral width, for example, can lead to signal foldover (cf. Chapter 1). Such a situation may occur if carbon atoms are present that resonate outside the 200 ppm spectral window that commonly is chosen for the purpose of obtaining survey spectra (this is indeed the case with, for example, ketone and aldehyde carbonyl carbon atoms). Another critical parameter is the pulse flip angle (for example, the use of an inappropriately large flip angle may result in suppression of some of the quaternary carbon resonances).

In addition to poor spectral quality due to improper instrument settings, the user should be watchful for spurious signals, as these can easily be mistaken for real peaks. Often, these 'ghost signals' arise from rf interference or from improper balancing of the quadrature receiver.

Also, ^{13}C n.m.r. spectra generally are obtained from sample solutions in common deuterated solvents (for example, $CDCl_3$, dimethylsulphoxide-d_6, benzene-d_6, etc.). Since the ^{2}H resonance serves to establish field frequency lock; the user should be aware that the deuterated solvents themselves contribute characteristic multiplets, and these patterns should be taken into account when interpreting the spectrum.

The task of spectral interpretation can then be divided into a number of steps. Depending upon the level of sophistication of the user and of the available instrumentation, two potential alternative routes may be considered. Figure 5.1 summarizes the steps in each of these routes in the form of a block diagram.

Steps 1 and 2 are unrelated to the specific type of spectroscopy that is being used. Elemental microanalysis normally provides the molecular formula from which the elemental composition and molecular mass are derived. Without the need for any additional information, simply knowing the molecular formula permits the user to derive a key parameter, the so-called '*double-bond equivalence*' (i.e. the degree of unsaturation). This quantity is defined as the sum of double bonds, rings, and *twice* the number of triple bonds. (Recall that every degree of unsaturation removes two hydrogens; hence, triple bonds each possess two degrees of unsaturation.) The presence of heteroatoms in the molecule, if any, has the following effects:

(1) Oxygen neither adds to nor reduces the number of hydrogens;
(2) Nitrogen adds one hydrogen per nitrogen atom;

(3) Halogens eliminate one hydrogen per halogen atom;

(4) The effect of divalent sulphur and selenium is the same as that of oxygen, while phosphorus behaves like nitrogen.

Step 3 is concerned with evaluation of the proton noise-decoupled ^{13}C n.m.r. spectrum. From a simple line count, information can be deduced regarding *molecular symmetry*. Whereas a line number that is less than the number of carbon atoms in the molecular formula might be due to accidental chemical shift degeneracy, it usually indicates instead that the molecule possesses at least one symmetry element.

Step 4 comprises an in-depth evaluation of the proton noise-decoupled ^{13}C n.m.r. spectrum in terms of the chemical shift values of resonances that correspond to saturated and unsaturated carbon atoms. Since spectral overlap may occur between resonances of carbon atoms that are hybridized differently, such an analysis does not always lead to unambiguous conclusions. In general, however, application of this analysis permits the number of double and triple bonds to be deduced; together with the information obtained in step 2, it is then possible to deduce the number of rings.

In step 5, we inquire about the number of hydrogens that are attached to each of the carbon atoms in the molecule. Carbon-13 n.m.r. spectroscopy is uniquely capable of providing this information directly. The user typically has various options at his disposal for deriving this information; (see discussion in Chapter 3). Traditionally, single frequency off-resonance decoupling (SFORD) has been used for this purpose. Alternatively, the user may resort to one of the polarization transfer experiments, for example INEPT or DEPT (see Chapter 3). The experiment of choice in this regard is probably DEPT, since it is the less sensitive of the two polarization transfer methods to variations in the carbon–hydrogen coupling constant.[1] In general, SFORD spectra afford the same information, although analysis of SFORD spectra is often less straightforward, particularly when the ^{13}C n.m.r. spectrum is cluttered.

A decision must be made at this stage in ^{13}C n.m.r. spectral analysis. If the information thus far provided is deemed adequate, the user can proceed to the final step (step 8). Thus, individual elements of structural information can now be combined in such a way as to afford a unique structure. However, such a situation is exceptional; more commonly, additional spectral information will be needed in order to establish the structure unambiguously.

Such information, if needed, is provided by recording and analysing the corresponding ^{1}H n.m.r. spectrum (i.e. step 7a). If the single-resonance ^{1}H n.m.r. spectrum cannot be analysed (perhaps because it is inordinately complex), spin–spin coupling patterns can be elucidated via application of homonuclear magnetic double resonance techniques or by application of two-dimensional techniques, (for example, COSY,[2] (step 7b)). Homonuclear nuclear

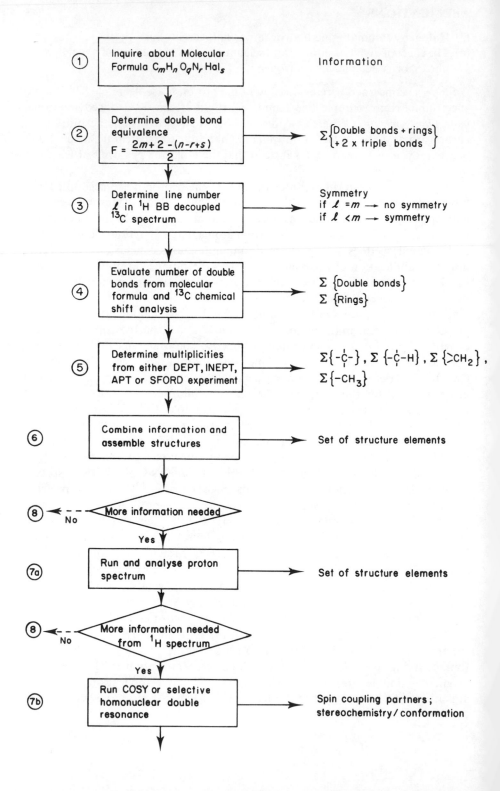

		Information
①	Inquire about Molecular Formula $C_mH_nO_qN_rHal_s$	Information
②	Determine double bond equivalence $F = \dfrac{2m+2-(n-r+s)}{2}$	$\Sigma \begin{Bmatrix} \text{Double bonds + rings} \\ \text{+ 2 x triple bonds} \end{Bmatrix}$
③	Determine line number l in ^1H BB decoupled ^{13}C spectrum	Symmetry if $l = m \rightarrow$ no symmetry if $l < m \rightarrow$ symmetry
④	Evaluate number of double bonds from molecular formula and ^{13}C chemical shift analysis	$\Sigma \{\text{Double bonds}\}$ $\Sigma \{\text{Rings}\}$
⑤	Determine multiplicities from either DEPT, INEPT, APT or SFORD experiment	$\Sigma\{-\overset{\mid}{\underset{\mid}{C}}-\}, \Sigma\{-\overset{\mid}{\underset{\mid}{C}}-H\}, \Sigma\{{>}CH_2\},$ $\Sigma\{-CH_3\}$
⑥	Combine information and assemble structures	Set of structure elements
⑧ ← - - No	More information needed	
	Yes ↓	
⑦a	Run and analyse proton spectrum	Set of structure elements
⑧ ← - - No	More information needed from ^1H spectrum	
	Yes ↓	
⑦b	Run COSY or selective homonuclear double resonance	Spin coupling partners; stereochemistry / conformation

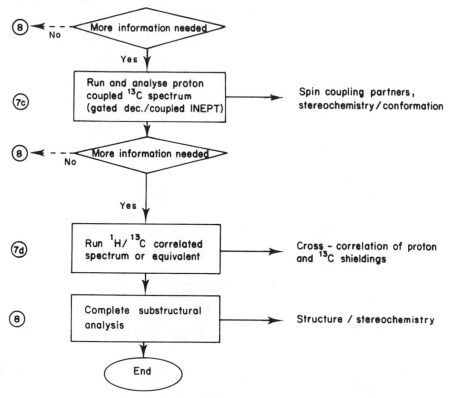

Fig. 5.1   Possible routes for structure analysis (for details see text).

Overhauser effect (NOE) experiments or their two-dimensional counterparts (NOESY)[3] are also useful in this context.

At this point, the information implicit in the ^{1}H n.m.r. spectrum has probably been exhausted. If additional information is required, the user can resort to analysis of the proton-coupled ^{13}C n.m.r. spectrum (step 7c), which can be recorded in a variety of ways.

An additional tool for ^{13}C n.m.r. spectral analysis is the cross-assignment of protons with the carbon atoms to which they are directly bonded. Such correlation techniques often are used to corroborate an otherwise ambiguous structural assignment. Once again, several alternative instrumental methods are available for this purpose (see discussion in Chapter 3). In principle, this information should be available from any double-resonance experiment (for example, a series of SFORD spectra, step 7b). However, in practice, it is often considerably more efficient to employ two-dimensional polarization transfer experiments (such as ^{13}C{^{1}H} correlated spectroscopy) for this purpose (step 7d). Given the information obtained from application of each of the above steps,

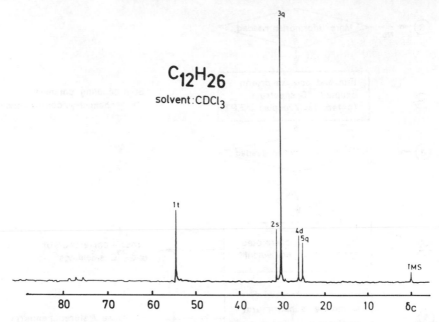

Fig. 5.2 20 MHz proton noise-decoupled ^{13}C spectrum of a hydrocarbon $C_{12}H_{26}$. The multiplicities are marked on top of each line (S = singlet, d = doublet, t = triplet, q = quartet).

the user should now be in a position to arrive at a unique structure (or at least a limited set of alternative, isomeric structures) in step 8.

The use of the guidelines summarized in Fig. 5.1 is illustrated with the following simple example. Figure 5.2 shows the proton noise-decoupled 20 MHz ^{13}C n.m.r. spectrum (100 ppm spectral width) of a hydrocarbon whose molecular formula is $C_{12}H_{26}$.*

A first inspection indicates that the spectral quality requirements outlined above are satisfied. The weak triplet at 76.9 is due to the solvent, $CDCl_3$.

(1) Molecular formula: $C_{12}H_{26}$.
(2) $F = (2 \times 12 + 2 - 26)/2 = 0$; consequently, the molecule possesses no double bonds, triple bonds, or rings.
(3) $l = 5$.
   There are only five resonances, but twelve carbon atoms. Acidental equivalence for such a small molecule is exceedingly unlikely. Hence, the molecule possesses symmetry. Among the various possibilities for the distribution of the twelve carbons among the five signals, four must be considered:

$$4:2:2:2:2$$
$$6:2:2:1:1$$
$$6:3:1:1:1$$

---

*Throughout this chapter, the lines originating from the solute molecules are numbered consecutively from high to low frequency in order to facilitate discussion.

and

$$8:1:1:1:1$$

(4) As already indicated by the molecular formula, the molecule contains no double or triple bonds. The most deshielded carbon atom ($\delta 54.32$) reflects heavy α-substitution for this carbon.

(5) The multiplicities (as given by the letters on top of each line), together with the most likely 6:2:2:1:1 carbon distribution based, on the observed intensities, suggest the following number of primary, secondary, tertiary, and quaternary carbon atoms:

$$(6 + 1) \times CH_3; \quad 2 \times CH_2; \quad 1 \times CH; \quad 2 \times C$$

Although no quantitative relationship holds under these spectral recording conditions (due to partial saturation and differential nuclear Overhauser effects), it is clear that the methylene resonance [$\delta 54.32(t)$] is substantially more intense than are the methine and highest-field methyl group signals ($\delta 26.21$ (d) and $\delta 25.29$ (q), respectively). However, the intensity of the quaternary carbon resonance ($\delta 31.39$ (s)) is comparable to that of each of the two highest field single-carbon resonances. Since quaternary carbon atom resonances generally appear attenuated because of their longer spin–lattice relaxation times, this signal is suspected of belonging to two carbon atoms. Since the ^1H n.m.r. spectrum obtained at this field strength (2.3 T) is not very informative, no ^1H—^{13}C correlation was attempted.

(6) The six equivalent methyl resonances indicate the presence of two equivalent t-butyl groups:

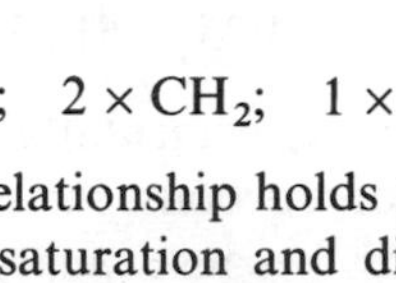

$$2 \times -\overset{\displaystyle \diagup CH_3}{\underset{\displaystyle \diagdown CH_3}{C-CH_3}}$$

Thus, resonances corresponding to two equivalent methylenes, one methine, and one methyl group remain:

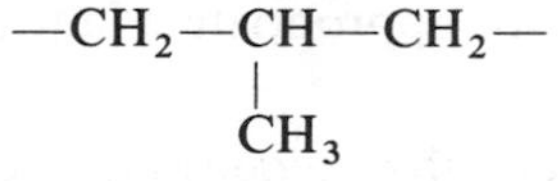

$$-CH_2-\underset{\displaystyle CH_3}{\overset{\displaystyle |}{CH}}-CH_2-$$

(7) The above structural elements can be mutually linked in only one way; hence the structure shown below is uniquely defined by the information contained in the ^{13}C n.m.r.:

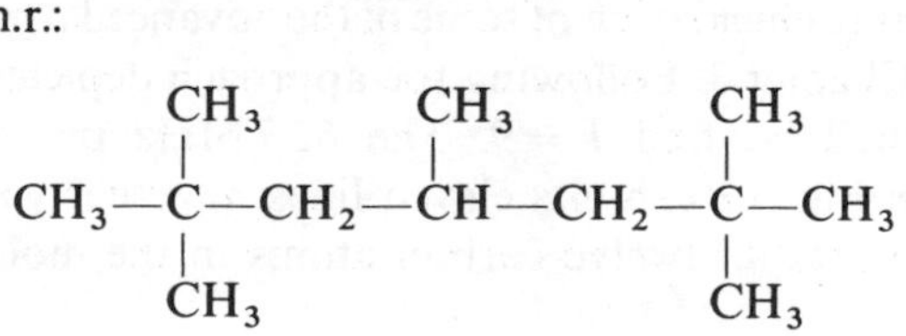

$$CH_3-\underset{\displaystyle CH_3}{\overset{\displaystyle CH_3}{\overset{\displaystyle |}{\underset{\displaystyle |}{C}}}}-CH_2-\underset{\displaystyle CH_3}{\overset{\displaystyle CH_3}{\overset{\displaystyle |}{CH}}}-CH_2-\underset{\displaystyle CH_3}{\overset{\displaystyle CH_3}{\overset{\displaystyle |}{\underset{\displaystyle |}{C}}}}-CH_3$$

No other structure is compatible with the observed line number and multiplicities unless accidental line degeneracies are admitted.

In order to increase the level of confidence in this structural assignment, the experimental ^{13}C n.m.r. chemical shifts can be compared with those predicted by application of Lindeman and Adam's additivity rules (see Chapter 2):

	C	C	C	C	4-CH$_3$
Observed	30.37	31.39	54.32	26.21	25.29
Predicted	29.78	31.75	51.83	24.24	21.59

Excellent agreement between experimental and calculated shifts generally is obtained for straight-chain hydrocarbons; increased branching leads to larger deviations. This probably occurs because highly branched molecules assume a preferred conformation that leads to different γ-gauche interactions than would be incurred by corresponding straight-chain hydrocarbons. Nevertheless, a reasonable qualitative fit is obtained that further supports the structural assignment.

Problem 1: The 25.2 MHz proton noise-decoupled and off-resonance decoupled ^{13}C n.m.r. spectra are given for a hydrocarbon, C_8H_{18}. Suggest a structure for this compound that is consistent with the given n.m.r. spectral information, and predict the ^{13}C n.m.r. chemical shifts of each carbon atom in your structure on the basis of additivity rules.

Problem 2: The 20 MHz proton noise-decoupled ^{13}C n.m.r. spectrum of a hydrocarbon is given, along with the multiplicities obtained from the corresponding off-resonance decoupled spectrum. The molecular formula of this hydrocarbon is not known. Suggest a structure for this compound that is consistent with the given n.m.r. spectral information, and corroborate your structure by comparing experimental ^{13}C n.m.r. chemical shift values with those predicted by application of the appropriate additivity rules.

A more demanding problem that requires the combined use of ^{13}C and corresponding ^{1}H n.m.r. spectral data is concerned with the structure of a synthetic amine whose molecular formula is $C_{12}H_{10}N_2$. This problem illustrates the power for structure elucidation of some of the advanced n.m.r. techniques that were discussed in Chapter 3. Following the approach depicted in Fig. 5.1, from application of step 2 we find $F = 9$. The 62.7 MHz proton-decoupled ^{13}C spectrum shown in Fig. 5.3 exhibits eleven lines; together with a resonance at $\delta 39.4$, these account for all twelve carbon atoms in the molecule. The lack of

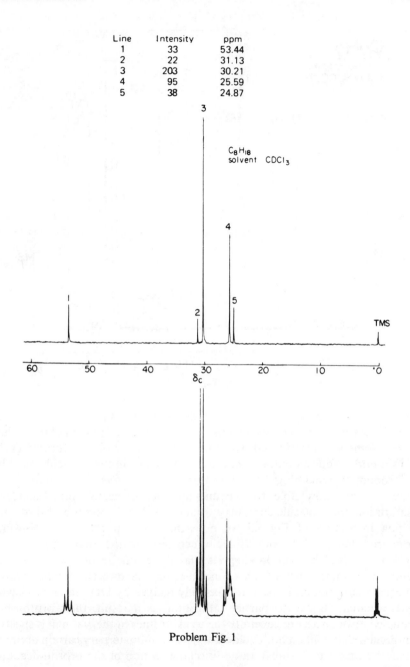

Line	Intensity	ppm
1	33	53.44
2	22	31.13
3	203	30.21
4	95	25.59
5	38	24.87

C_8H_{18}
solvent $CDCl_3$

Problem Fig. 1

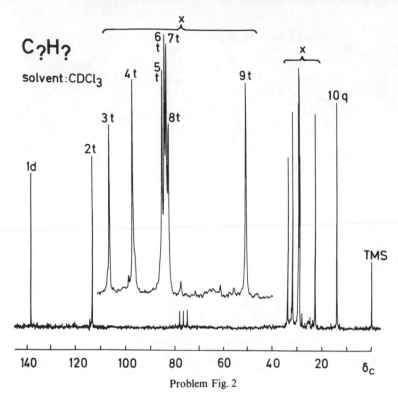

Problem Fig. 2

spectral degeneracy indicates that the molecule contains no symmetry element. From the chemical shifts of resonances 1–11, we conclude that all of these eleven carbon atoms are unsaturated (sp²). Since the number of unsaturated carbon atoms is odd, it follows that one of the double bonds in the molecule must be a C=N bond; the remaining ten carbon atoms comprise five C=C double bonds. Hence, we are now able to account for the molecule's nine degrees of unsaturation; the molecule contains a total of six double bonds and three rings.

Upon inspection of Fig. 5.3 we note that four of the eleven downfield resonances (lines 2, 3, 4, and 7) are much less intense than those of their neighbours. This observation suggests that these four resonances may correspond to quaternary carbon atoms. Since they have no directly attached protons, quaternary carbon atoms are less effectively relaxed by DD interactions which causes attenuation due to partial saturation; as a result, their proton-noise decoupled resonances are generally the weakest lines in the ^{13}C n.m.r. spectrum of molecules that contain both quaternary and non-quaternary carbon atoms (see Chapters 1 and 3 for details). In addition, inspection of the proton-decoupled INEPT spectrum (not shown) indicates that, among the non-quaternary carbon atoms, seven are methines (CH) and one is a methyl group (CH₃).

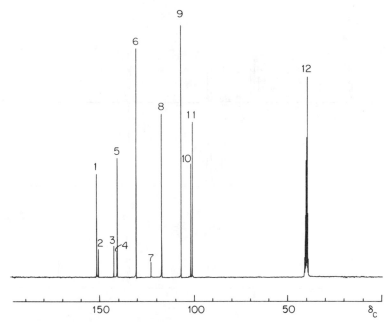

Fig. 5.3 Proton noise-decoupled ^{13}C nmr spectrum of a synthetic amine with molecular formula $C_{12}H_{10}N_2$ (courtesy Drs J. Cook and S. Wehrli).

Although we have probably not yet fully exploited all available information at this point (step 6) it is obvious that additional spectral information will be needed in order to assign a structure to the unknown sample. The answer to the question that is asked at the first decision point is decidedly 'yes'.

The corresponding ^{1}H n.m.r. spectrum of this compound is shown in Fig. 5.4. Inspection of this spectrum indicates the presence of seven resonances in the aromatic proton region. Chemical shift and spin–spin coupling information derived via inspection of Fig. 5.4 is summarized in Table 5.1.

The chemical shift of the resonance at $\delta 8.17$ is characteristic of an *ortho* proton in a pyridine ring that is coupled to one neighbouring (*meta*) proton ($J_{HH} = 4.2$ Hz). Since no other proton in the molecule is so highly deshielded it can be inferred that the other *ortho* position in the pyridine ring does not bear a hydrogen atom (i.e. the remaining *ortho* position is substituted). This conclusion is confirmed via examination of the corresponding ^{13}C n.m.r. spectrum (Fig. 5.3) which exhibits two downfield resonances ($\delta 151.2$ and 150.2) that correspond to *ortho* pyridine ring carbon atoms.

The ^{1}H n.m.r. resonance at $\delta 8.17$ is coupled to the resonance at $\delta 6.42$ (which partially overlaps another doublet that is part of the pattern at $\delta 6.40$). The fact that the resonance at $\delta 6.42$ is not further coupled indicates that the *para* position in the pyridine ring is substituted.

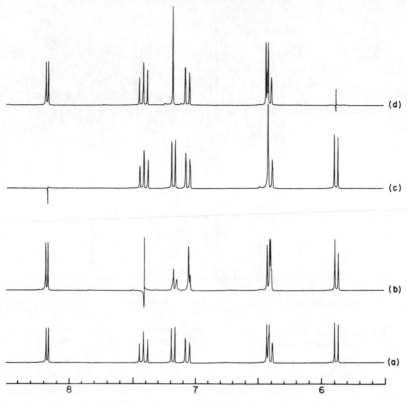

Fig. 5.4   250 MHz proton nmr spectrum of synthetic amine $C_{12}H_{10}N_2$ (courtesy Drs J. Cook and S. Wehrli). (a) Single resonance; (b – d) double resonance with irradiation at δ = 7.4 (b), δ = 8.2 (c), δ = 5.9 (d).

TABLE 5.1   Proton chemical shifts, spin–spin coupling constants, and signal multiplicities in compound $C_{12}H_{10}N_2$

δ_H	J_{HH}(Hz)	Multiplicity (Integral)
8.17	4.2	d(1H)
7.41	7.2	t(1H)
7.17	6.4	d(1H)
7.06	7.2;0.8	dd(1H)
6.42	4.2	d(1H)
6.40	7.2;0.8	dd(1H)
5.88	6.4	d(1H)
3.30	0	s(3H)

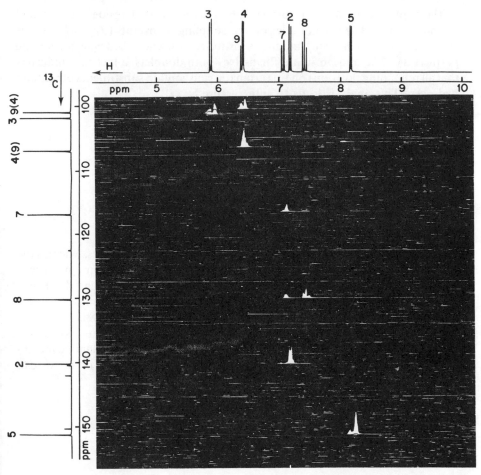

Fig. 5.5 ^1H—^{13}C correlated nmr spectrum of synthetic amine $C_{12}H_{10}N_2$ (5.9 T) (courtesy Drs J. Cook and S. Wehrli).

Since the second *ortho* position in this pyridine moiety is also substituted and since there exists no resonance with only long-range couplings, it follows that the second *meta* position must also be substituted. From this information, we can infer the presence of substructure **a**:

The triplet at $\delta 7.41$ must correspond to a proton that resides in a second aromatic ring. The magnitude of its coupling constant ($J_{HH} = 7.2$ Hz) is characteristic of *ortho* coupling in a substituted benzene. The triplet's coupling partners at $\delta 7.06$ and $\delta 6.40$ are themselves both doublets, a fact that indicates that all remaining positions on this aromatic ring contain substituents other than hydrogen. This analysis leads to substructure **b**:

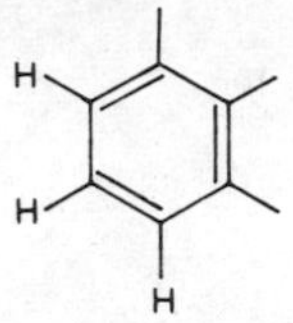

b

Substructures **a** and **b** together contain six quaternary carbon atoms, whereas the ^{13}C n.m.r. spectrum (Fig. 5.3) indicates the presence of only four quaternary carbons in the molecule. Hence, we deduce that two carbons are shared between these two substructures, with the consequence that the two rings must be fused.

Finally, the pair of doublets at $\delta 7.17$ ($J_{HH} = 6.4$ Hz) and at $\delta 5.88$ is characteristic of a vinyl group in which one of the sp^2-hybridized carbon atoms is situated α to nitrogen, (i.e. the one at $\delta 7.17$). By combining the resulting vinylamine moiety with three quaternary carbon atoms in either substructure **a** or **b**, we can construct another heterocyclic ring:

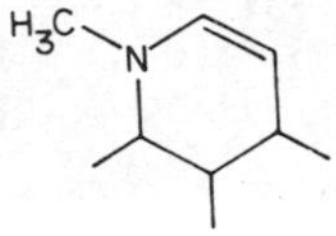

c

By fusing **a** with **b**, three alternative substructures (i.e. **d**, **e**, and **f**) can be generated: Substructure **f** can be discarded immediately, as it cannot be fused to a vinylamine moiety to generate a third six-membered ring.

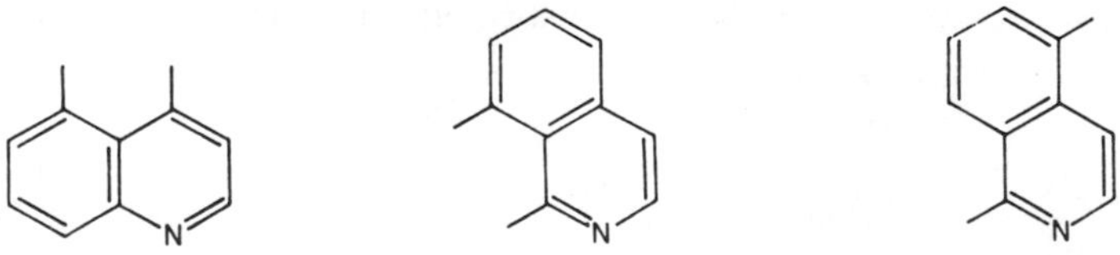

d

e

f

Each of the two remaining substructures (i.e. **d** and **e**) can be combined with the vinylamine moiety ($-CH=CH-N-CH_3$) in two different ways: combination with **d** affords **g** and **h**, whereas combination with **e** affords **i** and **j**.

g

h

i

j

Structure **i** can be ruled out, since the quaternary carbon atom that is flanked by two nitrogen atoms would be uniquely deshielded in a manner that is inconsistent with the observed ^{13}C n.m.r. chemical shift data.

Homonuclear NOE experiments performed via irradiation of the methyl resonance were found to afford a 20% enhancement of the intensity of each of the ^1H n.m.r. signals at $\delta 6.40$ and 7.17. The former signal has been shown to correspond to one of the benzene doublets; accordingly, this signal can be assigned to a benzenoid proton that is in close proximity to the $N-CH_3$ group. No such relationship exists in structure **g**; hence, we are left only with structural alternatives **h** and **j**.

The corresponding $^1H-{}^{13}C$ correlated n.m.r. spectrum (Fig. 5.5) provides the necessary information that will permit us to distinguish between structural alternatives **h** and **j**. Examination of Fig. 5.5 permits unambiguous assignment of the strategic carbon atom that is bonded to the olefinic proton that resonates at $\delta 5.88$. (This is the vinyl carbon atom which is β to nitrogen in substructure **c**.) The expected long-range coupling pattern for this carbon atom in **h** is a doublet of doublets (dd, due to spin–spin couplings $^3J_{CCCH}$ and $^2J_{CCH}$); the expected pattern indeed is observed (see Fig. 5.6, which shows an expansion of a coupled INEPT spectrum). By way of contrast, structure **j** would be expected to display only a doublet for this carbon atom (due to spin–spin coupling $^2J_{CCH}$). The foregoing arguments lead uniquely to the conclusion that the correct structure of this compound must be **h**.

Figure 5.7 shows the 25.2 MHz proton noise-decoupled spectrum of a thermal decomposition product[4] obtained from **1**. The molecular formula $C_{13}H_{20}O_2$ of the reaction product indicates the same elemental composition as that of the starting material. Consequently only a rearrangement has taken place, i.e. the

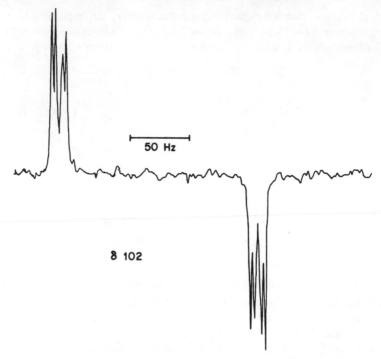

Fig. 5.6 Expansion of the proton-coupled INEPT spectrum of synthetic amine $C_{12}H_{10}N_2$ for the carbon resonating at $\delta = 102$.

sum of rings and double bonds remains unaltered. From the line number ($l = 6$) it can be concluded that one deals with a symmetric species, probably possessing a plane or centre of symmetry, or a twofold axis where one carbon is not affected by the symmetry operation. The chemical shift of 210.3 ppm suggests a ketone. Although the centre carbon in allenes also resonates in this region, this can be excluded because sp^2 carbons of allenes characteristically absorb between 75 and 90 ppm, a region which is empty in the present case. The line at 130.8 ppm is indicative of an olefinic double bond. The presence of only one such line furthermore implies equivalence of the two carbons. The region between 60 and 80 ppm where ether-type carbons are known to resonate is empty, which allows one to infer that both oxygens are part of (equivalent) carbonyl functions. The remaining lines in the high-field region all originate from sp^3-type carbons. The off-resonance data indicate that the methyl carbons are equivalent. A comparison of the methyl peak intensity with the highest field methylene resonance of 20.1 ppm suggests a 4:1 ratio for the two peaks. Four equivalent methyl carbons stipulate an arrangement of two symmetric sets of geminal methyl groups. Since

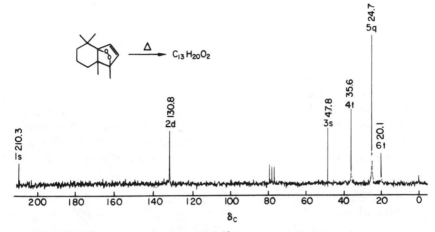

1

Fig. 5.7   25.2 MHz proton noise-decoupled ^{13}C spectrum of a thermal rearrangement product $C_{13}H_{20}O_2$ originating from a peroxide of known structure (see text).[4]

there is only one quaternary carbon resonance at 47.8 ppm, this must belong to the geminal methyl groups and thus represents a two-carbon resonance. The lines (in order of increasing field) therefore have to be assigned the following weights:

$$2:2:2:2:4:1$$

We may now list the structure elements:

$$2 \times \quad \underset{\diagup \diagdown}{\overset{H_3C \quad CH_3}{\underset{C}{\diagdown \diagup}}}$$

$$1 \times -CH{=}CH-$$
$$2 \times {>}C{=}O$$
$$2 \times -CH_2-$$
$$1 \times -CH_2-$$

The most likely symmetry element for the molecule is a plane of symmetry which bisects the single methylene carbon and the C=C double bond. Since $F = 4$ and the molecule possesses three double bonds, we have to account for one ring. This reduces the number of possible structures to four:

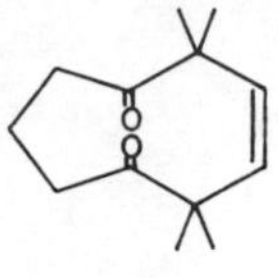

2

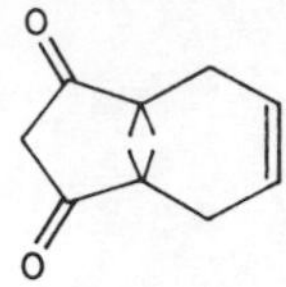

4

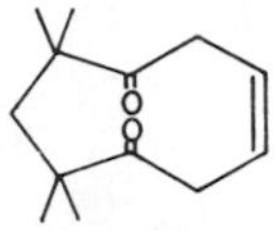

3

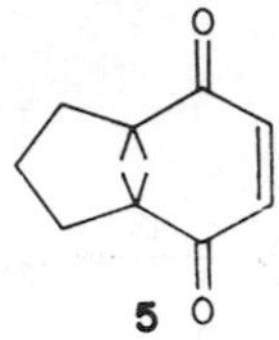

5

Structures **3** and **4** are immediately ruled out on the basis of the chemical shift of single methylene carbon which would be considerably less shielded in either structure. The discrimination between **2** and **5** is more difficult. A comparison with chemical shifts of α-trisubstituted saturated ketones shows that such a carbonyl would resonate at around 215 ppm. This clearly eliminates structure **2**.

Problem 3: Determine the structure of a molecule with molecular formula $C_{13}H_{17}NO$, whose 25.2 MHz proton noise-decoupled ^{13}C and 100 MHz proton n.m.r. spectra are given, and make a full assignment of the ^{13}C spectrum.

Note: The small amount of sample available did not permit the recording of the off-resonance decoupled ^{13}C spectrum.

Problem 4: The 25.2 MHz proton noise-decoupled spectrum of a compound is given whose structure is suspected to be[5]

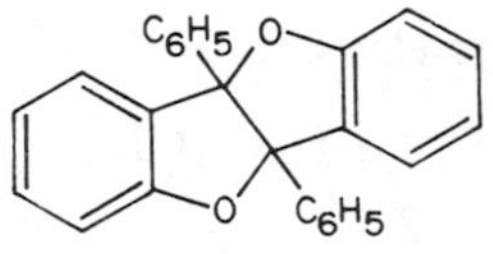

(1) How many lines would you expect and which is the symmetry of the proposed structure?
(2) Is the spectrum compatible with the given structure?
(3) If your answer to question (2) is negative, suggest an alternative structure and give reasons for your choice.

Problem 5: Hydrogenation of 'binor-s' (**I**, $C_{14}H_{16}$) over PtO_2 in HOAc at 40 atm. H_2 pressure affords a single compound, 'tetrahydrobinor-S' (**II**, $C_{14}H_{20}$).

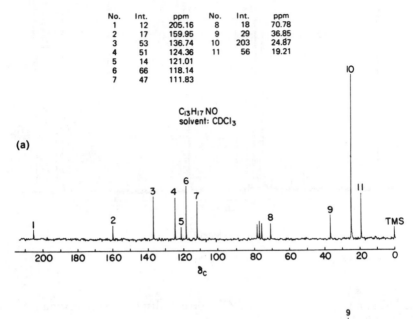

No.	Int.	ppm	No.	Int.	ppm
1	12	205.16	8	18	70.78
2	17	159.95	9	29	36.85
3	53	136.74	10	203	24.87
4	51	124.36	11	56	19.21
5	14	121.01			
6	66	118.14			
7	47	111.83			

$C_{13}H_{17}NO$
solvent: $CDCl_3$

(a)

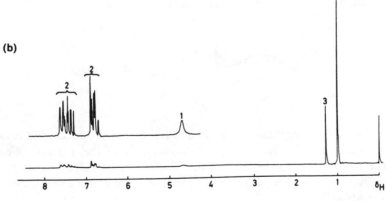

(b)

Problem Fig. 3.

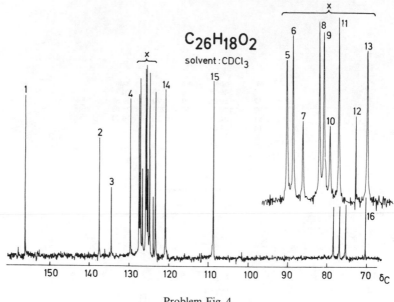

Problem Fig. 4.

The ^{13}C n.m.r. spectrum of **II** (CDCl$_3$ solvent) displays seven resonances: δ49.8 (d), 39.3 (d), 37.8 (d), 37.0 (d), 32.4 (t), 32.2 (t), and 24.1 (t).

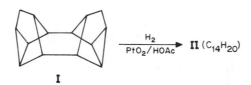

(1) In principle, hydrogenation of **I** can give rise to four products (that result via hydrogenolysis of its two cyclopropane rings). Draw the structure of each of these four possible products. Which one(s) can be ruled out as possible candidate structures for **II** on the basis of the C ^{13}n.m.r. spectral information given above?

(2) A contour plot of the carbon–carbon connectivity two-dimensional n.m.r. spectrum[6] of **II** is shown in Problem Fig. 5(b). The horizontal lines in this figure indicate carbon–carbon connectivities. Which *one* of the four possible structures for **II** that you drew in response to part (1), above, is consistent with the information given in the figure? Briefly justify your answer.

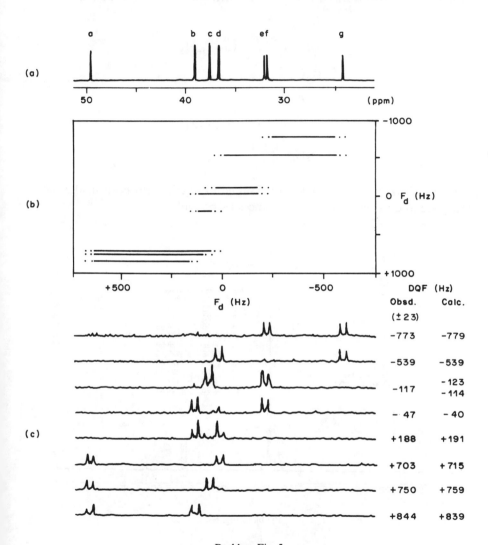

Problem Fig. 5
(a) ^{13}C nmr spectrum of tetrahydrobinor-S in CDCl$_3$. (b) Contour plot of the CCC two-dimensional nmr spectrum. The horizontal lines indicate the connectivities between the various AX (or AB) carbon doublets. (c) F_d traces at the calculated double quantum frequencies more clearly showing the connectivities between various carbon centres and the agreement between the calculated and observed DQFs. Reprinted with permission from V. V. Krishnamurthy, J. G. Shi and G. A. Olah, *J. Org. Chem.* **50**, 3005. Copyright (1985) American Chemical Society.

Problem 6: A hydrocarbon has been assigned the following structure:[7]

$$H_5C_6$$
$$\diagdown$$
$$C-C\equiv C-C_6H_5$$
$$\diagup\;\;\big|$$
$$H_5C_6\;\;\big|$$
$$\big|$$
$$H_5C_6$$
$$\diagdown$$
$$C-C\equiv C-C_6H_5$$
$$\diagup$$
$$H_5C_6$$

The 25.2 MHz ^{13}C spectrum is in obvious contradiction. Propose an alternative structure and assign as many resonances as possible.

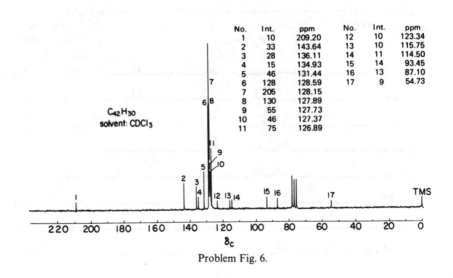

No.	Int.	ppm	No.	Int.	ppm
1	10	209.20	12	10	123.34
2	33	143.64	13	10	115.75
3	28	136.11	14	11	114.50
4	15	134.93	15	14	93.45
5	46	131.44	16	13	87.10
6	128	128.59	17	9	54.73
7	205	128.15			
8	130	127.89			
9	55	127.73			
10	46	127.37			
11	75	126.89			

$C_{42}H_{30}$
solvent: $CDCl_3$

Problem Fig. 6.

Problem 7: The 25.2 MHz proton noise-decoupled ^{13}C and the 100 MHz proton spectrum of a compound with the molecular formula $C_{20}H_{16}O_6$ are given. Determine its structure and assign all resonances.[8]

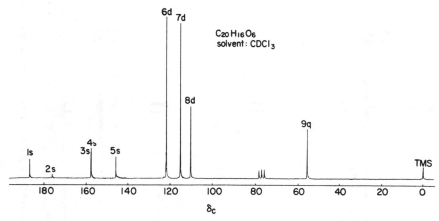

No.	Int.	ppm	No.	Int.	ppm
1	24	186.85	8	89	110.38
2	5	176.41	9	62	55.63
3	32	157.98			
4	39	157.83			
5	27	145.97			
6	200	121.81			
7	191	115.30			

$C_{20}H_{16}O_6$
solvent: $CDCl_3$

δ_C

Problem Fig. 7(a).

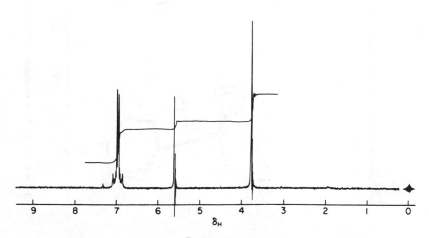

δ_H

Problem Fig. 7(b).

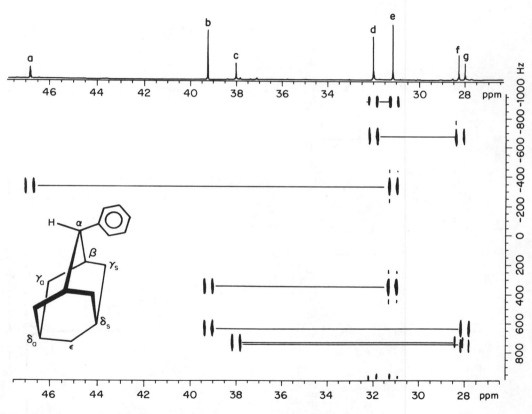

Problem Fig. 8   Two-dimensional INADEQUATE spectrum of the aliphatic part of 2-phenyladamantane.

Problem 8: Although a number of investigators have studied the ^{13}C n.m.r. spectra of substituted adamantanes, some ambiguity yet remains concerning chemical shift assignments of the *syn* and *anti* δ ring carbon atoms in 2-substituted adamantanes (for example, δ_s versus δ_a in 2-phenyladamantane (**I**, below)). This question has been resolved in the case of **I** via examination of its two-dimensional INADEQUATE spectrum, shown in Problem Fig. 8. Assign the aliphatic carbon resonances in **I** by using the carbon–carbon connectivity data contained in the two-dimensional INADEQUATE spectrum in conjunction with arguments based on chemical shift data contained therein.

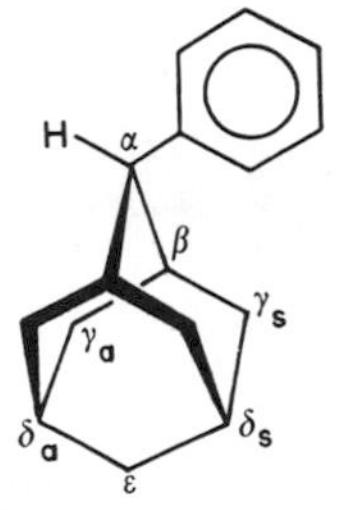

I

Problem 9: A compound with the molecular formula $C_{14}H_{20}O_2$ is characterized by its 60 MHz proton and 25.2 MHz proton noise-decoupled ^{13}C spectrum.[9] Furthermore, the ^{13}C signal multiplicities as obtained from the off-resonance decoupled spectrum are known.

(1) Determine the structure.
(2) Assign the ^{13}C spectrum.

Problem 10: Derive the structure of a naturally occurring hydrocarbon of molecular formula $C_{10}H_{16}$ with the aid of the 20 MHz proton noise and off-resonance decoupled ^{13}C spectra and the 100 MHz proton spectrum.

Note: Two further peaks were observed in the ^{13}C spectrum at 116.3(d) and 144.4 ppm (s).[10]

Provided that a high-resolution mass spectrometer is available and a molecular ion peak is obtained, the molecular mass and elemental composition can be determined unambiguously. In the following discussion we will study a problem where this information was not available. The compound was received for n.m.r. analysis with the remark that it was suspected to be an epoxide.

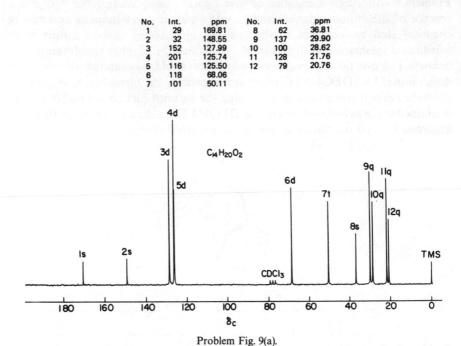

No.	Int.	ppm	No.	Int.	ppm
1	29	169.81	8	62	36.81
2	32	148.55	9	137	29.90
3	152	127.99	10	100	28.62
4	201	125.74	11	128	21.76
5	116	125.50	12	79	20.76
6	118	68.06			
7	101	50.11			

$C_{14}H_{20}O_2$

Problem Fig. 9(a).

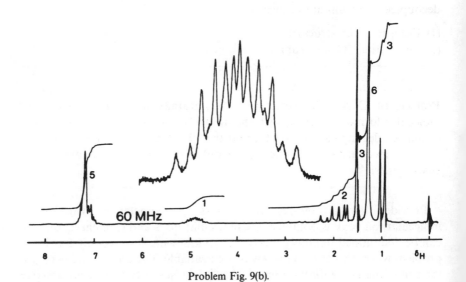

Problem Fig. 9(b).

The 300 MHz proton spectrum in Fig. 5.8(a)[9] yields the following information. From the normalized integrated intensities, a total of seventeen protons is obtained. Two of these are found to be exchanged upon addition of D_2O, which points to an NH_2 group.* The two pairs of lines centred at 7.30 and 7.80 ppm are characteristic of a *para*-substituted benzene and the chemical shifts suggest a tosyl group. This is supported by the singlet (3H) at 2.43 ppm. In the high-field region one finds two doublets at 0.87 ppm (3H, $J = 6.5$ Hz) and 1.00 ppm (3H, $J = 6.5$ Hz). This is typical of diastereotopic isopropyl methyl groups, whose non-equivalence is caused by attachment of the isopropyl methine carbon to an asymmetric carbon. The isopropyl proton resonance at 1.4 ppm exhibits further splitting due to a proton which resonates at 3.13 ppm ($J = 10$ Hz). The chemical

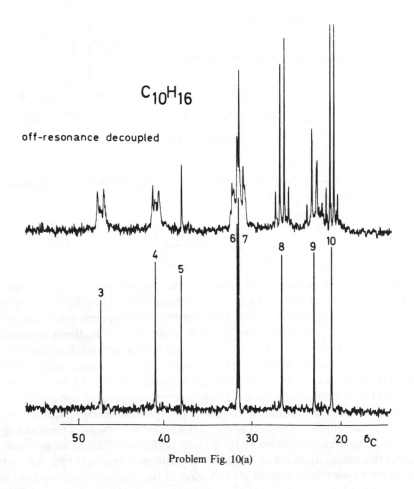

Problem Fig. 10(a)

*The integral was run after D_2O exchange.

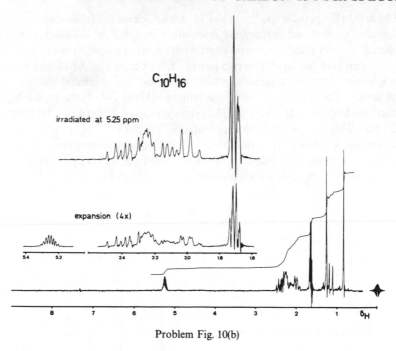

Problem Fig. 10(b)

shift of this latter is fairly characteristic of an epoxy proton. This suggests the following moiety to be present:

Since no further protons are observed, the second epoxy carbon can be assumed to be quaternary. This is the information available from the proton spectrum. The ^{13}C spectrum in Fig. 5.8(b) confirms the assignments made on the basis of the proton spectrum. Peaks 2, 3, 4, and 5 belong to the p-substituted benzene moiety while peaks 7 and 9 are assigned to the two methine carbons of $(CH_3)_2CHCHO$. Line 8, whose lowered intensity indicates a quaternary carbon, is identified as the second epoxy carbon. The high-field lines 10, 11, and 12, finally, are due to the three methyl carbons. The two remaining peaks 1 and 6 are both of low intensity and thus account for two more non-proton-bearing carbons, which shows the total number of carbon atoms in the molecule to be 14. The shielding of the carbon represented by line 1 (161.5 ppm) is characteristic of an sp^2 carbon attached to a hetero atom (O or N), while the position of line 6 (113 ppm) is rather unusual for a quaternary carbon. At this stage of the analysis the compound was also subjected to examination by induced electron emission spectroscopy (IEE) in

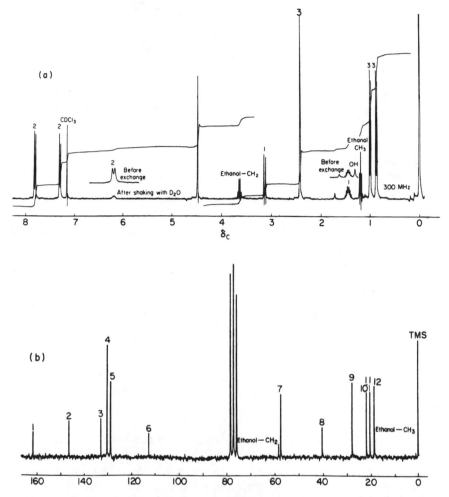

Fig. 5.8 NMR spectra of an unknown epoxide.[9] (a) 300 MHz proton spectrum recorded in deuterochloroform; (b) 25.2 MHz proton noise-decoupled ^{13}C spectrum of the same compound.

order to corroborate the presence of sulphur. The result indicated a 3:1 ratio of nitrogen and sulphur. Since halogen is absent, and on the assumption that no further oxygen atoms are present, the molecular formula of the compound is $C_{14}H_{17}N_3O_3S$, for which $F = 8$. In order to account for this unsaturation number, three additional double bonds and/or rings must be present, i.e. two carbons and two nitrogens have to be incorporated. Peaks 1 and 6 in the ^{13}C spectrum are typical of C=N and C≡N carbons; the latter could be confirmed by a strong laser Raman line at 2240 cm^{-1}. This allows one to list the following structure elements:

The high shielding observed for the quaternary aliphatic carbon (line 8) strongly suggests it to be substituted by carbon rather than nitrogen, which gives rise to the following structure **6**:

6

So far we have considered only proton noise and single-frequency off-resonance decoupled spectra as a basis for structure determination using ^{13}C n.m.r. In chapter 3 it was pointed out that proton-coupled single-resonance spectra may provide additional diagnostic criteria in the form of one-bond and long-range ^{13}C—^{1}H spin–spin coupling constants. The former, for example, have been shown to be applicable to probe carbon hybridization and to give information on ring size (cf. Chapter 2). Long-range couplings, on the other hand, enable one to establish the type of substitution and to differentiate *cis/trans* isomerism. The stereospecific nature of $^3J_{CCCH}$ often renders it the parameter of choice in conformational analysis.

The next example we wish to discuss in some depth concerns a simple problem in connection with structural isomerism where neither proton nor ^{13}C noise and off resonance decoupled spectra led to an unequivocal solution. A reaction yielded a cyclic sulphone[11] $C_6H_{10}SO_2$ which, on chemical grounds, could be identical with one of structures **7** to **13**.

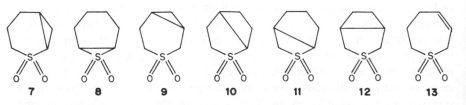

The proton coupled ^{13}C spectrum in Fig. 5.9 exhibits six distinct multiplets and thus immediately rules out the symmetric structures **8** and **12**. The absence of

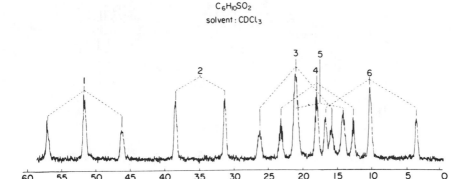

$C_6H_{10}SO_2$

solvent : $CDCl_3$

Fig. 5.9   25.2 MHz proton-coupled ^{13}C spectrum of a sulphone,[11] recorded by gating the noise decoupler (on during pulse delay, off during data acquisition).

olefinic resonances also eliminates **13**. The remaining structures would all give rise to the same signal multiplicities. However, the information theoretically inherent in ^{13}C spectra is not yet exhausted. It is well known (cf. Chapter 2) that cyclopropyl carbons exhibit characteristic values for their one-bond $^{13}C-^{1}H$ spin–spin coupling constants (160–170 Hz) whereas in cyclobutane and larger carbon rings $^{1}J_{CH}$ is much smaller (125–35 Hz). For this reason, the spectrum in Fig. 5.9 has been recorded in the single-resonance mode. The key feature in this spectrum is the triplet centred at 10.2 ppm corresponding to a coupling constant of 167 Hz. This clearly assigns this resonance to a cyclopropyl methylene carbon. A further confirmation of the correctness of this assignment is the unusually high shielding. This finding eliminates all structures except **7** and **9**. For the methine carbons resonating at 17.5 and 34.9 ppm, coupling constants of 172 and 177 Hz, respectively, are obtained. The large deshielding of the latter methine carbon can only be explained as being caused by attachment to the electronegative sulphone group and hence proves **7** to be the correct structure. This is further corroborated by the coupling constants and chemical shifts found for the methylene carbons pertaining to the six-membered ring: 132 Hz (18.1 ppm), 132 Hz (21.2 ppm), and 141 Hz (51.7 ppm). The largest coupling constant is associated with the least shielding. Both effects are a consequence of the electron-withdrawing nature of the sulphone group. In the case of structure **9**, there would be two strongly deshielded methylene carbons with both exhibiting an increased value of $^{1}J_{CH}$.

The following interpretation problem deals with structural isomerism in substituted aromatics. In this case it is symmetry that excludes proton n.m.r. as the problem-solving technique.

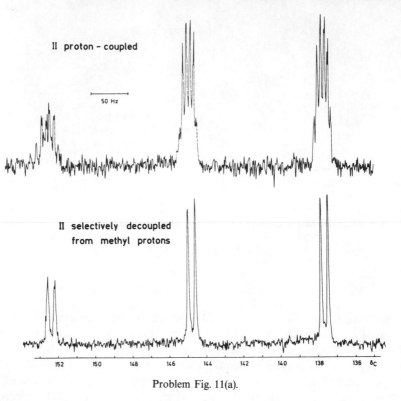

Problem Fig. 11(a).

Problem 11: Neither the proton nor the proton-decoupled ^{13}C spectra provide the criteria necessary for the distinction between the isomers of dimethylpyrazine. Assign the isomers 2,3- and 2,6-dimethylpyrazine to the proton-coupled and selectively decoupled ^{13}C spectra and determine the various spin–spin coupling constants.

Problem 12: Two-dimensional heteronuclear shift correlation spectroscopy via long-range couplings has been utilized to establish atomic connectivities in racemic acetate (**I**).

I

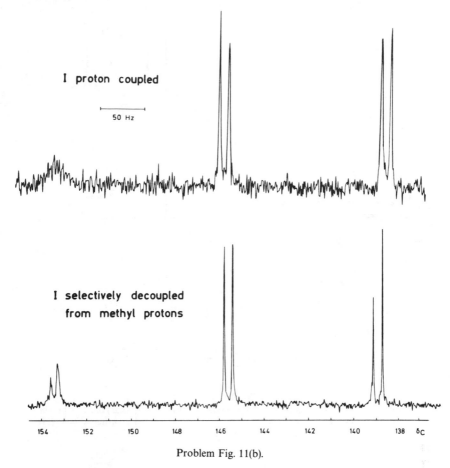

I proton coupled

50 Hz

I selectively decoupled
from methyl protons

154   152   150   148   146   144   142   140   138   δC

Problem Fig. 11(b).

A section of the ^1H, ^{13}C correlated (via long-range coupling) spectrum of racemic
I for the four olefinic carbon atoms and the two carbonyl groups in the high-field
proton region (which contains the six methyl group ^1H n.m.r. resonances) is
shown in Problem Fig. 1?. An additional, important observation (not shown in
the figure) involves the four cross peaks of carbon atoms **iii** and **vi** with methyl
resonances **b** and **e**: two of these, (**iii–b**) and (**vi–e**) are more intense than the other
two [(**iii–e**) and (**vi–b**)]. If it is assumed that the enhanced intensities for cross
peaks (**iii–b**) and (**vi–e**) reflect the fact that geminal coupling is more developed
than vicinal coupling under the experimental conditions employed, relevant
spectral assignments can then be made with a high degree of confidence.

On the basis of the information that appears in Problem Fig. 12 and in the
above discussion, assign the six carbon resonances (**i–vi**) and the six methyl

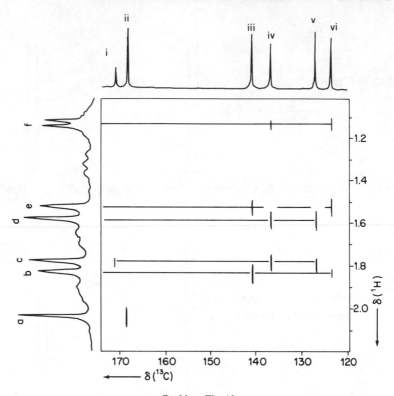

Problem Fig. 12
Section of ^{1}H–^{13}C correlated spectrum (via long-range couplings) of *rac*-1 for the olefinic carbon atoms and carbonyl groups and the methyl protons at 270 MHz proton frequency.[12]

resonances (**a–f**) to the corresponding carbon atoms (1, 2, 3, 5, 6, and 7) and methyl groups (*α–ζ*) in **I**.

Natural products

Structural determination of natural products represents a particularly challenging task. In synthetic chemistry, the history of the sample under investigation is normally known, i.e. the product is the result of a reaction carried out on a well-defined starting material. In contrast, the natural-product chemist is very often confronted with an entirely unknown sample, usually extracted from some organism. After mass spectrometry and elemental analysis have yielded the molecular formula, the sample is examined by ^{13}C and ^{1}H n.m.r., ms, infra-red and ultra-violet. Earlier, such problems were almost exclusively solved by chemistry alone by transforming the sample into less complex degradation products which were finally identified by comparison of their physical properties with those of known compounds. This lengthy procedure has been superseded by the advent of non-destructive spectroscopic analysis methods. Natural-products chemistry is probably one of the areas where n.m.r. has brought about most progress in that it has shortened the time required to determine an unknown structure by orders of magnitude.

The examples that follow have in common the problem that an unambiguous structure could not be derived without additional support from ^{13}C n.m.r.

The first example of a natural-product structure problem deals with a molecule which was isolated from a plant (*Gentiana cruciata*) and, based on elemental analysis and molecular ion peak (m.s.), assigned the molecular formula $C_6H_7O_3$. Its chemical and spectroscopic properties suggested the following formula **14**:

14

A reinvestigation[13] with the aid of ^{13}C and 100 MHz proton n.m.r. revealed several contradictions and permitted the correct structure to be assigned. In the subsequent discussion we will again adhere to the scheme outlined in Table 5.1. The molecular formula indicates an unsaturation number $F = 4$. The most striking feature of the ^{13}C spectrum in Fig. 5.10(a) is the fact that the lines occur as pairs. The relative intensities for the two lines forming the pair are roughly equal throughout the spectrum. These findings are typical of a mixture of two very similar constituents. The chemical shifts permit the following conclusions to be reached with regard to functional groups:

194.0; 191.1 ppm: ketonic, $\alpha\beta$-unsaturated carbonyl carbon

168.1; 168.7 ppm: ester (lactone) carbonyl or olefinic carbon β to a ketonic carbonyl function

159.3; 157.6 ppm: olefinic carbon bonded to oxygen or nitrogen

97.4; 97.0 ppm: olefinic or aliphatic carbon

63.5; 63.6 ppm: aliphatic carbon α to oxygen or nitrogen

36.3; 35.7 ppm: aliphatic carbon, carbon-substituted

No.	Int.	ppm	No.	Int.	ppm
1	62	194.05	10	198	63.48
2	24	191.09	11	209	36.28
3	24	168.68	12	116	35.70
4	34	168.08			
5	112	159.35			
6	65	157.62			
7	59	97.39			
8	36	96.95			
9	139	63.59			

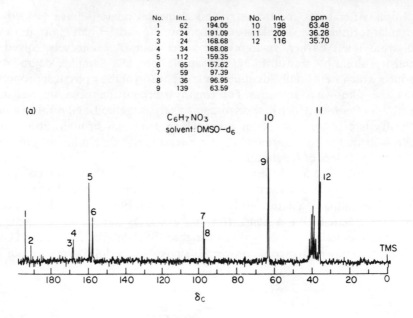

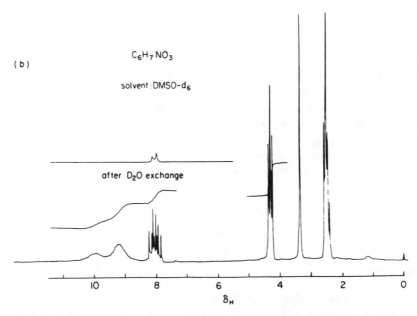

Fig. 5.10 NMR spectra of an unknown natural product $C_6H_7O_3N$, recorded in hexadeuterodimethylsulphoxide. (a) 25.2 MHz proton noise-decoupled ^{13}C spectrum; (b) 100 MHz proton spectrum.[13]

One of the two resonances giving rise to controversy is the one at 97.4 (97.0 ppm), because this corresponds to a position at which the sp^2 and sp^3 resonance regions overlap. Such unusual high-field shifts are found for olefinic carbons which are β to a lone-pair atom, with a significant contribution from a dipolar resonance structure.

$$\ddot{X}-C=C \leftrightarrow \overset{\oplus}{X}=C-\overset{\ominus}{C}$$

An inspection of the proton spectrum depicted in Fig. 5.8(b) is necessary at this stage. This displays two broad absorptions of unequal intensity at 10.0 and 9.2 ppm, respectively, which disappear upon addition of D_2O. At approximately 8.0 ppm one finds a complex multiplet which may be disentangled as two overlapping doublets of doublets ($J = 17$ and 9 Hz). Upon deuterium exchange these degenerate into broadened singlets of unequal intensity. This proves that the splitting observed for this olefinic proton is caused by coupling to the two protons represented by the broad peak, which is identified as belonging to NH_2. The broadening is not a consequence of a chemical exchange process taking place at intermediate rate in the n.m.r. time scale; rather it is due to coupling to ^{14}N. The latter could have been confirmed by heteronuclear decoupling, an experiment which was not performed in the present case. The non-equivalence of the NH_2 protons, which manifests itself in unequal coupling ($J^{trans} = 17$ Hz, $J^{cis} = 9$ Hz), must be caused by hindered rotation about the C—N bond (cf. also the next section). As mentioned earlier, the carbonyl ^{13}C resonances at 194.0 and 191.1 ppm are characteristic of an $\alpha\beta$-unsaturated ketone. This allows the following partial structure 15 to be derived:

15

In order to account for a double bond equivalence of 4, the structure must be assigned a ring. The two triplets centred at 4.2 and 2.5 ppm (the latter is somewhat masked by the solvent resonance) are interpreted in terms of the partial structure—CH_2—CH_2—O for which there is also evidence in the ^{13}C spectrum. The proton chemical shift of 2.5 ppm indicates this methylene to be adjacent to a

16   17

carbonyl carbon. The only missing carbon belongs to the carbonyl group of the lactone function. By incorporating this and connecting the two subunits, one arrives at the structures **16** and **17**.

The *cis/trans* isomers **16** and **17** are obviously stabilized through hydrogen bonds. The unusually high shielding found for the quaternary olefinic carbon points out the importance of the dipolar resonance hybrid **18**.

18

Moreover, the broadening observed for the NH proton signals which is ascribed to coupling to ^{14}N stipulates an electron deficiency at the nitrogen atom.

Problem 13: The infra-red spectrum of a sesquiterpene[14] with empirical formula $C_{15}H_{20}O_2$ suggests the presence of a furan ring, conjugated with a formyl group. From the 100 MHz proton spectrum, which is not shown in the figure, the partial structure

$$OHC(furan)CH_2-CH(CH_3)-\overset{|}{C}=CH-$$

could be determined. Derive the structure from the 25.2 MHz proton noise-decoupled ^{13}C spectrum. The signal multiplicities are given on top of each line.

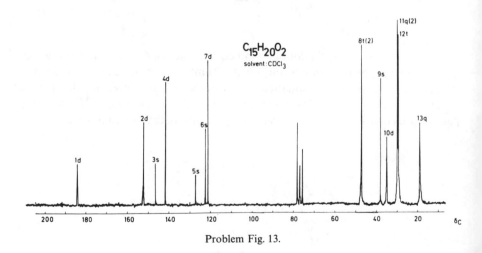

$C_{15}H_{20}O_2$

solvent : CDCl$_3$

Problem Fig. 13.

No.	Int.	ppm	No.	Int.	ppm
1	60	199.22	10	163	37.93
2	45	171.40	11	191	35.49
3	81	169.73	12	188	33.85
4	201	124.15	13	173	32.35
5	106	82.85	14	157	30.42
6	194	52.52	15	204	28.54
7	184	45.71	16	150	21.82
8	173	39.02	17	170	20.05
9	105	38.47	18	177	19.81
			19	167	17.37

(a)

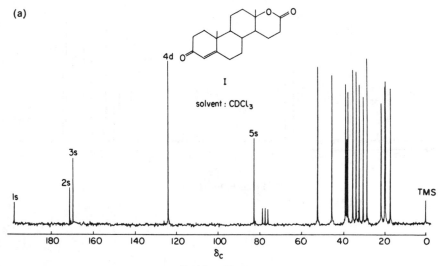

Problem Fig. 14(a).

EXPANSION I

(b)

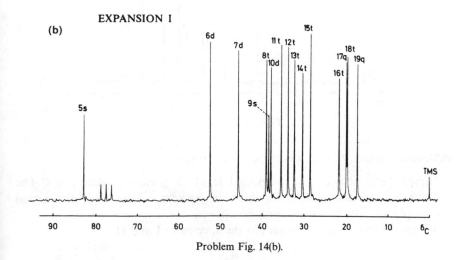

Problem Fig. 14(b).

No.	Int.	ppm	No.	Int.	ppm
1	45	186.20	11	107	39.01
2	28	171.14	12	119	37.97
3	75	167.69	13	105	32.28
4	106	154.73	14	100	32.08
5	103	128.23	15	93	28.51
6	127	124.30	16	117	23.42
7	68	82.61	17	204	20.15
8	78	50.95	18	113	18.71
9	112	45.67			
10	47	43.04			

(c)

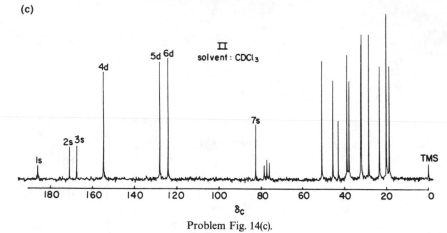

Problem Fig. 14(c).

(d)   **EXPANSION II**

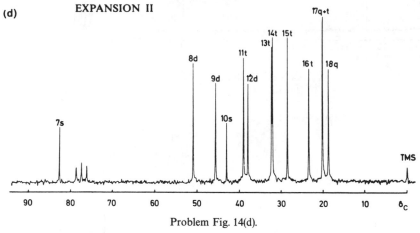

Problem Fig. 14(d).

Problem 14: A lactone[15] is known to have structure **I**.

(1) Derive the structure of a compound **II** which is closely related to **I**. The 25.2 MHz proton noise-decoupled ^{13}C spectra, along with the signal multiplicities, are given for both components.

(2) Assign the low-field resonances in the spectra of **I** and **II**.

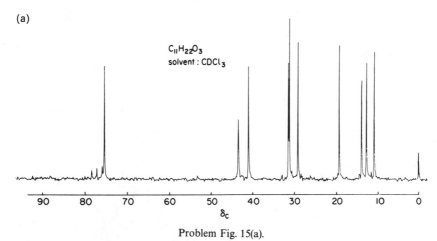

I

Problem 15: A naturally occurring aliphatic hydroxy acid[16] was studied by ^{13}C and proton n.m.r. The 25.2 MHz ^{13}C noise and off-resonance decoupled spectra

No.	Int.	ppm	No.	Int.	ppm
1	138	75.48	6	169	29.25
2	73	43.45	7	164	19.43
3	138	41.02	8	121	14.08
4	143	31.53	9	143	12.91
5	197	31.29	10	156	11.09

(a)

$C_{11}H_{22}O_3$
solvent : CDCl$_3$

δ_C

Problem Fig. 15(a).

(b)

off-resonance decoupled

δ_C

Problem Fig. 15(b).

(c)

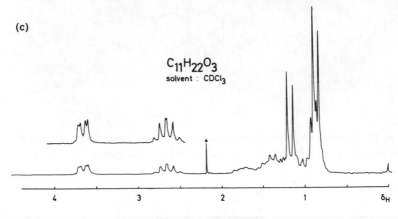

$C_{11}H_{22}O_3$
solvent : $CDCl_3$

Problem Fig. 15(c).

(d)

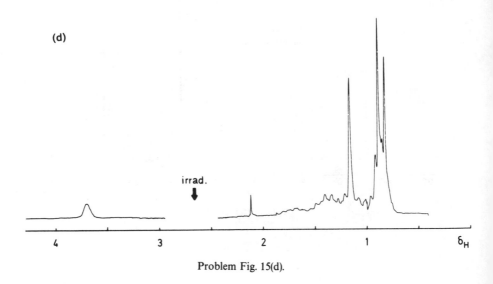

irrad.

Problem Fig. 15(d).

of the aliphatic region, together with the corresponding 100 MHz proton single and double resonance spectra, are given.

(1) Determine the structure.
(2) Eliminate uncertainties regarding assignment of the correct isomer by rationalizing the chemical shifts of the high-field signals by comparison with predicted values on the basis of Lindeman and Adams additivity rules (cf. Chapter 2).

Problem 16: A hydrocarbon[9] extracted from cod liver oil gave a molecular ion peak in the mass spectrum corresponding to a molecular mass of 268. Both the 300 MHz proton and the 25.2 MHz ^{13}C spectrum provided evidence of aliphatic carbons only. Determine the structure and corroborate your result by comparing the shifts with those computed with the aid of Lindeman and Adams additivity increments.

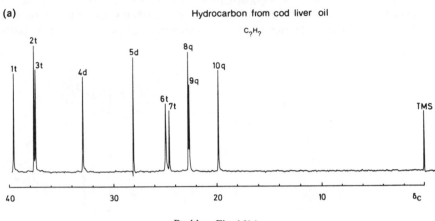

(a) Hydrocarbon from cod liver oil

$C_?H_?$

Problem Fig. 16(a).

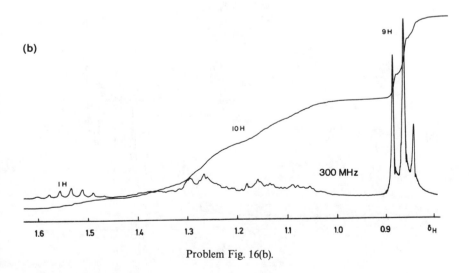

(b)

Problem Fig. 16(b).

Figures 5.11(a)–(d) show the 100 MHz single and double resonance spectra and the 25.2 MHz proton noise-decoupled spectrum of a natural product[17] whose molecular formula is known to be $C_{11}H_{16}O_3$, which corresponds to an unsaturation number of 4. The ^{13}C spectrum in Fig. 5.11(d) displays the expected number of peaks in accordance with the molecular formula. Four of the peaks (1,2,4,8) are of markedly lower intensity, which identifies them as quaternaries. Lines 1 and 2 (183.0, 172.1 ppm) suggest carbonyl carbons in acid, ester, or lactone functions. Line 3 at 112.5 ppm is typical of an olefinic carbon, while lines 4 and 5 (87.1 and 66.4 ppm) must be assigned to sp^3 carbons adjacent to oxygen. The remainder of the peaks (6–11) pertain to sp^3 carbons in non-heteroatom functions. Since, in the absence of nitrogen, olefinic resonances appear as pairs,

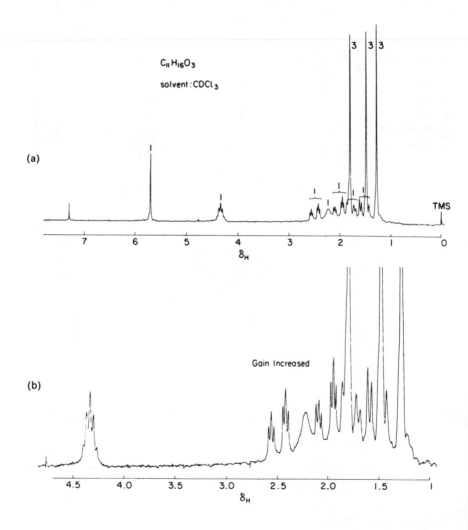

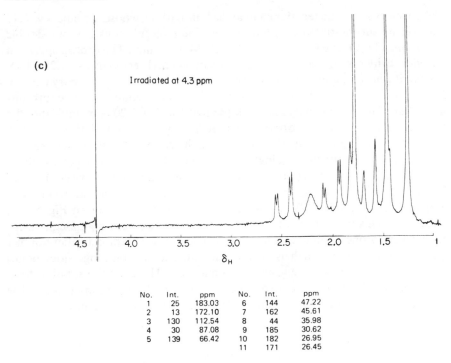

Irradiated at 4.3 ppm

No.	Int.	ppm	No.	Int.	ppm
1	25	183.03	6	144	47.22
2	13	172.10	7	162	45.61
3	130	112.54	8	44	35.98
4	30	87.08	9	185	30.62
5	139	66.42	10	182	26.95
			11	171	26.45

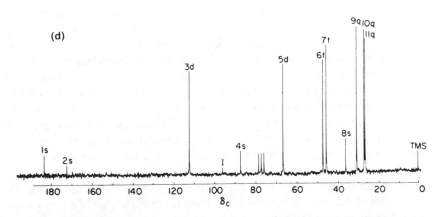

Fig. 5.11 NMR spectra of an unknown natural product $C_{11}H_{16}O_3$, recorded in deutero-chloroform. (a) 100 MHz single-resonance spectrum; (b) as (a), but high-field region expanded and at increased gain; (c) same as (b), but spin decoupled from the proton resonating at 4.3 ppm; (d) 25.2 MHz proton noise-decoupled ^{13}C spectrum.

peak 2 clearly has to be reassigned to an olefinic carbon, whose extreme low-field shift will be substantiated at a later stage. The molecule thus has two double bonds (one C=O, one C=C) and, since $F = 4$, two rings. The proton spectrum provides evidence of three non-equivalent methyl resonances (1.25, 1.45, 1.80 ppm) whose singlet multiplicity proves attachment to a quaternary carbon. The weak resonances between 1.4 and 2.7 ppm account for five protons altogether. The broad, only partly separated, line at 2.20 ppm indicates the presence of an alcoholic hydroxyl function. The quintuplet at 4.35 ppm (^{1}H, $J \sim 3.5$ Hz) must be due to an ether proton equally coupled to four other protons. Finally, at 5.7 ppm one finds a singlet (^{1}H) assignable to an olefinic proton. Since the proton at 4.35 ppm couples to the protons resonating between 1.4 and 2.7 ppm, irradiation at the position of the low-field line should result in a simplification of the high-field resonances. The partial spectrum in Fig. 5.11(c) reveals two AB systems ($J = 14$ Hz) originating from two pairs of non-equivalent methylene protons. Moreover, the two low-field doublets exhibit further doublet splitting ($J = 2.0$ Hz), which is readily identified as a long-range (four-bond) coupling as it occurs in zigzag conformations. These findings point to a conformationally rigid system to which these methylene protons belong. The data therefore suggest that the molecule contains the following moiety:

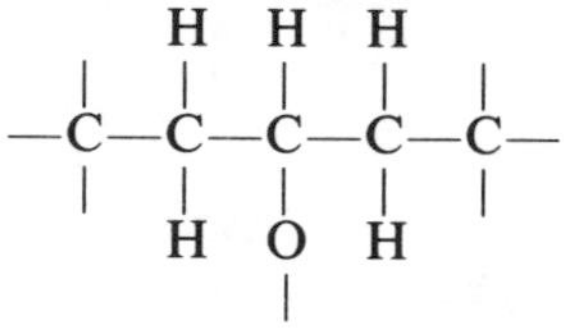

$$
\begin{array}{ccccc}
 & H & H & H & \\
 & | & | & | & \\
-C & -C & -C & -C & -C- \\
 & | & | & | & | \\
 & H & O & H & \\
 & & | & &
\end{array}
$$

Apart from the small long-range coupling, no further couplings occur for the methylene protons. It can be concluded from this that the methylene groups are each attatched to a quaternary carbon. The ^{13}C spectrum confirms the presence of the above structure element (peaks 3, 6, 7 for the three proton-bearing carbons). Since there are four quaternary carbons in the molecule, it is not clear which peaks are to be assigned to the carbons α to the methylenes, although the olefinic double bond can be excluded. This is because one of the olefinic carbons bears a proton which almost certainly would result in allylic coupling to one group of methylenes protons. However, the proton spectrum provides no evidence of such a coupling. The unusual deshielding exhibited by line 2 requires that this carbon be electron-deficient as a consequence of mesomeric interaction with a strongly electron-attracting group. In fact, chemical shift comparisons show that the chemical shift of this olefinic quaternary carbon is typical of that of a non-proton bearing carbon in β position to a carbonyl group:

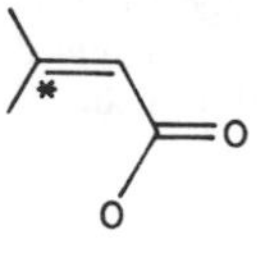

Consequently it is clear that the quaternary carbons adjacent to the methylene carbons are identical to those represented by peaks 4 and 8. There is only one way to link the two subunits:

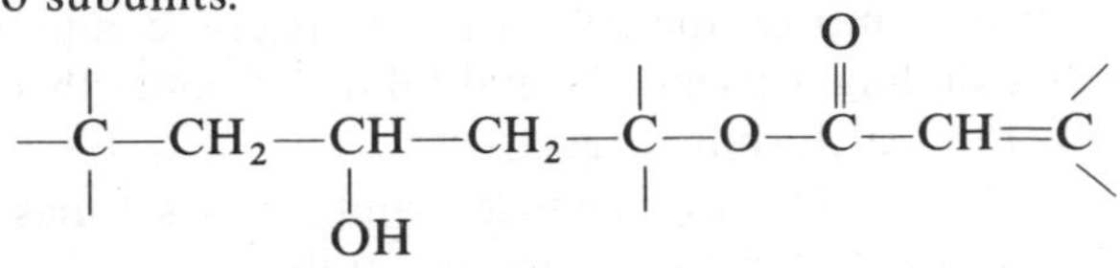

$$-\overset{|}{\underset{|}{C}}-CH_2-\underset{\underset{OH}{|}}{CH}-CH_2-\overset{O}{\underset{|}{C}}-O-\overset{\overset{O}{\|}}{C}-CH=C\overset{\diagup}{\diagdown}$$

19

Ring closure is now accomplished in a straightforward manner by excluding in advance versions with methyl groups in allylic position.

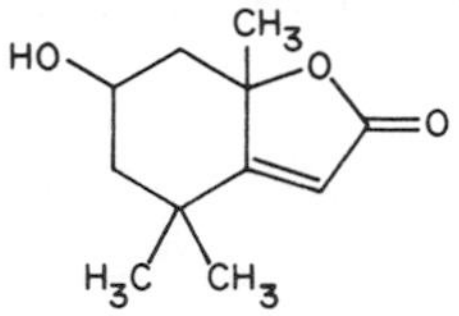

20

Problem 17: Longifolene, $C_{15}H_{24}$, is a tricyclic sesquiterpene found in the terpene oils from several *Pinus* species.[18] Analysis of its ^{13}C n.m.r. spectrum was aided via spectral editing with the DEPT pulse sequence (Problem Fig. 17(a)). Definitive assignment of the ^{13}C n.m.r. spectrum of longifolene follows from double-quantum coherence measurements (Problem Fig. 17(b)). (*Note*: Only the sp^3 carbon atoms in longifolene appear in Figs 17(a) and 17(b). In addition, there are two downfield resonances that correspond to the two vinyl carbon atoms: $\delta167.59$ (singlet in the off-resonance decoupled spectrum) and $\delta98.93$ (triplet in the off-resonance decoupled spectrum.

(1) Assign all resonances in the ^{13}C n.m.r. spectrum of longifolene on the basis of the spectral information presented in Problem Figs 17(a) and 17(b).
(2) Once all ^{13}C resonances have been assigned, it is then possible to make corresponding ^1H n.m.r. assignments via evaluation of the heteronuclear ^1H—^{13}C chemical shift correlation diagram for longifolene (Problem Fig. 17(c)). Assign as many of the ^1H resonances for longifolene as you can.

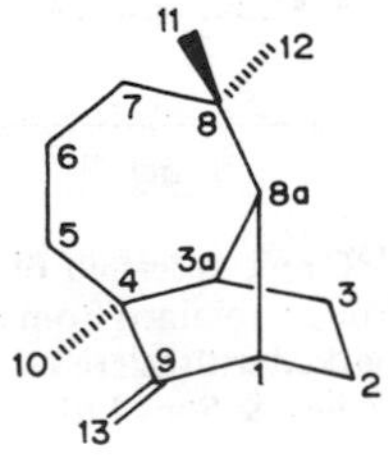

Stereochemistry, geometric isomerism

In Chapter 3 the techniques and parameters suited to probe stereochemistry have been outlined. These comprise the chemical shift (steric compression: γ and δ effects), spin–spin coupling between ^{13}C and ^{1}H or ^{13}C and other spin-$\frac{1}{2}$ nuclei), 2D spectroscopy, rare earth shift reagents, etc.

Syn–anti isomerism of imines, isoimides, and oximes[19] has been studied extensively by n.m.r. and the kinetic parameters for the interconversion have been determined. For a number of systems the free energy difference ΔG between the two isomers is close to zero, i.e. they co-exist at room temperature (cf. next section). Although the proton spectra may reflect the presence of two isomers, identification is often hampered due to signal overlap and the impossibility of rationalizing the shifts.

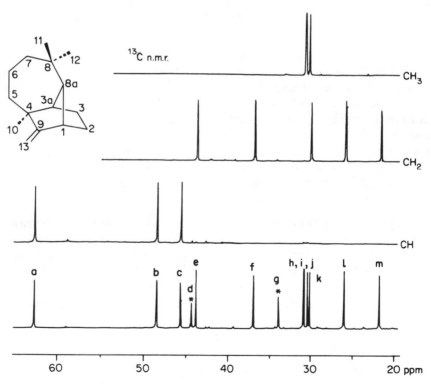

Problem Fig. 17(a) ^{1}H decoupled ^{13}C nmr spectrum of longifolene (sp^3 region only) and subspectra for CH, CH$_2$, and CH$_3$ groups obtained from experiments with the DEPT pulse sequence;[18] quaternary carbons are marked with asterisks. Reproduced by permission of John Wiley & Sons Ltd.

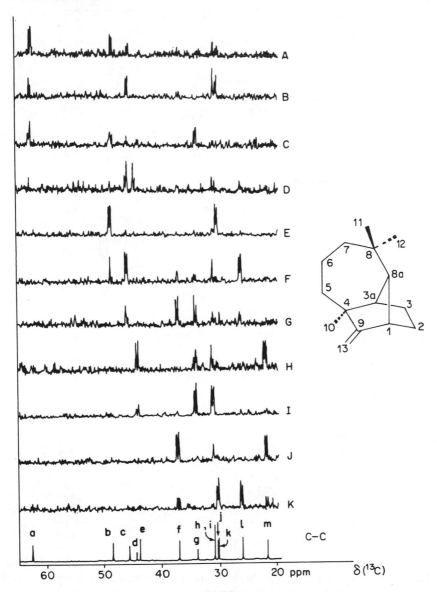

Problem Fig. 17(b)   ¹³C nmr double-quantum coherence spectra of longifoline determined at 100.6 MHz.[18]

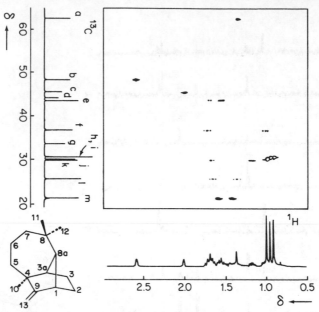

Problem Fig. 17(c)   400/100 MHz ^{1}H—^{13}C chemical shift correlation diagram for longifolene.[18]

Large steric compression shifts were reported for the syn isomer (22) of *N*-phenylmaleoisoimide:[20]

anti   **21**

syn   **22**

The determination of the configurations of a large number of ketoximes[21] was performed in the same way. Some representative examples are given below:

23

24

Structure **25**: with labels OH, N, 30.7, CH₃, 22.7, 42.0, 32.2, H₃C—C—CH₂—C—CH₃, CH₃

Structure **26**: with labels HO, N, 29.9, CH₃, 16.4, 49.2, 31.6, H₃C—C—CH₂—C—CH₃, CH₃

Structure **27**: 39.1, 35.5, 27.1, 34.9, 27.8, 42.0, N—OH

Structure **28**: 38.3, 35.6, 26.0, 37.2, 27.4, 38.5, N, HO

Again large γ-*gauche* effects are observed for carbons syn relative to the N—OH group (up to 7 ppm). This allows the two isomers to be distinguished from each other.

The identification of diastereomers, before the availability of ^{13}C n.m.r., had been a formidable task. The most common approach used to differentiate and assign stereochemistry involves proton n.m.r. through the angular dependence of the vicinal coupling constant. However, often the coupling information is not available because either the coupling does not exist or the critical lines in the spectrum are masked by superimposed signals. The following is an example where ^{13}C n.m.r. provided a clear answer to a complex stereochemical problem. The assignment of a *threo* and *erythro* configuration to the two diastereomeric 2-amino-3-mercapto-DL-butyric acids[22] **29** and **30** could not be made on the basis of the almost identical proton spectra:

Structure **29** (*threo*): COOH, H—C—NH₂, HS—C—H, CH₃

Structure **30** (*erythro*): COOH, H—C—NH₂, H—C—SH, CH₃

In order to obtain a conformationally rigid system, the two amino acids were converted to the diastereomeric DL-*N*-formyl-4-carboxythiazolidines **31** and **32**:

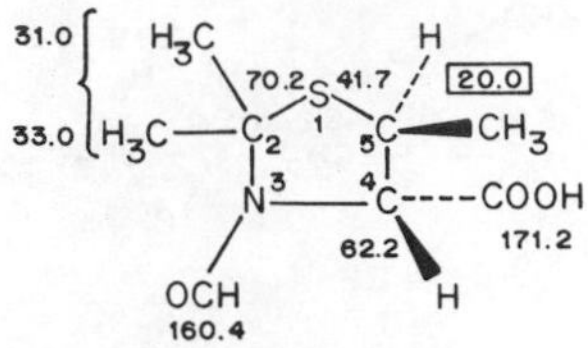

threo   **31** 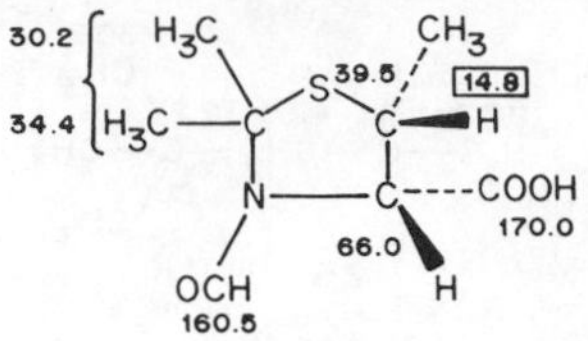   **32**   erythro

Because rotation around the C—N bond is slow in the n.m.r. time scale (cf. next section), two sets of signals were found in the ^{13}C spectra of **31** and **32** which were assigned to conformers **33** and **34**:

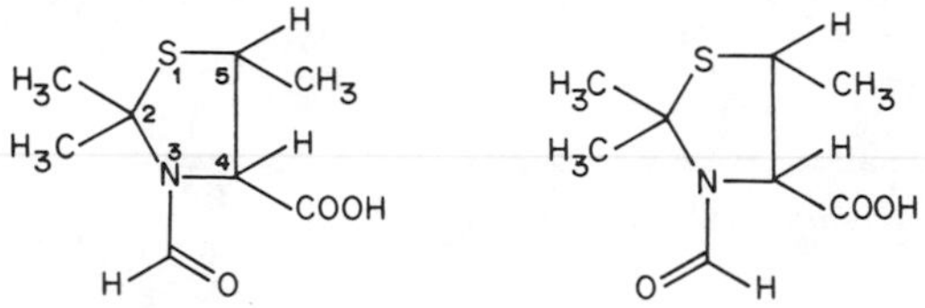

33   **34**

On the reasonable assumption that in the more highly populated isomer the carbonyl is *cis* relative to C-4, the two sets of lines could be assigned. The chemical shifts given for **31** and **32** are those assigned to conformer **33**. The only large chemical shift difference between **31** and **32** occurs for the C-5 methyl (20.0 versus 14.8 ppm). Whereas in **31** the carboxyl groups and C_5—CH_3 are *trans*, they are *cis* in **32**. The 5.2 ppm upfield shift for this carbon in **32** unambiguously proves that **32** has to be associated with the *erythro* configuration.

Problem 18:[23] The 25.2 MHz proton noise-decoupled ^{13}C spectrum of a sample labelled '1,3-dimethylcyclohexane' is given.

(1) Explain the number of lines in the spectrum.
(2) Make a full assignment.
(3) Rationalize the chemical shifts using appropriate additivity rules.

Problem 19: The reaction of equimolar amounts of trimeric 3,3-dimethyl-3H-indole and p-chlorobenzoyl chloride[24] in pyridine yielded two compounds of the

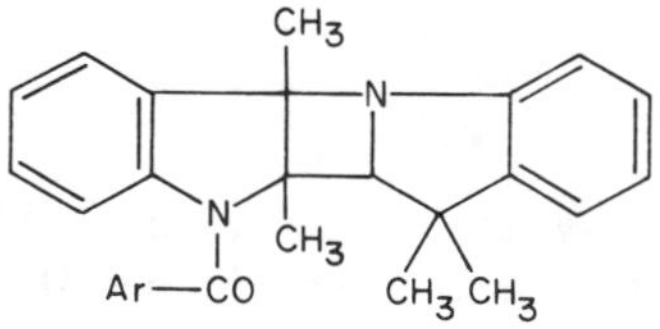

I 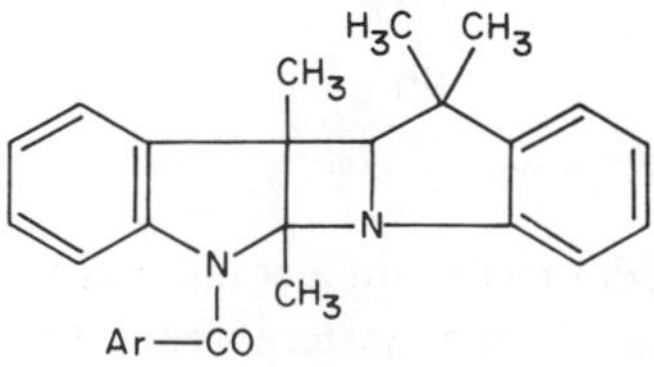   II

1,3 – Dimethylcyclohexane

solvent : acetone – d_6

Line	Intensity	ppm	Line	Intensity	ppm
1	70	44.74	6	73	27.44
2	33	41.40	7	99	26.69
3	194	35.29	8	207	22.93
4	74	33.94	9	37	21.01
5	184	33.08	10	74	20.71

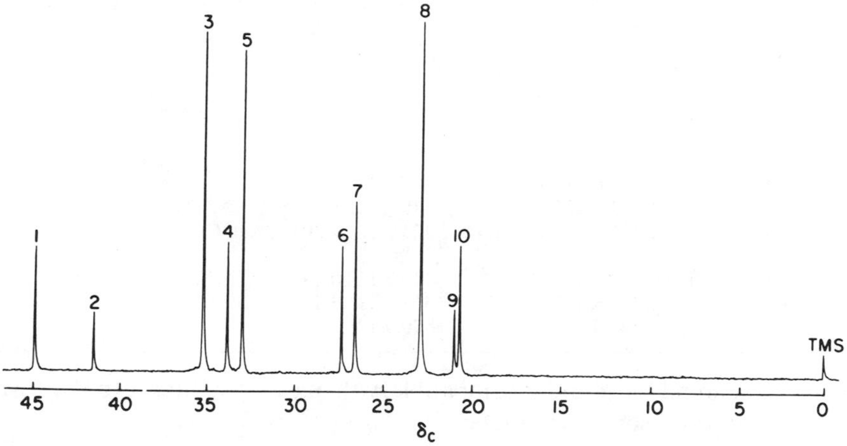

Problem Fig. 18.

molecular formula $C_{27}H_{25}ON_2Cl$. On the basis of infra-red, ultra-violet, and proton n.m.r. spectral data the two compounds were suggested to be stereoisomers of either **I** or **II**.

However, on mechanistic grounds, structures **III** and **IV** also merited examination.

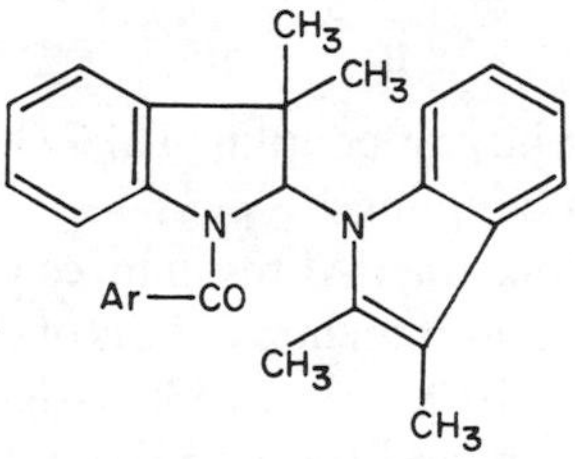

III 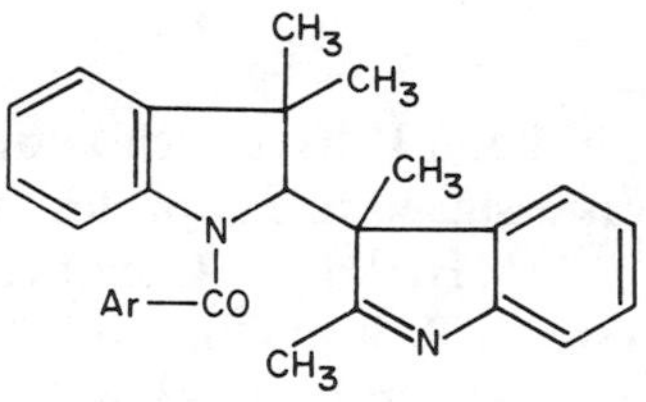   **IV**

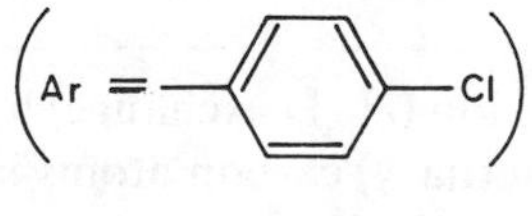

$$\left(Ar = \underset{}{\text{—}}\!\!\bigcirc\!\!\text{—} Cl \right)$$

| | Shieldings of isomers A and B | |
| | A | B |
Carbon type	δ_C	δ_C
CH$_3$	8.6, 10.1, 20.8, 31.9	8.7, 12.0, 20.5, 32.4
—CH(sp^3)	81.0	81.0
—C—(sp^3)	45.4	45.7
>C=O	169.1	169.8
=C(sp^2)—H	111.4, 116.6, 117.9 119.1, 121.4, 121.9 125.1, 127.3(2C), 128.2(3C)	107.2, 116.9, 117.8 119.3, 121.1, 121.9 125.0, 126.7(2C), 128.2(3C)
=C (sp^2)	107.6, 129.4, 131.5 133.6, 134.6, 135.6, 139.1, 141.0	110.2, 128.0, 130.4 133.9, 135.5, 136.6 139.0, 141.7

Are the data found for **A** and **B** compatible with one of structures **I** to **IV**? If yes, state what type of isomerism applies.

Problem 20: Considerable controversy has existed in the literature regarding the assignment of the methyl resonances in the ^{13}C n.m.r. spectra of angelic acid (**I**) and tiglic acid (**II**):

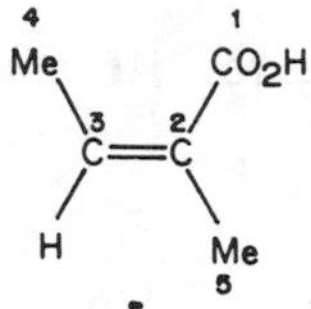

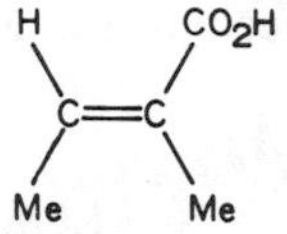

J_{HH} values in **I** and **II** have been known; the largest coupling (*ca* 7 Hz) in each instance was assigned to J_{34} proton–proton couplings.

The 400/100 MHz ^{1}H—^{13}C shift correlations (methyl resonances only) for **I** and **II** are given in Problem Fig. 10.[25] The ^{13}C n.m.r. chemical shifts of the methyl groups are as follows: **I**: δ15.93 and 20.22; **2**: δ10.95 and 13.83. Assign the ^{13}C methyl resonances (C$_4$ and C$_5$) and the ^{1}H resonances (H$_4$ and H$_5$) in **I** and in **II** on the basis of the information given.

Problem 21: Selective deuteration (H–D exchange) has been utilized for assignment of non-protonated (quaternary) carbon atoms in small- and intermediate-sized molecules. This method takes advantage of the fact that the signal intensity

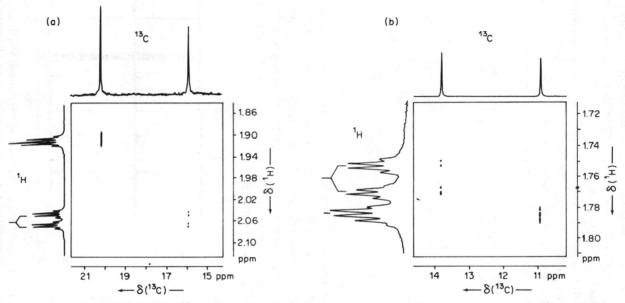

Problem Fig. 20   400/100 MHz ¹H—¹³C shift correlations for (a) angelic and (b) tiglic acid; methyl resonances only.[25] Reproduced by permission of John Wiley & Sons Ltd.

for non-protonated carbon atoms is greatly affected by the number of neighbouring protons. (s. Chapter 3) Therefore, replacement of nearest-neighbour protons
by deuterons is expected to *decrease* the intensity of the signal associated with the
quaternary carbon atom in question.

Information that is complementary to the H–D exchange experiment can be
garnered via examination of heteronuclear NOE in systems that contain
quaternary carbon centres. Here, saturation of neighbouring proton signals
should *increase* the intensity of the signal associated with the non-protonated
carbons.

Both of these n.m.r. techniques have been applied to the question of whether
compound **I** exists in the **E** form (**Ia**) or in the **Z** form (**Ib**).[26]

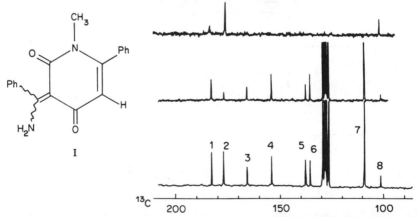

Relevant ^{13}C n.m.r spectra are given below. In addition, note that C_7 is a doublet
while C_1–C_6 and C_8 are singlets in the off resonance-decoupled ^{13}C n.m.r.
spectrum of **I**. On the basis of the given ^{13}C n.m.r. spectral information, deduce
the structure of **I**. Assign as many of the numbered ^{13}C n.m.r. resonances in **I** as
you can on the basis of chemical shift arguments and on the basis of peak
intensities in the ^{13}C n.m.r. spectra of **I**:

Problem Fig. 21   Proton-decoupled ^{13}C spectrum for **I** in MeOH (bottom), MeOD (centre), and the
difference heteronuclear NOE spectrum (top). ^{1}H irradiation at the frequency for the lowest field NH
proton. Reprinted with permission from Shapiro, Kolpak and Lemke, *J. Org. Chem*, **49**, 187.
Copyright (1984) American Chemical Society.

Cations, anions

In view of their importance as intermediates in solvolytic reactions and electrophilic substitution, carbocations have attracted great interest among chemists. Since the majority of carbocations and carbanions are short-lived species under normal conditions, previously they could only be studied indirectly. ^{13}C n.m.r. has finally afforded direct insight into the structure and electronic nature of organic cations and anions.[27] A very large number of stable carbocations and carbanions have been characterized by measuring their proton and ^{13}C spectra in SO_2ClF–SbF_5 or SO_2ClF–FSO_3H–SbF_5 solvent mixtures at low temperature. The chemical shifts of carbons in carbocations have been shown to vary over a wide range, depending on the ability of the system to delocalize the positive charge.

Problem 22: Predict the relative order of chemical shifts for the carbon formally bearing the positive charge and give reasons.[28]

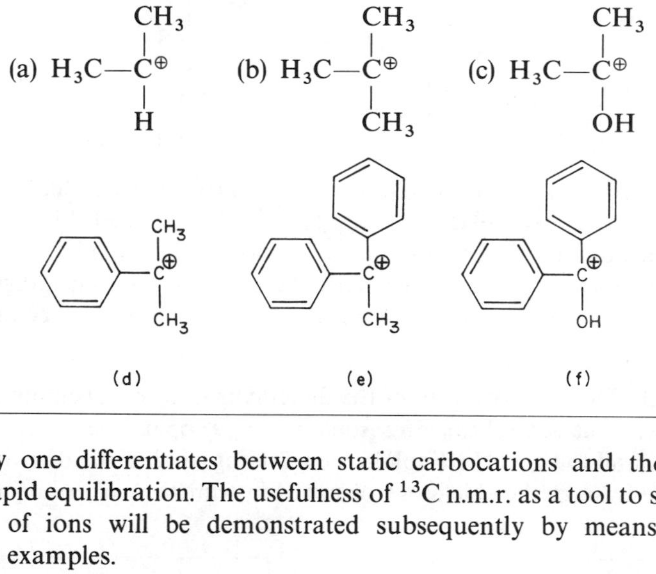

(a) $H_3C-\overset{\displaystyle CH_3}{\underset{\displaystyle H}{\overset{|}{\underset{|}{C}}}}{}^{\oplus}$
   (b) $H_3C-\overset{\displaystyle CH_3}{\underset{\displaystyle CH_3}{\overset{|}{\underset{|}{C}}}}{}^{\oplus}$
   (c) $H_3C-\overset{\displaystyle CH_3}{\underset{\displaystyle OH}{\overset{|}{\underset{|}{C}}}}{}^{\oplus}$

(d)   (e)   (f)

Basically one differentiates between static carbocations and those which undergo rapid equilibration. The usefulness of ^{13}C n.m.r. as a tool to study both categories of ions will be demonstrated subsequently by means of some illustrative examples.

The importance of arenium ions[29,30] as intermediates in electrophilic aromatic

35

TABLE 5.2[30] ^{13}C shieldings in pentamethylnitrobenzenium
and pentamethylbenzenium ions

Ion	C_4	$C_{3,5}$	$C_{2,6}$	C_1
36	205.8	142.5	181.4	98.7
37	192.2	133.1	191.9	98.4
38	197.7	142.7	184.3	99.4
39	200.5	145.5	181.5	99.9
40	191.9	139.5	193.8	57.7
41	187.0	129.7	202.3	57.1
42	185.9	139.7	196.8	58.7

substitution is generally acknowledged. Nuclear magnetic resonance finally established the nature of these transient species as σ complexes of type **35**, where E stands for electrophile. Table 5.2 lists the chemical shifts found in some pentamethylbenzenium and pentamethylnitrobenzenium ions[30] (**36–43**).

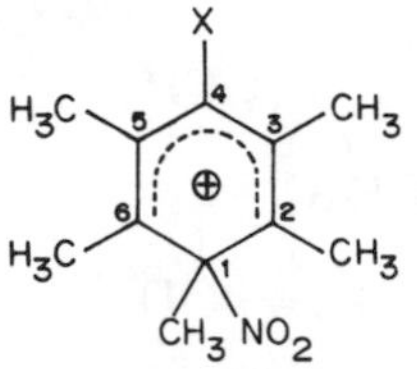

36 X=CH$_3$	**40** X=CH$_3$
37 X=F	**41** X=F
38 X=Cl	**42** X=Cl
39 X=Br	**43** X=Br

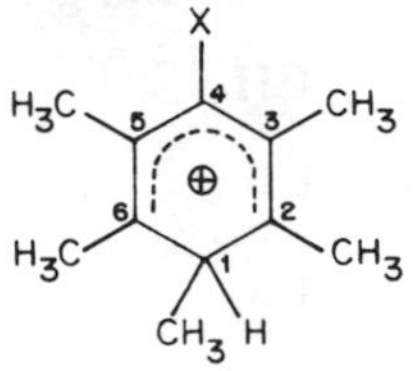

The large upfield shifts observed for C-1 clearly indicate that a change in hybridization has taken place from sp^2 to sp^3. The chemical shifts for the *ortho* and *para* carbons, on the other hand, are found at substantially lower field than those in pentamethylbenzene, whereas the *meta* carbon shieldings are little affected by ion formation, in accordance with MO charge density calculations.

Problem 23: The ^{13}C spectrum of the dimethylisopropylcarbenium ion affords two lines with the central carbon resonating at 198 ppm. The coupling constant $^1J_{CH}$ was found to be 65 Hz. Explain these findings and rationalize the chemical shift on the basis of the shieldings observed in trimethylcarbenium ion: 48 and 328 ppm.

A large number of carbocations are known to equilibrate rapidly between degenerate forms whether they belong to the classical trivalent (carbenium) ions

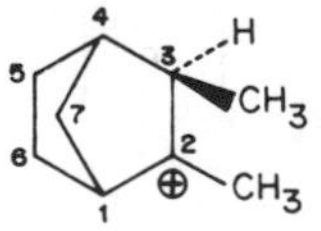

44

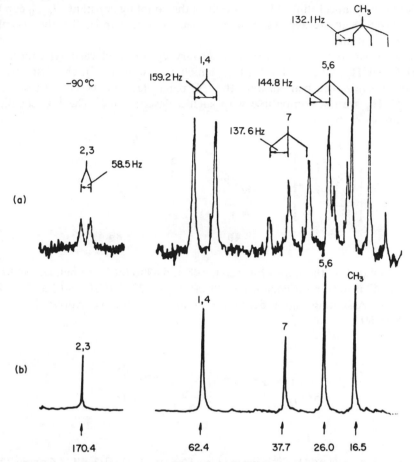

Fig. 5.12   25.2 MHz ^{13}C spectra of 2-methyl-exo-3-methyl-2-norbornyl cation at
−90°C.[32] (a) proton coupled; (b) proton noise-decoupled.

or to the bridged non-classical tetra- and pentavalent carbonium ions. The nature
of 2-methyl-*exo*-3-methyl-2-norbornyl cation **44** was disclosed from its ^{13}C and
proton spectra.[32] Figure 5.12 shows the proton coupled and noise-decoupled
^{13}C spectrum of the ion under conditions as indicated. The latter displays five
resonances. Chemical shifts and one-bond ^{13}C—H coupling constants point to a
rapidly equilibrating carbenium ion, i.e.

Both the chemical shift of $C_{2,3}$ as well as the coupling constant $^1J_{C_{2,3}H}$ can be rationalized as an average of the values expected for C-2 and C-3 in the case of a static carbenium ion.

In contrast to the secondary and tertiary cyclopropyl carbinyl cations c-$C_3H_5C^+HCH_3$ and c-$C_3H_5C^+(CH_3)_2$ which were shown to be characterized as partly delocalized carbenium ions, the ^{13}C n.m.r. data for the parent cation, c-$C_3H_5C^+H_2$ are not compatible with such a description.[33] The ^{13}C chemical shifts are given below:

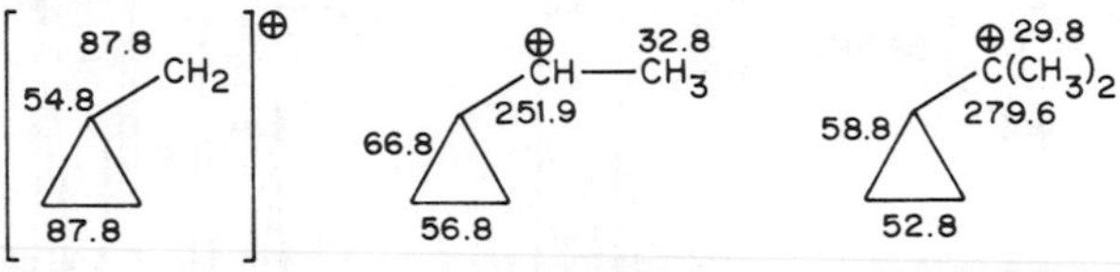

47   **48**   **49**

Extrapolation of the shifts observed for **48** and **49** to the hypothetical localized structure **47** results in the chemical shifts $\delta(CH_2^+) = 222$, $\delta(CH_2) = 61$ and $\delta(CH) = 74$ ppm. Assuming rapid degenerate rearrangement as expressed by the equilibria **50**,

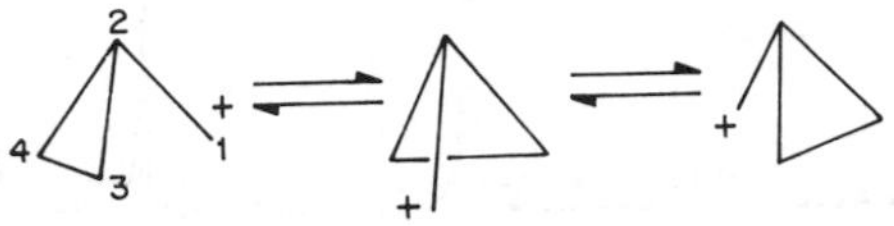

50

the chemical shift δCH_2 is estimated as $[2(61) + 1(222)]/3 = 114.6$ ppm. This value deviates substantially from the experimentally observed shift of 87.8 ppm, thus rendering such an equilibrium very unlikely. Analogous considerations equally permit one to exclude the classical cyclobutenium and allyl carbinyl cations. In fact the ^{13}C spectroscopic data could only be matched to the non-classical cyclopropylcarbonium ion which may be formed from **47** by bond delocalization via two-electron three-centre bond formation (**51**):

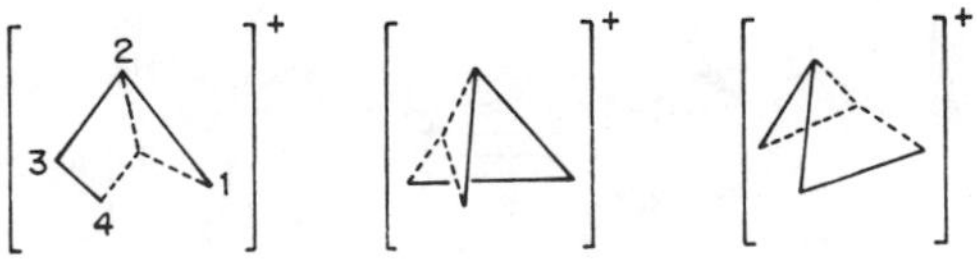

51

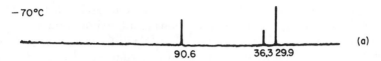

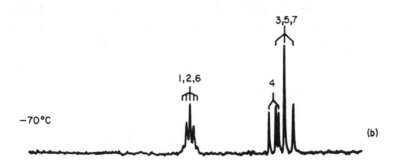

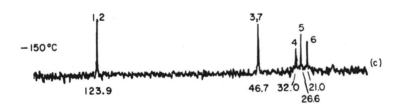

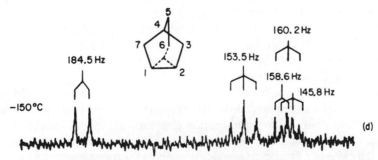

Fig. 5.13   25.2 MHz ^{13}C spectra of the 2-norbornyl cation. (a) − 70°C, proton noise-decoupled; (b) − 70°C, off-resonance decoupled; (c) − 150°C, proton noise-decoupled; (d) − 150°C, proton coupled.[34]

The [13]C n.m.r. spectrum of the 2-norbornyl cation, obtained by Olah and coworkers in 1973 under stable ion conditions in a low-nucleophilicity 'super-acid' solvent system,[34] is shown in Fig. 5.13. At $-70°C$, only three resonances were observed in the proton noise-decoupled spectrum (Fig. 5.13(a)). In the off-resonance decoupled spectrum at $-70°C$, the lowest field signal exhibited quintuplet multiplicity (Fig. 5.13(b)). These results are consistent with a car-benium ion that possesses threefold symmetry at $-70°C$. At $-150°C$, the equilibration process becomes slow on the n.m.r. time scale, and five resonances appear (Fig. 5.13(c)). The bridgehead carbon (C-6) absorbs at 21.0 ppm, a result which implies that very little positive charge resides at this position. In contrast, carbon atoms C-1 and C-2 resonate at 123.9 ppm.

Three processes have been identifed as being operational in the 2-norbornyl cation: (1) 6,2-hydride shift, (2) 3,2-hydride shift, and (3) Wagner–Meerwein rearrangement:[35]

The reader can readily verify the observed three-signal [13]C n.m.r. spectrum of this cation at $-70°C$ (Fig. 5.13(a)) by invoking concurrent, rapid Wagner–Meerwein rearrangements and 6, 2-hydride shifts in this system. At $-150°C$, the 6, 2-hydride shift becomes slow on the n.m.r. time scale, and the resulting five-signal spectrum (Fig. 5.13(c)) can be accounted for by invoking only rapid Wagner–Meerwein rearrangement of the 2-norbornyl cation at this temperature. The 3, 2-hydride shift in this system remains slow on the n.m.r. time scale at temperatures below $ca - 60°C$.[36] Thus far, all attempts to 'freeze out' the Wagner–Meerwein rearrangement (i.e. to render it slow on the n.m.r. time scale) have met with failure.

As an alternative to the 'equilibrating classical ion' model for the Wagner–Meerwein rearrangement, a number of investigators have interpreted the results of ^{13}C and [1]H n.m.r. studies as providing support for the existence of a σ-bridged non-classical 2-norbornyl cation that might be represented by a resonance hybrid (52c) of the two equilibrating 'classical' ions 52a–b:

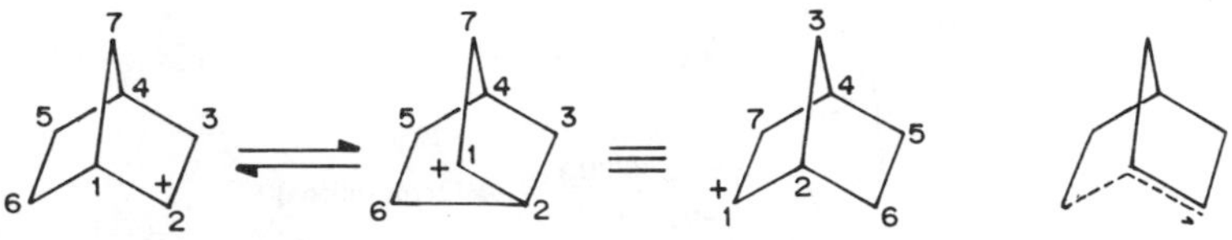

52a   **52b**   **52c**

It has been argued that equilibration of **52a** with **52b** should result in averaging of the ^{13}C chemical shifts of the carbon atom at the charge site (i.e. C-2, which bears a positive charge in **52a** and is uncharged in **52b**). The operation of non-classical charge delocalization (as in **52c**) is expected to result in a chemical shift at the charge site that is upfield of this 'average' value.

This interpretation, however, has not gone unchallenged.[37] First, Kirmse[38] has pointed out that the choice of model system upon which the 'average chemical shift' is based is critical to this argument. Second, and perhaps more important, such arguments are of necessity based upon the assumption that n.m.r. chemical shifts correlate in a simple and direct way with charge densities in carbenium ions[38]. While this statement certainly has merit, it belies the complexity of the situation.[39] Nevertheless, much useful information has been obtained from ^{13}C and ^{1}H chemical shift studies that pertains to the structure of carbenium ions.[40,41]

Problem 24: The diol **I** is found to transform into a stable dication[42] in SbF$_5$– SO$_2$ solution at $-70°$C whose ^{13}C spectrum shows four resonances: 10.6 ppm (CH$_3$), -2 ppm (CH$_3$), 126.3 ppm ($-\overset{|}{\underset{|}{C}}-$) and 22.5 ppm ($-\overset{|}{\underset{|}{C}}-$). Propose a structure.

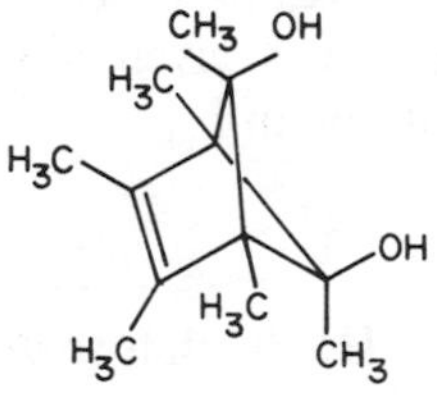

I

Problem 25: Considerable controversy has accompanied the question of the detailed structure of the 2-norbornyl carbocation. However, this question can be resolved readily in the case of the corresponding 9-pentacyclo[4.3.0.0^{2,4}.0^{3,8}.0^{5,7}]nonyl carbocation (**I**, produced when the corresponding alcohol is dissolved in 'magic acid'):

There are three possible structural situations that might be adopted by cation **I**:

(1) The cation might have a static structure, **Ia**.
(2) The cation might be represented in terms of rapidly equilibrating classical ions (**Ia⇌Ib⇌**etc.).
(3) The cation might adopt a non-classical structure, **Ic**.

The ^{1}H and ^{13}C NMR spectra of **I** are given below:

(1) ^{1}H n.m.r. spectrum: $\delta 2.25, 2.38, 3.97$ (broad singlets, relative intensities 1:1:1);
(2) ^{13}C n.m.r. spectrum: $\delta 29.55$ ($^1J_{CH} = 204\,Hz$); $\delta 31.43$ ($^1J_{CH} = 179\,Hz$); $\delta 41.28$ ($^1J_{CH} = 185\,Hz$).

Which structure, **Ia**, **Ia⇌Ib**, or **Ic**, best fits the given n.m.r. data for the carbocation? Utilize the $^1J_{CH}$ values given above to assign the ^{13}C n.m.r. resonances to specific carbon atoms in **I**.

Problem 26: Two-dimensional $\{^2H\}^{13}C$ shift correlation experiments have been employed recently for n.m.r. analysis of deuterated or partially deuterated sites. Through the use of this technique, n.m.r. analysis of the partially ^{2}H-labelled products that result from deuterium scrambling during the course of chemical reactions has been shown to be capable of providing information of mechanistic import.

An interesting application of this technique to the study of the detailed mechanism of decomposition of 3, 3-dideuteriobicyclo [2.1.1]hexyl-2-diazonium ion in aqueous medium has been reported:

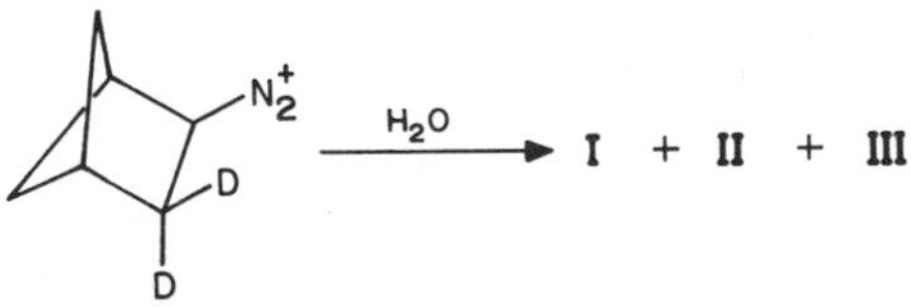

The product of this reaction has been shown to be a mixture of *three* deuterated bicyclo[2.1.1]hexan-2-ols, (**I**, **II**, and **III**). The ^{13}C n.m.r. chemical shift values for the six carbon atoms in bicyclo[2.1.1]hexan-2-ol are listed below:*

Carbon atom	Chemical shift (ppm)
C-1	46.2
C-2	72.0
C-3	38.2
C-4	38.3
C-5	34.9 (*syn* to OH)
C-6	39.2 (*anti* to OH)

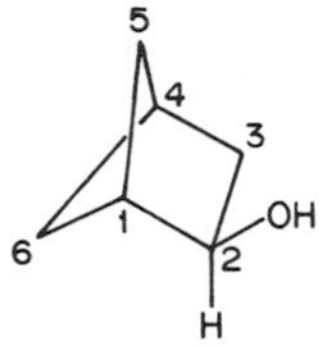

The one-dimensional $\{^2H\}^{13}C$ polarization transfer experiment (Problem Fig. 26(a)) affords three ^{13}C signals due to partially deuterated carbons. A fourth (downfield signal can also be seen in this figure; however, the two-dimensional $\{^2H\}^{13}C$ correlation experiment is not sufficiently sensitive to permit correlation of this ^{13}C peak with corresponding ^2H signals (see Problem Fig. 26(b)). The ^2H n.m.r. spectrum of the product shows six absorption signals, 1-6 (Problem Fig. 26(c)).

(1) Do the CD_2 units in the starting bicyclo[2.2.1]hexyl-2-diazonium ion remain intact in each of the three product bicyclo[2.2.1]hexan-2-ols, or are they further scrambled?

(2) Suggest structures for the three reaction products that are consistent with the information in Problem Figs 26(a)–(c) and with the given ^{13}C n.m.r. chemical shift data.

(3) Assign the peaks in ^2H n.m.r. spectrum of the product mixture to each of the three deuterated bicyclo[2.1.1]hexan-2-ols that are formed in this reaction by making use of the information contained in the two-dimensional $\{^2H\}^{13}C$ shift correlation (Problem Fig. 26(b)). (Note that definite assignment of the ^2H n.m.r. spectrum is of particular value for quantitative analysis of the product mixture since it is this spectrum that affords the most accurate integration results).

---

*The ^{13}C n.m.r. chemical shifts for individual carbon atoms in bicyclo[2.1.1]hexan-2-ol were determined by Kirmse and Zellmer.[43] The chemical shift values for C-3 and C-6 in this compound that are reported in the above table require that the corresponding assignments previously reported by Stothers and Tan[44] be reversed.

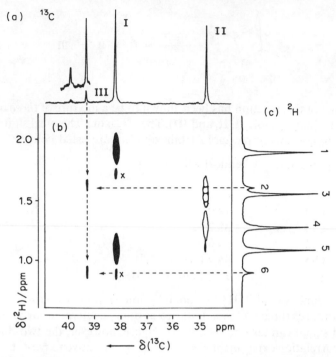

Problem Fig. 26 Analysis of a mixture of bicyclo[2.1.1]hexanols (I-III). (a) Result of a one-dimensional {²H}¹³C polarization transfer experiment using polarization transfer pulse sequence with ¹H and ²H decoupling; (b) Two-dimensional ²H, ¹³C shift correlation with continuous ¹H broadband decoupling; Signals marked with (x) are t_1 quadrature detection artefacts. (c) ²H spectrum (61.4 MHz) of I-III. Reprinted with permission from Wesener and Günther, *J. Am. Chem. Soc.*, **107**, 1537. Copyright (1985) American Chemical Society.

Halonium ions were postulated to occur as intermediates in electrophilic halogenation reactions. A number of stable alkyl and dialkylhalonium ions could be detected at low temperature in superacid media.[45–47] It is now well established that the positive charge in halonium ions is mainly localized on the halogen atom. In the ethylene bromonium[45] ion the ¹³C chemical shift is 75.6 ppm. Based on a comparison with the chemical shift of the isopropyl carbenium ion (318 ppm) one would expect about 310 ppm for the CH_2 carbon carrying the charge in the hypothetical ion **53** whereas the sp^3 carbon α to Br would be expected to have a shift of approximately 39 ppm. From this, an average of 175 ppm would result in marked contrast to the observed value. The one-bond ^{13}C—H coupling constant of 185 Hz is close to the one in ethylene oxide which further supports the cyclic structure **54**.

In much the same way, ^{13}C n.m.r. lends itself to the study of carbanions, in particular their charge density distribution.[48] The latter is known to be related to product distribution and kinetic parameters in reactions involving protonation. In conjugated dienyl anions[48] (**55–57**), it was found that the odd-numbered carbons all resonate at substantially higher field in accord with MO calculations, predicting increased electron charge for these centres:

55	**56**	**57**

5.3   DYNAMIC PROCESSES, CONFORMATIONAL ANALYSIS

In addition to structure elucidation, ^{13}C n.m.r. has proved its potential for the examination of various kinds of transient processes such as chemical exchange, intramolecular rearrangements as they occur in tautomeric equilibria, valence tautomerism, and other rapid isomerization processes. A second area where ^{13}C n.m.r. has been utilized successfully concerns the study of molecular conformation in solution, whether one deals with stable conformers or molecules where rapid interconversion occurs at ambient temperature. Both aspects of conformational analysis will be considered in this section.

The study of dynamic processes in general requires the analysis of the spectra at several temperatures.[49] A dynamic process is said to be slow in the n.m.r. time scale if the nucleus under investigation exchanges between two environments at a rate which is small compared with the frequency separation in the two environments. Two separate signals can be discerned only if this condition is satisfied. The exchange rate is controlled by the temperature of the sample. Although the majority of temperature-dependent n.m.r. studies carried out so far used the proton, this nucleus suffers from two deficiencies. First, proton spectra are complicated by spin–spin coupling which produces frequent overlap and thus renders line shape studies difficult. Second, the proton chemical shift differences of a nucleus in only slightly different environments are small. This stipulates very low temperatures in order to reach coalescence (critical temperature below which the two lines begin to separate). The experiment often fails, however, because the coalescence temperature is beyond technical feasibility. Both problems are usually alleviated by ^{13}C n.m.r. No complex spin multiplets occur because ^{13}C spectra are recorded under proton noise-decoupling conditions. Moreover, owing to the sensitivity of ^{13}C shieldings to minor structural changes, the chemical shift difference exhibited by a carbon in different environments is generally larger than for corresponding protons. The practical consequence of this property is the higher coalescence temperatures generally observed for ^{13}C which enables the study of rapid equilibria not amenable to proton n.m.r. examination.[27]

The information which may be obtained from temperature-dependent n.m.r. is qualitative and quantitative and comprises three areas:

(1) Characterization of structure (conformation) of the species undergoing exchange. This requires that the data be collected in the slow-exchange limit, i.e. at temperatures below coalescence. Except for the different experimental conditions, the problems are similar to those outlined in the previous section.

(2) Determination of the relative thermodynamic stability in terms of the free energy ΔG. This again requires slow-exchange spectra from which the molar ratio of the two components is measured by integration. The method consequently is confined to small values of ΔG and K, the thermodynamic stability constant. If the concentration of the two exchanging components are [A] and [B] according to the equilibrium

$$A \overset{k_A}{\underset{k_B}{\rightleftharpoons}} B \qquad (5.1)$$

the equilibrium constant K may be defined as

$$\frac{k_B}{k_A} = K = [B]/[A] = \exp(-\Delta G/RT) \qquad (5.2)$$

k_A and k_B in Eqn (5.2) denote the rate constants in s^{-1}. Since ΔG is temperature-dependent through the entropy term, it is often mandatory to determine ΔS as well in order to be able to extrapolate ΔG to ambient temperature. Denoting the populations of A and B as p_A and p_B, respectively, the chemical shift of carbon C_i, δC_i^{av} is a weighted average of the corresponding chemical shifts δC_{iA} and δC_{iB}:

$$\delta C_i^{av} = p_A \delta C_{iA} + p_B \delta C_{iB} \qquad (5.3)$$

or

$$\delta C_i^{av} = p_A \delta C_{iA} + (1 - p_A)\delta C_{iB} \qquad (5.4)$$

Measurement of δC_i^{av} for two different temperatures T_1 and T_2 thus permits one to determine $\Delta G(T_1)$ and $\Delta G(T_2)$ from which ΔS is obtained by insertion into the Helmholtz equation:

$$\Delta G = \Delta H - T\Delta S \qquad (5.5)$$

(3) Derivation of kinetic parameters. The determination of the kinetic parameters in terms of rate constants and activation energies requires a line shape analysis, i.e. experiments have to be performed near the coalescence temperature. For a detailed discussion of the procedures, the reader is referred to the pertinent reviews.[49] At temperatures above coalescence, the rate constants k_A and k_B for a simple two-site exchange process are given by Eqns (5.6a) and (5.6b):

$$k_A = \frac{p_A p_B^2 4\pi v_{AB}^2}{\delta v} \qquad (5.6a)$$

$$k_B = \frac{p_A^2 p_B 4\pi v_{AB}^2}{\delta v} \qquad (5.6b)$$

In Eqns (5.6), v_{AB} represents the chemical shift difference of an exchanging nucleus in environments A and B, and δv is the line width corrected for effective transverse relaxation. For degenerate processes (interchange between equal energy forms) $p_A = p_B$ and Eqns (5.6a) and (5.6b) simplify to

$$k = \pi v_{AB}^2/2\delta v \qquad (5.7)$$

At the coalescence temperature, the rate becomes

$$k = \pi v_{AB}/\sqrt{2} \qquad (5.8)$$

which is restricted to cases in which $p_A = p_B$. However, accurate rate constants near the coalescence temperature can only be derived from a computer-matching of the experimental spectra.

Below the coalescence temperature, finally, where separate lines are observed, the rate constant may be approximated by Eqns (5.9a) and (5.9b):

$$k_A = \pi \delta v_A \qquad (5.9a)$$
$$k_B = \pi \delta v_B \qquad (5.9b)$$

where δv_A and δv_B are the corrected linewidths.

The reaction rate constant k_r is related to the enthalpy and entropy of activation, ΔH^* and ΔS^*, respectively, through the Eyring equation

$$k_r = (\kappa kT/h)\exp(-\Delta G^*/RT)$$
$$= (\kappa kT/h)\exp(\Delta S^*/R)\exp(-\Delta H^*/RT) \qquad (5.10)$$

In Eqn. (5.10), κ represents the transmission coefficient, k the Boltzmann constant; the remaining quantities have their usual significance.

Intramolecular rearrangements, fluxional molecules

Imidazole is known to undergo rapid tautomeric equilibration between equal energy tautomers as depicted in **58**.

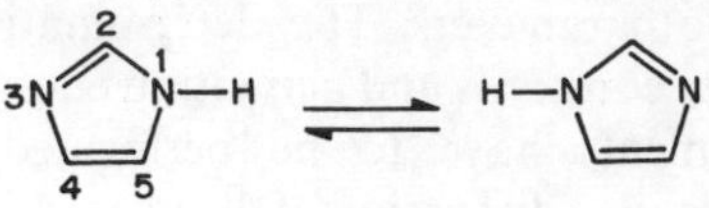

58

The ^{13}C spectrum at ambient temperature consequently consists of only two resonances. For a long time it was an unanswered question which was the predominant tautomer in the amino acid L-histidine which possesses the imidazole moiety:

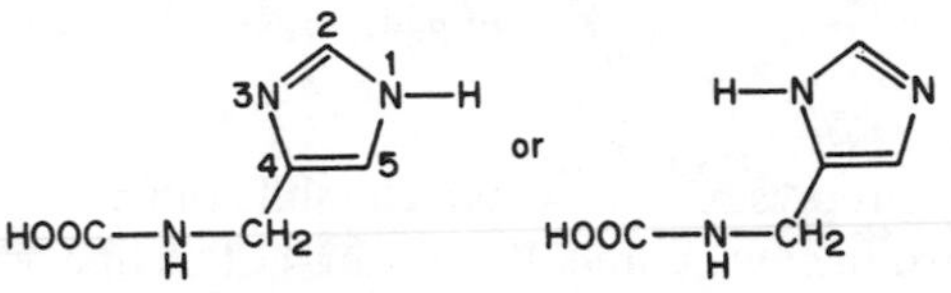

59   **60**

Conclusive evidence that **59** prevails over **60** was provided by the pH dependence of chemical shifts. The titration shifts for imidazole carbons in histidine were found to be identical in sign and similar in magnitude to those in 1-methylhistidine. This clearly establishes the ^{1}H tautomer to be predominant in basic solution.

The 22.6 MHz ^{13}C spectrum of tropolone acetate[51] at $-70°C$ consists of nine resonances which could be assigned as follows:

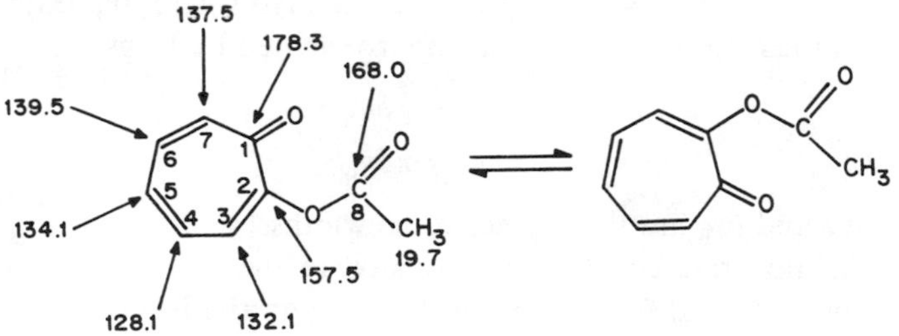

61   **62**

At elevated temperatures, fast acetyl migration should result in averaging of C-1 and C-2, C-3 and C-7, and C-4 and C-6. Because the two interchanging forms are equivalent, their populations are equal and therefore the expected shifts of the exchanging nuclei are just the mean of the shifts observed in the slow-exchange limit. At the line widths observed at 50°C, only three lines could be discerned at 167 ppm (C-1, C-2, C-8), 134 ppm (C-3, C-4, C-5, C-6, C-7), and 19.8 ppm (CH$_3$). Line shape analysis in the temperature range between -70 and $-10°C$ yielded for the free energy of activation $\Delta G^* = 10.8$ k cal/mol.

The fluxional nature of a large number of inorganic and organometallic molecules is well documented. The structure of dicyclopentadienyltricarbonyldirhodium[52] was shown by X-ray crystallography to be **63**. At room

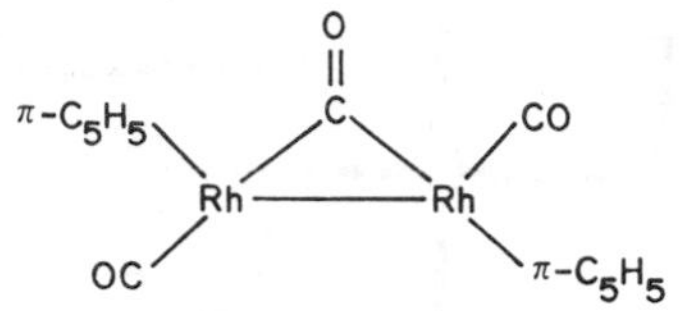

63

temperature, the carbonyl region of the ^{13}C spectrum exhibits only a triplet ($J_{Rh-C} = 43$ Hz) due to equal coupling to the two $^{103}RH*$ spins. This behaviour suggested rapid exchange of bridging and terminal carbonyls rather than an equilibrium between a bridging and non-bridging moiety. At $-80°C$ this process was found to become slow, which gave rise to the observation of a triplet at 231.8 ppm ($J_{Rh-C} = 45$ Hz) and a doublet at 191.8 ppm ($J_{Rh-C} = 83$ Hz). From these data the weighted average coupling constant can be predicted as $(0 + 83 + 45)/3 = 42.7$ Hz, which is in good agreement with the experimental figure.

A somewhat more complex situation was found for dicyclopentadienyltetracarbonyldiiron **64** and **65**:

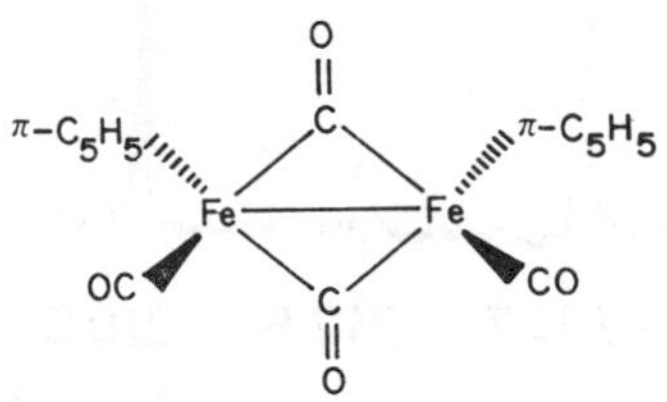

64 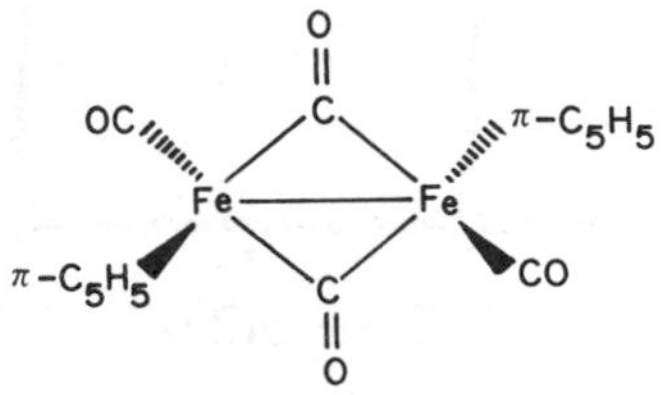   **65**

The temperature-dependent ^{13}C spectra of the carbonyl region[53] (Fig. 5.14) afforded insight into the dynamic processes occurring in this molecule. The high-temperature spectrum reveals only one resonance at 243 ppm which broadens with decreasing temperature. At $-35°C$, well below the coalescence temperature, three lines appear at 274.7, 242.6, and 210.5 ppm. When the temperature is lowered further, the centre line broadens and finally vanishes at $-85°C$. These data provide clear evidence of three different time-dependent processes, which are suggested to originate from an interchange of bridging and terminal carbonyls occurring for both the *cis* (**64**) and *trans* isomer (**65**) and from *cis–trans* interconversion. However, the two equilibria differ in their activation energy. The chemical shifts of the bridging carbonyl, on the one hand, and terminal carbonyls,

$*^{103}Rh$ has 100% abundance and spin $I = 1/2$.

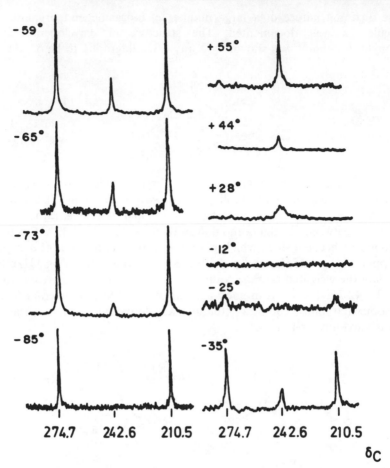

Fig. 5.14 Temperature-dependent ^{13}C spectra of carbonyl carbons in dimeric cyclopentadienyliron dicarbonyl.[53] The original CS_2-based chemical shift scale was converted to TMS standard by using the relationship $\delta\,TMS = 192.4 - \delta\,CS_2$.

on the other, were assumed to be identical in the two isomers under the experimental conditions used. At the low-temperature limit no interconversion occurs. Hence two resonances are observed which are assignable to the bridging (274.7 ppm) and terminal (210.5 ppm) carbons. As the temperature is raised, the kinetically less stable (*trans*) isomer begins to interconvert, thus resulting in the centre line at 242.6 ppm. At $-35°C$ the *cis* isomer equally starts to undergo rapid intramolecular conversion. The disappearance of the centre line at $-12°C$ is due to *cis–trans* isomerization becoming fast, finally leading to a single-line spectrum.

Conformational equilibria, configurational inversion

Considerable effort has been devoted to the characterization of the conformation of carbocycles and heterocycles. One of the early examples examined by low-temperature ^{13}C n.m.r. is methylcyclohexane,[54] which was known to exist primarily as the equatorial conformer. This molecule is favourable for ^{13}C examination at low temperature because of the predicted large chemical shift differences for C-3, 5 and the methyl carbon in the two conformers. The low-temperature 63.1 MHz ^{13}C spectrum in Fig. 5.15 reveals, besides the five strong resonances due to the equatorial conformer, two weak lines at the high-field end of the spectrum. The chemical shifts are in agreement with those predicted for carbons 3 and 5, and C(1)—CH$_3$ in the axial form.[55] From the intensity ratio observed for corresponding lines, an equilibrium constant of approximately 100 was estimated in accord with the accepted value of 1.6 kcal/mol by which the equatorial form is more stable than its axial counterpart. Because of the steric interactions between protons C(3)—H and C(5)—H with the axial substituent, the equatorial form is generally favoured, i.e. ΔG is positive for the equatorial-to-axial conformational change. An interesting exception to this rule was found for cyclohexylmercuric derivatives[56] which exist preferentially in their axial forms. Unambiguous assignment of the axial and equatorial forms could be made on the basis of the vicinal ^{199}Hg—^{13}C coupling constant, which obeys a Karplus relationship (cf. Chapter 2). The 63.1 MHz ^{13}C spectra recorded at $-90°$C displayed separate resonances for the two conformers from which $\Delta G = -0.2$ to -0.3 kcal was obtained. The anomalous behaviour observed for these compounds is ascribed to the exceptionally long carbon–mercury bond resulting in

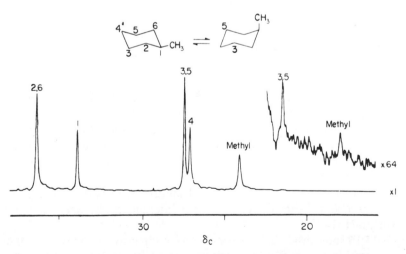

Fig. 5.15   63.1 MHz proton noise-decoupled ^{13}C spectrum of neat methylcyclohexane at $-110°$C.[54]

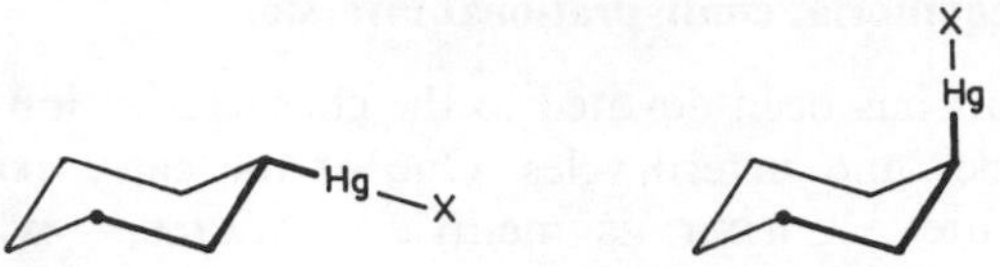

66

trans

$^3J_{HgC} = 275$ Hz

67

gauche

$^3J_{HgC} = 70$ Hz

lowered steric repulsion. This interpretion is corroborated by the absence of a steric compression shift for $C_{3,5}$ of the axial conformer.

Substitution of a hydrogen by the bulky *t*-butyl group locks the cyclohexane ring in a conformation in which the *t*-butyl substituent is equatorial. Although the molecule may still undergo rapid equilibration, the population of the conformer with the *t*-butyl substituent in axial position becomes negligibly small.

So far, we have only dealt with chair conformations, which in most situations constitute the lowest energy form of six-membered rings. Exceptions are expected in those instances where 1,3-*syn*-diaxial interactions occur in both chair conformations. An analysis of the ring carbon chemical shifts in methyl-substituted 1,3-dioxanes[57] provides clear evidence of the presence of a non-chair conformation. Although the ring carbon shieldings could be fitted extremely well with a set of additive increments (similar to those used for methylcyclohexanes), large deviations between observed and predicted chemical shifts were found for *trans*-2,4,4,6-tetramethyl-1,3-dioxane:

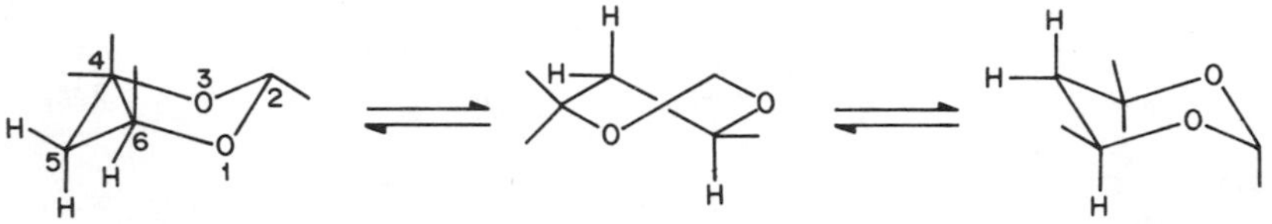

68   **69**   **70**

These are for **68**, $\Delta\delta = 3.7$ (C-2), 5.5 (C-4), 1.6 (C-5), and 3.6 ppm (C-6) with the observed shielding being smaller throughout than the predicted values. This points to the existence of the twisted conformer **69**, for which the high-field proton spectrum provided definitive proof.

In contrast to the previously discussed methylcyclohexane, the 1,1,*cis*-1,2, *trans*-1,3, and *cis*-1,4-dimethylcyclohexanes interconvert between equal energy conformers **71–74**.[58,59]

Figure 5.16 shows the ^{13}C spectra of *cis*-1,4-dimethylcyclohexane (**74**) in the fast and slow exchange limit [(a) and (c)] and near coalescence (b). Since the frequency differences between C-2 and C-3, C-1 and C-4 and CH$_3$ (axial) and CH$_3$ (equatorial) are different, their coalescence temperatures are all different. This is

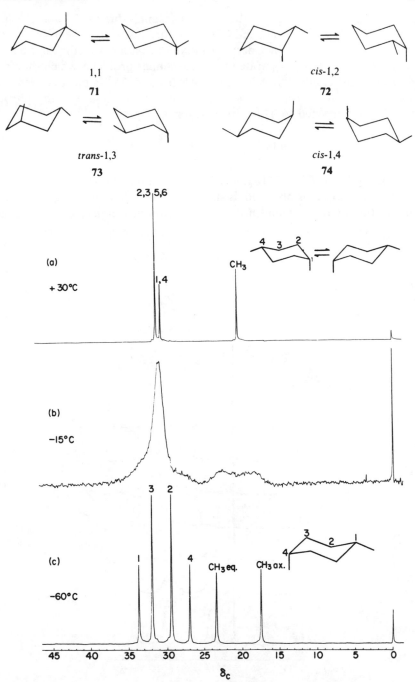

Fig. 5.16   Temperature-dependent 25.2 MHz ^{13}C spectra of 1, 4-dimethylcyclohexane, 90% in hexadeuteroacetone. (a) $+30°$C (fast exchange limit); (b) $-15°$C (near coalescence); (c) $-60°$C (slow exchange limit).

clearly recognized from Fig. 5.16(b), where the signals due to C-2 and C-3 have coalesced, whereas CH_3 (axial) and CH_3 (equatorial) still give rise to two (partly overlapping) resonances. When deriving the kinetic parameters, the three pairs of carbons can be treated as separate two-site exchange problems which should all lead to the same result. The calculation of ΔH^* and ΔS^* is most conveniently performed through a least-squares linear regressions analysis. For this purpose Eqn. (5.10) is converted to its logarithmic form:

$$R \ln \frac{hk_r}{\kappa k T} = -\Delta H^*/T + \Delta S^* \tag{5.11}$$

By plotting $R \ln [hk_r/(\kappa k T)]$ against $1/T$, a straight line with slope $-\Delta H^*$ and intercept ΔS^* should be obtained. It should be noted that due to the degenerate nature of the interconversion in the above-mentioned molecules, $p_A = p_B$, and therefore $k_A = k_B = k_r$.

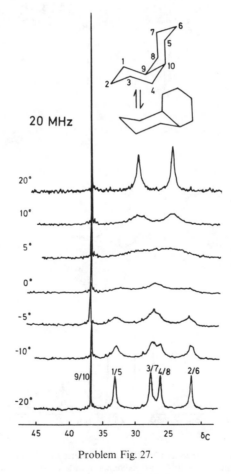

Problem Fig. 27.

Problem 27: The temperature-dependent proton noise-decoupled ^{13}C spectra of *cis*-decalin are given. Determine ΔH^* and ΔS^* with the aid of Eqns (5.6), (5.9) and (5.11).

Note: The corrected linewidths may be determined by using the width of the line which is not affected by the exchange process. Insert $\kappa = 0.5$ for the transmission coefficient. Use the following constants:

$$R = 1.985 \times 10^{-3} \, \text{kcal} \, (\text{gmol})^{-1} \, \text{deg}^{-1}$$
$$k/h = 2.21 \times 10^{10} \, \text{s} \, \text{deg}^{-1}$$

The conformations of several large carbocycles have been studied by low-temperature high-field ^{13}C n.m.r.[60-63] High magnetic fields are imperative for the examination of such molecules because of the small shielding differences for the exchanging carbons, associated with very low energy barriers for interconversion.

Strain energy calculations suggested slightly higher stability of the twisted boat-chair (TBC) over the twisted chair-boat (TCB) conformation in cyclononane.[60]

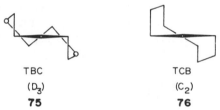

TBC	TCB
(D_3)	(C_2)
75	**76**

The 63.1 MHz ^{13}C spectrum at $-162°C$ exhibits two resonances whose areas reflect a 2:1 ratio which is consistent with the D_3 symmetry of the TBC (**75**) but not with a TCB (**76**) conformation. The conformational process occurring is assumed to involve conversion of the TBC into the boat-chair (BC) **77** and back into an enantiomeric TBC conformer:

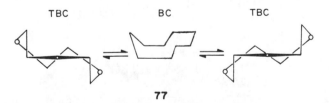

TBC   BC   TBC

77

The dynamics of conformational interconversions are often characterized by more than one rate process occurring. *Cis, cis*-1,5-cyclooctadiene[62] is an illustrative example of such a case. Possible conformations which have to be considered for this molecule are the boat (B) **78**, twist boat (TB) **79**, skew (S) **80**,

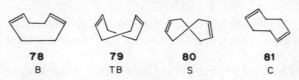

78	79	80	81
B	TB	S	C

and chair (C) **81** forms. Whereas the 60 MHz proton spectrum remained invariant down to $-150°C$, both the high-field proton and ^{13}C spectrum reflected the presence of dynamic processes. The ^{13}C spectrum of the methylene carbons converted from a singlet at ambient temperature to a 1:1 doublet below $-176°C$. This result precludes the B, S, or C forms as the major conformer and is consistent only with the C_2 symmetry of the twist boat form which undergoes a spectral process with $\Delta G^* = 4.2 \pm 0.2 \, kcal \, mol^{-1}$ at $-176°C$. The 251 MHz proton spectrum reveals two further spectral processes. The methylene proton resonance converts from a broadened singlet at $-165°C$ to an asymmetric doublet at $-168°C$ whose low-field component further splits at $-178°C$. The activation barriers associated with these processes were determined to be $\Delta G^* = 4.9 + 0.1 \, kcal \, mol^{-1}$ and $4.4 \pm 0.1 \, kcal \, mol^{-1}$, respectively. These observations are consistent with the TB as the ground conformation.

The lowest energy conformation of cyclotetradecane[63] was assumed to be the rectangular diamond lattice form **82** whose C_{2h} symmetry give rise to four observationally distinguishable carbons in the ratio 1:2:2:2. The low-temperature 63.1 MHz ^{13}C spectrum in Fig. 5.17 displays three resonance lines

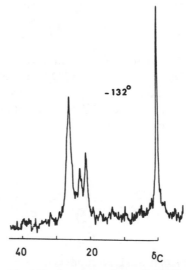

Fig. 5.17 Low-temperature 63.1 MHz proton noise-decoupled ^{13}C spectrum of cyclotetradecane, 1.2% in CS_2/CH_2 $CH_2{=}CHCl$.[63]

82

whose areas reflect a ratio of 4:1:2, which agrees with the suggested conformation provided the observed degeneracy is considered to be accidental.

The potential energy barriers to rotation about single bonds in general are very low (usually in the order of a few cal/mol) and only in exceptional circumstances could slow-exchange limit spectra be obtained that permitted the determination of the rotamer populations.[49] However, even for rapidly rotating groups, the residence times for the three lowest energy staggered rotamer conformations may be different as reflected by chemical shifts and coupling constants.

5.4 TWO-DIMENSIONAL TECHNIQUES FOR THE ELUCIDATION OF EXCHANGE NETWORKS

Elucidation of complex exchange phenomena involving exchange between multiple sites has been aided significantly by the application of two-dimensional n.m.r. methods.[64] The basic experiment follows the general pattern for two-dimensional n.m.r. as outlined in Chapter 3. During a preparation period, transverse magnetization is created via a 90° pulse; this is followed by an evolution period, t_1, during which the spins accumulate phase according to their characteristic chemical shift. A second 90° pulse then returns the y' component of the transverse magnetization to the longitudinal axis (Fig. 5.18(a)). During the subsequent mixing period of fixed duration, t_m, nuclear spins undergo exchange processes which manifest themselves via generation of cross peaks in the two-dimensional map. The two-dimensional spectral data array is generated in the usual manner: t_1 is stepped through several values, and this is followed by a double Fourier transformation.

Consider the simplest case of two groups of spins, A and B, which resonate at frequencies v_A and v_B, respectively, and which are exchanging slowly (i.e. $k \ll |v_A - v_B|$). Spins that belong to one of the two spin groups (for example, B) accumulate a phase angle $\phi_B = 2\pi v_B t_1$ during period t_1; this results in longitudinal magnetization $M_{zB} = M_{oB} \cos 2\pi v_B t_1$ following the second 90° pulse. The resulting amplitude modulation causes resonance to occur in the t_1 dimension at frequency $v_1 = v_B$. During time interval t_2 the spins that do not exchange precess at frequency v_B, thereby resulting in a resonance at $v_2 = v_B$. This effect explains the occurrence of the diagonal peak B that is unrelated to the exchange process (Fig. 5.18(b)). If exchange is present, then during the mixing period t_m some of the spins A and B exchange while retaining their modulation in the t_1 domain. Spins labelled B now precess at frequency v_A, thereby resulting in resonance at $v_1 = v_B$

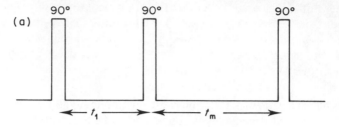

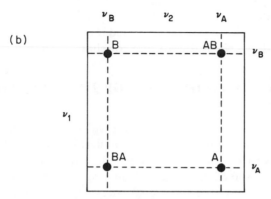

Fig. 5.18 Principle of two-dimensional exchange spectroscopy. (a) Pulse sequence diagram: t_1 is the evolution time and t_m the (usually constant) mixing time; (b) schematic illustration of two-dimensional map for chemical exchange between two species labelled (a) and (b). The off-diagonal peaks labelled AB and BA are indicative of chemical exchange.

and $v_2 = v_A$ (cf. the off-diagonal peak labelled AB in Fig. 18(b)). The remaining peaks, B and AB, can be explained in analogous fashion.

It is evident that in order for significant exchange to occur during the mixing period, the condition $t_m \gtrsim 1/k$ must hold. That is, the duration of the mixing period must be of the same order or longer than the lifetime of the spins at each site.

One advantage of studying exchange mechanisms by ^{13}C n.m.r. spectroscopy is the absence of cross peaks due to homonuclear spin–spin coupling.[65] An additional advantage of this technique *vis-à-vis* classical one-dimensional coalescence experiments is the fact that it is not necessary to perform experiments at temperatures that are near the coalescence temperature (where signal intensities of the exchange-broadened peaks are often very low). The other alternative method, i.e. saturation transfer,[66] is often very time consuming, since frequency-selective perturbation of various spectral lines is required.

Two-dimensional exchange experiments are particularly well suited for the study of multi-site exchange processes. The elucidation of complex exchange networks (for example, in *cis*-decalin and in bullvalene) through the use of ^{13}C two-dimensional n.m.r. exchange experiments has recently been demonstrated.[67]

Two-dimensional ^{13}C n.m.r. spectroscopy has been utilized for quantitative analysis of slow, intramolecular three-site exchange in a ruthenium cluster.[68] This was accomplished via detailed examination of the carbonyl ^{13}C region of the two-dimensional exchange spectrum of ^{13}CO-enriched $[Ru_3(\mu\text{-}H)(CO)_9(MeCCHCMe)]$ (**83**, below) as a function of the mixing time, t_m.

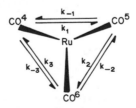

83

The one-dimensional ^{13}C n.m.r. spectrum of **83** in $CDCl_3$ solution at 299 K displays four resonances: one at $\delta 197.7$ (relative intensity 3, CO groups at Ru^2 which undergo rapid, localized exchange), and three additional resonances, each displaying relative intensity 2, at $\delta 195.9, 192.3,$ and 190.4 that correspond to the remaining CO groups (i.e. $(4, 4'), (5, 5'),$ and $(6, 6')$; note that 4, 5, and 6 undergo slow, localized exchange at Ru^3, and $4', 5',$ and $6'$ undergo slow, localized exchange at Ru^1).[69] In the two-dimensional experiment, no correlation peaks were observed that might indicate that exchange was occurring between CO moieties on Ru^2 and those on Ru^1 and Ru^3, while all possible correlation peaks interconnecting resonances due to exchange among $(4, 5, 6)$ or $(4', 5', 6')$ were observed. It was therefore concluded that CO exchange in **83** is an example of localized, three-site exchange in which populations in the three sites are equal:[68]

Six off-diagonal correlation peaks were obtained from the two-dimensional exchange spectrum of **83** for each t_m value employed. Analysis of the results in terms of relaxation and exchange behaviour[64,68] revealed that the three rate constants for three-site exchange in this system are essentially equal. This result suggests that CO exchange in **83** occurs via a concerted process that simultaneously involves all three CO groups.[68]

In principle, quantitative measurement of exchange rates requires observation of magnetization transfer as a function of mixing time. This can be accomplished by obtaining a set of m two-dimensional spectra at different mixing times t_m. Thus, in addition to the usual independent variables (i.e. evolution time (t_1) and detection time (t_2)), a third time variable (t_m) is introduced, resulting in a *three*-dimensional n.m.r. experiment.[64,70,71]

It is possible to reduce the potentially time-consuming three-dimensional experiment to a special form of a two-dimensional experiment by combining two of the three variables which are then incremented simultaneously. In practice, this is accomplished by setting $t_m = \kappa t_1$. The scaling parameter, κ, generally falls in the range between 10 and 100, depending upon exchange rates and the quality of spectral resolution. An accordion-like 'motion' results from application of the pulse sequence.

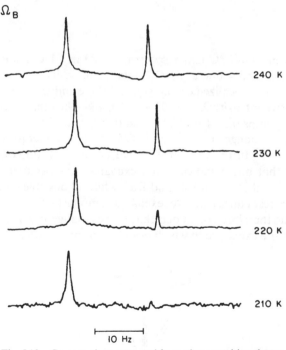

Fig. 5.19 Cross sections extracted from phase sensitive absorption mode accordion spectra of *cis*-decalin obtained at four different temperatures. The diagonal peaks (left) and cross peaks (right) show characteristic line shapes. At low temperatures, cross peaks vanish because antiphase components approach the same linewidth.[70] Reprinted with permission from Boden Hausen and Einst, *J. Am. Chem. Soc.* **104**, 1304. Copyright (1982) American Chemical Society.

Application of two-dimensional accordion spectroscopy to the study of ring inversion in *cis* decalin is shown in Fig. 5.19[70] This process is an example of a two-site exchange problem, $A \rightleftharpoons B$, with equal populations and for which $T_1^A = T_1^B$. In such an example, *diagonal peaks* are observed which reflect residual magnetization at sites A and B. The line shapes of diagonal peaks result from a *sum* of two Lorentzians whose integrated intensities are equal but whose widths differ. In addition, *cross-peaks* are also observed which result from exchange between sites A and B (i.e. these peaks result from interconversion of magnetization M_z^A and M_z^B that occurs during mixing interval t_m). The line shapes of cross peaks are determined by the *difference* between these same two Lorentzians.[70]

The rate of two-site exchange in *cis*-decalin can be determined via analysis of the accordion line shapes. Note the broad, positive pedestal associated with the low-field (diagonal) peak and the negative pedestal associated with the high-field (cross-peak) peak in the accordion spectra shown in Fig. 5.19. Activation parameters for the two-site exchange process can be determined via a simple Eyring plot; the results thereby obtained[70] are in good agreement with the corresponding values that were determined by the method shown in Problem 27 above.

5.5 ROTAMER EQUILIBRIA

The chemical shift is less suited as a means for the study of the favoured conformation of freely rotating groups because it is too sensitive to substituent effects. The majority of conformational studies of this kind have been performed by making use of the angular dependence of the vicinal coupling constant $^3J_{HCCH}$. The use of similar relationships for vicinal coupling involving ^{13}C and ^1H or ^{13}C and a spin-$\frac{1}{2}$ nucleus such as ^{31}P, ^{19}F, ^{199}Hg is described in Chapters 2 and 3 in connection with a discussion of stereochemical assignment techniques.

In substituted ethanes of the type $CR^1H_2CR^2R^3H$ the vicinal spin–spin coupling constants J_{AC} and J_{BC} of the three-spin system formed by the three magnetically non-equivalent protons are weighted averages of the coupling constants in the three minimum potential energy rotamers which are illustrated by the Newman projections **84–86**.

In rotamer **86** the two vicinal coupling constants J_{AC} and J_{BC} are *gauche*, in **84** and **85** they are *gauche* and *trans*, respectively. By defining rotamer populations

p_I, p_{II}, and $p_{III} (= 1 - p_I - p_{II})$, the vicinal coupling constants J_{AC} and J_{BC} can be expressed as

$$J_{AC} = p_I J^t + p_{II} J^g + (1 - p_I - p_{II}) J^g \tag{5.12}$$

and

$$J_{BC} = p_I J^g + p_{II} J^t + (1 - p_I - p_{II}) J^g \tag{5.13}$$

As has been pointed out in a study of side-chain conformation in amino acids and peptides,[72] the conventional approach fails if, for example, geminal protons cannot be assigned or only one vicinal coupling constant occurs. In the former case, only the fractional population p_{III} can be calculated from the sum of Eqns (5.12) and (5.13). This situation may be illustrated with aspartic acid (**87**),

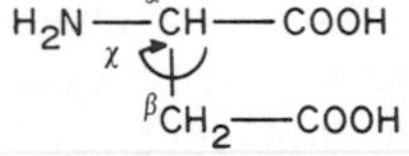

87

whose β-CH$_2$s cannot be assigned. In this case the side-chain conformation in terms of the relative populations of the three staggered rotamers can be determined by additionally exploiting the angular dependence of the three-bond coupling constants $^3J_{CO-C_\beta H_A}$ and $^3J_{CO-C_\beta H_B}$, which are a function of the dihedral angle χ. The rotamers **88, 89**, and **90** can be represented by the corresponding Newman projections (projection along the C_α—C_β bond):

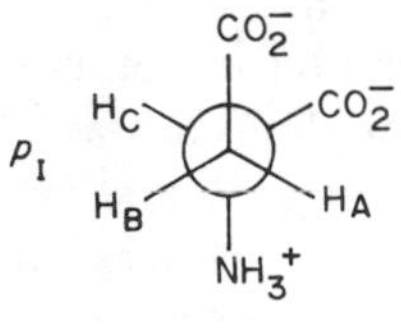

88 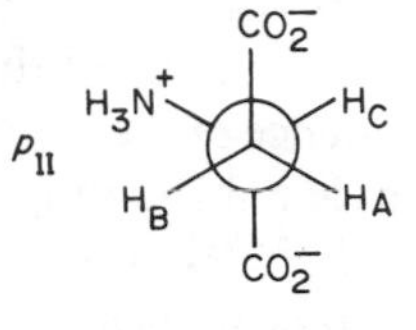   **89** 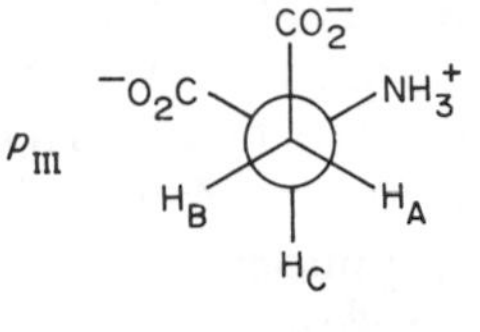   **90**

From the observed ^{1}H—^{1}H coupling constants ($^3J_{C_\alpha H-C_\beta H} = 9.4$ and 4.2 Hz) and by inserting $J_{HH}^g = 2.56$ Hz and $J_{HH}^t = 13.6$ Hz, the approximate rotamer populations are calculated to be either (a) or (b):

(a) $p_I = 0.15, p_{II} = 0.62, p_{III} = 0.23$

(b) $p_I = 0.62, p_{II} = 0.15, p_{III} = 0.23$

The proton coupled ^{13}C spectrum of the α-carboxyl carbon shown in Fig. 5.20 reveals a doublet of triplets whose larger splitting arises from the geminal coupling to the α-proton and the triplet originates from nearly equal coupling to C_β—H$_A$ and C_β—H$_B$. The averaged coupling constants $J_{CO-C_\beta H_A}$ and $J_{CO-C_\beta H_B}$ can be expressed by Eqns (5.14) and (5.15):

$$J_{CO-C_\beta H_A} = p_I J_{CH}^g + p_{II} J_{CH}^g + (1 - p_I - p_{II}) J_{CH}^t \tag{5.14}$$

$$J_{CO-C_\beta H_B} = p_I J_{CH}^t + p_{II} J_{CH}^g + (1 - p_I - p_{II}) J_{CH}^g \tag{5.15}$$

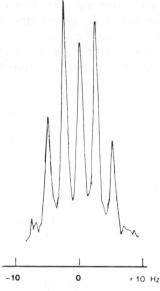

Fig. 5.20 Proton-coupled 25.2 MHz
^{13}C spectrum of the α-carboxyl
carbon in aspartic acid in
D$_2$O at pD 11.[72]

where J_{CH}^g and J_{CH}^t now represent the vicinal CCCH coupling constants. The sum of $^3J_{CO-C_\beta H_A}$ and $^3J_{CO-C_\beta H_B}$ in aspartic acid is 5.1 Hz. Insertion of this value into the sum of Eqns(5.14) and (5.15) yields, for possibility (a):

$$0.38 J_{CH}^t + 1.62 J_{CH}^g = 5.1 \tag{5.16}$$

and for possibility (b):

$$0.85 J_{CH}^t + 1.15 J_{CH}^g = 5.1 \tag{5.17}$$

In alanine, where, for reasons for symmetry, $p_I = p_{II} = p_{III}$, it was noted that

$$J_{CH}^t + 2J_{CH}^g = 12.6 \tag{5.18}$$

Combination of Eqn. (5.18) with (5.16) and (5.17), respectively, gives reasonable values only for assumption (a):

$$J_{CH}^t = 11.9 \text{ Hz}, \quad J_{CH}^g = 0.4 \text{ Hz}$$

Hence the correct fractional populations are as given for (a) with rotamer II dominating.

The kinetics of rotameric interconversions are accessible to the n.m.r. experiment only in those exceptional cases where, for some reason, rotation is hindered, and ΔG^* is raised above the critical limit so that spectra can be obtained under intermediate exchange rate conditions.

One group of compounds satisfying the aforestated condition is characterized by heavy substitution of the rotating carbon. 9-Chloromethyltriptycene[73] illustrates such a case of sterically hindered rotation (**91, 92**).

91   **92**

Although the proton spectra were found to be temperature-dependent, their complexity prevented precise kinetic data from being extracted. The temperature-dependent ^{13}C spectra of the aromatic carbon region displayed in Fig. 5.21 exhibit only six signals in the fast exchange limit, due to averaging of carbons 1, 5, 11; 2, 6, 12, etc. in rings a_1, a_2, and a_3 (**92**). In the slow exchange limit, the symmetry of the molecule is lowered from C_{3v} to C_s. In the region of the quaternary carbon resonances, for example, four lines at an intensity ratio of 2:2:1:1 are now observed. From line shape measurements, the following parameters could be derived for the barrier to rotation: $\Delta H^* = 13.5 \pm 0.7\,\text{kcal mol}^{-1}$, $\Delta S^* = 1.8$ e.u.

High barriers to rotation also occur if the rotation is hindered for electronic reasons. In amides and related compounds, the electronic structure of the C—N bond may be described by resonance hybrids **93** and **94** which imply a partial

93   **94**

double-bond character of the C—N bond. For the same reason, carbonyl carbons conjugated to unsaturated systems are stabilized in the two conformations for which the π-electron systems are coplanar. The lowest energy conformers of 2-furaldehyde, for example, are those with the C=O bond parallel to the aromatic ring (**95, 96**).[74] The slow-exchange limit ^{13}C spectrum for this

95   **96**

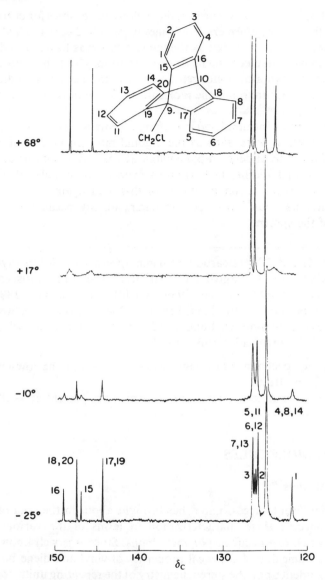

Fig. 5.21 Temperature-dependent 25.2 MHz proton noise-
decoupled ^{13}C spectra of 9-chloromethyltriptycene.[73]

compound recorded at $-90°C$ exhibits two sets of resonances with an
approximate intensity ratio of 9:1. Assignment of lines to the two conformers was
made on the basis of the expected steric compression shift for C-3 in the anti form
(**96**). The chemical shift differences for the remaining carbons are small, which is

indicative of only minor differences in the π-electron densities for corresponding carbons in the two conformers. In N-methylpyrrole-2-carbaldehyde, the C-3 resonance was fond to broaden as the temperature was lowered and finally to sharpen again without changing its original position. This behaviour points to a large value for the syn/anti conformational equilibrium constant, indicating that over a large temperature range the molecule predominantly exists in a preferred conformation.

A further form of intramolecular exchange processes is peculiar to systems having lone electron pairs, such as the trigonal nitrogen or phosphorus. Most amines are known to undergo rapid configurational inversion through a planar transition state. In aziridine, the barriers to inversion are usually sufficiently high to permit their examination by n.m.r. In this area again ^{13}C n.m.r. proved distinctly advantageous[74] over proton n.m.r., mainly because of the greater simplicity of the spectra.

Problem 28: The 90.5 MHz carbon-13 n.m.r. spectra of 2, 3-dimethylbutane at $-100°$C and at $-180°$C are given. At $-100°$C, the spectrum consists of only two lines (intensity ratio 1:2 (downfield versus upfield)). Note that at $-180°$C each of these signals separates into two lines. The mutual separation, $\Delta\delta$, between the two low-field lines (signal numbers **1** and **2**) is 2.71 ppm, and their intensity ratio was found to be 1:2 (downfield versus upfield).

(1) Explain the spectrum and assign resonances **1–4** in the low-temperature ($-180°$C) spectrum.
(2) Quantify the rotameric conformations which are present at $-180°$C.

5.6 MACROMOLECULES[75]

Synthetic polymers[76]

The most significant success n.m.r. has brought about in the area of synthetic polymers concerns their characterization in terms of the various forms of isomerism which can occur in polymer chains. Since it is well known that the physical, mechanical, and chemical properties of vinyl and diene polymers are strongly dependent upon the stereochemistry of the repeating units, considerable effort in the recent past has been put into stereochemical sequence analysis. The majority of polymer studies carried out until about 1975 used proton n.m.r., although this nucleus has some severe disadvantages. First, the chemical shift differences between nuclei of the same structural unit, but being in different stereochemical environments, are usually small. The demands for high field (spectrometers with superconducting magnets) are thus rather stringent. Second, the analysis of the generally complex n.m.r. patterns may be complicated further

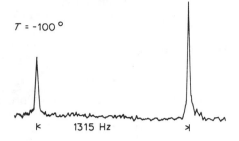

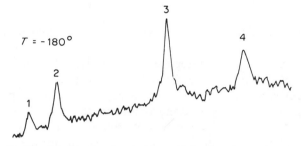

Problem Fig. 28 Reprinted with permission from Lunazzi *et al.*,
J. Am. Chem. Soc., **99**, 4573. Copyright (1977) American Chemical
Society.

by spin–spin coupling. This is usually circumvented by using specifically deuterated polymers, whose synthesis, however, may be a tedious and costly problem. Third, the broadening caused by dipolar spin–spin relaxation associated with slow molecular motion is much more severe for protons than for ^{13}C. This becomes understandable by taking into consideration that the transverse relaxation rate due to dipolar relaxation is proportional to the squares of the interacting magnetic moments (cf. Chapter 4). It is for this reason that high-resolution ^{13}C spectra can be obtained from highly viscous solutions, in some cases even from solids.

Stereoregularity of polymers

The majority of polymers are derived from substituted ethylene. For a vinyl derivative of the type

$$\begin{array}{ccc} A & & H \\ \diagdown & & \diagup \\ & C{=}C & \\ \diagup & & \diagdown \\ H & & H \end{array}$$

polymerization usually results in head-to-tail addition,

$$\cdots-\underset{\underset{\displaystyle H}{|}}{\overset{\overset{\displaystyle A}{|}}{C}}-\underset{\underset{\displaystyle H}{|}}{\overset{\overset{\displaystyle H}{|}}{C}}-\underset{\underset{\displaystyle H}{|}}{\overset{\overset{\displaystyle A}{|}}{C}}-\underset{\underset{\displaystyle H}{|}}{\overset{\overset{\displaystyle H}{|}}{C}}-\underset{\underset{\displaystyle H}{|}}{\overset{\overset{\displaystyle A}{|}}{C}}-\underset{\underset{\displaystyle H}{|}}{\overset{\overset{\displaystyle H}{|}}{C}}-\cdots$$

but occasionally head-to-head and tail-to-tail addition may occur:

$$\cdots-\underset{\underset{\displaystyle H}{|}}{\overset{\overset{\displaystyle A}{|}}{C}}-\underset{\underset{\displaystyle H}{|}}{\overset{\overset{\displaystyle H}{|}}{C}}-\underset{\underset{\displaystyle H}{|}}{\overset{\overset{\displaystyle H}{|}}{C}}-\underset{\underset{\displaystyle H}{|}}{\overset{\overset{\displaystyle A}{|}}{C}}-\underset{\underset{\displaystyle H}{|}}{\overset{\overset{\displaystyle A}{|}}{C}}-\underset{\underset{\displaystyle H}{|}}{\overset{\overset{\displaystyle H}{|}}{C}}-\cdots$$

A further form of structural isomerism in a polymer chain may result from cross-linking.

More important is the stereoregularity of a polymer. In all vinyl polymers (except polyethylene and those formed from symmetrically di- and tetra-substituted ethylene), chiral carbon centres are introduced. Depending on the relative configuration of neighbouring asymmetric carbons, basically two types of stereoregularity can be differentiated (**97, 98**). In a purely isotactic polymer chain (**97**) the relative handedness of all asymmetric carbon centres is the same, whereas in a syndiotactic chain (**98**) it varies alternately. Aside from these 'pure' forms of configurational sequences, various degrees of regularity occur. A polymer with no configurational preference is termed atactic.

isotactic

97

syndiotactic

98

In preceding sections we have seen that carbons in different stereochemical environments may readily be distinguished. Projected on a vinyl polymer of the general type

$$(-CHA-CH_2)_n$$

the methylene carbon will have different shielding depending on whether the adjacent tertiary carbon atoms have the same or different configuration:

meso(m)

racemic(r)

We may henceforth use these abbreviations for the representation of repeating units. Two consecutive repeating units are referred to as a dyad, with the relative configuration at the asymmetric centres denoted r (racemic) or m (meso). This constitutes the basis for the notation of configurational sequences in vinyl polymers. By analogy, three adjacent repeating units are denoted as a triad. The methine carbon shielding differs, depending on whether the carbon is in an rr, mm, or rm/mr triad:

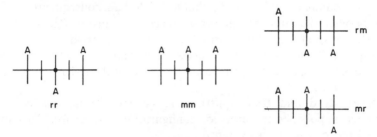

Although rm and mr are indistinguishable, both must be taken into account when relative line intensities are considered.

Figure 5.22 shows an early ^{13}C spectrum of isotactic and atactic polypropy-

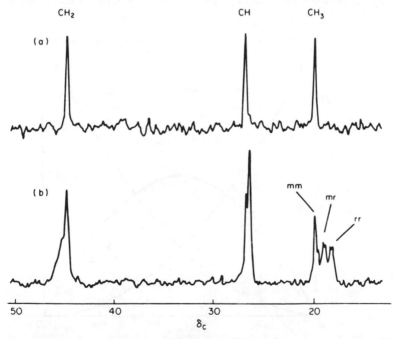

Fig. 5.22 25.2 MHz proton noise-decoupled ^{13}C spectra of polypropylene.[77] (a) Isotactic, 5% in o-dichlorobenzene at 60°C; (b) atactic, 30% in o-dichlorobenzene at 60°C.

lene,[77] with the latter exhibiting triad splittings for the methyl resonance. The probability for the occurrence of various triads is a function of the statistics which underlies the chain propagation process. If one assumes that the probability for the generation of a meso dyad can be expressed by one single parameter, which may be denoted P_m, the process is governed by Bernoullian statistics. Such a process may be compared with that of reaching into a large jar of balls marked 'm' or 'r' with P_m being the proportion of 'm' balls. The probability of taking out an 'm' ball at random, of course, is P_m, that for an 'r' ball, consequently, is $1 - P_m$. The probability of an mm, rr, and mr sequence thus must be P_m^2, $(1 - P_m)^2$, and $2(P_m - P_m^2)$, respectively. The normalized theoretical triad probabilities are plotted against P_m in Fig. 5.23. If a process is Bernoullian, then the proportion of the areas due to mm, rr, and mr/rm peaks lie on a vertical line dependent only on a single value of P_m.

However, dyad and triad splittings in general do not suffice to fully characterize a polymer in terms of configurational sequence. This may be visualized by considering a sequence

$$\dots mm\ rr\ mm\ rr\ mm \dots$$

which may be generated by a non-Bernoullian propagation mechanism. It immediately becomes clear that by observing only dyad and triad intensities such

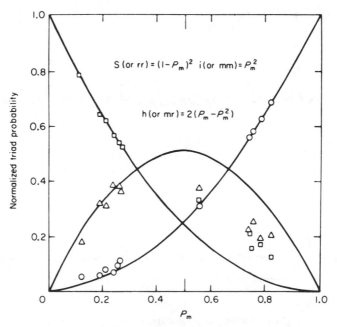

Fig. 5.23   Normalized Bernoullian triad probabilities as a function of P_m, the probability for the formation of a meso dyad.[75a]

a polymer cannot be distinguished from one which is randomly atactic. The large chemical shift dispersion, along with the superior resolution found in the ^{13}C spectra of macromolecules, renders possible the observation of higher configurational sequences, i.e. tetrads and pentads. In Table 5.3 the dyad, triad, tetrad, and pentad sequence designations are listed together with their probabilities.

The observation of tetrad and pentad placements in the ^{13}C spectra of vinyl polymers recorded at 23 kG or less has been reported in a number of cases, whereas the observation of corresponding configurational fine structure in proton n.m.r. requires much higher magnetic field.

The general relationship for the number of observationally distinguishable n-ads, $N(n)$ is given by Eqn. (5.19):[78]

$$N(n) = 2^{n-2} + 2^{m-1} \qquad (5.19)$$

where $m = n/2$ if n is even and $m = (n-1)/2$ if n is odd. Accordingly, the number of possible types of sequences is 2 for dyads, 3 for triads, 6 for tetrads, and 10 for pentads. It should be emphasized in this context that the propagation mechanism is immaterial with respect to the number of observationally different sequences.

In view of their tremendous economic importance, a great deal of effort has been put into the determination of the microtacticity of such widespread synthetic polymers as polyvinylchloride (PVC), polypropylene, poly-(methylmethacrylate) etc. As an illustrative example, we may discuss some data from earlier literature devoted to the determination of tetrad concentration in PVC[79]. Figure 5.24 shows the 25.2 MHz spectrum of a low molecular weight PVC clearly revealing tetrad fine structure for the methylene as well as pentad splittings for the methine resonance. All six observationally distinguishable tetrad placements show up as separate lines in the high-field CH_2 region. The assignments were made by fitting the observed methylene signal intensities to Bernoullian probabilities (cf. Table 5.3). The best fit was obtained for $P_m = 0.45$.

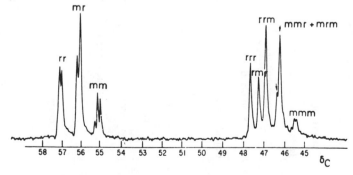

Fig. 5.24 25.2 MHz proton noise-decoupled ^{13}C spectrum of a low molecular weight polyvinylchloride, 10% (w/v) in o-dichlorobenzene at 60°C.[79]

TABLE 5.3 Designations and Bernoullian probabilities for n-ad placements ($n = 2, 3, 4, 5$) in vinyl polymers of the type $(-CH_2-CHR-)_n$[75a]

Designation		CHR Projection	Bernoullian probability	Designation		CH$_2$ Projection	Bernoullian probability
Triad	Isotactic, mm		P_m^2	Dyad	meso, m		P_m
	Heterotactic, mr		$2P_m(1-P_m)$		racemic, r		$(1-P_m)$
	Syndiotactic, rr		$(1-P_m)^2$	Tetrad	mmm		P_m^3
Pentad	mmmm (isotactic)		P_m^4		mmr		$2P_m^2(1-P_m)$
	mmmr		$2P_m^3(1-P_m)$		rmr		$P_m(1-P_m)^2$
	rmmr		$P_m^2(1-P_m)^2$		mrm		$P_m^2(1-P_m)$
	mmrm		$2P_m^3(1-P_m)$		rrm		$2P_m(1-P_m)^2$
	mmrr		$2P_m^2(1-P_m)^2$		rrr		$(1-P_m)^3$
	rmrm (heterotactic)		$2P_m^2(1-P_m)^2$				
	rmrr		$2P_m(1-P_m)^3$				
	mrrm		$P_m^2(1-P_m)^2$				
	rrrm		$2P_m(1-P_m)^3$				
	rrrr (syndiotactic)		$(1-P_m)^4$				

The dot in the projections represents the substituent R.

It is somewhat surprising that there is little similarity of chemical shift among carbons within r-centred tetrads, on the one hand, and those of m-centred tetrads, on the other.

A common use of ^{13}C n.m.r. spectroscopy for determining polymer microstructure consists of developing correlations between structure and some spectral property, such as ^{13}C n.m.r. chemical shifts. Empirical correlations (which usually require comparison of ^{13}C chemical shifts in the polymer with corresponding shifts in model compounds or in stereoregular polymers of known configuration) as well as theoretical models have proved useful in this regard. In 2, 4-dichloropentane,[80] which served as a model compound for the assignment of chemical shifts in PVC, the methylene carbon in the racemate is less shielded than that in the meso isomer (Fig. 5.25). This was interpreted in terms of γ-*gauche* and γ-*trans* interactions. Since, in contrast to 1, 4-*trans* orientation, a *gauche* orientation of substituents invokes additional shielding of the affected proton-bearing carbon (cf. Chapter 2), it was postulated that in racemic 2, 4-dichloropentane the rotamer populations allowed less γ-*gauche* interaction than in the meso stereoisomer. Proton n.m.r. data imply a preference for the rrr tetrad in PVC to assume a planar zigzag (all *trans*) conformation. Therefore, the resonance pertaining to this form would be expected at lowest field. Consequently, the methylene carbon in an rrr tetrad should expeience more *trans* interactions than in any other tetrad. In contrast, in the mmm tetrad the lowest energy conformation should give rise to more *gauche* interactions involving the methylene carbon than in any other tetrad. For this reason, the methylene carbon in an mmm tetrad is predicted to resonate at highest field. Both results are consistent with the experimental findings.

Problem 29: Assign the three lines belonging to the mm triad in the spectrum of atactic PVC (Fig. 5.24) in terms of pentad placements. (*Note*: Use signal heights to approximate the relative areas.)

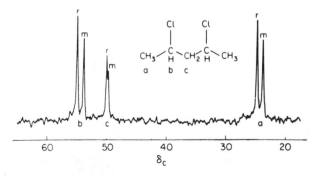

Fig. 5.25   22.6 MHz proton noise-decoupled ^{13}C spectrum of a mixture of meso and racemic 2, 4-dichloropentane.[80]

The foregoing analysis of microtacticity in PVC required the use of a model compound for assigning ^{13}C n.m.r. chemical shifts. More recently, two-dimensional n.m.r. has been utilized to make *absolute* tacticity assignments in PVC.[81,82] A combination of several two-dimensional n.m.r. techniques often is capable of providing sufficient information to permit complete assignment of stereochemical sequences in synthetic polymers and biopolymers, thereby obviating the need for chemical shift arguments. One important advantage of the use of two-dimensional n.m.r. techniques in this regard lies in the fact that multiplets that overlap when presented in one-dimensional n.m.r. spectra often can be isolated with suitable two-dimensional n.m.r. techniques. Hence, better effective spectral resolution generally can be obtained in the two-dimensional n.m.r. spectrum than in either the corresponding ^{13}C or proton n.m.r. spectrum alone.[81]

The carbon–proton shift-correlated two-dimensional n.m.r. spectrum of PVC is shown in Fig. 5.26.[82] The methine carbon (downfield) region displays three resonances (mm, rr, and mr triads) which display pentad fine structure. The centre triad can be assigned to the mr (= rm) sequence, since its chemical shift should be an average of the mm and rr triads.[82]

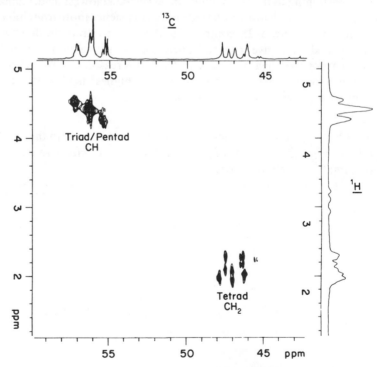

Fig. 5.26 500 MHz two-dimensional ^{1}H—^{13}C chemical shift correlated spectrum of a solution of PVC recorded in 1, 2, 4-TCP at 90°C.[82]

The methylene carbon (up-field) region contains at least six signals that correspond to tetrad sensitivity. When attempting to assign the signals in the methylene carbon region we can make use of the property that the methylene protons are magnetically non-equivalent in an m dyad (*heterosteric*), while those in a r dyad are equivalent (*homosteric*; see **99**, below). The reader should verify

99

that of the six possible methylene carbon tetrad sequences, only two (rrr and mrm) contain a carbon atom that correlates with a single proton resonance (i.e. the methylene protons on the central carbon atoms that are involved in these two tetrads are equivalent and thus provide a singlet resonance).

Another tool for the configurational assignment of proton and ^{13}C resonances in vinyl polymers is the heteronuclear RELAY experiment.[83-90] This experiment combines the features of homonuclear COSY and heteronuclear chemical shift-correlation experiments.[89] In the conventional heteronuclear chemical shift-correlation experiment, INEPT-type transfer of magnetization occurs from a proton to the directly attached ^{13}C nucleus (to which it is scalar-coupled; see relevant discussion in Chapter 3). While the RELAY spectrum contains this same information, it provides additional information from coherence transfer between more remote nuclei which are mutually scalar-coupled. As an example, consider the stepwise transfer of magnetization: (1) from proton H_a to another proton H_b which resides on a neighbouring carbon atom C_b followed by (2) polarization transfer from H_b to its directly attached carbon atom C_b. The observation of this polarization transfer sequence provides information that pertains to long-range connectivities in the molecule. A variety of connectivity sequences have been elucidated via two-dimensional correlation spectroscopy; these include, *interalia*, $H_a \rightarrow X \rightarrow H_b$[91,92] and $X \rightarrow H_a \rightarrow H_b$[93] where, typically, $X = {}^{13}C$.

The traditional heteronuclear RELAY experiment[83,84] is a relatively sensitive to the ^1H-^1H coupling constant. An improvement in this respect is the heteronuclear RELAY experiment with spin-locking.[82,90]

In this experiment *net* magnetization transpose between mutually scalar-coupled nuclei is achieved via Hartmann–Hahn cross polarization,[94] an experiment analogous to solid-state cross polarization which proceeds via a dipole–dipole coupling pathway (cf. Section 3.10). Consider a homonuclear (for example, proton–proton) cross-polarization experiment performed with a single coherent rf field. In the ideal case where two scalar-coupled protons, H_A and H_B,

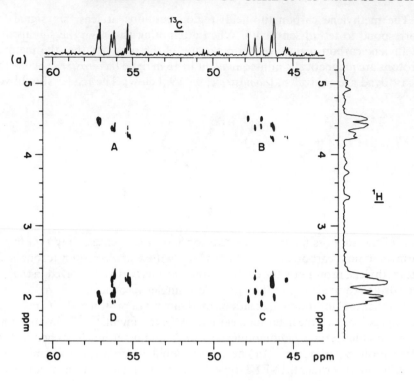

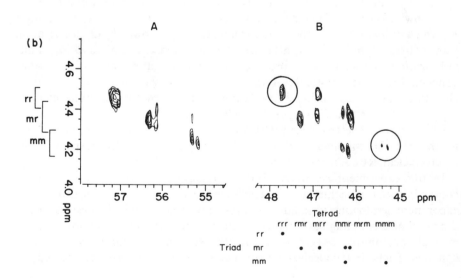

(c)

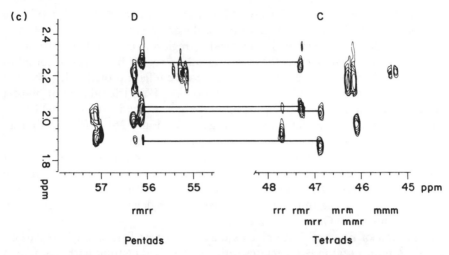

Pentads   Tetrads

Fig. 5.27   (a) 500 MHz two-dimensional heteronuclear spin-lock RELAY spectrum of a 15% solution of PVC in 1, 2, 4-TCB at 90°C. Regions A and C appear in the chemical shift correlated spectrum (Fig. 5.26) while regions B and D result from relayed magnetization.[82] (b) Expansion of regions A and B from Fig. 5.27(a). Tetrad assignments are consistent with experimental spectrum. (c) Expansion of regions C and D in Fig. 5.27(a). The horizontal lines connect a single carbon pentad resonance to tetrad resonances *mrr* and *rmr*, which combine to give only one possible pentad sequence: *rmrr*.[82] Reprinted with permission from Crowther, Szeverenyi and Levy, *Macromol.* **19**, 1333. Copyright (1986) American Chemical Society.

experience identical effective rf field strengths, a perfect Hartmann–Hahn match[94] is achieved that results in exchange of spin-locked magnetization at a rate $1/J_{AB}$. Small Hartmann–Hahn mismatches that arise when mutually scalar-coupled nuclei experience slightly different effective rf field strengths can be minimized via phase-alternation of the spin lock field along the $\pm x$ axis.[90,94]

The two-dimensional heteronuclear spin-lock RELAY spectrum of PVC is shown in Fig. 5.27(a).[82] Four regions, A-D, are observed in the spectrum shown in Fig. 5.27(a). Two of these regions (A and C) involve correlation of methylene and methine carbons with their own methylene and methine protons, respectively. However, in regions B and D, the following correlations apply, respectively:

(1) Methylene carbons are correlated with methine protons; here, transfer of magnetization occurs from the methine protons to the methylene carbons via three-bond proton–proton coupling by means of homonuclear Hartmann–Hahn type cross polarization.[94]

(2) Methine carbons are correlated with methylene protons in a analogous way wherein magnetization transfer occurs from the methylene protons to the methine carbons. Peak assignments can then be made on the basis of observed peak multiplicities in the two-dimensional n.m.r. spectra.

Careful examination of regions A and B in Fig. 5.27(a) (see expansion of these regions in Fig. 5.27(b)) reveal that, of the peaks contained in region B, only two tetrads do *not* correlate with the mr triad in region A (see circled contours in Fig. 5.27(b), region B). Hence, these two tetrads in region B must correspond to mmm and rrr (since these do not contain the rm configurational unit). These in turn correspond to the methylene carbon atoms in PVC that absorb at highest and at lowest field, respectively. Since all carbon atoms in the low-field peak correlate only to single methylene hydrogens (see Fig. 5.26), it follows that all methylene protons in this tetrad are equivalent, and, therefore, that this peak must be due to the rrr tetrad. The methylene carbon atom resonance at highest field, therefore, can be assigned to the mmm tetrad. (These conclusions are consistent with those discussed earlier in connection with the anticipated effects of *trans* versus *gauche* interactions involving methylene mmm and rrr tetrad carbon atoms in PVC.)

Inspection of Figure 5.27(b) reveals that the highest field methylene carbon tetrad (i.e. mmm, region B) correlates with the high-field methine carbon signal in region A. Symmetry considerations require that an mmm methylene carbon tetrad correlate only with an mm methine triad. Hence, the high-field methine carbon signal can be assigned to the mm triad, and the remaining (low-field) signal must correspond to the rr triad.

Four methylene carbon tetrads (i.e. rmr, rrm, mrm, and mmr) remain to be assigned. (Recall that all four of these unassigned methylene carbon tetrads correlate with the mr methine carbon triad). Two of the unassigned methylene tetrads also correlate with the rr and mm methine triads; hence, these tetrads must be rrm and mmr, respectively. The remaining two unassigned tetrads can be distinguished simply by recalling that, of these two tetrads, only mrm contains a methylene carbon atom that correlates with a single methylene proton (Fig. 5.26). Once the mrm tetrad has been identified, the lone remaining tetrad must be rmr, and the assignment of the methylene carbon resonances in the ^{13}C n.m.r. spectrum of PVC is now complete. It is clear from this example that the two-dimensional n.m.r. techniques that have been developed for the study of small molecules indeed can be applied to the task of rendering detailed stereochemical assignment in macromolecules.

The above assignments are summarized in Figure 5.28.[81] Note that the improved spectral resolution obtained at 125 MHz (cf. the corresponding 25.2 MHz ^{13}C n.m.r. spectrum of PVC shown in Fig. 5.24) permits clean mutual separation and firm assignment of the mmr and mrm tetrad signals. The above conclusions require reversal of previous assignments that were made for these two methylene carbon tetrad signals.[79,95]

Finally, expansion of the C and D regions in Fig. 5.27(a) is shown in Fig. 5.27(c).[82] Assignment of peaks in one of the mr-centred methine carbon pentads is illustrated therein. The horizontal lines in this figure indicate that this single pentad resonance correlates only with tetrad resonances mrr and rmr. The

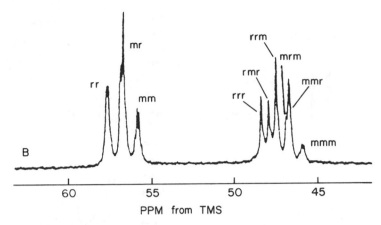

Fig. 5.28 125 MHz ^{13}C nmr spectrum of PVC at 65°C.[81] Reprinted with permission from Mirau and Bovey, *Macromol.*, **19**, r10. Copyright (1986) American Chemical Society.

latter can combine in only one way: i.e. to afford the rmrr pentad sequence, which is thereby assigned unequivocally.

Diene polymers

Geometrical isomerism as it occurs in diene polymers, such as polybutadiene and polyisoprene, has been extensively studied by both proton and ^{13}C n.m.r. However, whereas in polybutadiene the vinyl proton seems to be insensitive to *cis–trans* isomerism and only minor shielding differences were found for the methyl protons, both vinyl and methyl carbon shielding reflect the configurational environment.[96,97] The methylene carbons in *cis*-polybutadiene are more highly shielded by 5.3 ppm in comparison with their *trans* counterparts. This may be accounted for by 1, 4 steric interactions in the former configuration. Under sufficiently high resolution, the olefinic region of the spectrum displays four separate resonances as shown in Fig. 5.29(a).[97] The assignment is easily made with a polybutadiene whose *cis/trans* ratio differs from one. A comparison of the relative intensities with those of the methylene carbon signals then immediately shows that the two lower field lines at 130.85 and 130.75 ppm are associated with carbons at a *trans* double bond, while the resonances at 130.25 and 130.10 ppm belong to *cis* olefinic carbons. The small splittings indicate that the chemical shift of these carbons is further affected by the configuration of the carbons in neighbouring units similar to the effects observed in vinyl polymers.

In principle, four observationally distinguishable sequences are present for each *cis* and *trans* carbon (**100–103** and **104–107**, respectively). Since δa and $\delta a'$,

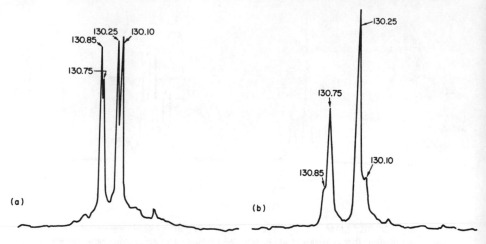

Fig. 5.29   25.2 MHz proton noise-decoupled ^{13}C spectra of polybutadiene (olefinic region). (a) In *n*-heptane, 51% *cis*, 49% *trans*; (b) in hexadeuterobenzene, 57.5% *cis*, 38% *trans*, 4.5% 1, 2 linkages.[97]

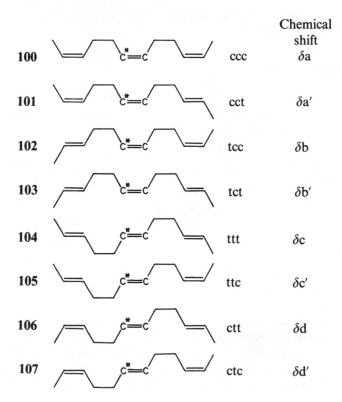

			Chemical shift
100		ccc	δa
101		cct	δa′
102		tcc	δb
103		tct	δb′
104		ttt	δc
105		ttc	δc′
106		ctt	δd
107		ctc	δd′

and δb and δb' etc. are expected to be more similar than, for example, δa and δb, or δa' and δb', only four olefinic resonances typically are observed. An assignment of all four resonances on a 1:1 basis can be achieved from a polymer in which either *cis* or *trans* linkages prevail. In Fig. 5.29(b) the olefinic region of a predominantly *cis*-1, 4-polybutadiene is shown. Since in such a polymer (57.5% *cis* and 38% *trans*), ctc + ctt sequences are statistically more frequent than ttt + ttc sequences, the lower intensity line at 130.85 ppm is assigned to the latter.* Similar arguments may be applied to the assignment of the line due to *cis*-centred sequences. The following are the designations of olefinic resonances in 1, 4-polybutadiene:

δ_C	Assignment
130.85	ttt, ttc
130.75	ctc, ctt
130.25	ccc, cct
130.10	tct, tcc

The microstructure in the spectra of the 1:1 *cis/trans*-1, 4-polybutadienes does not permit distinction between a regular sequence of the type

$$\ldots ccttcctt \ldots$$

and a random distribution. This again demonstrates the need for the observation of higher configurational sequences.

The olefinic region in the spectrum of the polymer shown in Fig. 5.29(b) exhibited two further weak signals at 145.2 and 112.4 ppm. The chemical shift of the latter resonance is characteristic of a terminal methylene. Hence the two resonances can be associated with the α and β carbons in 1, 2-linked units (**108**).

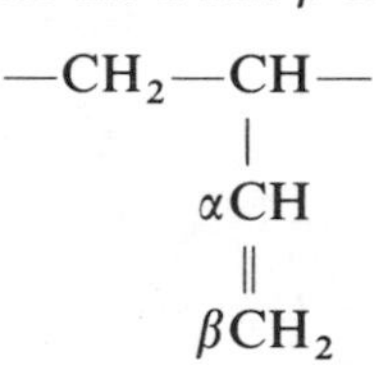

$$-CH_2-CH-$$
$$|$$
$$\alpha CH$$
$$\|$$
$$\beta CH_2$$

108

A third area in polymer research concerns the characterization of copolymers in terms of sequence distribution. Because of the potential applicability of ethylene–propylene copolymer as a general-purpose rubber, considerable effort has been invested in the study of this polymer. Its physical properties are found to be critically dependent upon the fraction of propylene units. The inability of proton chemical shifts to discriminate between various types of methine and methylene protons in an aliphatic carbon chain precludes

---

*This assignment is not in accord with that given in Ref. 97.

TABLE 5.4 Predicted chemical shifts in fundamental ethylene–propylene copolymer sequences[98]

Type	Copolymer sequence	Carbon	δC
A	C—C—C—C—C—C—C—C—C	A	29.96
B	(branched propylene unit)	B-m	19.63
		B-t	32.52
		B-α	36.91
		B-β	27.27
		B-γ	30.21
C	(branched sequence)	C-m	20.12
		C-t	30.45
		C-α'	43.86
		C-α	37.16
		C-β	27.27
		C-γ	30.21
D	(branched sequence)	D-m	19.63
		D-t	32.52
		D-α'	34.22
		D-α	36.91
		D-β	27.27
		D-γ	30.21
E	(branched sequence)	E-m	19.63
		E-t	32.52
		E-α'	37.16
		E-β'	24.58
		E-α	36.91
		E-β	27.27
		E-γ	30.21
F	(branched sequence)	F-m	20.12
		F-m'	20.61
		F-t	30.45
		F-t'	28.38
		F-α'	44.11
		F-α	37.16
		F-β	27.27
		F-γ	30.21
G	(branched sequence)	G-m	16.64
		G-t	37.06
		G-α	34.22
		G-β	27.52
		G-γ	30.21

TABLE 5.4 (contd.)

Type	Copolymer sequence	Carbon	δC
H	$\begin{array}{c} \quad\ \text{C} \qquad\quad \text{C}^{m'}\ \text{C}^{m} \\ \quad\ | \qquad\quad\ |\quad\ | \\ \text{C—C—C—C—C—C—C—C—C—C—C—C} \\ \qquad\qquad\ \alpha'\ \ \iota'\ \ \iota\ \ \ \alpha\ \ \ \beta\ \ \ \gamma \end{array}$	H-m	16.64
		H-m'	16.64
		H-t'	37.06
		H-t	37.06
		H-α'	31.53
		H-α	34.22
		H-β	27.52
		H-γ	30.21
I	$\begin{array}{c} \quad\ \text{C} \qquad \text{C}^{m'}\ \text{C}^{m} \\ \quad\ | \qquad\ |\quad\ | \\ \text{C—C—C—C—C—C—C—C—C—C—C—C} \\ \qquad\qquad \alpha'\ \ \iota'\ \ \iota\ \ \ \alpha\ \ \ \beta\ \ \ \gamma \end{array}$	I-m'	17.13
		I-m	16.64
		I-t'	34.99
		I-t	37.06
		I-α'	41.17
		I-α	34.22
		I-β	27.52
		I-γ	30.21

proton n.m.r. from being a suitable technique. On the other hand, we have seen that ^{13}C shieldings are sensitive to the number of $\alpha, \beta,$ and γ carbons (cf. Chapter 2). The fundamental ethylene–propylene copolymer sequences including both head-to-tail as well as head-to-head and tail-to-tail linkages of monomer units are compiled in Table 5.4[98] along with the chemical shifts predicted with the aid of Lindeman and Adams additivity rules. These data show, for example, that head-to-head sequences (defined as two neighbouring propylene units with adjacent tertiary carbons) can easily be identified from their characteristic methyl shieldings. Provided the peak areas are proportional to the corresponding carbon concentrations (which can be attained under suitable experimental conditions), the ethylene/propylene molar ratio can be calculated from the relative methyl carbon intensity $I(\mathrm{CH}_3)/\sum_i I_i$ where $\sum_i I_i$ represents the total peak area:

$$I(\mathrm{CH}_3)/\sum_i I_i = P/(2E + 3P) \qquad (5.20)$$

In Eqn. (5.20) E and P denote the ethylene and propylene mole fractions, where $P = 1 - E$.

Problem 30: Assign all peaks in the spectrum of an ethylene-propylene copolymer and determine the molar ratio of propylene units.

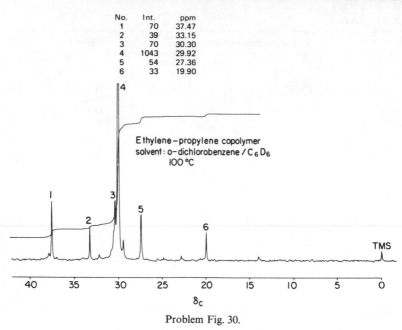

No.	Int.	ppm
1	70	37.47
2	39	33.15
3	70	30.30
4	1043	29.92
5	54	27.36
6	33	19.90

Ethylene–propylene copolymer
solvent: o–dichlorobenzene / C_6D_6
100 °C

Problem Fig. 30.

Compositional fine structure has been observed for a variety of other copolymer systems such as ethylene oxide–propylene oxide,[99] styrene–butadiene,[100] ethylene–vinyl acetate,[101] and methylmethacrylate–methacrylic acid[102] copolymers. The analysis of such systems is often complicated due to the presence of both compositional and configurational microstructure, unless the polymer is stereochemically uniform. The methylmethacrylate–methacrylic acid (MMA–MAA) copolymer is a rather special case because of the similarity of the repeating units. Here it emerges that the carboxyl carbon shielding in particular is sensitive to the neighbouring unit. In fact the resonance of this carbon reflects compositional triads.

Problem 31: Assign all six lines in the carbonyl region of an MMA–MAA copolymer in terms of compositional triads.[102]

Ring-opening Polymerization[103]

Cycloalkenes and bicycloalkenes commonly undergo ring-opening polymerization in the presence of a suitable olefin metathesis catalyst:

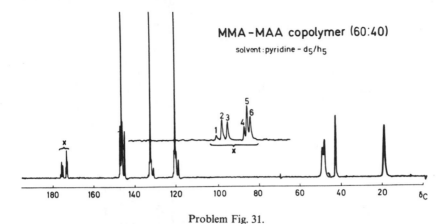

Problem Fig. 31.

The mechanism of ring-opening polymerization is believed to involve an intermediate metallocarbene species.[104]

Ring-opened polymers (for example, poly-(1, 3-cyclopentylenevinylenes) (**110**), prepared via olefin metathesis of substituted norbornenes (**109**)), have received considerable attention in recent years:[105]

109   **110**

Note the following structural features of polymer **110**: (1) the *cis* relationship between carbon–hydrogen bonds at C_1 and C_4, (2) *cis, trans* (c, t) isomerism is possible at each carbon–carbon double bond, (3) m, r ring dyad tacticity is possible, and (4) for the case of R = H substrate monomer **109** lacks a symmetry plane, and the possibility of head–head (HH), head–tail (HT), and/or tail–tail (TT) polymerization must be considered. Carbon-13 n.m.r. spectroscopy affords valuable information regarding these structural features in ring-opened polymers derived from substituted norbornenes.[105]

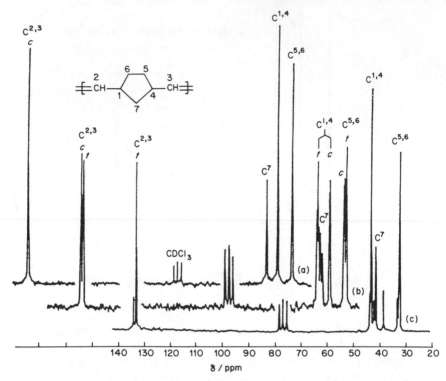

Fig. 5.30   22.6 MHz ^{13}C nmr spectra of poly(1, 3-cyclopentylenevinylene). Spectra (a) and (b) are offset 40 ppm and 20 ppm, respectively. (a) ReCl$_5$ catalyst; (b) WCl$_6$/EtAlCl$_2$ catalyst.[94] Reproduced by permission of Hüthigs Wepf Verlag.

Carbon-13 n.m.r. spectra of **110** prepared via reaction of norbornene with a variety of olefin metathesis catalyst systems are shown in Fig. 5.30.[106] The n.m.r. spectrum shown in Fig. 5.30(a) is quite simple; the lack of fine structure reflects the fact that reaction of norbornene with ReCl$_5$ catalyst affords the corresponding all-*cis*, fully syndiotactic polymer **110**. In contrast to this result, all of the non-olefinic carbon atoms in the spectra shown in Figs 5.30(b) and (c) display considerable fine structure. Since the chemical shift of each of these carbon atoms is affected by the *cis/trans* nature of neighbouring carbon–carbon double bonds in the polymer, the presence of extensive fine structure in these spectra attests to the lack of double-bond stereoregularity in the polymers formed via reaction of norbornene with WCl$_6$-EtAlCl$_2$ and with ruthenium-based catalyst systems, respectively. In addition, the presence of fine structure in the olefinic carbon region in the n.m.r. spectra shown in Figs 5.30(b) and (c) reflects the mutual influence of neighbouring double bonds whose nearest respective olefinic carbon atoms are separated by at least four carbon–carbon bonds.[107]

Determination of tacticity (m, r dyads) in **110** (R = H) has not yet been carried out due to the small shift differences for m and r dyads in this system.[108] This problem was solved originally via examination of ^{13}C n.m.r. spectra of ring-opened polymers derived from optically active 5-methyl and 5, 5-dimethylnorbornenes.[109] Ivin and coworkers[106] have shown that, in the event of stereospecific propagation of ring-opening polymerization of norbornene via metallocarbene intermediates, all-*cis* polymer should be syndiotactic (rr, **111**), whereas all-*trans* polymer should be isotactic (mm, **112**).

111

112

Syndiotactic polymers that are formed via ring-opening polymerization of unsymmetrically substituted norbornenes (for example, **109**, R = H) contain the HH, TT sequence; corresponding isotactic polymers contain an all-HT sequence.[109] Hence, up to eight signals can be observed in the olefinic carbon region in ^{13}C n.m.r. spectra of ring-opened polymers derived from unsymmetrically substituted norbornenes (i.e. *cis* and *trans* double bonds in the polymer each produce a set of four resonances due to HH, TT, HT, and TH combinations).

The use of optically active substrates for making tacticity assignment in ring-opened polymers is illustrated in Fig. 5.31[110,111] The olefinic carbon region of the ^{13}C n.m.r. spectrum of all-*cis* polymer produced via reaction of ReCl$_5$ with racemic *exo*-5-methylnorbornene (($\pm$)-**109**, (R = CH$_3$)) is shown in Fig. 5.31(a).[111] Here, the methyl groups are distributed randomly, and the four olefinic carbon resonances that correspond to HH, TT, HT, and TH sequences are equally intense. The olefinic carbon chemical shift assignments that accompany Fig. 5.31(a) were made on the basis of anticipated γ- and δ-methyl substituent effects.[111]

The corresponding ^{13}C n.m.r. spectrum of all-*cis* polymer obtained via reaction of ReCl$_5$ with (+)-**109** (R = CH$_3$) is shown in Fig. 5.31(b). A clear

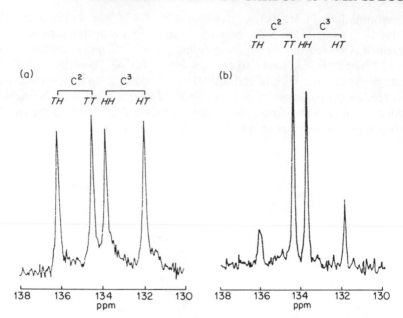

Fig. 5.31 22.6 MHz ^{13}C nmr spectra (olefinic carbon region) of ring-opened polymers synthesized from (a) ($\pm$)-**109**($R = CH_3$) and (b) (+)-**109**($R = CH_3$).[111]

preference for TT, HH sequences is apparent in this spectrum. The relatively small contribution of peaks due to TH, HT sequences (i.e. 13.5%) can be ascribed to polymer produced via concomitant reaction of $ReCl_5$ with (−)-**109** ($R = CH_3$) that was present as an impurity in the starting material (along with the corresponding, desired (+)-enantiomer). After correcting for the presence of (−)-enantiomeric impurity in the substrate, the resulting ^{13}C n.m.r. spectrum (Fig. 5.31(b)) is consistent with the conclusion that $ReCl_5$-promoted ring-opening polymerization of (+)-**109** (R═CH_3) affords the corresponding all-*cis*, fully syndiotactic polymer.[111] The mechanistic significance of this observation has been discussed elsewhere by Ivin and coworkers.[110]

Biopolymers

Among the molecules studied by n.m.r. biopolymers probably represent the greatest challenge to the spectroscopist. Over the past years a large effort has been expended in the domain of protein and polynucleotide characterization. For a detailed discussion of the recent literature on this subject, the reader is referred to the pertinent reviews.[75a,75b,112]

The simplest model compounds used in the analysis of protein spectra are the amino acids whose ^{13}C spectra have been assigned.[113,114] While remote side-chain carbons are found to be relatively insensitive to the type of amide linkage, the α and β carbons undergo greater shielding changes when a peptide bond is formed. The characteristic shifts induced upon formation of specific amide bonds reflect the so-called secondary structure observed in peptides and proteins. A prominent feature of ^{13}C chemical shifts in amino acids and peptides is their pH dependence.[115] Ionization studies, therefore, have been used successfully to tackle assignment problems in amino acids and peptides. Small peptides can formally be considered as oligomeric models for proteins. In order to gain some picture of the relative shifts expected when an amino acid is incorporated into a peptide chain, we may consider the chemical shifts of a central residue denoted X in a pentapeptide of the type Gly–Gly–X–Gly–Gly.[116] Table 5.5 lists the chemical shifts in the free amino acid X and those in the peptide. Measurement of the spectra at various pH values by making use of the characteristic pK values of the different ionizable groups in the central residues aided in the assignment of closely spaced lines. From these data it is seen that upon incorporation of the amino acid into the peptide chain, the α and β carbons experience average shifts of approximately -1 and $+1$ ppm, respectively.

The potential of ^{13}C n.m.r. in the study of complex peptides may be exemplified with oxytocin,[117] a cyclic nonapeptide hormone, whose spectrum could be deciphered with the help of specially prepared small peptides.

The aliphatic region of the 25.2 MHz spectrum of oxytocin (111) is depicted in Fig. 5.32 along with assignments to α-, β-, γ-, and δ-carbons in the corresponding

TABLE 5.5 Comparison of chemical shifts in free amino acids X and in the pentapeptide Gly–Gly–X–Gly–Gly obtained at neutral pH[116]

X	Carbon	δ peptide	δ free	δ peptide $-$ δ free
Serine	CO	171.9	171.8	$+0.1$
	Cα	55.4	56.3	-0.9
	Cβ	60.8	60.0	$+0.8$
Threonine	CO	172.0	172.3	-0.3
	Cα	58.7	60.1	-1.4
	Cβ	66.6	65.8	$+0.8$
	Cγ	18.3	19.1	-0.8
Aspartic acid	CO	173.5	173.8	-0.3
	Cα	51.2	52.0	-0.8
	Cβ	38.2	36.2	$+2.0$
	Cγ	177.2	177.1	$+0.1$
Asparagine	CO	172.4	173.4	-1.0
	Cα	50.1	51.3	-1.2
	Cβ	36.0	34.9	$+1.1$
	Cγ	174.1	174.4	-0.3
Glutamic acid	CO	173.9	174.3	-0.4
	Cα	53.8	54.6	-0.8
	Cβ	26.9	26.7	$+0.2$
	Cγ	33.0	33.3	-0.3
	Cδ	181.0	180.8	$+0.2$

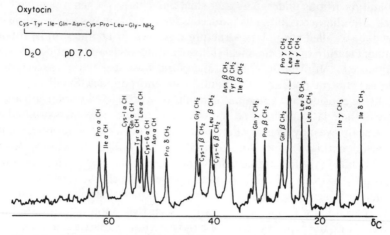

Fig. 5.32 High-field region of the 25.2 MHz proton noise-decoupled ^{13}C spectrum of oxytocin in D_2O at pD 7.0.[117] The chemical shifts are given relative to external TMS.

Oxytocin

111

($\equiv$ Cys-Tyr-Ile-Gln-Asn-Cys-Pro-Leu-Gly-NH$_2$)

amino acid residues. The α-carbons characteristically appear in the region between 50 and 65 ppm. The α-amino group of the terminal cysteine moiety can be titrated. From examination of the long-range pH-induced shifts it was concluded that changing from basic to acidic medium produces a conformational transition at the isoleucyl residue in oxytocin concomitant with alteration of the disulphide dihedral angle.[117-119]

Combinations of homo- and heteronuclear two-dimensional n.m.r. techniques have proved valuable for assigning ^{13}C and proton n.m.r. chemical shifts in peptides. In general, carbonyl carbons in peptides resonate within a relatively narrow window and are thus difficult to assign. As demonstrated previously for vinyl polymers, two-dimensional chemical shift correlation is a powerful tool for enhancing the effective spectral resolution. An illustration of this capability is the assignment of peptide carbonyls;[120,121] the basic experiment has been discussed in Section 3.5. The distinguishing feature in the present application of the technique is the transfer of magnetization via two- and three-bond long-range couplings involving peptide protons. Note that this differs from the original experiment wherein magnetization transfer is induced between directly bonded nuclei. In this way, the connectivities established by the observation of magnetization transfer due to $^{2}J_{CH}$ and $^{3}J_{CH}$ scalar couplings in a peptide permit complete assignment of its proton n.m.r. spectrum or vice versa (see below). Additionally, such connectivities can be used to establish the amino acid sequence in peptides.

The pulse sequence for heteronuclear ^{13}C—^{1}H correlation spectroscopy is shown in Figs 3.40 and 3.41. An important prerequisite in this sequence is the proper choice of the two short, fixed delays (Δ_1 and Δ_2) needed to prevent mutual cancellation of antiphase components (see Section 3.5). Typically, a compromise value is employed for these delays that approximates J-values for the carbon–proton scalar couplings of interest. Since the delays are proportional to $1/J_{CH}$, the use of long-range couplings results in longer delay times.

As an example, two-dimensional heteronuclear correlation of α-hydrogen and carbonyl ^{13}C resonances has been used to assign the proton and ^{13}C n.m.r. spectra of a protected tetrapeptide, $CH_3CO-Thr-Phe-Thr-Ser-NH_2$ (**112**).[121] The four α-protons in **112** are circled and numbered sequentially in the structure shown below:

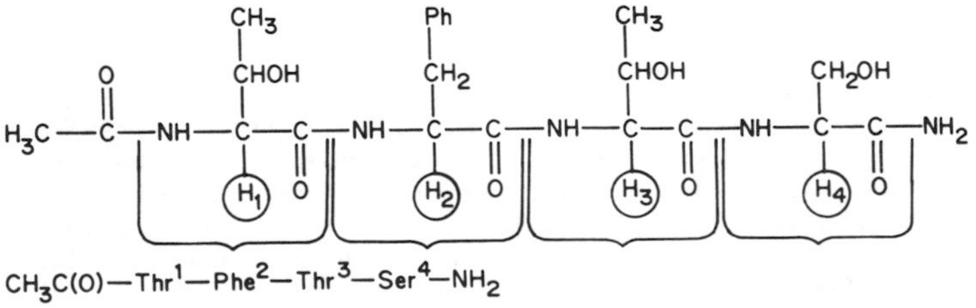

112

The two-dimensional ^{13}C—^1H heteronuclear chemical shift correlation spectrum of **112** is shown in Fig. 5.33.[121a] An expansion of the α-proton region, including relevant proton n.m.r. chemical shift assignments for **112**,[121b] is displayed in the ω_1 dimension of Fig. 5.33. It should be noted that the α-proton in the *N*-terminal amino acid residue (i.e. Thr[1]) lacks $^3J_{CH}$ coupling to a carbonyl carbon atom. In addition, it is useful to note that the carbonyl carbon atom in the *C*-terminal amino acid residue (i.e. Ser[4]) lacks $^3J_{CH}$ coupling to an α-proton. Hence, cross peaks in the two-dimensional heteronuclear correlation spectrum

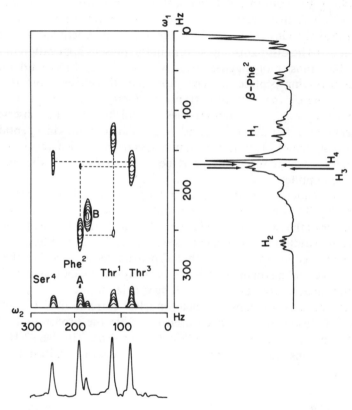

Fig. 5.33   The 270 MHz two-dimensional heteronuclear chemical shift correlation spectrum of a 0.16 M solution of **112** in DMSO-d_6 at 30°C. In the ω_1 dimension: the α region of the proton spectrum with assignments given for α-protons.[121b] In the ω_2 dimension: the carbonyl region of the ^{13}C decoupled spectrum with assignments given for the carbonyl carbon atoms.[121a] A and B are fold-over peaks. The dotted line shows the connectivities of the amino acids. Reproduced by permission of VCH Verlagsgesellschaft.

that correspond to the α-proton in the N-terminal residue and to the carbonyl carbon atom in the C-terminal residue can be identified readily (see below).

On the other hand, H_2, for example, displays two sets of cross peaks that correspond to $^2J_{CH}$ (to the carbonyl carbon in Phe^2) and $^3J_{CH}$ (to the carbonyl carbon in Thr^1). A compromise setting was used for Δ_1 and Δ_2, and spin–spin couplings were found to range between 1 and 7 Hz. The relative intensities of these two cross peaks are very different (see Fig. 5.33). Finally, it should be noted that, in the case of compound 112, it happens that the anticipated $^3J_{CH}$ between H_1 and the N-acetyl carbonyl group in the CH_3CO–Thr^1 residue is unresolved.[121] Given this information, the ^{13}C and ^1H n.m.r. peak assignments for 112, shown in Fig. 5.33, can be made as follows:

(1) Of the various α-protons in 112, all display cross peaks that arise from two couplings ($^2J_{CH}$ and $^3J_{CH}$) except for that which resonates at highest field (which has a single cross peak that arises from $^2J_{CH}$ coupling). Accordingly, this signal must correspond to H_1 in 112, and the carbonyl carbon atom with which it correlates in Fig. 5.33 must reside in the CH_3CO–Thr^1 residue.

(2) Turning next to the α-proton in 112 which resonates at lowest field, the fact that the apparent multiplicity of this peak is a double doublet (i.e. the X part of an ABX system) suggests that this proton is situated adjacent to a methylene group that contains non-equivalent protons. Of the three α-protons in 112 that remain to be assigned, only H_2 and H_4 are so situated. Since this low-field α-proton signal can be seen to correlate via magnetization transfer arising from $^3J_{CH}$ (low-intensity cross-peak) with the carbonyl carbon atom in the CH_3CO–Thr^1 residue, this proton signal must be due to H_2, and the other carbonyl carbon atom with which it correlates via magnetization transfer arising from $^2J_{CH}$ (prominent cross peak) must reside in the Phe^2 residue.

(3) The two α-protons in 112 that remain unassigned are indicated by arrows in Fig. 5.33. Of these two protons, the one that resonates at lower field is seen to correlate with the carbonyl carbon atom in the Phe^2 residue. The intensity of the contour that establishes this correlation is weak. Since H_4 does not couple with the carbonyl in Phe^2 this proton must be assigned to H_3. Since this α-proton is also strongly scalar-coupled ($^2J_{CH}$, prominent intensity contour) to another carbonyl carbon atom, that carbon resonance to which H_3 is so correlated must be due to the carbonyl carbon atom in the Thr^3 residue.

(4) The only remaining α-proton resonance, therefore, can be assigned to H_4, and the lone remaining carbonyl carbon resonance must be due to the carbonyl carbon atom in the Ser^4–NH$_2$ residue.

Another, somewhat more complicated case in point is cyclosporin A (113, an immunosuppressive cyclic undecapeptide):

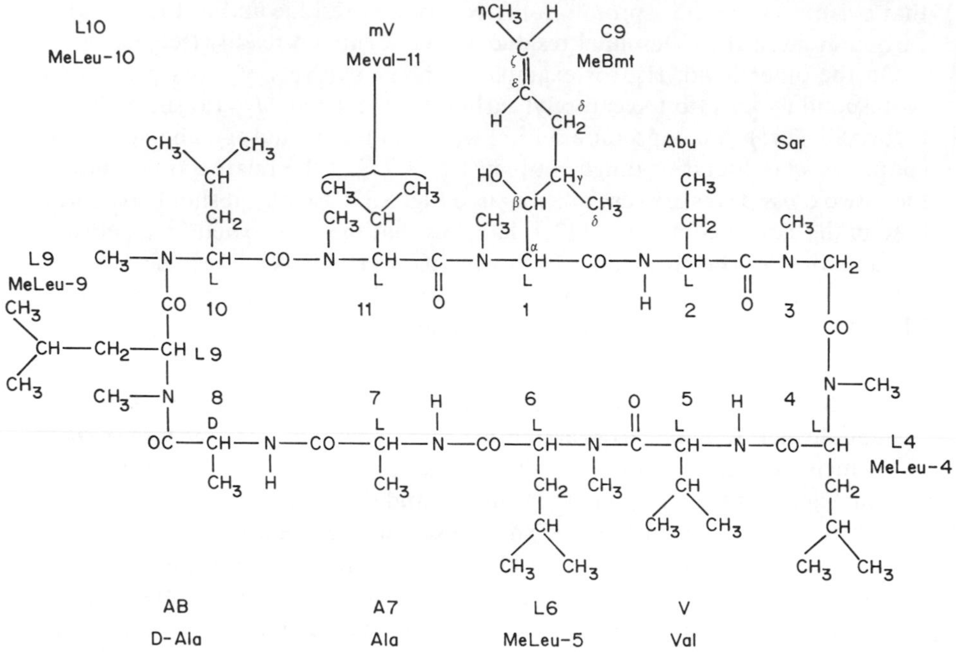

L10
MeLeu-10

mV
Meval-11

ηCH₃ H
C9
MeBmt

Abu   Sar

L9
MeLeu-9

AB
D-Ala

A7
Ala

L6
MeLeu-5

V
Val

113*

The carbon–proton (heterocorrelated) n.m.r. spectrum of **113** is shown in Fig. 5.34.[120] Here, all of the carbonyl carbons can be assigned via observation of their couplings to directly attached protons and to remote protons by using normal and relayed carbon–proton correlated techniques, respectively. The heteronuclear couplings between carbonyl carbon atoms and as many as four different kinds of protons (i.e. NH or CH₃N, two α-protons and one β-proton) appear as cross peaks in the spectrum shown in Fig. 5.34. Optimization of cross peaks that are due to relatively small long-range ^{1}H—^{13}C couplings (i.e. of the order of ca 5–10 Hz) can be accomplished simply by increasing the delays Δ_1 and Δ_2 in the basic pulse sequence for heteronuclear shift correlation from the normal 'compromise' values of 3.5 and 2.2 ms, respectively, to ca 60 and 40 ms, respectively, as discussed in the previous example.[86,120]

The eleven carbonyl carbons in cyclosporin A are correlated in this spectrum with four kinds of protons (see circled protons in the peptide substructure shown in Fig. 5.34). As in the preceding example, this application of two-dimensional-shift correlation spectroscopy affords information regarding connectivities between adjacent amino acid residues. Importantly, the *direction* of connectivity

*For abbreviations of amino acid residues, see Ref. 120.

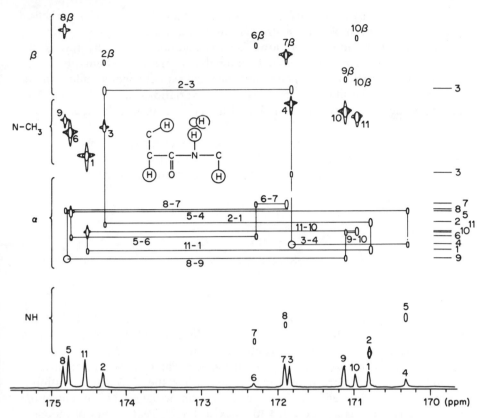

Fig. 5.34 Carbon–proton hetero-correlated nmr spectrum of **113** showing region of carbonyl carbon resonances and α, β, and N-CH$_3$ peptide proton resonances. Pulse timing parameters were optimized for 5 Hz ^{13}C—^{1}H coupling constants. Numbers refer to peptide residues (**113**).[120] Reprinted with permission from Kessler, Loosli and Oschkinat, *Helv. Chim. Acta*, **68**, 661 (1985).

is also determined, since couplings of carbonyl carbon to N$\underline{\text{H}}$ or NC$\underline{\text{H}}_3$ protons occur only to that amino acid which follows in the sequence.[121] Additional applications of two-dimensional n.m.r. spectroscopy to conformational analysis of peptides have been reviewed recently.[125] The reader is referred to the original literature[120] for a more detailed discussion of proton and ^{13}C n.m.r. chemical shift assignments in cyclosporin A.

Proteins

Proteins in their native state occur as a right-handed α-helix, i.e. they exist in a distinct conformation. The α-helix as it occurs in nature, as well as in synthetic polypeptides, consists of approximately 3.6 residues per turn. Since the transition of the protein from its native state to a random coil conformation results in the loss of its biological activity, much of the research on proteins in solution has been focused on the study of the helix–coil transition process. In order to lose its helical conformation, intramolecular hydrogen bonds are disrupted. Hence helix–coil transitions can be induced by lowering the solution pH or by raising the temperature. The process is found to be reversible within certain limits and can be pursued on the basis of the concomitant chemical shift changes. In addition, the conformational transition is accompanied by variations in the dynamic properties of the protein which are reflected in the relaxation behaviour (cf. next section).

Synthetic poly-α-L-amino acids[126–128] provide excellent models since they exhibit a behaviour which is in many respects typical of an α-helix. An example is poly-γ-benzyl-L-glutamate (PBLG)[126] (114), whose proton spectra had already

$$(-\overset{\alpha}{C}H—CO—NH—)_n$$
$$\beta CH_2$$
$$\gamma CH_2$$
$$\delta CO—O—CH_2—\text{(phenyl: 1,2,3,4)}$$

114

been studied earlier as a function of added trifluoroacetic acid (TFA). In Fig. 5.35 the ^{13}C spectra of PBLG are depicted as a function of TFA concentration. Progressively increasing TFA concentration results in signal displacements that reflect the conformational changes taking place. The α-carbon resonance, for example, is shifted more than 3 ppm upfield when the TFA concentration is increased from 3% (helix) to 29% (coil). At intermediate acid concentrations where the two forms are expected to co-exist, the C_α peak shows pronounced broadening which could be interpreted as resulting from chemical exchange at a rate near coalescence. However, the phenomenon could as well result from molecular weight polydispersity (different chain length of the polymer moieties), since it is known that the helix–coil transition process for short chains occurs at lower acid concentration. It is also recognized from Fig. 5.35 that the lines—in particular the α-carbon resonances—have narrowed considerably in the random coil polymer, which is a consequence of the enhanced chain mobility in this conformation (cf. Chapter 4 and next section). Similar results were obtained for analogous systems poly-L-methionine[127] and poly-L-lysine.[128]

The development of large-sample diameter nmr probes has made it feasible to observe single carbon resonances in natural proteins. The sensitivity problem in these cases is usually one of concentration limitation. The limited solubility of these macromolecules means that the measurements must be carried out at concentrations as low as 10^{-2} molar and less. Because of the large number of nuclei present in nearly equal magnetic environments, signal overlap occurs, and the spectra observed therefore represent the envelopes arising from a large number of overlaying signals.

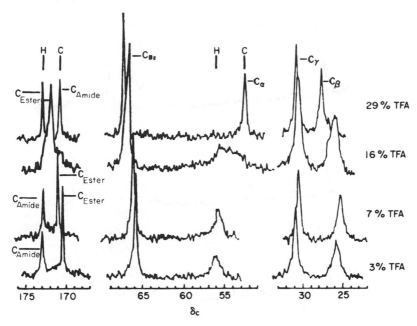

Fig. 5.35 25.2 MHz proton noise-decoupled ^{13}C spectra of poly-γ-benzyl-L-glutamate in CDCl$_3$-TFA:[126] 3%, 7% TFA, helix; 16% TFA, helix and coil; 29% TFA, coil.

One of the early proteins studied by ^{13}C n.m.r. is ribonuclease A,[129,130] a molecule with molecular mass 13 700 consisting of 124 amino acid residues. It contains eighteen basic and ten acidic amino acids and the amino acid chain is cross-linked via four disulphide bridges. The spectra of ribonuclease in its native state (Fig. 5.36(a))[130] exhibit relatively little fine structure, which is accounted for by two factors. First, the lines are broadened because of efficient spin–spin relaxation due to slow overall reorientation of the macromolecule. Second, slight chemical shift differences occur for carbons which are almost identical but lose this near-equivalence in the folded form. By lowering the pH, the protein disrupts and transforms into a denatured form. The spectrum recorded at pH 1.46 (Fig. 5.36(d)) shows a much larger degree of fine structure since the peaks have narrowed considerably and some chemical shift non-equivalence has been removed in this flexible form of the molecule. A number of tentative assignments could be made on the basis of a chemical shift comparison with amino acids and small peptides. A higher degree of reliability of such assignments is probably attained when the protein carbon shifts are compared with those observed in synthetic polypeptide oligomers whose sequences are identical with those found

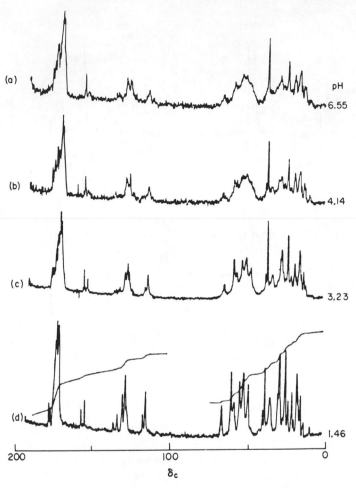

Fig. 5.36 15.1 MHz proton noise-decoupled ^{13}C spectra of ribonuclease A, 0.02 M in H_2O at various pHs.[130] The chemical shift scale was altered from CS_2 to TMS standard using the relationship $\delta_C^{TMS} = 192.4 - \delta_C^{CS}2$.

in some portion of the protein. Table 5.6[130] lists the tentative assignments for the carbon chemical shifts convering the region between *ca* 40 and 175 ppm in bovine pancreatic ribonuclease A.

Similar spectra were obtained from other proteins such as hen egg white lysozyme[131] and the heme proteins.[132] The latter are formally characterized by consisting of a porphyrin iron unit which is linked to a peptide chain. One member of this class of proteins is cytochrome *c*, which is widely distributed in living organisms (plants and animals). Although numerous different species are

TABLE 5.6   Partial chemical shift assignments in the spectrum of denatured bovine pancreatic ribonuclease A, recorded at pH 1.46[130]

$\delta_C{}^a$	Assignment	$\delta_C{}^a$	Assignment
176.9	glutamic acid C^δ		serine C^α
175.9	aspartic acid C^γ		tyrosine C^α
177–169	—CONH—, COOH		phenylalanine C^α
156.4	arginine C^ζ		glutamine C^α
154.2	tyrosine C^ζ		methionine C^α
135.9	phenylalanine C^γ		arginine C^α
133.2	histidine $C^{\epsilon 1}$	50.8–51.6	glutamic acid C^α
129.9	tyrosine $C^{\delta 1}$, $C^{\delta 2}$		lysine C^α
128.1	phenylalanine $C^{\epsilon 1}$, $C^{\epsilon 2}$		leucine C^α
	phenylalanine $C^{\delta 1}$, $C^{\delta 2}$		histidine C^α
	histidine C^γ		aspartic acid C^α
127.5	tyrosine C^γ		cystine C^α
126.7	phenylalanine C^ζ		asparagine C^α
116.9	histidine $C^{\delta 2}$	39.6	alanine C^α
115.1	tyrosine $C^{\epsilon 1}$, $C^{\epsilon 2}$	42.4	glycine C^α
66.6	threonine C^β		
	serine C^β		
	proline C^α		
60.6–58.8	valine C^α		
	isoleucine C^α		
	threonine C^α		

^{a}In ppm from TMS, converted from CS_2 standard using the relationship $\delta_C^{TMS} = 192.4 - \delta_C^{CS_2}$.

13   14   15   18   17 18   19
—Lys–Cys–Ala–Gln–Cys–His–Thr

115

known, the variations are confined to the sequence of the 124 amino acid residues. However, all variants have the cysteine residues 14 and 17 in common, through which the polypeptide chain is bonded to the heme group. A sequence, for example, such as encountered in horse-heart cytochrome c is shown in **115**. The hexaco-ordinated iron atom of the heme group is bound to four nitrogens while one of the axial ligands is known to be an imidazole nitrogen of the histidine residue 18 and the other is the sulphur of methionine 80 (see **116**):

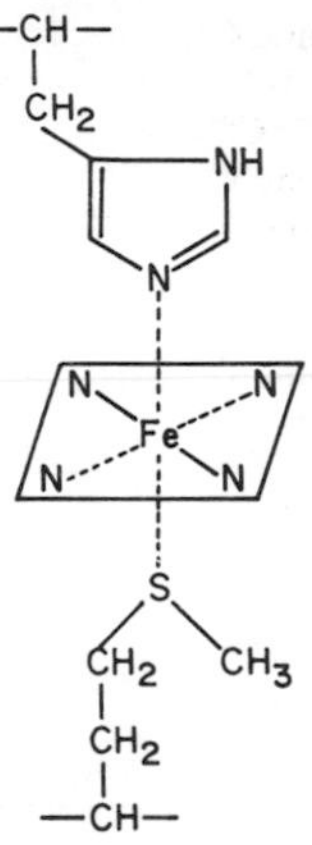

116

The role of cytochrome c is to serve as an electron carrier in the oxidative degradation of nutrients. The molecule is capable of doing so because the heme iron can occur in different valence states. In ferrocytochrome c, the iron is divalent and the complex is diamagnetic ($S = 0$), whereas in ferricytochrome c, the iron is trivalent and the complex is paramagnetic ($S = \frac{1}{2}$). Figure 5.37[132] displays the aromatic region of cytochrome c in both redox states as well as an intermediate state resulting from fast exchange between the two species. In order to identify non-protonated resonances, off-resonance proton noise-decoupling was used. Whereas in normal off-resonance decoupling the proton decoupler frequency is applied in the coherent mode (cf. Chapter 3), off-resonance noise-decoupling leads to very much broadened resonances for all proton-bearing carbons while leaving quaternary carbon resonances unaffected. The experiments proved that all narrow lines belong exclusively to quaternary carbons. The spectrum of the aromatic region in ferrocytochrome c in Fig. 5.37(c) shows twenty-four single-carbon resonances (peaks 2, 4–6, 8–10, 12, 13, 15–26, 28–30) and six two-carbon resonances, all due to quaternary carbons. This division into single and two-carbon resonances was made on the assumption of equal NOE for all carbons of this type. Although no individual assignment can as yet be made for most of these peaks, they are known to belong to eighteen quaternary carbons in the side-chains of twelve aromatic amino acid residues (four tyrosines, four phenyl-

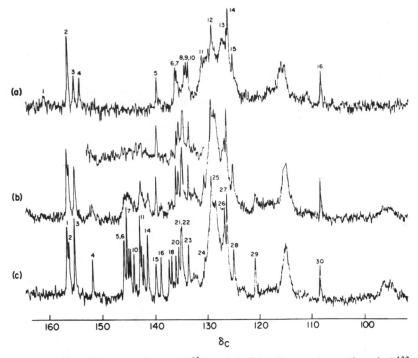

Fig. 5.37   15.1 MHz proton-decoupled ^{13}C spectra of cytochrome c (aromatic region).[132] The peaks numbered consecutively are due to quaternary carbons. (a) Ferricytochrome c, 8 mM; (b) 9:1 mixture of ferrocytochrome c and ferricytochrome c; (c) ferrocytochrome c.

alanines, three histidines, one tryptophan), sixteen quaternary carbons of the haem moiety and two ζ-carbons of arginine.

Since the two redox forms co-exist, some chemical shift averaging caused by fast electron transfer is observed, and because the fast exchange limit conditions apply for most of the carbons (cf. Section 5.3), the resonances remain sharp (Fig. 5.37(b)), with the exception of those carbons which are close to the paramagnetic centre and therefore experience large contact shifts (cf. Chapter 3). These comprise the porphyrin carbons as well as C^γ of histidine 18 (peak 29). The broadening may be due to deviations from the fast exchange conditions as well as hyperfine coupling. Figure 5.38[132] represents a chemical shift correlation diagram obtained by plotting the chemical shifts as a function of the relative oxidation state. This was realized by suitably mixing ferri- and ferrocytochrome c.

Recently, two-dimensional n.m.r. spectroscopic techniques have been employed for the study of proteins. The first such applications made use of specifically ^{13}C-enriched proteins.[133,134] However, it was found subsequently that sensitivity-related problems could be overcome by limiting data collection to

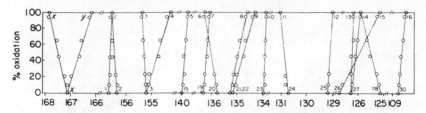

Fig. 5.38   Chemical shifts of some fast-exchanging quaternary carbon resonances observed in spectra of mixtures of the two redox forms of horse-heart cytochrome c.[132] The numbering is the one used in Fig. 5.37 except for the lines marked x and y, which refer to carbonyl signals.

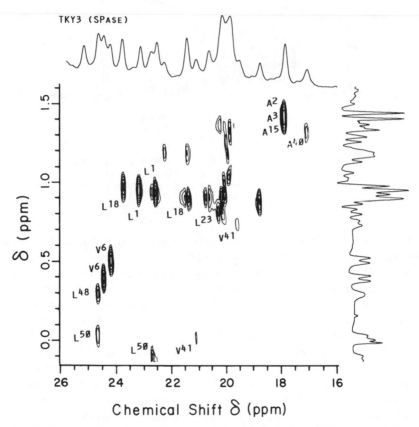

Fig. 5.39   200 MHz proton–carbon correlated spectrum of the methyl region of turkey ovomucoid third-domain protein. The single-letter code (A = alanine, L = leucine, V = valine) specifies the amino acid and the associated numbers refer to position of amino acid residuales in the protein.[135] Reproduced with permission from Westler, Ortiz-Polo and Markley, *J. Magn. Res.*, **58**, 352 (1984).

a relatively small portion of the spectrum.[135] The intensity of cross peaks (produced via transfer of proton spin polarization to carbon) is proportional to the number of protons attached to a given carbon atom; hence, sensitivity can be maximized by focusing attention on the 'protein methyl group region.

The natural-abundance carbon–proton chemical-shift correlated two-dimensional n.m.r. spectrum of the methyl region of turkey ovomucoid third domain (OMTKY3, a serine proteinase inhibitor obtained via cleavage of egg white ovomucoid) is shown in Fig. 5.39.[135] OMTKY3 contains 26 methyl groups, *ca* 20 of which are resolved in the contour plot shown in this figure. Carbon resonances were then assigned by correlation to proton signals whose respective assignments had been made previously via one- and two-dimensional proton n.m.r. experiments with avian ovomucoid third domains.[135]

Paralleling ^{13}C studies of proteins, this has proved equally successful in probing a second class of important biopolymers, the nucleic acids. While early studies were mainly related to the assignment of resonances in the spectra of the monomer units which constitute the basis of nucleic acids, i.e. of the nucleosides and nucleotides,[136] the ^{13}C spectra of synthetic polynucleotides serving as model nucleic acids have later been reported.[137] The first natural-abundance ^{13}C spectrum of a nucleic acid is that of yeast transfer ribonucleic acid (tRNA).[138] The spectrum could be partially assigned by chemical shift comparison with the nucleotides, in a manner similar to that described for the proteins.

5.7 SOLID-STATE [13]C N.M.R. APPLICATIONS

Elements of the theory of cross-polarization magic angle spinning (CP/MAS) have been discussed in Chapter 3. Some applications of the CP/MAS method will now be presented and briefly discussed.

Lippmaa and coworkers[139] recently have itemized several general conclusions regarding chemical shifts in the [13]C n.m.r. spectra of solid samples that are crucially important for the interpretation of solid-state n.m.r. spectra:

(1) Isotropic chemical shifts are approximately the same for samples in the (neat) liquid and in the solid state in the absence of specific interactions in the solid state.

(2) While molecular inversion symmetry in the liquid state is maintained in the solid state, other types of molecular symmetry can be lost.

(3) Rotational freezing in the solid state results in non-equivalence (i.e. groups like *t*-butyl which are enantiotopic in solution may become diastereotopic in the solid state).

(4) Intramolecular non-bonded interactions predominate over intermolecular non-bonded interactions in the solid state.

(5) Intermolecular hydrogen bonding in the solid state results in loss of molecular symmetry.

The above generalizations should be borne in mind when considering the problems that follow.

Problem 32: Isotropic ^{13}C chemical shifts are given below for maleic acid and for fumaric acid, both in the solid state and in solution.[139] Comment briefly on the data in terms of anticipated differences between solid- and solution-state maleic acid *vis-à-vis* solid- and solution-state fumaric acid.

| Compound | Physical state | [13]C isotropic chemical shifts (ppm) | | | |
		C_1	C_2	C_3	C_4
Maleic acid	(a) Powder sample, 20°C	171.5	143.3	135.6	175.4
Maleic acid	(b) 10% solution in ethanol	168.3	131.5	131.5	168.3
Fumaric acid	(a) Powder sample, 20°C	172.7	138.3	138.3	172.7
Fumaric acid	(b) 10% solution in ethanol	167.3	134.5	134.5	167.3

Problem 33: [13]C n.m.r. spectra of crystalline naphthazarin B (I) have been obtained by using cross-polarization magic angle spinning (CP/MAS) techni-

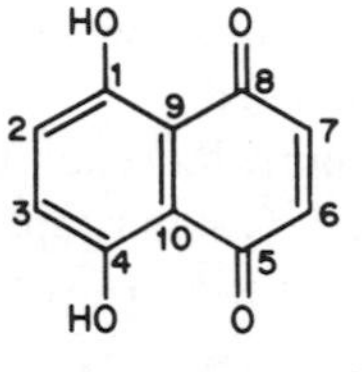

Naphthazarin B (I)

ques. The spectra thereby obtained at + 25°C and at − 160°C are shown in the figure.[140] Peak assignments which correspond to numbered carbon atoms in the structure are as follows: carbons 1, 4, 5, and 8: $\delta 173$; carbons 2, 3, 6, and 7: $\delta 134$; carbons 9 and 10: $\delta 112$ (+ 25°C). Account for the dramatic change in the [13]C n.m.r. spectrum of solid naphthazarin B that occurs when the temperature is lowered from + 25°C to − 160°C.

Problem 34: (1) The [13]C n.m.r. chemical shifts and spectral assignments for the carbon atoms in semibullvalene (CCl_2F_2 solution) obtained at − 95°C and at − 160°C are given below:[141a]

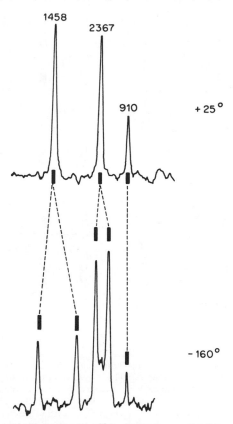

Problem Fig. 33 ^{13}C nmr spectra of solid naphthazarin B obtained using CP/MAS techniques (600 scans, 30 Hz line broadening, cross-polarization time 1 ms, and recycle time 3 s): at 25°C and − 160°C.[140] Reprinted with permission from Shiau *et al.*, *J. Am. Chem. Soc.* **102**, 4546. Copyright (1980) American Chemical Society.

Temperature (°C)	Chemical shift (ppm)	Assignments
− 95	50.7	1, 5
	87.2	2, 4, 6, 8
	121.7	3, 7
− 160	42.2	2, 8
	48.0	1
	53.1	5
	121.7	3, 7
	131.8	4, 6

Semibullvalene

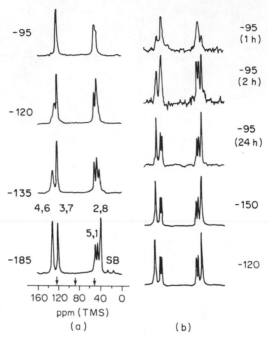

Problem Fig. 34   (a) ^{13}C CPMAS spectra of semibullvalene cooled into the solid state showing reversible
dynamic behaviour.[141b] The arrows show the position of
the three resonances for the average spectrum obtained in
solution.[141a] The peaks marked SB are the upfield
spinning side bands of the vinylic carbon resonances.
(b) Spectra of the same sample during annealing at
$-95°$C. After annealing for several hours, the spectra no
longer change with temperature cycling. Reprinted with
permission from Macho, Miller and Yannoni, J. Am.
Chem. Soc. **105**, 3735. Copyright (1983) American Chemical Society.

Account mechanistically for the observed temperature dependence of the
^{13}C n.m.r. spectrum of semibullvalene.

(2) The corresponding ^{13}C CP/MAS spectrum of solid semibullvalene obtained between $-95°$C and $-185°$C is shown in the figure.[141b] Compare the
^{13}C CP/MAS n.m.r. data for solid semibullvalene with the corresponding
^{13}C n.m.r. data obtained for a CCl_2F_2 solution of semibullvalene (given in part
(1), above). What is the significance of the observed difference between the
appearance of the ^{13}C n.m.r. spectrum of semibullvalene in solution at $-95°$C and
the appearance of the corresponding ^{13}C CP/MAS n.m.r spectrum of this
compound at the same temperature?

An interesting addition to solid-state ^{13}C n.m.r. spectroscopy, is nutation
n.m.r. spectroscopy, introduced by Yannoni and coworkers.[142] This tech-

nique exploits homonuclear (^{13}C—^{13}C) dipole–dipole couplings for measuring carbon–carbon bond lengths in the solid state. Since the probability that two ^{13}C labels will occur in the same molecule is only 1.2×10^{-4}, it is necessary to employ specific isotopic enrichment. The two mutually coupled carbon atoms generate a doublet whose width is governed by Eqn. (3.31). If it were not for chemical shift broadening (caused by chemical shift anisotropy), the splitting could be measured simply from the Fourier transform of the free induction decay in the usual way. In a transient nutation, chemical shift effects virtually are absent because the magnetization is forced to precess around the (much weaker) B_1 field at resonance. The spins then precess slowly at frequency $\omega_N = \gamma_C B_1$, modulated by spin–spin coupling. Such a precession is also denoted 'forced', since it is driven by the B_1 field which is applied during the data-collection period. Linewidths in nutation spectra are sensitive to the homogeneity of B_1.

Application of the nutation n.m.r. technique takes advantage of the fact that dipole–dipole splitting between two ^{13}C nuclei varies inversely with the cube of the internuclear distance. Nutation n.m.r. spectroscopy has been utilized to determine internuclear carbon–carbon bond distances within ca 1% accuracy. It should be noted that there is no requirement that the ^{13}C nuclei in question be mutually bonded in order for the nutation n.m.r. technique to be applicable to determining internuclear distances.[142,143]

The nutation n.m.r. technique has been applied to studies of polymerization mechanisms.[144] Figure 5.40(a) depicts ^{13}C n.m.r. nutation spectra of poly(phenylacetylene) at 77 K; the poly(phenylacetylene) sample had been prepared via reaction of $Ph^{13}C\equiv^{13}CH/Ph$—$C\equiv CH$ with tetra(n-butoxy)titanium-triethylaluminum in toluene. By way of contrast, Fig. 5.40(b) depicts the corresponding ^{13}C n.m.r. nutation spectrum of another sample of poly(phenylacetylene) that had been prepared via the corresponding reaction of $Ph^{13}C\equiv^{13}CH/PhC\equiv CH$ with pentachloromolybdenum–tetraphenyltin in toluene. The simulated spectrum (dotted curves in Fig. 5.40(a)) corresponds to 91% of the contiguous ^{13}C nuclei being separated by 1.36 Å and 9% by 1.48 Å. In the case of the second sample of poly(phenylacetylene), the results of theoretical spectral simulation (dotted curves in Fig. 5.40(b)) indicated that 88% of the contiguous ^{13}C nuclei are separated by 1.48 Å and 12% by 1.36 Å. The distances 1.36 Å and 1.48 Å correspond to carbon–carbon double and single bonds, respectively, in polyenes. The results indicate that most of the contiguous ^{13}C nuclei in the first poly(phenylacetylene) sample are linked by carbon–carbon double bonds, whereas in the second poly(phenylacetylene) sample the contiguous ^{13}C nuclei are singly bonded. These results are consistent with the following mechanistic schemes:

(1) The titanium-initiated polymerization of phenylacetylene proceeds via direct insertion of phenylacetylene into a titanium–carbon bond (Mechanism A, Fig. 5.41);

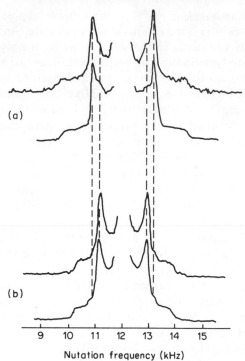

Nutation frequency (kHz)

Fig. 5.40 ^{13}C nmr nutation spectra of poly (phenyl-acetylene) at 77°K. The cross-polarization time was 0.5 ms and the recycle time 1 s (144 000 transients). For details see text.[144] Reprinted with permission from Katz *et al.*, *J. Am. Chem. Soc.* **105**, 7787. Copyright (1983) American Chemical Society.

(2) The molybdenum-promoted polymerization of phenylacetylene may instead follow an olefin metathesis mechanism, proceeding via a metallocyclic intermediate (Mechanism B, Fig. 5.41).[144]

5.8 MECHANISTIC STUDIES[145]

The elucidation of reaction mechanisms in synthetic and biogenetic studies was mainly accomplished until recently by means of radioactive labelling using either ^{14}C or ^{3}H. In order to determine where and to what extent the label of the substrate or precursor molecule had been incorporated into the product molecule, the latter had to be converted to a number of fragments by step-wise degradation. From the radioactivity measured for the degradation products it could be established whether a particular site in the molecule was labelled or not. The use of ^{13}C or deuterium labels, on the other hand, has the advantage that the often tedious degradation procedures become unnecessary because the ^{13}C

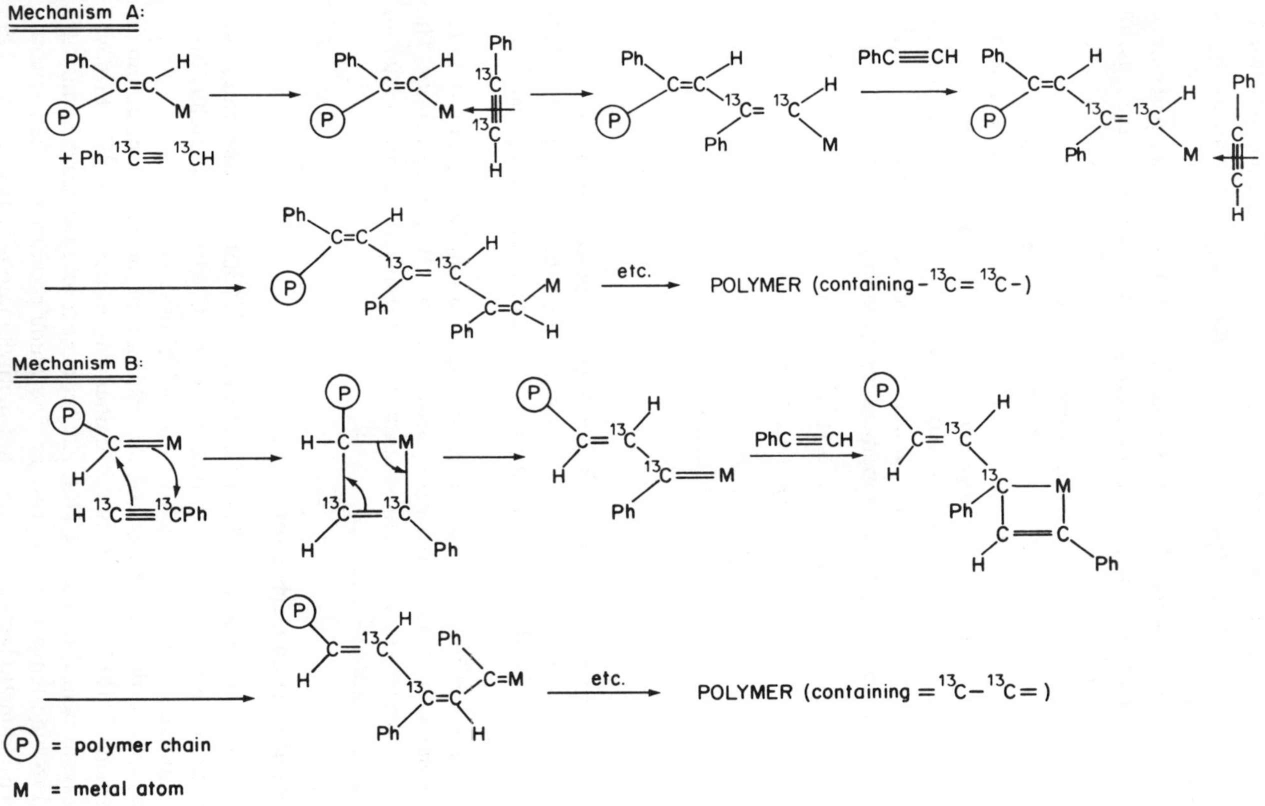

Fig. 5.41 Reaction mechanisms derived from the spectra in Fig. 5.40.

spectrum affords direct insight into the label distribution of the product. Its only drawback is the requirement of a relatively high percentage of incorporation, a condition which in biosynthetic problems is not always fulfilled. This, however, is partly overcome by the possibility of using a high level of enrichment in the substrate molecules.

An important question which has to be clarified beforehand is the percentage enrichment required. In Chapter 3, for example, we have seen that useful information is available from $^{13}C-^{13}C$ spin–spin coupling. At natural abundance, the ^{13}C satellites arising from coupling between adjacent carbons in isotopomers containing two ^{13}C atoms have intensities on the order of 0.5% relative to the centre peak. Hence satellite peaks are only detectable for exceptionally high signal-to-noise ratios and therefore some degree of enrichment in general will be required. Let us assume the relative enrichment values for two carbons C_A and C_B in a compound to be a and b. Then the probabilities for the occurrence of the four possible isotopomers are as follows:

$$^{12}C_A{}^{12}C_B: p_1 = (1-a)(1-b)$$
$$^{13}C_A{}^{12}C_B: p_2 = a(1-b)$$
$$^{12}C_A{}^{13}C_B: p_3 = (1-a)b$$
$$^{13}C_A{}^{13}C_B: p_4 = a.b$$
$$\overline{\sum_i p_i = 1}$$

Thus for an enrichment to 10% at both sites A and B one obtains for the average satellite line intensities $p_4/(2p_2) = 0.055$ or 5.5% of the centre peaks. This corresponds to about 46% of the intensity of a natural abundance peak, as can easily be verified.

For quantitative determinations the experiments should be carried out under conditions where the peak areas are proportional to the ^{13}C concentrations. The experimental requirements for satisfying these conditions are described later, in the section on quantitative analysis.

Elucidation of reaction mechanisms

Whereas early reaction mechanism studies based on ^{13}C-labelled substrates used proton n.m.r., where molecules with incorporated ^{13}C give rise to satellite lines, this technique has been superseded by direct ^{13}C observation.

An example is the elucidation of the mechanism of ring contraction in phenylcarbene[146] (117). For this purpose phenyldiazomethane-[^{13}C] (118) was prepared from benzaldehyde-^{13}CHO, 92% enriched, and the reaction mixture subjected to high-temperature pyrolysis. The crude product was subsequently treated with dimethylamine to yield 6-dimethylamino-6-methylfulvene (119), whose ^{13}C spectrum in Fig. 5.42 indicates that complete label scrambling has

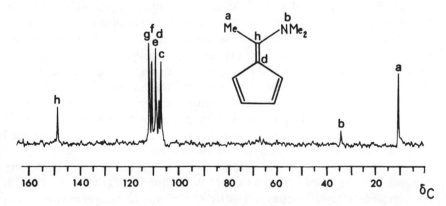

117   **118**   **119**

taken place. The low-intensity peak (marked b) in Fig. 5.42 originates from the *N*-methyl carbons and thus represents natural-abundance ^{13}C. Since this degenerate peak is just marginally observable, all remaining lines must result from ^{13}C incorporation. A plausible mechanism which accounts for these data involves a degenerate phenylcarbene rearrangement via a bicyclic intermediate (**120**) and the ring-expanded cycloheptatrienylidene (**121**) involving hydrogen shifts. Hence

120

121

before ring contraction occurs, a pre-equilibrium is established in which phenylcarbene rapidly interconverts with a bicyclic species and with cyclo-

Fig. 5.42 22.6 MHz proton noise-decoupled ^{13}C spectrum of 6-dimethylamino-6-methylfulvene obtained from a labelled sample as described in the text.[146]

heptatrienylidene. A quantitative evaluation of the ^{13}C data showed that the degree of incorporation was indeed the same for all carbons in 6-N,N-dimethylamino-6-methylfulvene.

Several mechanistic schemes have been postulated to account for the various products (and the distribution of ^{13}C label) in the phenylcarbene rearrangement which were reviewed critically.[147,148]

Evidence for the occurrence of a degenerate rearrangement in 2-acetoxycyclohexanone[149] has been established with the aid of a [1-^{13}C]-enriched sample. Upon treatment of **122** with acetic acid and potassium acetate at 142°C, the ^{13}C spectrum revealed equal distribution of the label among positions 1 and 2. This result is in accordance with the postulated formation of the symmetrical intermediate **123** which leads to an exchange of the oxygen functions. The spectrum in Fig. 5.43 shows the region of carbons 3, 4, 5 and 6. Although no ^{13}C had been incorporated at these sites, the signals reflect the labelling of C-1 and C-2 in that satellites appear arising from one- and two-bond ^{13}C − ^{13}C coupling. Heating of **122** in CH$_3$COOH/CH$_3$COOK at 218°C resulted in complete label scrambling among all ring carbon centres. Exchange of the oxygen functions between positions α and α' (C-2/C-6 in the first step) was suggested to involve the allylic rearrangement of enol **124** as indicated.

(122)   (123)   (124)

In Chapter 3 we pointed to the characteristic effects which substitution of hydrogen by deuterium produces on the ^{13}C spectra. Isotopic substitution was shown to result in characteristic shifts as well as spin–spin coupling involving deuterium. Although deuterium tracers may potentially be studied by proton n.m.r. and mass spectroscopy, this approach often fails because the proton spectrum is not analysable. The mass spectral data, on the other hand, afford information regarding the level of deuterium incorporation, but unless the fragmentation pattern is understood it does not permit any conclusions as to which sites had been deuterated. However, ^{2}H n.m.r. spectroscopy is hampered by the small chemical shift dispersion of this nucleus.

An example illustrating the potential of ^{13}C n.m.r. for probing deuterium incorporation is discussed subsequently.[150] When humulene (**125**) is treated with D$_2$SO$_4$, an acid-catalysed cyclization with concomitant deuterium incorporation occurs, leading to product **126**. From the mass spectrum an average deuterium incorporation of 1.85 deuterium atoms per molecule is indicated. In Fig. 5.44(a)

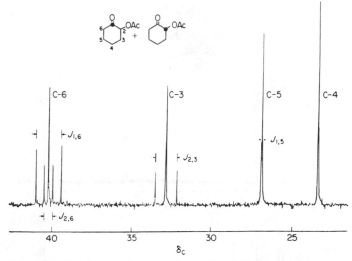

Fig. 5.43 Methylene carbon signals in the 25.2 MHz proton noise-decoupled ^{13}C spectrum of a mixture of [1-^{13}C]- and [2-^{13}C]-2-acetoxyclohexanone showing fine structure due to one- and two-bond ^{13}C—^{13}C couplings.[149]

125

126

the fully assigned ^{13}C spectrum of the protio compound is shown. In the spectrum of the deutero species in Fig. 5.44(b), 1:1:1 triplets appear at positions assignable to C-3(-5), C-8(-10), and C-12(-15), thus proving that monodeuteration has taken place at these carbon centres. Since only a fraction of the molecules are deuterated, the triplets have superimposed singlets, originating from molecules bearing only protons at these sites. For the same reason the resonances due to C-1, 7 and C-4 show up as 'doublets' of unequal intensity arising from different isotopomers. The observed small upfield shift of one component is a geminal isotope effect. (cf. Section 3.9). The C-9 resonance even appears in the form of a 'triplet' caused by a superposition of the spectra containing CH_2, CHD and CD_2 at carbon 8. The labelling pattern found provided experimental evidence of the cyclization mechanism suggested earlier.

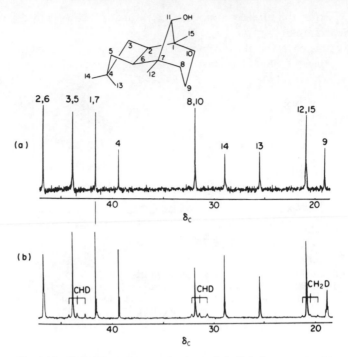

Fig. 5.44　25.2 MHz proton noise-decoupled ^{13}C spectra of **126**. (a) Protio isotopomer; (b) formed from cyclization of **125** in D_2SO_4.[150]

^{13}C labelling in biosynthesis[145,151]

A common approach in the elucidation of biosynthetic pathways leading to secondary metabolites consists of feeding specifically labelled precursors to appropriate cultures. If the label used is ^{13}C, the metabolite under investigation is isolated and subjected to ^{13}C n.m.r. analysis. Provided the natural abundance ^{13}C spectrum has previously been assigned, the label distribution can be determined from a comparison with the spectrum of the labelled sample. Conversely, this technique allows one to assign the spectrum from a known biosynthetic pathway. This aspect of the method has previously been dealt with in

127

Chapter 3. From the already abundant literature[145,151] we will discuss a few illustrative examples.

In order to prove the polyacetate origin of the antibiotic lactone asperlin (**127**), [2-¹³C]acetate (61% enriched) was fed to growing cultures of *Aspergillus*

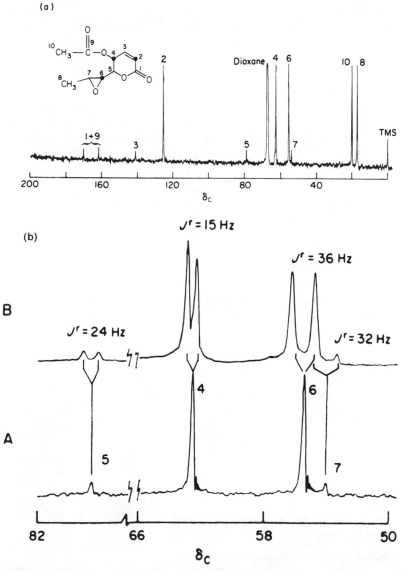

Fig. 5.45 25.2 MHz ¹³C spectra of asperlin, biosynthesized from [2-¹³C]acetate.[152] (a) 200 ppm survey spectrum, proton noise-decoupled; (b) high-field region, proton noise-decoupled (A) and off-resonance decoupled, irradiation at 3.70 ppm (B).

nidulans and the isolated product examined by ^{13}C n.m.r.[152] The ^{13}C spectrum of the labelled antibiotic (recorded in the continuous-wave mode) displayed in Fig. 5.45(a) exhibits enhanced intensity for five of the ten carbon signals. A total assignment of the spectrum could be achieved with the aid of multiplicities and magnitudes of the residual splittings observed in the single-frequency off-resonance decoupled spectra. Since the spectrum was recorded using the dioxane solvent resonance as a homonuclear lock signal, the decoupler had been centred at the dioxane proton frequency. Consequently the residual splittings progressively increase with increasing proton chemical shifts with respect to that of dioxane ($\delta H = 3.70$ ppm). The partial off-resonance spectrum of the lower field aliphatic region in Fig. 5.45(b) clearly indicates the differences in the reduced multiplets pertaining to carbons 4, 5, 6, and 7. The chemical shifts of the attached protons are: $H_4, \delta 5.22$; $H_5, \delta 3.90$; $H_6, \delta 3.05$; $H_7, \delta 2.93$ and the residual splittings J^r are as indicated in Fig. 5.45(b). Since, according to Eqn. (3.9), (p. 102) the residual splitting J^r also depends on the actual ^{13}C—^{1}H one-bond coupling constant $^1J_{CH}$ and the decoupler power level $\gamma H_2/2\pi$, its calculation requires the knowledge of these quantities. For the ^{13}C-enriched carbons $^1J_{CH}$ could be measured from the ^{13}C satellites and the expected residual splittings calculated. Using $\gamma H_2/2\pi = 21\,00$ Hz, one obtains for C_4 and C_6 values for J^r of 15.5 and 32 Hz in good agreement with the experimentally observed values. The labelling pattern found verifies the tetraacetyl origin of the eight-carbon epoxy-γ-lactone moiety in the antibiotic.

Problem 35: The proton noise-decoupled ^{13}C spectrum, including off-resonance information, obtained from a natural abundance sample of thermozymocidin triacetyl-γ-lactone is given.[153]

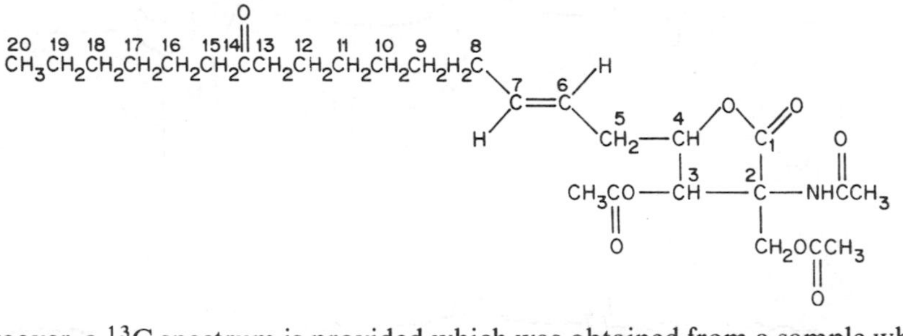

Moreover, a ^{13}C spectrum is provided which was obtained from a sample which was isolated from a culture broth which had previously been fed with [1-^{13}C]acetate (90%).

(1) Establish the labelling pattern.
(2) Decide whether acetate participates in the biosynthesis of the serine moiety of the molecule (serine = $HOCH_2CH(NH_2)COOH$).

No.	Int.	ppm	No.	Int.	ppm
1	26	212.09	11	81	62.59
2	29	172.85	12	145	42.88
3	32	170.53	13	101	32.53
4	57	170.17	14	83	32.28
5	46	169.46	15	93	31.69
6	66	135.20	16	204	29.01
7	72	123.67	17	171	23.88
8	89	81.55	18	118	22.61
9	77	72.14	19	99	20.60
10	83	63.36	20	104	20.36
			21	85	14.04

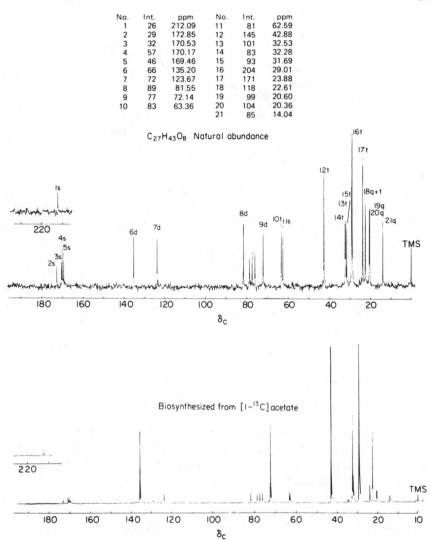

Problem Fig. 35.

We have seen previously that a completely unambiguous assignment of the spectrum represents the prime task in order to delineate a labelling pattern on the grounds of the ^{13}C spectrum. In the subsequent example this is achieved with a combination of residual splittings and $^{13}C—^{13}C$ coupling data. In the biosynthesis of dihydrocoriolin C (**128**), which belongs to a class of tricyclic sesquiterpene antibiotics, the biosynthesis had been assumed to proceed via humulene (**129**). Three possible cyclization routes which all have protonation at

128   **129**

C-10 in common can be considered. In order to gain an insight into the biogenesis, coriolin was biosynthesized from $[1, 2\text{-}^{13}C]$acetate precursor.[154] The potential advantages of dilabelling have already been outlined in Chapter 3. The three conceivable pathways would lead to the label distribution as indicated by structures **130, 131** and **132** where the lines in bold type indicate bonds formed from a

130   **131**   **132**

$$R_1 = R_3 = COCH_3, R_2 = \overset{\displaystyle OCOCH_3}{\underset{\displaystyle |}{COCHC_6H_{13}}}$$

dilabelled acetate precursor. However, this does not imply that all remaining carbons are unlabelled. Because of label dilution, ^{13}C—^{13}C couplings are not observable unless the bond in question has been formed from a dilabelled acetate unit. The anticipated labelling patterns differ according to the specific intra-molecular rearrangements peculiar to each of the pathways. For the present discussion, it is sufficient to know the predicted results as expressed by the three structures **130, 131**, and **132**. Table 5.7 lists the chemical shift and spin–spin coupling assignments, from which it immediately can be inferred that the labelling scheme **130** applies. It is, for example, readily recognized that a ^{13}C—^{13}C coupling is expected to occur twice between a singlet and a quartet carbon (C_3—C_{12} and C_{11}—C_{15}) only for scheme **130**. A further argument in favour of the mechanism leading to **130** is the presence of a ^{13}C—^{13}C coupling involving a quaternary and a methine carbon (C_6—C_7). The magnitude of 26 Hz is typical of epoxy carbons.

Apart from mono- and dilabelled acetate, specifically labelled propionate and butyrate precursors have been fed to the appropriate cultures. In the antibiotic **133** $[1\text{-}^{13}C]$butyrate incorporation experiments proved unequivocally that all

TABLE 5.7 ^{13}C chemical shifts and ^{13}C—^{13}C spin–spin coupling constants in dihydro-coriolin C triacetate[154]

Position		$^1J_{cc}$(Hz)
C_1	(80.8, d)–C_2 (51.0, d)	42
C_3	(47.1, s)–C_{12}(13.1, q)	38
C_4	(64.5, s)–C_{13}(44.9, t)	30
C_5	(71.7, d)	—
C_6	(60.9, d)–C_7 (73.5, s)	26
C_8	(72.7, d)	—
C_9	(41.3, d)–C_{10}(37.6, t)	33
C_{11}	(43.9, s)–C_{15}(21.4, q)	34
C_{14}	(26.5, q)	—
$C_{1'}$	(170.0, s)–$C_{2'}$ (72.4, d)	65
$C_{3'}$	(31.3, t)–$C_{4'}$ (25.2, t)	34
$C_{5'}$	(28.8, t)–$C_{6'}$ (31.5, t)	35
$C_{7'}$	(22.5, t)–$C_{8'}$ (14.0, q)	34

Note: In parentheses: chemical shift in ppm, residual splitting (s = singlet, d = doublet, t = triplet, q = quartet).

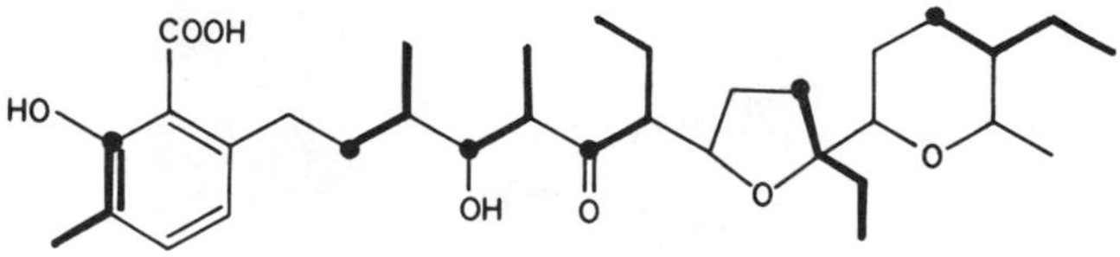

133

ethyl groups were butyrate-derived, whereas all methyl groups except one have propionate as a precursor.[155]

An interesting experiment reported recently was carried out with the purpose of establishing the origin of C-2 in cephalosporin (**134**). Since valine had clearly

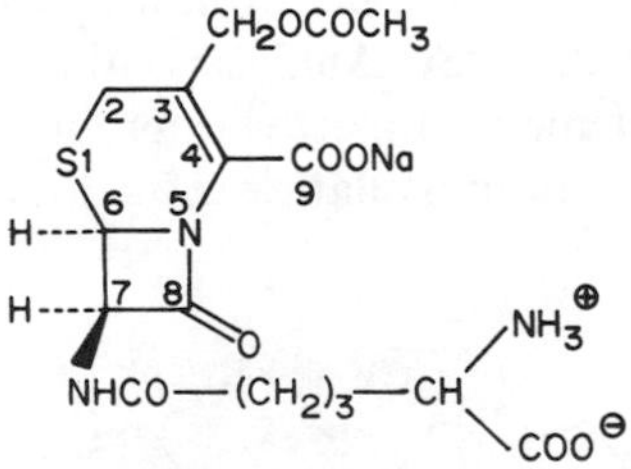

134

been shown to participate as a substrate in the biosynthesis of β-lactam antibiotics, it was of interest to obtain detailed information regarding the fate of the isopropyl group.[156] For this purpose (2RS, 3S)-[4-^{13}C]valine (135) was synthesized and the appropriate culture incubated with this precursor. The

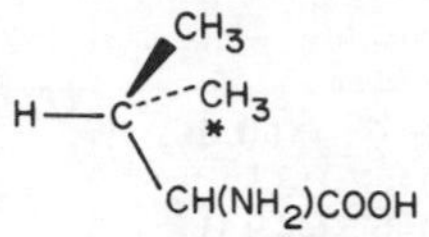

(3S)-[4-^{13}C]valine

135

antibiotic resulting from the biosynthesis involving the chiral substrate **135** exhibited C-2 to be the exclusive site of enrichment.

A useful extension of the doubly ^{13}C labelled acetate incorporation technique for the elucidation of biosynthetic pathways involves *in vivo* incorporation of uniformly ^{13}C labelled glucose (i.e. [U-^{13}C$_6$]glucose) into metabolites. In this way, biosynthetic steps that involve cleavage and/or rearrangement of an initially intact ^{13}C—^{13}C bond in the uniformly labelled substrate can be identified readily. When [U-^{13}C$_6$]glucose was fed to cultures of *Streptomyces* UC5319, pentalenolactone (136) was produced.

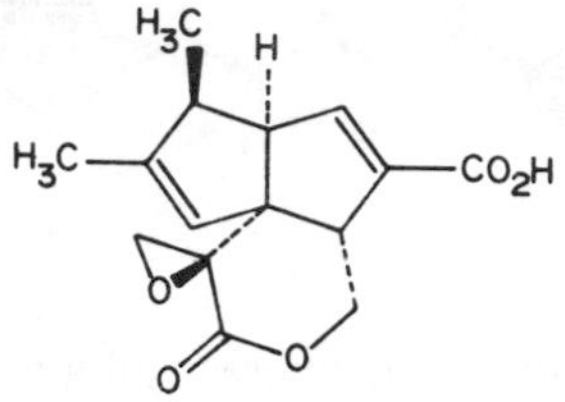

136

It was shown that in this system, [U-^{13}C$_6$]glucose serves as a precursor to [1,2-^{13}C$_2$]CoA (a co-enzyme) which then promotes cyclization of farnesyl pyrophosphate, ultimately affording **136**. Analysis of the ^{13}C n.m.r. spectrum of biosynthetically labelled pentalenic acid methyl ester (derived from the initially formed metabolite **136**) led to the postulation of a mevalonoid biosynthetic pathway as illustrated below.[157]

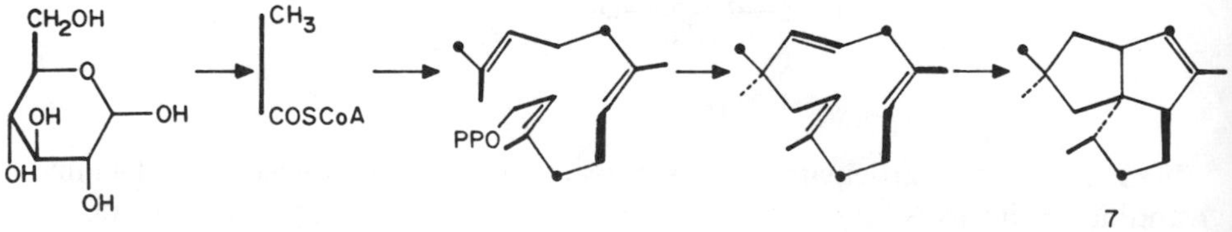

7

More recently, ^{13}C n.m.r. studies of *in vivo* incorporation of $[U-^{13}C_6]$glucose have been utilized to investigate the biosynthesis of additional glucose meta-bolites. In one such study, $[U-^{13}C_6]$glucose, when fed to *Streptomyces flocculus*, culture afforded ^{13}C-labelled streptonigrin (137).

137

The biosynthetic pathway leading to streptonigrin was elucidated via analysis of $^{13}C—^{13}C$ spin–spin couplings in the ^{13}C n.m.r. spectrum of the material thereby produced.[158]

As in numerous other examples, the application of two-dimensional n.m.r. techniques has significantly enhanced the utility of ^{13}C n.m.r. spectroscopy for elucidating biosynthetic pathways. For example, the mechanism of biosynthesis of ravidomycin (138, an antitumor antibiotic produced by *Streptomyces ravidus*) has been investigated by two-dimensional INADEQUATE techniques (see Chapter 3, p. 165). Carbon-13 n.m.r. chemical shift assignments for 138 have been reported recently.[159]

In order to determine the pattern of incorporation of intact two-carbon subunits, ^{13}C-labelled 138 was produced from doubly labelled $[1, 2-^{13}C]$acetate via bacterial fermentation. Carbon–carbon connectivities in labelled 138 thereby obtained were then analysed via double-quantum two-dimensional n.m.r. techniques.

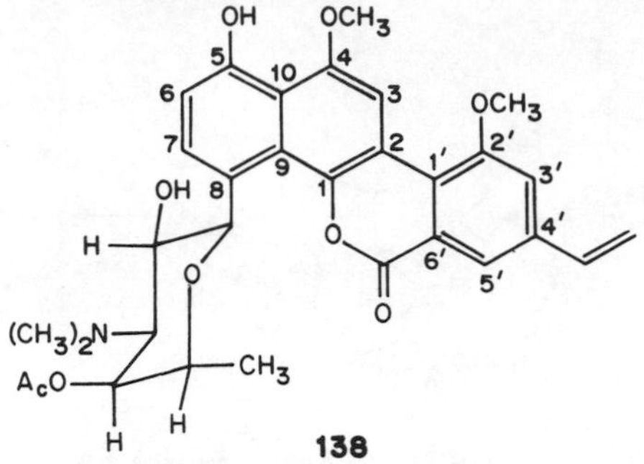

138

A partial two-dimensional INADEQUATE spectrum of labelled **138** produced as described above is shown in Fig. 5.46.[160] The carbon–carbon connectivities for the ravidomycin aglycone established from the two-dimensional n.m.r. spectrum shown in Fig. 5.46 are indicated by the bold-face lines in structure **139**.

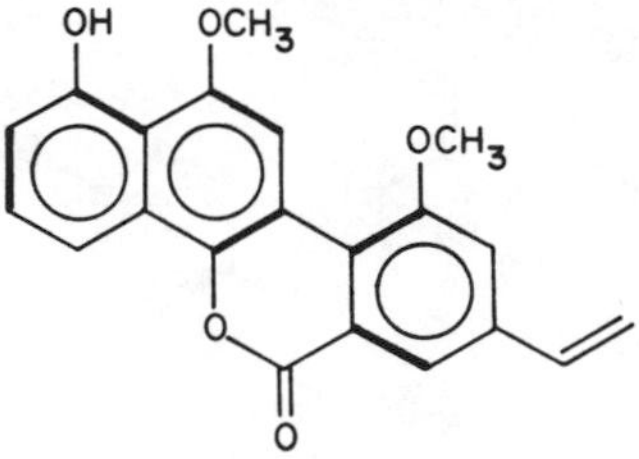

139

Double-quantum coherence n.m.r. spectroscopy has also been utilized recently to investigate the mechanism of biosynthesis of mevinolin (**140**, produced by *Aspergillus terreus*; mevinolin is an inhibitor of cholesterol biosynthesis in humans).[161]

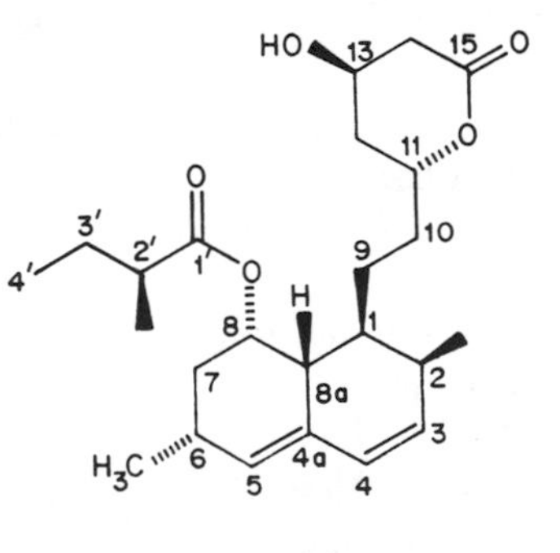

140

The above studies serve to illustrate the value of the two-dimensional INADEQUATE technique for unambiguous detection of intact multicarbon

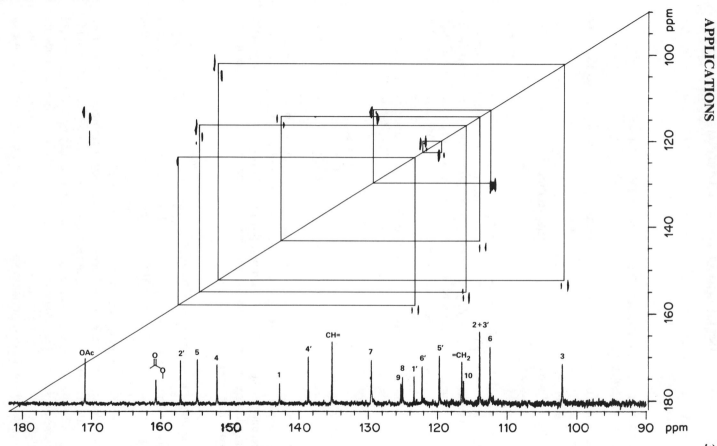

Figure 5.46 Partial two-dimensional INADEQUATE spectrum of [1,2-^{13}C]acetate-labelled revidomycin. Straight lines connecting off-axis peaks indicate coupled ^{13}C paris.[159]

units. Accordingly, extensive future exploitation of this technique for the elucidation of biosynthetic pathways is anticipated.

5.9 SPIN–LATTICE RELAXATION AND NUCLEAR OVERHAUSER STUDIES

The basics of spin–lattice relaxation, including an introduction to the most common mechanisms governing ^{13}C relaxation, have been given in Chapter 4. Although relaxation phenomena primarily reflect the dynamics rather than the structure of molecules, there is an intimate relationship between the two. This may be visualized by considering that molecular motion in the simplest case is described by a single correlation time. In the case of the most common mechanism, dipole–dipole relaxation, the relevant quantity is the rotational correlation time τ_c, which is a function of frictional and inertial effects. Hence, it is understandable that a bulky molecule, for example, will be hindered in its reorientation due to friction between solute and solvent molecules (and at high concentration also due to friction between the solute molecules themselves). Moreover, the moments of inertia of such molecules will be larger than for their smaller counterparts. Both effects tend to decelerate molecular motion. Molecules possessing functional groups with labile protons are known to associate with increasing concentration and thus again affect reorientation rates. These are two examples illustrating how molecular motion is related to structure. In the forthcoming discussion we will preferentially stress the structural and stereochemical aspects of relaxation, while the treatment of dynamic and mechanistic phenomena will be confined to a minimum (cf. Chapter 4). Moreover, emphasis will be placed on dipolar relaxation, which has been shown to be the dominant mechanism for the relaxation of proton-bearing and quaternary carbons in medium- and large-sized organic molecules. For a more profound coverage of the dynamic aspects of ^{13}C relaxation, the interested reader is referred to the pertinent reviews.[162,163]

Applications to structure assignment[164]

This section will outline how spin–lattice relaxation data may be exploited as an additional assignment criterion. It thus complements the techniques discussed in Chapter 3. Among all carbons in a ^{13}C n.m.r. spectrum, the quaternaries are undoubtedly the most difficult ones to assign. While single-resonance spectra in small molecules may provide the necessary criteria for unambiguous assignment, this method usually fails in larger molecules for reasons of complexity of the spin–spin coupling patterns. In such cases relaxation data may provide unequivocal arguments. Besides the designation of quaternary carbons, relaxation times are equally suited for the assignment of protonated carbons, particularly in those instances where the off-resonance spectrum cannot be disentangled because of numerous signal overlaps. Furthermore, relaxation times may serve to discriminate carbons which undergo internal reorientation from those which belong to the rigid backbone of the molecule.

Quaternary carbons

It has been pointed out in Chapter 1 that the signal intensities in a pulsed spectrum do not necessarily reflect their true values. This is due to relaxation rate differentials between individual carbons of a molecule. Since the pulse intervals, dictated by sensitivity considerations, do not permit complete recovery of the magnetization at the end of the acquisition period, the more slowly relaxing carbons will give rise to lowered signal intensities. Although this may be considered a disadvantage in quantitative studies, it can be beneficial from a diagnostic point of view. In particular, quaternary carbons whose dipolar relaxation is less efficient (r_{CH}^{-6} term, Eqn. (4.15)) can in general immediately be assigned without recourse to single-frequency decoupling techniques (cf. Fig. 1.11).

Furthermore, we have seen that the observed relaxation rate can generally be expressed as the sum of individual contributions (Eqn. (4.38)) with one term usually dominating ($1/T_1^{DD}$). If the correlation time τ_c is short enough and r_{CH}^{-6} is relatively small—a situation that often applies to quaternary carbons in small-to-intermediate-sized molecules—dipolar relaxation may be quite inefficient and there will thus be competition from other mechanisms. Under these circumstances $T_1^{DD} > T_1^{obs}$ and therefore the NOE is below its maximum value. On this assumption it can be shown that η is a function of $\sum_i r_{CH_i}^{-6}$; it decreases with

141

increasing CH distance. In pyrene (**141**) for example, the quaternary carbons exhibit incomplete NOE ($\eta < \eta_0$). The spectrum in Fig. 5.47 illustrates a relative NOE measurement carried out by applying $90°$ pulses using sufficiently long pulse intervals to ensure complete relaxation of all carbons between the pulses.[164] Assuming full dipolar relaxation for the proton-bearing carbons, the NOE enhancement factors for the two sets of quaternary carbons (11, 12, 13, 14 and 15, 16) can be determined from a comparison of their peak areas with those of the proton-bearing carbons. The two carbons 15 and 16, buried in the interior of the molecule, with no geminal protons, exhibit a lower NOE ($\eta = 0.5$) than the four peripheral quaternary carbons 11–14 ($\eta = 1.25$). Remembering that $1/T_1 = 1/T_1^{DD} + 1/T_1^O$, where $1/T_1^O$ signifies the relaxation rate arising from other than dipolar contributions, then Eqn. (4.40) can be modified by replacing $1/T_1^{DD}$ with Eqn. (4.15), which leads to

$$\eta = \eta_0 \frac{\hbar^2 \gamma_C^2 \gamma_H^2 \sum_i r_{CH_i}^{-6} \tau_c}{\hbar^2 \gamma_C^2 \gamma_H^2 \sum_i r_{CH_i}^{-6} \tau_c + 1/T_1^O} \tag{5.21}$$

Hence, under these circumstances, the fractional NOE, η, becomes dependent on $\sum_i r_{CH_i}^{-6}$, and for a given value of the non-dipolar relaxation contribution, η decreases with decreasing magnitude of $\sum_i r_{CH_i}^{-6}$, as is apparent from the results in Fig. 5.47. However, this behaviour is rather exceptional and is only observed for relatively small molecules. For larger molecules, the tumbling slows down and the DD relaxation totally prevails over SR and CSA contribution.[165] As a consequence, $1/T_1^O$ in Eqn. (5.21) becomes negligibly small and $\eta = \eta_0$, i.e. all carbons experience full NOE independent of the number and distance of nearest hydrogens. The method described is then no longer feasible for assignment purposes.

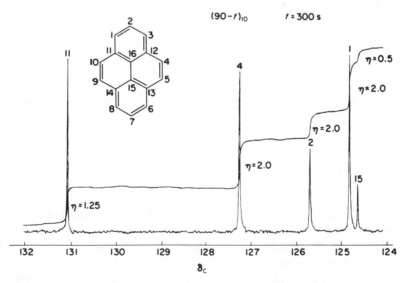

Fig. 5.47 Integrated ^{13}C spectrum and nuclear Overhauser enhancement factors η in pyrene. The spectrum was obtained from a degassed sample under conditions as indicated.[164]

Problem 36: A series of inversion-recovery T_1 spectra are given, showing the region of the two quaternary carbons in benzonitrile.[164] The spectra were obtained by plotting the difference $S_\infty - S(t)$ of the signal amplitudes against τ whose values are given on the right-hand side of each trace. Furthermore, the completely relaxed spectra of the same region are given with no proton decoupling and with decoupling, respectively.

(1) Determine the T_1s of the two carbons using the procedure described in Chapter 4.
(2) Determine the NOE and T_1^{DD}.
(3) Assign the two lines to the two quaternary carbons.

In codeine (**142**), the average relaxation time of the methine carbons was found to be short $(T_1(CH) \approx 150\,\text{ms})$, thus indicating a long rotational correlation

142

time.[166] Under these circumstances, the relaxation even of the quaternary carbons is expected to be predominantly dipolar, which was experimentally corroborated by the full NOE of all quaternary carbons. In this case, the measured relaxation rates, according to Eqn. (4.15), are proportional to the sum of the inverse sixth power of the C—H internuclear distances. This provides a unique means of assigning these carbons to the particular resonances. Among the four quaternary aromatic carbon signals, those at lowest field are (for chemical shift reasons) assigned to the oxygen-bearing ones; however, an individual assignment on this basis is not possible. Taking into account that C-14 has one proton in α position, whereas C-13 has none, the latter is anticipated to relax more slowly. Hence the line at 146.6 ppm $(T_1 = 8.9\,\text{s})$ belongs to C-13 and that at 141.7 ppm $(T_1 = 5.2\,\text{s})$ to C-14. Analogous arguments apply to C-3 and C-12 with three and no α protons, respectively. It is interesting to note that C-12 relaxes faster than C-13 (5.2 versus 8.9 s), which can be explained by the close proximity of C-12 to the methylene protons at C-15. Chemical shift and relaxation data of the quaternary carbons in **142** are compiled in Table 5.8.

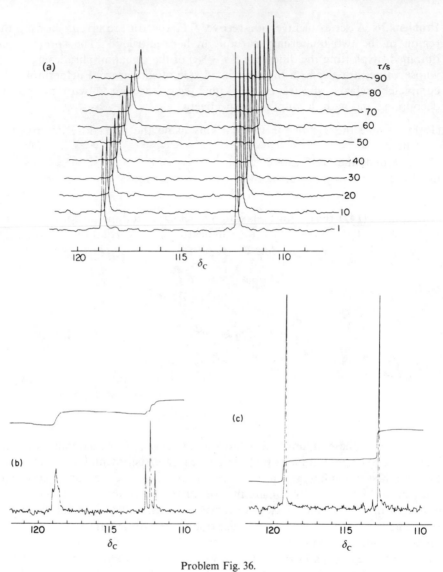

Problem Fig. 36.

Problem 37: The alkaloid reserpine has a total of 11 quaternary carbons, which were shown to be predominantly relaxed by DD interaction.[166] It was found, moreover, that the relaxation for all carbons could be described by a single correlation time (except methyl carbons).

TABLE 5.8 Chemical shifts and relaxation times of quaternary carbons in codeine (142)[a,166]

Carbon	$T_1(s)$	δ_C(ppm)
11	1.53	42.9
3	1.82	127.0
14	5.2	141.7
13	8.9	146.6
12	5.1	130.9

[a] At 25.2 MHz and 30°C, 1.6 molar in $CDCl_3$.

The chemical shifts and spin–lattice relaxation times for the quaternary carbons are listed below. Assign these carbons using chemical shift and relaxation arguments.

$T_1(s)$	δ_C(ppm)
2.9	156.5
2.5	136.8
1.6	130.9
1.6	108.2
3.8	122.5
7.5	125.6
4.9	153.3
4.9	153.3
2.4	173.2
5.6	165.9
12.8	142.8

An extension of one-dimensional heteronuclear NOE experiments discussed in Chapter 4 is the HOESY[167] experiment. HOESY has been utilized for structural elucidation and/or conformational analysis of small organic molecules[167,168] and of larger, biologically important molecules (for example, natural products[169,170] and peptides[171-173]). In order to ensure that the magnetization transfer occurs via a dipolar mechanism (rather than via scalar interactions, the HOESY pulse sequence[167,168] effectively eliminates scalar coupling between the two types of spins (i.e. ^1H and ^{13}C). This experiment lends itself to elucidation of the mechanism of relaxation, i.e. to the determination of the fractional contribution of the DD mechanism to overall nuclear relaxation.

The observation of NOE permits connectivities to be established; for example, molecular fragments that are connected through a common quaternary carbon often can be identified readily by using this technique. Similar information is accessible through the use of heteronuclear RELAY techniques. An important distinction between NOE and RELAY experiments is that the former involves (through-space) dipole–dipole interactions between neighbouring nuclei whereas the latter is transmitted via scalar coupling between neighbouring nuclei. Thus, the NOE is potentially capable of providing information regarding internuclear distances that has bearing upon the three-dimensional structure of the molecule under investigation (cf. also p. 387, 388).[174]

An example that illustrates the use of two-dimensional HOESY is provided by a study[167] of dipolar interactions between quaternary carbons and protons in camphor. A contour plot of the HOESY spectrum of camphor is displayed in Fig. 5.48. The proton chemical shift assignments shown in the F_1 dimension in Fig. 5.48 were established via analysis of the COSY spectrum of camphor. Connectivities established via analysis of scalar couplings in the COSY spectrum were found to be fully consistent with those established via analysis of dipole–dipole interactions in the HOESY experiment.[167]

Let us illustrate the assignment process for the case of the three methyl groups in camphor (Fig. 5.48). Note that carbon h generates cross peaks only with the protons in methyl groups B and C; hence, we can immediately assign the resonance that corresponds to the protons A in methyl group a. Additionally, this observation permits assignment of the resonance that belongs to quaternary carbon i (note cross peak (A, i) in Fig. 5.48). The fact that both protons B and C afford cross peaks with quaternary carbon h establishes the connectivity $(CH_3)_B$–C_h–$(CH_3)_C$.

Problem 38: The HOESY spectrum of fluoranthene (**143**) is shown below.[167] In addition, the following T_1 values have been measured for the nine magnetically non-equivalent ^{13}C nuclei in fluoranthene (unassigned as presented here): 2.1, 2.2, 2.4, 2.4, 2.5, 11.6, 15.1, 15.6, and 16.1 s. Based upon this information, assign all of the ^{13}C and proton resonances in fluoranthene.

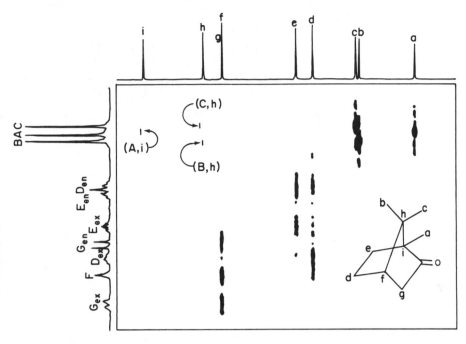

Fig. 5.48 Contour plot of the HOESY spectrum of camphor. Spectral width in the F_2 dimension (^{13}C) was 5400 Hz and in the F_1 dimension was 730 Hz. Cross peaks arise from dipolar coupling between quaternary carbons and nearby protons.[167] Reprinted with permission from Yu and Levy, *J. Am. Chem. Soc.* **106**, 6533. Copyright (1984) American Chemical Society.

Since the NOE varies with the inverse sixth power of internuclear distance, carbon–proton internuclear distances in principle can be determined via a HOESY experiment in dipolar-interacting systems. This approach has been utilized successfully to define hydrogen-bonding patterns in, for example, rifamycin S (a macrocyclic natural product),[169,170] valinomycin (a cyclic dodecadepsipeptide),[171] and the [Val³] analogue of the fungal tetrapeptide HC toxin.[175] It has been pointed out[176] that more than one carbonyl group may be in sufficiently close proximity to a given NH group to produce similar NOE effects. Nevertheless, this technique continues to be employed as a tool for determining molecular conformation in biomolecules.

Proton-bearing carbons: the effect of internal motion

We have seen in Chapter 4 that, for a molecule which undergoes isotropic tumbling, the relaxation rate for dipolar relaxation is given by Eqn. (4.15). If contributions from protons apart from those directly attached are ignored, which

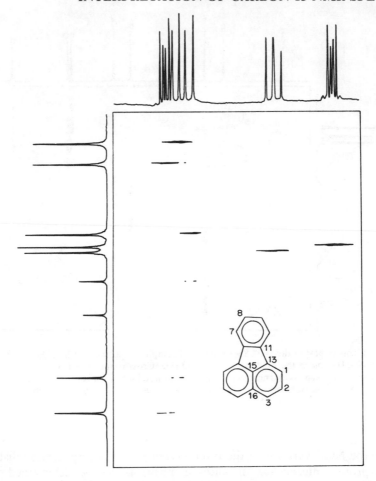

Problem Fig. 38 The contour plot of a HOESY spectrum of fluoranthene.[167] Spectral acquisition parameters are the following: 2 s for mixing
time, 2315 Hz in $F_2(^{13}C)$ dimension, and 64 spectra (128 scans each) were
accumulated with a t_1 increment to make 240 Hz in the F_1 dimension. The
dipolar interactions between three quaternary carbons and nearby protons
clearly show up as cross peaks labelled in parentheses.

turns out to be a valid approximation, Eqn. (4.15) simplifies to

$$1/T_1^{DD} = \hbar^2 \gamma_C^2 \gamma_H^2 N r_{CH}^{-6} \tau_c \qquad (5.22)$$

where N represents the number of directly bonded protons.

If these conditions apply, differentiation between methyl, methylene, and
methine protons should be straightforward. Although single-frequency off-
resonance decoupling (in principle) provides the same information, the analysis

of such spectra is often hampered because of crowding of the signals and the concomitant overlaps. The simple scheme breaks down (1) as soon as the molecule as a whole undergoes anisotropic motion, i.e. if the reorientation around some geometric axis is favoured, or (2) if the molecule possesses groups which are subjected to fast internal motion. The simplest case of the latter situation can be established for a methyl group attached to the rigid backbone of the molecule. The relaxation time of such a carbon can no longer be described in terms of the correlation time controlling the overall motional behaviour of the molecule. If the internal rotation is fast compared to the tumbling rate of the whole molecule, a lengthening of T_1 of the carbon in question is always observed. Basically, this behaviour is generally found for carbons which can undergo unhindered internal motion such as those belonging to a flexible side chain bonded to a slowly diffusing larger molecule. If treated with care, this provides one with an outstanding diagnostic tool. Often it is not even necessary to determine precise relaxation times and it suffices to run one single inversion-recovery experiment which provides the relative order of T_1s. The spectrum in Fig. 5.49 illustrates the potential of the method. It shows a single inversion-recovery trace of cholestane (144) obtained by choosing $\tau = 0.5$ s. This spectrum

144

displays weakly negative lines for the methine carbons pertaining to the backbone of the molecule, whereas the corresponding methylene carbons, which relax twice as fast as the methines, appear positive, with all of them exhibiting approximately the same intensity. The methyl carbons which, in the absence of internal motion, should relax three times as fast as a methine carbon, obviously have long T_1s as evidenced by their intense inverted lines. It is further noted that carbons 22, 23, 24, and 25 all exhibit less efficient relaxation than their backbone counterparts, with their T_1s progressively increasing towards the chain end. Therefore, if the multiplicities are known, the additional knowledge of (relative) T_1 values should enable one to make specific assignments and, in general, to distinguish between backbone and side-chain carbons.

The best-investigated case of internal motion is that of a rapidly spinning methyl group.[177] It has been shown that in some instances the methyl group rotation can become so fast that spin rotation may compete with dipolar relaxation.[178] In toluene, for example, where the methyl group is essentially a free

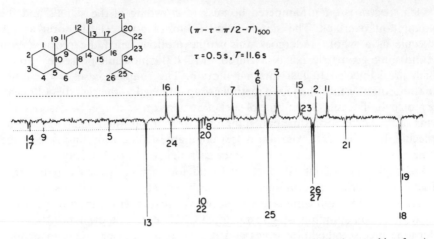

Fig. 5.49 Inversion-recovery ^{13}C spectrum of cholestane (*ca* 1-molar in deuterochloroform) under conditions as indicated. It is to be noted that backbone methine carbons appear slightly negative (dotted line), whereas corresponding methylene signals are positive, indicative of their faster relaxation. The deviation found for C-3, which is relaxed more completely than other backbone methylenes, is a consequence of the slightly anisotropic motion of this molecule.

rotor (≈ 0 kcal for a sixfold barrier), substantial contributions from spin–rotation were observed.[179] In contrast, in *o*-xylene,[180] methyl carbon relaxation is dipolar, which is ascribed to the fact that the two adjacent methyl groups have a threefold barrier to rotation of approximately 2 kcal mol^{-1}. A comparison of the dipolar relaxation times T_1^{DD} obtained by dissecting the experimental T_1s in *o*-xylene (**145**) and mesitylene (**146**) using NOE data, is illustrative of this behaviour.

145   **146**

It would, however, be incorrect to associate less efficient dipolar relaxation of methyl carbons with increased steric compression. The fallacy of such an interpretation is substantiated by the results obtained for hemimellitene (**147**)

147

whose T_1^{DD} data are given in the formula.[180] It is obvious here that the sterically perturbed 2-CH_3 exhibits a longer T_1^{DD}, thus pointing to increased motion for this carbon. The reason for this apparent contradiction lies in the fact that, unlike in toluene, the 2-methyl carbon has a sixfold energy profile, which is readily recognized if one considers the minimum energy rotamers (**148** and **149**).

148 **149**

The 1-CH_3 in conformer **148** and the 3-CH_3 in conformer **149** are both in minimum energy configurations relative to the 2-CH_3 and have threefold energy profiles. The 2-CH_3, on the other hand, has the same steric energy in both forms although it has been rotated through an angle of 120°. Hence it does not possess a threefold but a sixfold barrier to rotation.

Further examples documenting the relationship between conformational energy wells and spin–lattice relaxation times in rotating groups are found in 1-methylnaphthalene (**150**) and 9-methylanthracene (**151**).[181] In **150** methyl group rotation is hindered due to steric interaction with the peri hydrogen, and hence T_1 is short ($T_1(CH_3)/T_1(CH) \approx 1$), whereas no such minimum energy rotamer exists

150 **151**

for **151**, leading to rapid spinning ($T_1(CH_3)/T_1(CH) \approx 3$). Calculations[165] show that in the case where group rotation is very much faster than overall reorientation, $T_1(CH_3)/T_1(CH) = 3$, provided that the methine carbon belongs to the rigid backbone of the molecule. On the other hand, for a completely locked methyl group, $T_1(CH_3)/T_1(CH) = \frac{1}{3}$, i.e.

$$\tfrac{1}{3}T_1(CH) \leqslant T_1(CH_3) \leqslant 3T_1(CH)$$

Locked   Free rotor

An analysis of the motional behaviour of the angular methyl carbons in androstanes, in terms of the correlation time τ_c for overall motion, and τ_G accounting for methyl group rotation, reveals some interesting features.[182] The computational procedure required to separate group rotation from overall diffusion is discussed in Ref. 165. While the activation energy for rotation around the C—C bond in ethane is known to be 3.6 kcal mol^{-1}, corresponding to $\tau_G = 8 \times 10^{-11}$ s, the T_1s found for C-18 and C-19 in the substituted androstanes correspond to a τ_G of the order of 10^{-12} s, thus pointing to significantly lower barrier to rotation in these compounds. It is interesting that, in most derivatives, τ_G(C-18) > τ_G(C-19). Some τ values, together with T_1 s of backbone and CH$_3$ carbons from which the correlation times were derived, are listed in Table 5.9 for some substituted androstanes.

The methyl carbon T_1 data are clearly related to the number of 1–3 diaxial interactions with axial hydrogens, in that each interaction lowers the barrier to rotation of the particular methyl. From **152** it is recognized that C-19 has five

152

axial protons in 3-position whereas C-18 has only three. It is noteworthy that ring A and B substitution leaves τ_G(C-19) practically unaffected (1.7 ± 0.2 ps). This holds equally for the 17-ketone. This is understandable considering the fact that the number of diaxial interactions remains unaltered. Introduction of a keto function at C-11 removes one axial hydrogen and probably renders the interaction with C$_8$—H less severe because of the concomitant distortion of ring

TABLE 5.9   Spin–lattice and correlation time data in substituted androstanes[a,182]

| Substituent | NT_1^b | | | τ_G(C-18)^d | τ_G(C-19)^d |
	CH$_1$CH$_2$^{c}	C-18	C-19		
2-β-OH	2.0	7.7	8.3	1.7	1.5
2-CO	2.0	7.3	8.6	1.9	1.4
3-β-OH	2.2	9.0	13.2	1.5	0.5
3-α-OH	2.1	8.0	12.0	1.7	0.6
3-CO	2.7	9.6	12.6	1.5	0.8
4-α-OH	2.5	8.6	12.7	1.7	0.7
4-β-OH	2.0	7.7	7.8	1.8	1.7
4-CO	2.2	8.1	8.7	1.7	1.5
11-CO	2.3	5.4	13.8	4.7	0.5
16-CO	2.1	4.5	11.8	7.2	0.6
17-CO	2.5	8.8	13.8	1.7	0.6
Δ^5-ene	2.4	8.4	6.8	1.8	2.9

^{a} 1-molar in CDCl$_3$.
^{b} NT_1 represents the product of the number of directly bonded protons and T_1(in s).
^{c} Average for backbone CH$_2$ and CH carbons.
^{d} In ps.

C. This is reflected by the increase of τ_G(C-18) in this compound from 1.7 to 4.7 ps. The correctness of this interpretation is borne out by the data obtained for C-19, where τ_G(C-19) = 0.6 + 0.1 ps for the 3α-, 3β-, and 4α- alcohols as well as the 11-, 16-, and 17-ketones, whereas a significantly higher value was obtained for all other derivatives. The effect is particularly conspicuous for the Δ^5 compound, where the introduction of the double bond not only removes the axial hydrogen C_6—H, but also lowers the steric interaction with C_2—H and C_4—H because of ring distortion. The resultant stabilization of the methyl carbon C-19 manifests itself in a considerably shortened relaxation time ($NT_1 = 6.8$ s against 13.8 s in the 17-ketone), which corresponds to a τ_G roughly six times longer than in the ring D-substituted systems.

Anisotropic overall motion

It was pointed out earlier that the simple equation describing dipolar relaxation in terms of a single correlation time τ_c, as expressed by Eqns (4.15) or (5.22), is only valid on condition that the molecule tumbles in an isotropic fashion, i.e. if there is no preference for rotation around some specific geometrical axis. This condition is often violated and anisotropic rotational diffusion is well established.[179,183] A classical example is that of monosubstituted benzenes (**153**). Since the moment of inertia relative to the twofold axis is less than that relative to an axis perpendicular to C_2, rotation around the former is favoured. Depending on the

153

mass, polarity, and bulkiness of R, rotation around the C_2 axis may be five to twenty times faster than rotation around a perpendicular axis. This should result in a different relaxation rate for carbon 4 as compared to carbons 2, 3, 5, and 6 because the C—H internuclear distance vector in the former is aligned with the main axis for reorientation.[183] Hence the local field H_{loc} produced by proton C_4—H is not modulated and cannot give rise to relaxation of this carbon. In other words, the relaxation of the *para* carbon is only affected by the slower motion around the perpendicular axis. Since in the motional narrowing limit, less motion is always equivalent to more efficient relaxation, the *para* carbon generally exhibits a shorter T_1 than the *ortho* and *meta* carbons. Some representative T_1 values of proton-bearing carbons in monosubstituted benzenes are given below (**154–161**) (in seconds).[179]

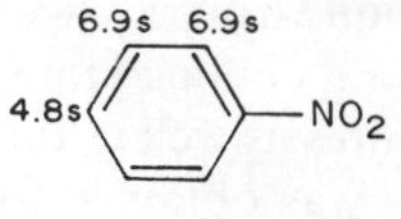

6.9s   6.9s

4.8s ⟨benzene ring⟩—NO₂

154

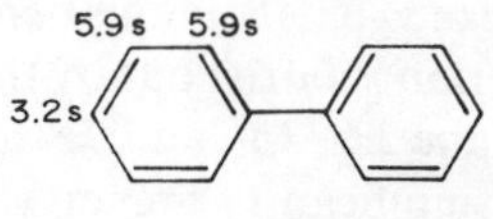

5.9s   5.9s

3.2s ⟨biphenyl⟩

155

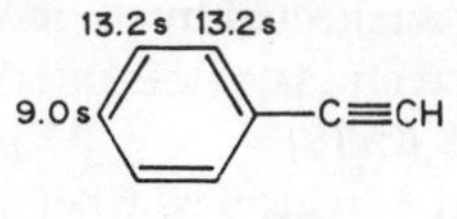

13.2s   13.2s

9.0s ⟨benzene ring⟩—C≡CH

156

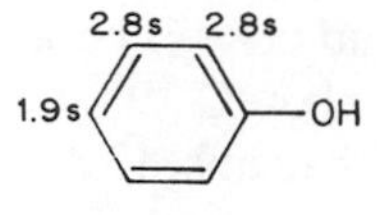

2.8s   2.8s

1.9s ⟨benzene ring⟩—OH

157

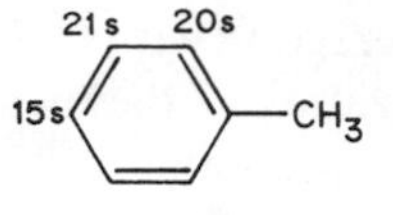

21s   20s

15s ⟨benzene ring⟩—CH₃

158

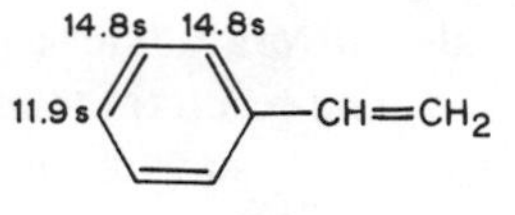

14.8s   14.8s

11.9s ⟨benzene ring⟩—CH=CH₂

159

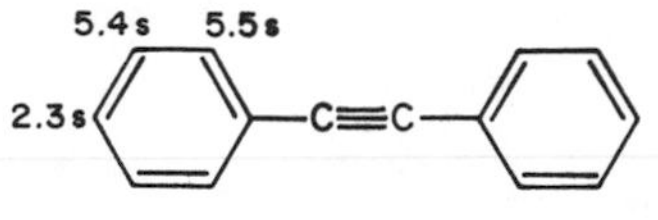

5.4s   5.5s

2.3s ⟨benzene ring⟩—C≡C—⟨benzene ring⟩

160

5.3s   5.2s

1.1s ⟨benzene ring⟩—C≡C—C≡C—⟨benzene ring⟩

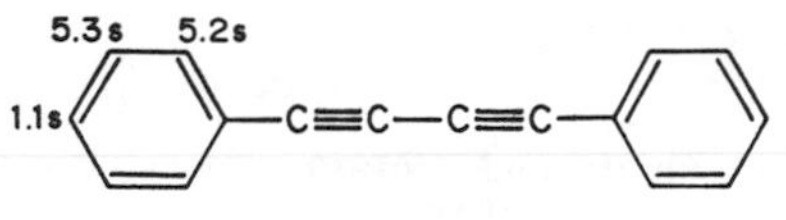

161

It has been shown that, from the ratio $T_1(o, m)/T_1(p)$, the ratio of the rotation rates for the reorientation with respect to the two axes can be determined.[179] Particularly conspicuous are the data found for diphenyldiacetylene (**161**), a rod-shaped molecule where $T_1(o, m)/T_1(p) = 5$. This unusually large value is a consequence of the distinctive shape of the molecule which renders reorientation around the short axis especially unfavourable.

In unsymmetrically disubstituted benzenes, motional anisotropy may be used to assign their resonances. This is demonstrated with two *ortho*-substituted molecules, **162** and **163** whose motion is only slightly anisotropic, but significant enough to aid the assignment.[164]

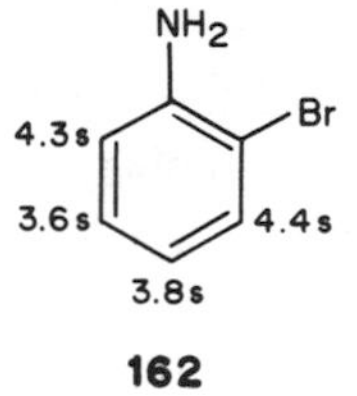

NH₂

4.3s

3.6s   Br

     4.4s

3.8s

162

COOCH₃

3.2s

2.8s   OH

     3.2s

2.2s

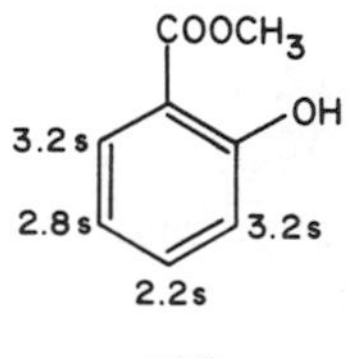

163

Problem 39: The proton noise-decoupled ^{13}C spectrum of 3-bromobiphenyl in DMSO-d₆, together with individual spin–lattice relaxation times for proton-bearing carbons, is given.[179] Assign all lines with the aid of relaxation and chemical shift data.

Motional anisotropy as evidenced from ^{13}C relaxation data has also been observed for larger molecules. Violation of Eqn. (5.22) is always a clear indication that the overall reorientation of a molecule is anisotropic. In **164**, for example,

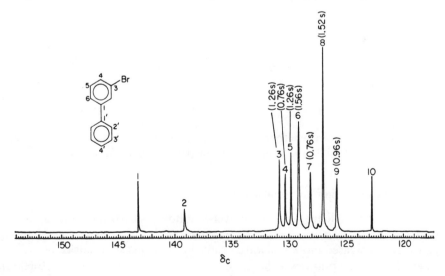

164

T_1 s of proton-bearing carbons 6 and 8 were found to be 75 and 130 ms, respectively,[184] pointing to preferred rotation around the long axis of the molecule. Figure 5.50 shows a selection of some inversion-recovery T_1 spectra recorded with the objective of measuring the relaxation times of the protonated carbons in **164**. The assignment of the two lines to carbons 6 and 8 was made by means of single-resonance spectra. It should also be noted that these T_1 s are almost two orders of magnitude shorter than those in some of the monosubstituted benzenes discussed earlier, which is due to the larger size of the molecule as well as strong intermolecular association due to hydrogen bonding.[185,186]

A quantitative treatment of anisotropic motion requires an analysis of the data

Problem Fig. 39.

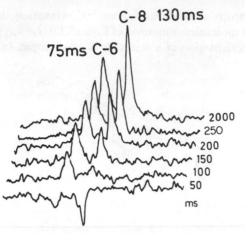

Fig. 5.50   Selection of inversion-recovery ^{13}C spectra of **164**, showing signals due to C-6 and C-8.[184]

in terms of the rotational diffusion tensor[187] which replaces the single correlation time sufficient for the case of isotropic reorientation. In the case of an asymmetric top molecule (for example, decalin, **165**), the three principal values R_1, R_2, and R_3 of the rotational diffusion tensor are necessary to describe the motion of the molecule.[188] Although the mathematics involved is well established and suitable computer programs are available,[189] there are often no straightforward solutions

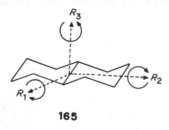

165

because it is uncertain whether the diffusional principal axis system coincides with that defined for the inertia tensor (for details see Chapter 4.4).

Dipolar relaxation in macromolecules

So far, we have implicitly assumed that the motional narrowing limit conditions apply, i.e. $(\omega_H + \omega_C)\tau_c \ll 1$ and consequently $1/T_1^{DD} \propto \tau_c$. Under these limiting conditions, increased motion always results in increased spin–lattice relaxation times. This assumption is safe up to molecular masses of about 1000. In macromolecules, however, motion can become so slow that the tumbling rate $1/\tau_c$ becomes comparable with ω_C and ω_H and one approaches the turning point of the curve which relates T_1 to τ_c in Fig. 4.3.[190-192] At the other extreme finally, where $(\omega_H + \omega_C)\tau_c \gg 1$, the relaxation rate increases linearly with $1/\tau_c$. In order to relate experimental T_1's to a correlation time, Eqn. (4.13) has therefore to be employed. Since this equation has two solutions with only one being physically meaningful, a further parameter is needed to obtain the correct value of τ_c. This, for example, can be achieved from a measurement of T_2, which, contrary to T_1, decreases monotonically when the motion is slowed down. Although T_2 data are not as easily amenable as T_1's, spin–spin relaxation times in the region of interest are sufficiently short to give rise to linewidths far beyond instrumental broadening. This often enables one to estimate T_2 from the linewidths observed. A further parameter that is dependent upon motion is the NOE. Whereas in the motional narrowing limit $\eta_0 = 0.5(\gamma_H/\gamma_C) = 1.988$ (Eqn. (4.18)) in the case where DD relaxation prevails over other mechanisms, η becomes a function of τ_c, ω_C and ω_H when motion is slow enough. At 23.4 kG, η drops to 1.54 at $\tau_c = 10^{-9}$ s and to 0.21 at $\tau_c = 10^{-8}$ s and finally, in the limit where $(\omega_H + \omega_C)\tau_c \gg 1$, it assumes a value of 0.15 (cf. Fig. 4.4). This is readily verified by substituting the transition probabilities in Eqn. (4.18) by the appropriate explicit expressions involving the spectral densities (Eqns (4.7)–(4.10)). Hence the statement that a full NOE is expected whenever relaxation is dominated by the DD mechanism is not of general validity.

These predictions found their first experimental confirmation in a study on the random-coil transition of poly-γ-benzyl-L-glutamate (PBLG) (166) (cf. page 338). Table 5.10 lists relaxation and NOE data obtained for the α-carbon in helical PBLG for three different molecular masses and compares them with

TABLE 5.10   Comparison of experimental linewidths and NOE data of the α-carbons in helical PBLG with computed values[a] obtained from spin–lattice relaxation times measured at 14 kG[a,191]

	7000[b]	17 000[b]	46 000[b]
Experimental T_1	35	39	44
Computed τ_c(ns)[c]	24(1.2)	27(1.6)	32(1.3)
η calculated[d]	0.2(1.5)	0.2(2.6)	0.2(2.7)
η observed	0.0	0.2	0.1
$(\pi T_2)^{-1}$ calculated[e]	38(9.5)	42(8.7)	48(7.6)
$(\pi T_2)^{-1}$ observed[f]	28	37	48

[a]Computed values inconsistent with experimental data are given in parentheses.
[b]Molecular mass.
[c]From Eqn. (4.13).
[d]From Eqn. (4.18).
[e]From Eqn. (4.17).
[f]Experimental full linewidth at half height.

computed data.[191] In contrast to observations on smaller molecules, T_1 is found to increase with increasing molecular mass. This is a clear indication that the motional narrowing limit conditions do not apply in this case. This is also borne out by the NOE, which is close to zero. It should be noted, however, that the equations used for computation 4.13, 4.17, and 4.18) are not strictly valid because the molecule's diffusion is probably not quite isotropic.

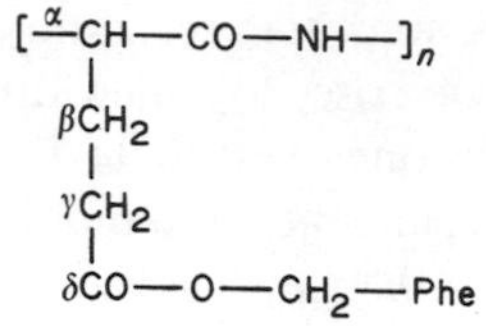

$$[-\overset{\alpha}{C}H-CO-NH-]_n$$
$$|$$
$$\beta CH_2$$
$$|$$
$$\gamma CH_2$$
$$|$$
$$\delta CO-O-CH_2-Phe$$

166

Reduced NOEs have also been reported for some synthetic high polymers such as polypropylene oxide[193] and polyisoprene[194] studied as solids. It may appear amazing at first that high-resolution ^{13}C n.m.r. spectra can be obtained at all from solid elastomers, in particular because these do not yield proton spectra under the same conditions. However, it has to be pointed out that dipolar coupling between ^{13}C and ^1H is weaker by a factor of about 16 since this interaction is proportional to the squares of the magnetogyric ratios γ, as is evident from Eqn. (4.17). It should also be mentioned that solid polymers normally do not have the properties of genuine solids in that their relaxation appears to be dominated by segmental motion. It has, however, been found that even in relatively non-viscous polymer solutions the rotational diffusion can become slow enough for the NOE to deviate substantially from 2 ($= \eta_0$). This clearly indicates that the macroscopic solution viscosity is not a relevant factor.

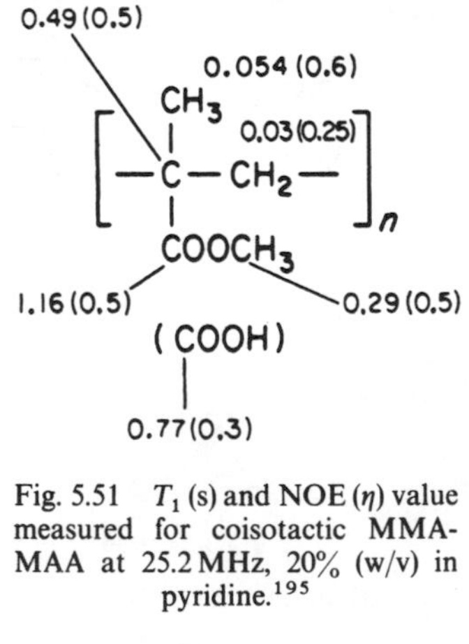

Fig. 5.51 T_1 (s) and NOE (η) value measured for coisotactic MMA-MAA at 25.2 MHz, 20% (w/v) in pyridine.[195]

An example is coisotactic methylmethacrylate-methacrylic acid copolymer (MMA–MAA), whose NOE and T_1 data obtained from 10% solutions point to the fact that the motional narrowing limit conditions do not apply[195] (Fig. 5.51). The value of $30 \pm 10\,\text{ms}$ for the methylene carbon, which would correspond to approximately 60 ms for a methine carbon of the same effective correlation time, is close to the possible minimum for dipolar relaxation (cf. Fig. 4.3). This equally holds for the NOE. Relaxation of the two methyl carbons on the other hand, is expected to be governed by a considerably shorter effective correlation time, which is a consequence of the internal motion, to which these carbons are subjected. However, it is interesting to note that even these carbons suffer a reduction of the NOE. For carbons undergoing internal motion both η and T_1 were shown to depend in a complicated manner on the correlation times for overall reorientation and group rotation.[190]

These results have important consequences in view of the applicability of ^{13}C n.m.r. as a quantitative method to the study of high polymers. Unless the necessary precautions are taken in the form of NOE-suppression techniques and the use of adequate recovery times between pulses, grossly distorted data may be obtained.

5.10 QUANTITATIVE ANALYSIS

Until relatively recently, ^{13}C n.m.r. spectroscopy was used only infrequently as a tool for quantitative analysis. To some extent, this situation reflects the fact that the techniques which are employed routinely to increase sensitivity concomitantly perturb ^{13}C nuclear energy level populations from their respective equilibrium values. Individual T_1 and NOE values can vary widely among ^{13}C nuclei in a given molecule; therefore, it is not surprising to find that, unless special precautions are taken, peak areas in ^{13}C n.m.r. spectra in general are not proportional to the number of ^{13}C nuclei present in the molecule.[200]

Three effects are considered to contribute to perturbing signal intensities in ^{13}C pulsed spectra:[200,201]

(1) The large spread of ^{13}C nuclear relaxation times in conjunction with the usually short pulse repetition times (i.e. short in comparison to T_1 values) leads to preferential saturation of the more slowly relaxing nuclei. Some implications of this phenomenon have been discussed in Chapters 1 and 4.

(2) The NOE observed in proton noise-decoupled spectra does not always reach its theoretical maximum. In small molecules this may result from the operation of other relaxation mechanisms that compete with dipole–dipole relaxation. Alternatively, in the case of macromolecules the motional narrowing limit conditions may be violated. The latter lowers the NOE in spite of the fact that no other relaxation mechanisms are present to any significant extent (cf. Chapter 4 and previous section).

(3) The power level of the exciting r.f. pulses does not satisfy the condition $\gamma H_2/2\pi > F/2$ (Eqn. 1.14 p. 16) which results in a non-uniform power spectrum. This means that resonances remote from the r.f. carrier are excited to a lesser extent than those near it. (We assume that the carrier frequency is placed at the centre of the spectrum.)

How can we get around these problems? The effect of unequal relaxation times can be overcome in different ways:

(1) Lowering the pulse flip angle to very small values (a few degrees) has the effect of perturbing the spin populations only very slightly, i.e. a large fraction of the equilibrium longitudinal magnetization is retained.
(2) A pulse delay may be introduced between the pulses long enough to result in complete repolarization of all nuclei. Since the re-establishment of the longitudinal magnetization to 99% of its equilibrium value following a 90° pulse requires a waiting time of $5T_1$, this method suffers from a severe loss of sensitivity. The same is also true of method (1).
(3) Dissolving a paramagnetic additive introduces a powerful relaxation mechanism which shortens the T_1 s of all carbons so that relaxation time differentials become insignificant. Deviations from this behaviour may be encountered for carbons which are buried at the interior of the molecule. A further drawback is the accompanying contamination of the sample which might be undesirable.

The experienced spectroscopist is usually in a position to assess whether there are carbons that may be relaxed by other than the dipolar mechanism, provided he knows the molecule of interest. Above molecular masses of about 300, the NOE is usually found to be complete (including quaternaries[165,166]) unless hydrogens are completely absent. In smaller molecules this may not be the case. Again, three ways offer themselves for eliminating the NOE:

(1) Operation without proton decoupling. The drawback of this method lies in the reappearance of $^{13}C—^1H$ spin–spin couplings which may cause severe overlap of resonances. Moreover, the sensitivity enhancement brought about by multiplet collapse is not realized.
(2) The decoupler is gated in such a way that it is on during data acquisition and off during a pulse delay, as outlined in Chapter 4. Again the sensitivity suffers, but it is a relatively safe way to remove the NOE on condition that the off period for the decoupler duty cycle is long enough.
(3) Instead of magnetically suppressing the NOE this can be achieved by chemical means as discussed in Chapter 4. 'Short-circuiting' the dipolar relaxation path through addition of paramagnetic species quenches the NOE. It has previously been pointed out that this method is inappropriate whenever T_1 s are short ($\leqslant 1$ s). Without introducing severe broadening of lines, (T_2 effect) T_1 may be shortened to about 0.1 s. Since the residual NOE is

proportional to the ratio T_1/T_1^{DD}, a tenfold reduction of T_1 lowers the NOE to 10%. On the other hand, if a typical T_1 is 10 s and a reduction to 0.1 s is achieved, the residual NOE is 1% which is below the integral accuracy and thus negligible.

As far as the r.f. power requirements are concerned it can be shown that the power demand $\gamma H_1/2\pi = F$ is usually adequate and should result in intensity errors of less than 2% across the spectral width of F Hz.[202] For a 10 k Hz spectrum (typical spectral width at 4.7 tesla) therefore, $\tau_{90°} = 25\ \mu s$ is satisfactory. Since the practical pulse flip angles are $<90°$, even less power may be tolerable. Should this requirement not be fulfilled, however, a calibration curve is necessary which relates the relative intensities to the frequency offset from the carrier.

An application of industrial importance involves the quantitation of petroleum fractions; an optimized experimental protocol for this purpose was evaluated recently.[203,204] The authors[204] demonstrated that a combination of gated decoupling (see Chapter 4) and the addition of relaxation-enhancing paramagnetic species (for example, $Cr(acac)_3$) is most appropriate. Criteria of relevance are: (1) total scan time to achieve adequate S/N, (2) solubility of the paramagnetic agent, (3) the concentration of the deuterated solvent required to achieve a stable field/frequency lock, and (4) the extent to which line broadening that necessarily accompanies the application of this protocol can be tolerated.

It is obvious that the gating delay, T_d (off-phase of the decoupler, cf. Section 4.5) that is needed for adequate NOE suppression can be reduced as the concentration, Q, of the paramagnetic additive is increased. At the same time, the lines are broadened due to shortening of T_2. The combination of T_d and Q for which the integral ratio of the two carbons with different relaxation characteristics was (within experimental error) equal to the theoretical value was determined for different concentrations of the paramagnetic species, $Cr(acac)_3$, in toluene solution.

A plot of T_d versus $[Q]$ appears as curve (a) in Fig. 5.52. Along with T_d, the mean linewidth, Δv, of solute signals was measured at different Q concentrations, thereby affording the plot of Δv versus $[Q]$ shown in curve (c), Fig. 5.52. The shaded area in Fig. 5.52 represents the region of T_d and corresponding $[Q]$ values within which quantitative measurement can be assured. The minimum scan time obviously is reached when $T_d = 0$, since for the scan time $T_s = (t_a + T_d)n$ holds, where t_a is the acquisition time and n is the number of transients. This situation is reached when $[Q] = 0.1$ mole per litre of relaxation reagent. At this limiting a value, the line width is approximately 7 Hz, a value which may be too large to permit resolution of two or more neighbouring resonances of interest. Hence, the choice of 'optimal conditions' depends very markedly upon the nature of the problem under consideration.

The use of $C_2D_2Cl_4/Fe(acac)_3$ as the solvent-relaxation reagent pair has been found to be particularly efficient for crude oil and petroleum product analysis via

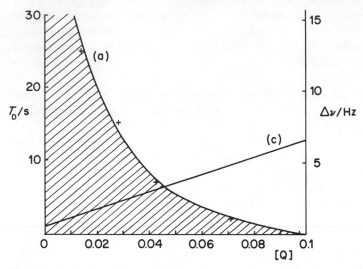

Fig. 5.52   Optimal (limiting) value of the pulse delay T_d for which the integral ratios in toluene approach the theoretical value, plotted against the concentration of Cu(acac)$_3$ in $C_2D_2Cl_4$, curve (a); curve (c) represents the linewidth, $\Delta\nu$.[204] Reprinted with permission Gillet *et al.*, *Anal. Chem.*, **52**, 813, Copyright (1980) American Chemical Society.

^{13}C n.m.r. spectrometry.[204] It was demonstrated via control experiments that potential complications introduced through association of the relaxation reagent with sample molecules (which would produce non-uniform sensitivity enhancement along the entire frequency range employed) in fact do not occur.[204]

Problem 40: The proton noise-decoupled ^{13}C spectrum of a mixture of four different carboxylic acid esters is given. In order to quench the NOE, a suitable amount of Cr(acac)$_3$ had been added to the solution. Moreover, the spectrum was obtained using sufficiently strong pulses ($\tau_{90°} = 15\,\mu s$) so that errors due to pulse imperfections are negligibly small.

(1) Assign all lines.
(2) Determine the composition of the mixture (in mole per cent) from the integrated carbonyl carbon region.

Applications

Quantitative analysis of mixtures of organic compounds is performed routinely by using a variety of analytical techniques. Until very recently, ^{13}C n.m.r.

No.	Int.	ppm	No.	Int.	ppm
1	60	171.03	10	48	118.69
2	21	168.19	11	44	64.59
3	37	167.25	12	205	60.51
4	31	161.19	13	47	53.11
5	95	133.49	14	115	52.17
6	44	132.79	15	46	26.00
7	40	131.10	16	146	20.85
8	179	130.16	17	195	14.39
9	183	129.02			

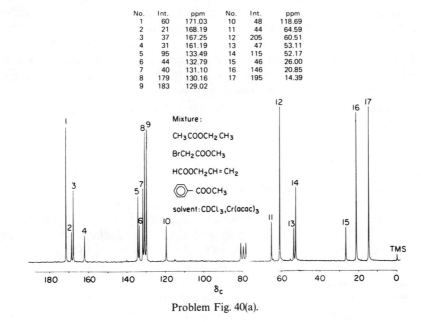

Mixture:

$CH_3COOCH_2CH_3$

$BrCH_2COOCH_3$

$HCOOCH_2CH=CH_2$

⬡—$COOCH_3$

solvent: $CDCl_3$, $Cr(acac)_3$

Problem Fig. 40(a).

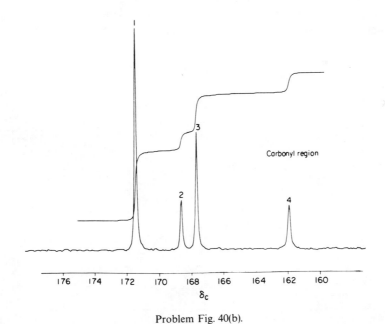

Carbonyl region

Problem Fig. 40(b).

spectroscopy has seldom been competitive with more sensitive and less expensive analytical techniques such as gas and liquid chromatography, ultra-violet, infrared, mass spectrometry, etc. In addition, the use of n.m.r. spectroscopy in this regard has been limited historically to those mixtures which contain comparable amounts of the constituents to be analysed. However, the strength of the n.m.r. method lies in its selectivity combined with its power to detect as well as to identify an unknown sample. With the advent of modern, highly sensitive high-field n.m.r. instrumentation, quantitative applications of ^{13}C n.m.r. spectroscopy for characterizing both solid and liquid mixtures have become more commonplace. Several examples have been reported wherein this technique has been applied to quantitative analysis of mixtures in agriculture and food technology[200] and in fuel-related industries (for example, analysis of solid fossil fuels[205] and crude oils[206]).

A large number of fats and oils are composed essentially of triglycerides of C_{18} fatty acids that contain between zero and three C=C double bonds; (for simplicity, these fatty acids are referred to by the abbreviations 18:0, 18:1, 18:2, and 18:3, respectively). The analytical procedure routinely used to quantify such mixtures is gas chromatography, (g.c.). In this procedure, the triglycerides are subjected to methanolysis or ethanolysis, and the resulting mixture of fatty acid methyl or ethyl esters is injected into the gas chromatograph. Peaks in the gas chromatogram are then integrated and identified according to their respective characteristic retention times. This method, although well established, suffers from several drawbacks: (1) a chemical step is required, (2) the method is destructive, and (3) considerable skill is required to identify structurally similar constituents which differ only slightly in their respective retention times.

TABLE 5.11   Peak assignments in the spectrum of linseed oil (Fig. 5.53).[207]

Peak	Assignment[a]	Integral
A	$C_1(a,b,c,d)$	102
B	$C_{16}(a)$	54
C	$C_{9,10}(c)$; $C_{9,13}(b)$; $C_9(a)$	122
D	$C_{10,12}(b)$; $C_{10,12,13}(a)$	195
E	$C_{15}(a)$	54
F	$C_{1',2',3'}(a,b,c,d)$	87
G	$C_2(a,b,c,d)$	96
H	$C_{16}(c,d)$	31
I	$C_{16}(b)$	14
J	$C_{4-7}(a)$; $C_{4-7,15}(b)$; $C_{4-7,12-15}(c)$; $C_{14-15}(d)$	561
K	$C_8(a)$; $C_{8,14}(b)$; $C_{8,11}(c)$	120
L	$C_{11,14}(a)$; $C_{11}(b)$	125
M	$C_3(a,b,c,d)$	96
N	$C_{17}(b,c,d)$	43
O	$C_{17}(a)$	54
P	$C_{18}(a,b,c,d)$	99

[a] $a = 18{:}3$; $b = 18{:}2$; $c = 18{:}1$; $d = 18{:}0$.

In order to assess the accuracy of ^{13}C n.m.r. spectroscopy as an alternative to g.c. methods, linseed oil (which contains mainly the fatty acids 18:0, 18:1, 18:2, and 18:3, including a small amount of 16:0) was investigated.[207]

Although the NOE is probably complete for all carbons in a molecule such as a C_{18}-fatty acid triglyceride, a fairly large spread of T_1 values is expected due to increased internal motion towards the end of the chain. For this reason, the sample was run as a solution which was 0.04 M in Cr(acac)$_3$. From the spectrum in Fig. 5.53, in which peaks or groups of peaks are labelled consecutively from low to high field, sixteen measurable quantities can be determined in the form of peak integrals that are related to the relative amounts of the four fatty acids. Line assignments and integral values are given in Table 5.11. If a, b, c, d denote the relative molar ratios of the fatty acids 18:3, 18:2, 18:1, and 18:0, it is readily recognized that these are related to peak areas as follows:

Parameter	Obtained from	Value
a	B, E, O, or average	54
b	I	14
c	$(C - a - 2b)/2$	20
d	$H - c$	11

The composition of the oil was subsequently determined in three different ways: (1) from the above starting set; (2) from a least-squares fit of all sixteen integral values to the values calculated from the parameter set (a, b, c, d) by iterating to minimum error; and (3) from g.c. analysis. The result of the procedure is given in Table 5.12. The data listed in Table 5.12 show remarkably good results obtained from the starting set, which are, of course, improved when a least-squares fit is performed, including all available integral values. Thanks to the resolving power of ^{13}C n.m.r., systematic errors due to peak overlaps are less likely to occur.

Another frequent analytical problem concerns the determination of the degree of unsaturation, usually expressed by the so-called iodine number. This is defined as the number of centigrams of iodine that will react with one gram of a fat:

$$\text{Iodine no.} = \frac{\text{Equivalent mass } (I_2)}{\text{Equivalent mass fat}} \times 100 \qquad (5.25)$$

Figure 5.54[201] shows the ^{13}C spectrum of a sample of corn oil, doped with 0.1 M Cr(acac)$_3$. Corn oil is similar to linseed oil in that it consists of a mixture of fatty acid triglycerides (mostly 18:1 and 18:2). The average molecular formula of the mixture is $C_{57}H_{(110-N)}O_6$, where N is the number of olefinic carbon atoms. The four major regions of the spectrum pertain to carbonyl, olefinic, glyceride, and remaining carbons. Since there are three carbonyl and three glyceride-type

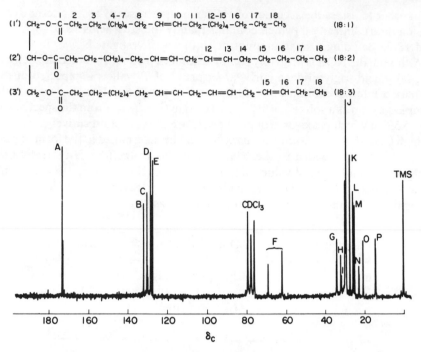

Fig. 5.53 Proton noise-decoupled ^{13}C spectrum of linseed oil in deuterochloroform with 0.04 M Cr(acac)$_3$.[207] The peak assignments are given in Table 5.12. The formula represents a hypothetical triglyceride of oleic (18:1), linoleic (18:2), and linolenic (18:3) acids.

TABLE 5.12 Composition of linseed oil determined with different methods

Component	Starting set	Least-squares fit	g.c. analysis
18:3(a)	54.5	55.6	55.3
18:2(b)	14.1	14.2	15.3
18:1(c)	20.2	20.0	19.3
18:0(d)	11.1	10.2[a]	9.9[b]

[a] n.m.r. gives the sum of all saturated acids.
[b] g.c. actually gives 6.0% 16:1 and 3.9% 18:0.

carbons, the sum of the olefinic region and the remaining carbons must equal 57. By dividing the sum of integrals by 57, the area per carbon is obtained. In this case, a correction factor 1.25 was used to account for the effect of low ^1H power. The average number of olefinic carbon atoms was found to be 9.0, corresponding to 4.5 double bonds, giving an iodine number of 129, in excellent agreement with a value of 128.8 determined by titration.

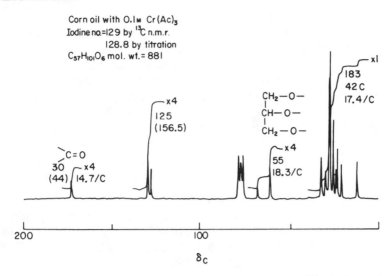

Fig. 5.54 ^{13}C spectrum of corn oil doped with 0.1 M Cr(acac)$_3$.[201] The integral values in parentheses were corrected to compensate for low pulse power.

The assessment of optical purity poses a particular challenge to the analytical chemist. N.m.r., notably of ^{19}F (due to its large chemical shift range), has been utilized in this regard with some success. An alternative approach focuses upon the use of chiral relaxation reagents using ^{13}C n.m.r.

Recently, Hofer and coworkers[208] have reported that tris[d, d-dicampholylmethanato]gadolinium (III) [Gd(dcm)$_3$, **167**], a chiral shift reagent, interacts with enantiomeric methylephedrines (**168**) and with enantiomeric ephedrines (**169**) to afford diastereoisomeric complexes in each case. Often, ^{13}C spin–lattice relaxation times (T_1) for carbon atoms in one of a pair of diastereoisomeric complexes differ dramatically from the T_1 value of the corresponding carbon atoms in the other diastereoisomeric lanthanide–substrate complex. This observation provides the basis for quantitative determination of enantiomeric excess, as illustrated in the example below.

167

168 R=CH$_3$
169 R=H

Figure 5.55 shows ^{13}C inversion recovery spectra of $(+) - $ **168** and $(-) - $ **168** obtained in the presence of **167** $(3 \times 10^{-3} \, \text{M})$ and recorded under identical experimental conditions. The gross difference between relaxation times of corresponding carbon atoms in the diastereoisomeric complexes **167**/$(+) - $ **168** and **167**/$(-) - $ **168** is apparent. It was also shown that the difference, $\Delta\Delta T_1$, between enantiomeric relaxation times depends upon the concentration of **167** (see Fig. 5.56).

Enantiomeric excess can be determined via graphical analysis of the expected biexponential relaxation behaviour.[209] Since the initial recovery curve primarily reflects relaxation of the more rapidly relaxing species, the measured rate at the larger recovery intervals is close to that of the more slowly relaxing species.

Problem 41: A plot of $\log(I_\infty - I_t)$ versus time of the ^{13}C n.m.r. signal at $\delta 72.4$ obtained for a mixture of **167**/$(+) - $ **168** and **167**/$(-) - $ **168** is shown below. The measured value of I_∞ is 5.65. Using these data, calculate the enantiomeric excess in the mixture of $(+) - $ **168** and $(-) - $ **168**.

Some progress in quantitative applications of ^{13}C n.m.r. spectroscopy have also been reported for the study of solid samples. A problem of particular complexity is the study of the composition of wood products. In the past, chemical modification of such materials was often required in order to obtain

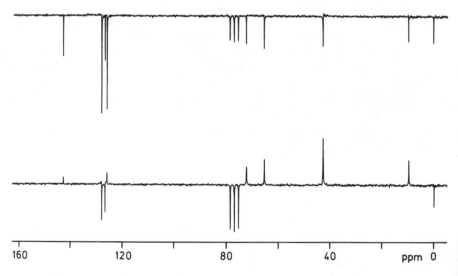

Fig. 5.55 ^{13}C inversion-recovery spectra of 1.87M $(+)$-methylephedrine (above) and 1.87 M $(-)$-methylephedrine (below) in chloroform-d containing 3×10^{-3} M Gd(dcm)$_3$ ($\tau_1 = 0.5$ s; $T_{rep} = 10$ s).[208] Reprinted with permission from Hofer, Keuper and Renker *Tetrahedron Lett.*, **25**, 1141. Copyright (1984) Pergamon Press Ltd.

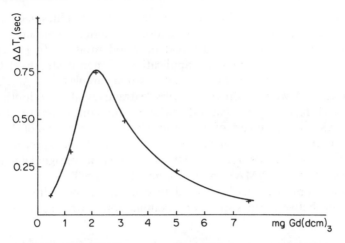

Fig. 5.56 Difference of relaxation times, $\Delta\Delta T_1$, between $(+)$- and $(-)$-ephedrine as a function of Gd(dcm)$_3$ concentration. Shown are values for the substituted aromatic carbon.[208] Reprinted with permission from Hofer, Keuper and Renken, *Tetrahedron Lett.*, **25**, 1141. Copyright (1984) Pergamon Press Ltd.

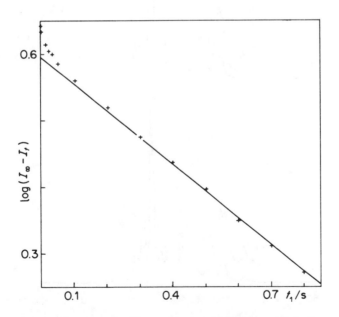

Problem Fig. 41 Plot of $\log(I_\infty - I_t)$ versus time for methyleph-edrine. (Shown are the intensities of the signal at $\delta = 72.4$ ppm).[209] Reprinted with permission from Hofer and Keuper, *Tetrahedron Lett.*, **25**, 5631. Copyright (1984) Pergamon Press Ltd.

soluble fractions for analysis. When lignin analysis was sought, some uncertainty typically remains as to whether such fractions are indeed representative of the lignin content of the unmodified wood or wood product. In principle, this difficulty can be circumvented by application of (non-destructive) CP/MAS techniques to the study of lignin content of wood samples.

The process of wood pulping removes 'extractives' (for example, phenols, terpenes, and resin acids) that otherwise would contribute to the overall ^{13}C CP/MAS n.m.r. spectrum of the wood sample. Some notion of the extent of this contribution can be gained via comparison of the spectra shown in Fig. 5.57.[210] It has been demonstrated that accurate relative signal intensities can be obtained for ^{13}C CP/MAS spectra of intact wood.[210]

This same approach has been used to compare carbohydrate, lignin, and protein ratios between grass species by examination of whole grass samples.[211]

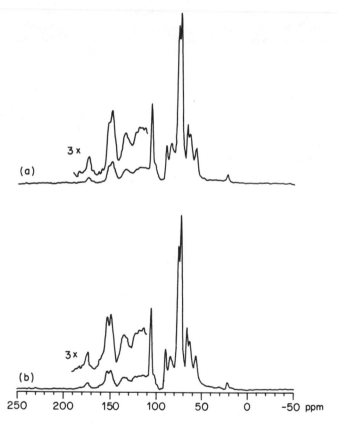

Fig. 5.57 ^{13}C CP/MAS spectra of southern pine wood. (a) Ground wood, (b) extractives-free wood.[210] Reprinted with permission from Haw, Maciel and Sehroder, *Anal. Chem.*, **56**, 1323. Copyright (1984) American Chemical Society.

Application of the CP/MAS technique is somewhat cumbersome at present; higher field strengths are needed to improve the signal-to-noise ratio (S/N) and to permit the use of decreased spectral acquisition times. At present, the CP/MAS technique is used in conjunction with near-infra-red reflectance spectroscopy and wet chemical methods for quantitative analysis of the components of whole grass samples.

Another possible application of ^{13}C CP/MAS n.m.r. spectroscopy involves quantitative analysis of solid fossil fuel samples. Carbon-13 n.m.r. spectra of such samples are exceedingly complex, and they are often poorly defined (i.e. low S/N) and poorly resolved due to the large number of spectral signals that result from their various carbonaceous organic components. An elegant spectral editing technique that permits considerable spectral simplification employs dipolar dephasing (DD).[212] The DD technique is a derivative of the CP/MAS technique that was introduced in Chapter 3. A delay, T_{DD}, between the cross-polarization period and the ^{13}C signal acquisition time is introduced before the application of the ^{1}H decoupling pulse. During this delay period, methylene and methine ^{13}C signals dephase more rapidly than do quaternary and methyl ^{13}C signals, thereby effectively suppressing the former. The resultant spectral simplification has enabled the reliable estimation of such quantities as (1) the aromaticity factor (i.e. the ratio of aromatic to aliphatic carbon atoms, f_a)[213,214] and (2) the ratio of protonated to non-protonated aromatic carbon atoms.[215,216]

The quantitative reliability of the ^{13}C CP/MAS n.m.r. technique for analysing coals and pitches has been investigated recently.[217] The results thereby obtained were compared with those from a corresponding experiment which utilized MAS combined with a simple 90° carbon pulse sequence with sufficiently long delays to permit complete recovery of ^{13}C magnetization within the repetition time of the experiment. The latter conditions were shown to provide truly quantitative experimental ^{13}C n.m.r. conditions. Significant discrepancies were observed between quantitative results obtained via CP/MAS techniques and those achieved by the latter procedure, thereby indicating that caution should be exercised when interpreting quantitative data that have been obtained via applications of ^{13}C CP/MAS n.m.r. techniques.[217] Other investigators have sounded similar warnings.[218,219]

Recently, the first application of a solid-state two-dimensional ^{1}H—^{13}C heteronuclear shift correlation experiment to the analysis of a whole coal has been reported.[220] Semi-quantitative analysis was aided by introducing intensity correction factors that take into account differences in T_1 values among the different types of ^{13}C atoms. The rationale of the two-dimensional approach is an improvement in spectral resolution that results from development of the spectrum in two dimensions, thereby permitting separation of the aliphatic region of the CP/MAS spectrum into functional categories.[220] Although the need for study of additional model compounds is apparent, the results of this first attempt to apply two-dimensional n.m.r. techniques to the study of solid fuel samples augurs well for the future of quantitative applications of ^{13}C n.m.r. spectroscopy.

In vivo applications of ^{13}C n.m.r.

In 1974, Richards and Moon first demonstrated that high-resolution spectra of phosphorus-31 can be obtained from living cells, thereby opening up new avenues for the study of cellular metabolism *in vivo*. In subsequent years, the field expanded very rapidly and was soon complemented by metabolic experiments, not only in intact cells but also in tissues, perfused organs and entire laboratory animals, and, finally, in humans. These exciting developments were made possible by the availability of large-bore high-field magnets and by the introduction of the surface coil by Ackerman *et al.* in 1981.[221] The nucleus of choice in most of this work has been and continues to be ^{31}P. Phosphorus provides a natural label in the form of the phosphate groups found in several of the high-energy metabolites such as adenosine triphosphate (ATP), adenosine diphosphate (ADP), and phosphocreatine (PCr) as well as inorganic phosphate resulting from cleavage of phosphate groups by hydrolytic reactions.

In this new field of medical n.m.r, which is currently being investigated as an adjunct to clinical magnetic resonance imaging,[222] ^{13}C n.m.r. spectroscopy has played only a secondary role. Nevertheless, it may provide information complementary to that obtainable by ^{31}P n.m.r. spectroscopy, as not all metabolites of interest, of course, contain phosphorus.

One motivation for the preference of ^{13}C over ^{31}P as an *in vivo* spectroscopic nucleus is the larger chemical shift range (200 ppm versus ~ 25 ppm for ^{31}P biological phosphates). Moreover, ^{13}C enables the use of enriched substrates, thus permitting delineation of metabolic pathways much as in biosynthetic studies in plants and bacteria, discussed earlier. For a detailed review of *in vivo* spectroscopy, the reader is referred to specialized reviews.[223,224]

Technical requirements

One of the major technical challenges for *in vivo* spectroscopy in animals and humans is the sheer size of the object to be studied. In recent years, whole-body superconducting magnets with field strengths up to 2 T and bore diameters of 1 m have been developed for clinical magnetic resonance imaging.[225] The purpose of operating at high magnetic field was to combine clinical imaging with spectroscopy such that the image essentially serves as a means for spatial localization from which the region of interest can be determined. Subsequently, spectral analysis using spatially localized methods is conducted.[226] This approach is still very much in the experimental stage, although it has become possible today to obtain spectra from spatially localized image-prescribed slices of tissue[227] or even from individual volume elements (voxels).[228] The techniques used involve magnetic field gradient-mediated localization, whereby, for example, a slice of tissue is excited by a combination of a field gradient perpendicular to the slice of interest in conjunction with a narrow-banded rf pulse. For a rf pulse (typically a sinc pulse) of bandwidth $\Delta\omega$ and gradient of amplitude G, a slice of tissue of width $\Delta r = \Delta\omega/(\gamma G)$ is excited perpendicular to the gradient orientation.

While most laboratory spectra of ^{13}C are obtained with simultaneous proton decoupling, this typically is not possible in larger objects such as humans. Since biologic tissues are electrically conductive, eddy-current-induced losses become the dominant source of resistance, thereby lowering the quality factor of the coil.[229] The rf power necessary to achieve decoupling over a significant volume would cause tissue heating and is, therefore, not practical. Therefore, most spectra reported to date in humans have been obtained without decoupling.

The third potential obstacle for *in vivo* ^{13}C spectroscopy is the low detection sensitivity, which typically mandates the use of enriched substrates. This poses a cost problem because of the need for very substantial ^{13}C quantities required by label dilution (in particular, for human applications). Nevertheless, it offers some opportunities as well, since it allows the use of site-specific enrichment. In this way, the fate of the carbon involved in metabolic breakdown or synthesis can be followed while minimizing background interference that arises from molecules which contain the isotope at natural abundance.

Applications to the study of cell metabolism in bacteria, animals, and humans

An interesting application of the use of enriched substrates has been reported in conjunction with the study of the redox status of the di- and triphosphopyridine nucleotide pools in *E. coli* bacteria.[230] It is well known that many metabolic processes such as reconstitution of ATP by oxidative phosphorylation critically depend upon the redox potential of the pyridine nucleotide pools. Therefore, the suggestion has been made to characterize the redox status in terms of the catabolic reduction charge (CRC), expressed as the ratio [NADH]/([NADH] + [NAD$^+$]). Likewise, the anabolic reduction charge (ARC) can be expressed as the ratio [NADPH]/([NADPH] + [NADP$^+$]), where NAD$^+$ is nicotinamide adenine dinucleotide and NADP$^+$ is nicotinamide adeninedinucleotidephosphate. Since NADH and NADPH have very similar spectroscopic properties, it is more convenient to define a mean reduction charge (MRC), defined as ([NADH] + [NADPH])/[total pyridine nucleotides]. The spectroscopic approach to determining this quantity via linewidth measurement techniques is based on chemical shifts that are sensitive to the redox status. One of the resonances that is sensitive to the oxidation state of NADH is C_2 of the nicotinic acid moiety. By incorporating [2-C 13] nicotinic acid into a culture of *E. coli* bacteria, enriched NAD$^+$/NADH is produced. Figure 5.58(a) shows a proton-decoupled C-13 N.M.R. spectrum of a culture of *E. coli* bacteria obtained in the way described. This spectrum shows two resonances, one from [2-^{13}C] nicotinate and one from

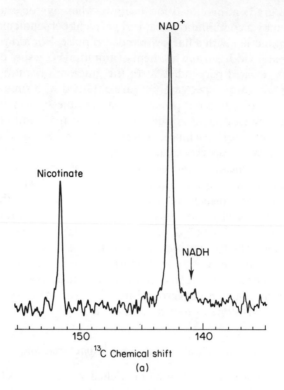

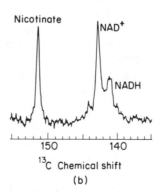

Fig. 5.58   (a) ^{13}C nmr spectrum of *E. coli* cultured on a medium containing [2-^{13}C] nicotinic acid. Resonances correspond to free nicotinate and to oxidized pyridine nucleotide. (b) Spectrum from a similar culture grown under aerobic conditions and observed for a total of 4 h under anaerobic conditions. Note the signal resulting from [2-^{13}C] nicotinic acid in reduced nucleotide (NADH).[230]

NAD$^+$. Figure 5.58(b) displays the corresponding spectrum obtained under anaerobic conditions and shows that part of NAD$^+$ has been reduced to NADH, resulting in a peak at high field relative to that of NAD$^+$. The mean reduction charge can be calculated from an analysis of the two signals pertaining to NAD$^+$ and NADH. It is interesting to note that the exchange between oxidized and reduced species is obviously sufficiently slow to give rise to separate resonances.

Perhaps the first successful ^{13}C n.m.r. spectra obtained from intact mammalian tissues were reported by Alger et al.[231] at 20 MHz ^{13}C frequency from various anatomic locations in a rat as well as from the human forearm. However, most of the signals detected arose from the fatty acid triglycerides which are very abundant in subcutaneous tissue. The presence of the olefinic carbon resonances enables quantification of the polyunsaturated fats and thus may permit the study of nutritional fat deficiencies and abnormalities in fatty acid metabolism. Reception of these signals is enhanced due to the characteristic sensitivity profile of surface coils which provide high sensitivity at the surface, decreasing rapidly with increasing depth. Figure 5.59 shows the natural abundance ^{13}C spectrum at 16.1 MHz (1.5 T) obtained by placing a 6.5 cm round surface coil on the temple of a normal human subject.[226] The spectrum which was recorded without proton noise decoupling shows spin coupling multiplets from fatty acid and glyceryl carbons as well as from the N(CH$_3$) moiety of choline (the latter probably arising from membrane phospholipids).

Administration of [1-^{13}C] glucose into the stomach of a live rat permitted the study of the time course of glycogen production by the liver.[231] While at first only the resonances of the α and β anomeric carbons at 92.3 and 96.8 ppm could be

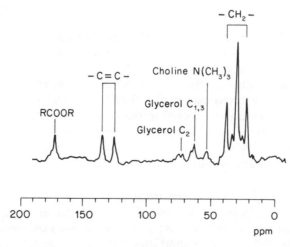

Fig. 5.59   Proton-coupled natural abundance ^{13}C spectrum obtained with a surface coil placed on the temple of a normal human subject.[226] Reproduced with permission from Bottomley et al., *Radiology*, **150**, 443 (1984).

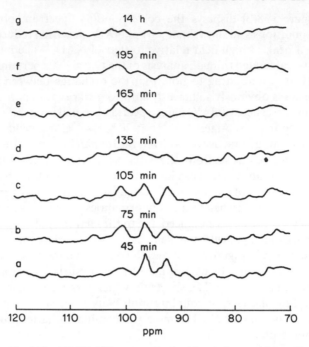

Fig. 5.60   20 MHz ^{13}C spectra obtained by placing a surface coil over the abdomen of a rat after feeding D-[1-^{13}C] glucose. The time between glucose feeding and spectral acquisition is provided in each spectrum. The signals at 101, 96.8, and 92.3 ppm arise from the C-1 carbons of glycogen and the beta and alpha anomers of D-glucose.[231] Reproduced by permission of the author.

detected, an additional resonance at 101 ppm emerged 75 min after administration of glucose. This resonance which first increased and later decreased again was ascribed to C-1 in glycogen (Fig. 5.60).

A key cell fuel is glucose, which is metabolized first by the glycolytic pathway and subsequently by the Krebs cycle to afford CO_2, which is then exhaled by the lungs. While glycolysis can proceed in the absence of oxygen, the Krebs cycle requires oxygen. In the case of oxygen deficiency (ischemia), the metabolic end product is lactic acid, which causes acidosis and, ultimately, cell death. The ability to detect and quantify lactic acid, therefore, is crucial for the evaluation of hypoxic insult.

Rothman et al.[232] recently demonstrated that lactic acid build-up in the ischemic rat brain can be studied quantitatively by combining site-specific ^{13}C enrichment and a novel indirect detection scheme. The experiment employed is based on a spectral editing method illustrated in Figure 5.61(a). This method employs a proton spin echo sequence for which the interpulse interval $\tau = 1/J_{CH}$ is selected; here, J_{CH} represents the one-bond ^{13}C—H coupling constant. In a first

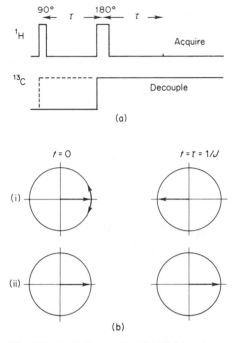

Fig. 5.61 (a) Pulse sequence for indirect detection of ^{13}C used to suppress isotopomers from C-12 containing species as described in the text. Data are collected by alternately turning the decoupler on and off and subtracting resulting FIDs. (b) The evolution of the transverse proton magnetization for spins bonded to ^{13}C (i) and ^{12}C (ii).

acquisition cycle, the ^{13}C decoupler is off during the defocusing period of the spin echo sequence. During this period, the protons dephase at a rate J_{CH}, each dephasing by π and $-\pi$ radians, respectively so that they are both aligned along the $-y$ axis of the rotating frame at time $t = \tau$ (Fig. 5.61(b)).

Since the ^{13}C decoupler is turned on at this point, the spins will retain their phase relationship. In a second acquisition cycle, this sequence is repeated with the ^{13}C decoupler always on. Therefore, the proton spins will not dephase by J modulation and will be aligned along the $+y$ axis once data acquisition starts. Therefore, by subtracting (b) from (a) (or conversely), the signals add coherently. By contrast, the species ^{12}C—H will not undergo J modulation and, therefore, upon subtraction the signals due to this species will cancel.

The purpose of the spin echo pulse is to rephase defocusing effects from magnetic field inhomogeneity and chemical shift. $^{1}J_{CH}$ for the methyl protons in lactate is 127 Hz, thus demanding $\tau = 7.8$ ms. In this way, the authors were able to

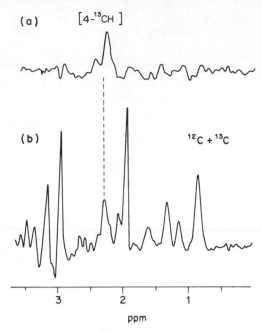

Fig. 5.62 (a) ^{13}C difference spectrum obtained from
a rat brain after infusion of [1-^{13}C] glucose prior to
death. (b) Spectrum of the sum of both ^{12}C and ^{13}C
species serving as a reference to calculate the lactate
concentration from a comparison of the peak intensity
to the creatine methyl resonance.[232]

measure the lactate concentration in the intact rat brain following infusion of
[1-^{13}C] glucose. This was achieved by comparing the (indirectly detected)
^{13}C difference spectrum with the (^{12}C + ^{13}C) subspectrum (Fig. 5.62).

Recently, Shulman *et al.* were able to obtain natural-abundance ^{13}C spectra of
the human liver *in vivo* at 2 T field strength and to isolate the resonances due to
glycogen, thereby demonstrating that the study of cellular metabolism by
^{13}C n.m.r. is feasible.[233]

Whether or not ^{13}C n.m.r. as a diagnostic modality is practical will become
evident during the years ahead. There is little doubt, however, that the technique
will have a place as a medical research tool which will further expand the scope of
^{13}C n.m.r.

'There is a great future in store for you and me, my boy, a great future.'[234]

References

1 (a) D. M. Doddrell, D. T. Pegg, and M. R. Bendall, *J. Magn. Res.* **48**, 323 (1982); (b)
   D. T. Pegg, D. M. Doddrell, and M. R. Bendall, *J. Chem. Phys.* **77**, 2745 (1982).

2 (a) W. P. Aue, E. Bartholdi, and R. R. Ernst, *J. Chem. Phys.* **64**, 2229 (1976); (b) A. Bax and R. Freeman, *J. Magn. Reson.* **42**, 164 (1981); (c) A. Bax and R. Freeman, *J. Magn. Reson.* **44**, 542 (1981).

3 (a) J. Jeener, B. H. Meier, P. Bachmann, and R. R. Ernst, *J. Chem. Phys.* **71**, 4546 (1979); (b) S. Macura and R. R. Ernst, *Mol. Phys.* **41**, 95 (1980).

4 W. Skorianetz and G. Ohloff, *Helv. Chim. Acta* **58**, 771 (1975).

5 P. Scribe and J. Rigaudy, Private communication.

6 A. Bax, R. Freeman, and T. A. Frenkiel, *J. Am. Chem. Soc.* **103**, 2102 (1981).

7 P. Capdevielle and J. Rigaudy, Private communication.

8 M. Maumy, Private communication.

9 J. N. Shoolery, *Applications Note* No. 5, Varian Assoc., Palo Alto, California, 1973.

10 J. Jautelat, J. B. Grutzner, and J. D. Roberts, *Proc. Nat. Acad. Sci. US* **65**, 288 (1970).

11 J. Kuszmann, P. Sohar, G. Horvath, and Zs. Mehesfalvi-Vajna, *Tetrahedron* **30**, 3905 (1974).

12 H. Kessler, C. Griesinger, and J. Lautz, *Angew. Chem., Internat. Edit.* **23**, 444 (1984).

13 S. Ghosal, R. K. Chaudhuri, M. P. Tiwari, A. K. Singh and F. W. Wehrli. *Tetrahedron Lett.* 403 (1974).

14 G. Magnusson and S. Thorsen, *Tetrahedron* **30**, 1431 (1974).

15 N. S. Bhacca and F. W. Wehrli, Unpublished results.

16 S. Altenburger, Private communication.

17 (a) S. Ghosal and F. W. Wehrli, Unpublished results; (b) R. Hodges and A. L. Porter, *Tetrahedron* **20**, 1463 (1964).

18 P. Joseph-N-Nathan, R. L. Santillan, P. Schmitt, and H. Günther, *Org. Magn. Reson.* **22**, 450 (1984).

19 H. Kessler, *Angew. Chem., Int. Edit.* **9**, 219 (1970).

20 C. K. Sauers and H. M. Relles, *J. Am. Chem. Soc.* **95**, 7731 (1973).

21 G. E. Hawkes, K. Herwig, and J. D. Roberts, *J. Org. Chem.* **39**, 1017 (1974).

22 S. Toppet, P. Claes, and J. Hoogmartens, *Org. Magn. Res.* **6**, 48 (1974).

23 F. W. Wehrli, *Nuclear Magnetic Resonance Spectroscopy of Nuclei Other than Protons* (Eds T. Axenrod and G. Webb), John Wiley, New York, 1974.

24 V. Dave, J. B. Stothers, and E. W. Warnhoff, *Tetrahedron Lett.* 4229 (1973).

25 P. Joseph-Nathan, J. R. Wesener, and H. Günther, *Org. Magn. Reson.* **22**, 190 (1984).

26 M. J. Shapiro, M. X. Kolpak, and T. L. Lemke, *J. Org. Chem.* **49**, 187 (1984).

27 For reviews, see (a) N. K. Wilson and J. B. Stothers, *Top. Stereochem.* **8**, 1 (1974); (b) J. B. Stothers in *Topics in Carbon-13 Nuclear Magnetic Resonance* (Ed. G.C. Levy), Vol. 1, Wiley-Interscience, New York, 1974, pp. 229–86.

28 G. A. Olah and P. W. Westerman, *J. Am. Chem. Soc.* **95**, 7530 (1973).

29 (a) G. A. Olah, H. C. Lin, and Y. K. Mo, *J. Am. Chem. Soc.* **94**, 3667 (1972); (b) G. A. Olah, G. D. Mateescu, and Y. K. Mo, *J. Am. Chem. Soc.* **95**, 1865 (1973).

30 G. A. Olah, H. C. Lin, and D. A. Forsyth, *J. Am. Chem. Soc.* **96**, 6908 (1974).

31 G. A. Olah, R. J. Spear, P. W. Westerman, and J.-M. Denis, *J. Am. Chem. Soc.* **96**, 5855 (1974).

32 G. A. Olah and G. Liang, *J. Am. Chem. Soc.* **96**, 189 (1974).

33 G. A. Olah, C. L. Jenell, D. P. Kelly, and R. D. Porter, *J. Am. Chem. Soc.* **94**, 146 (1972).

34 G. A. Olah, G. Liang, G. D. Mateescu and J. L. Riemenschneider, *J. Am. Chem. Soc.* **95**, 8698 (1973).

35 See: (1) P. D. Bartlett, *Nonclassical Ions*, W. A. Benjamin, Inc., New York, 1965, and references cited therein.

36 M. Saunders, P. von R. Schleyer, and G. A. Olah, *J. Am. Chem. Soc.* **86**, 5680 (1964).

37 For a review, see H. C. Brown, *Tetrahedron* **32**, 179 (1976).

38 W. Kirmse, *Top. Current Chem.* **80**, 89 (1979).

39 For reviews, see (a) D. G. Farnum, *Adv. Phys. Org. Chem.* **11**, 123 (1975); (b) G. M. Kramer, *Adv. Phys. Org. Chem.* **11**, 177 (1975); (c) R. N. Young, *Progr. NMR Spectrosc.* **12**, 261 (1979).

40 G. L. Nelson and E. A. Williams, *Progr. Phys. Org. Chem.* **12**, 229 (1976).

41 For a listing of ^{1}H and ^{13}C chemical shifts in bicyclic carbenium ions, see Table 8 in A. P. Marchand, *Stereochemical Applications of NMR Studies in Rigid Bicyclic Systems*, Verlag Chemie, Inc., Deerfield Beach, Florida, 1982, pp. 45–54.

42 H. Hogeveen and P. W. Kwant, *J. Am. Chem. Soc.* **96**, 2208 (1974).

43 H. Günther, Personal communication.

44 J. B. Stothers and C. T. Tan, *Can. J. Chem.* **55**, 841 (1977).

45 G. A. Olah and A. M. White, *J. Am. Chem. Soc.* **91**, 5801 (1969).

46 G. A. Olah and E. G. Melby, *J. Am. Chem. Soc.* **94**, 6220 (1972).

47 G. A. Olah, P. W. Westerman, E. G. Melby, and Y. K. Mo, *J. Am. Chem. Soc.* **96**, 3565 (1974).

48 R. B. Bates, S. Brenner, C. M. Cole, E. W. Davidson, G. D. Forsythe, D. A. McCombs, and A. S. Roth, *J. Am. Chem. Soc.* **95**, 926 (1973).

49 For general reviews of the subject, see, for example, (a) G. Binsch, *Topics in Stereochemistry* (Eds E. L. Eliel and N. L. Allinger), Vol. 3, Wiley-Interscience, New York, 1968; (b) I. O. Sutherland, *Annual Reports on Nuclear Magnetic Resonance Spectroscopy* (Ed. E. F. Mooney), Vol. 4, Acadmic Press, London and New York, 1971.

50 W. F. Reynolds, I. R. Peat, M. G. Freedman, and J. R. Lyerla, Jr, *J. Am. Chem. Soc.* **95**, 328 (1973).

51 S. Masamune, A. V. Kemp-Jones, J. Green, D. L. Rabenstein, M. Yasunami, K. Takase, and T. Nozoe, *J.C.S., Chem. Commun.* 283 (1973).

52 J. Evans, B. F. G. Johnson, J. Lewis, and J. R. Norton, *J.C.S., Chem. Commun.* 79 (1973).

53 O. A. Gansow, A. R. Burke, and W. D. Vernon, *J. Am. Chem. Soc.* **94**, 2550 (1972).

54 F. A. L. Anet, C. H. Bradley, and G. W. Buchanan, *J. Am. Chem. Soc.* **93**, 258 (1971).

55 D. K. Dalling and D. M. Grant, *J. Am. Chem. Soc.* **89**, 6612 (1967).

56 F. A. L. Anet and J. Krane, *Tetrahedron Lett.* 3255 (1974).

57 (a) F. G. Riddell, *J. Chem. Soc.* (*B*) 331 (1970); (b) G. M. Kelly and F. G. Riddell, *J. Chem. Soc.* 1030 (1971).

58 D. K. Dalling, D. M. Grant, and L. F. Johnson, *J. Am. Chem. Soc.* **93**, 3678 (1971).

59 H.-J. Schneider, R. Price, and T. Keller, *Angew. Chem.* **83**, 759 (1971).

60 F. A. L. Anet and J. J. Wagner, *J. Am. Chem. Soc.* **93**, 5266 (1971).

61 S. Masamune, K. Hojo, G. Bigam, and D. L. Barenstein, *J. Am. Chem. Soc.* **93**, 4966 (1971).

62 F. A. L. Anet and L. Kozerski, *J. Am. Chem. Soc.* **95**, 3407 (1973).

63 F. A. L. Anet, A. K. Cheng, and J. J. Wagner, *J. Am. Chem. Soc.* **94**, 9250 (1972).

64 See, for example A. D. Bax, *Two-Dimensional Nuclear Magnetic Resonance in Liquids*, Delft University Press, R. D. Reidel, Dordrecht, Holland, 1982.

65 S. Macura, Y. Huang, D. Suter, and R. R. Ernst, *J. Magn. Reson.* **43**, 259 (1981).

66 J. H. Noggle and R. E. Schirmer, *The Nuclear Overhauser Effect: Chemical Applications*, Academic Press, New York, 1971.

67 Y. Huang, S. Macura, and R. R. Ernst, *J. Am. Chem. Soc.* **103**, 5327 (1981).

68 G. E. Hawkes, L. Y. Lian, E. W. Randall, K. D. Sales, and S. Aime, *J. Magn. Reson.* **65**, 173 (1985).

69 S. Aime, L. Milone, D. Osella, M. Valle, and E. W. Randall, *Inorg. Chim. Acta*, **20**, 217 (1976).

70 G. Bodenhausen and R. R. Ernst, *J. Am. Chem. Soc.* **104**, 1304 (1982).

71 S. Macura and R. R. Ernst, *Mol. Phys.* **41**, 95 (1980).

72 P. E. Hansen, J. Feeney, and G. C. K. Roberts, *J. Magn. Res.* **17**, 249 (1975).
73 (a) Yu. K. Grishin, N. M. Sergeyev, O. A. Subbotin, and Yu. A. Ustynyuk, *Mol. Phys.* **25**, 297 (1973); (b) For a review on dynamic n.m.r. spectroscopy see, for example, M. Oki, *Application of Dynamic NMR Spectroscopy to Organic Chemistry*, VCH Publishers, Deerfield Beach, Florida, 1985.
74 (a) D. J. Chadwick, G. D. Meakins, and E. E. Richards, *Tetrahedron Lett.* 3183 (1974); (b) B. P. Roques, S. Combrisson, and F. W. Wehrli, *Tetrahedron Lett.* 1047 (1975).
75 For a general introduction and review on the application of n.m.r. to macromolecules see, for example, (a) R. A. Bovey, *High-Resolution Nuclear Magnetic Resonance of Macromolecules*, Academic Press, New York, 1972; (b) F. A. Bovey *Chain Structure and Conformation of Macromolecules*, Academic Press, New York, 1982; (c) J. C. Randall, *Polymer Sequence Determination*, Academic Press, New York, 1977.
76 For reviews on ^{13}C n.m.r. of synthetic polymers see, for example, (a) V. D. Mochel, *Rev. Macromol. Chem.*, **C8**, 289 (1972); (b) J. Schaefer, *Topics in Carbon-13 Nuclear Magnetic Resonance* (Ed. G. C. Levy), Vol. 1, Wiley-Interscience, New York, 1974; (c) A. E. Tonelli and F. C. Schilling, *Acc. Chem. Res.* **14**, 233 (1981).
77 L. F. Johnson, F. Heatley, and F. A. Bovey, *Macromolecules* **3**, 175 (1970).
78 Ref. 75(a), p. 79.
79 C. J. Carman, *Macromolecules* **6**, 725 (1973).
80 C. J. Carman, A. R. Tarpley, Jr, and J. H. Goldstein, *J. Am. Chem. Soc.* **93**, 2864 (1971).
81 P. A. Mirau and F. A. Bovey, *Macromolecules* **19**, 210 (1986).
82 M. W. Crowther, N. M. Szeverenyi, and G. C. Levy, *Macromolecules* **19**, 1333 (1986).
83 P. H. Bolton, *J. Magn. Reson.* **48**, 336 (1982).
84 P. H. Bolton and G. Bodenhausen, *Chem. Phys. Lett.* 139 (1982).
85 A. Bax, *J. Magn. Reson.* **48**, 336 (1982).
86 G. A. Morris, *Magn. Reson. Chem.* **24**, 371 (1986).
87 H. Kessler, M. Bernd, H. Kogler, J. Zarbock, O. W. Sorensen, G. Bodenhausen, and R. R. Ernst, *J. Am. Chem. Soc.* **105**, 6944 (1983).
88 H. Kessler, W. Bermel, and C. Griesinger, *J. Magn. Reson.* **62**, 573 (1985).
89 S. K. Sarkar and A. Bax, *J. Magn. Reson.* **63**, 512 (1985).
90 A. Bax, D. G. Davis, and S. K. Sarkar, *J. Magn. Reson.* **63**, 230 (1985).
91 M. A. Delsuc, E. Guittet, N. Trotin, and J. Y. Lallemand, *J. Magn. Reson.* **56**, 163 (1984).
92 D. Neuhaus, G. Wider, G. Wagner, and K. Wüthrich, *J. Magn. Reson.* **57**, 164 (1984).
93 L. D. Field and B. A. Messerle, *J. Magn. Reson.* **62**, 453 (1985).
94 D. G. Davis and A. Bax, *J. Am. Chem. Soc.* **107**, 2820 (1985).
95 N. Liu, S. N. Tong, and J. L. Kownig, *J. Appl. Polym. Sci.* **25**, 2205 (1980).
96 M. W. Duch and D. M. Grant, *Macromolecules* **3**, 165 (1970).
97 F. Conti, A. Segre, P. Pini, and L. Porri, *Polymer* **15**, 5 (1974).
98 Y. Tanaka and K. Hatada, *J. Polym. Sci.* **11**, 2057 (1973).
99 E. B. Whipple and P. J. Green, *Macromolecules* **6**, 38 (1973).
100 A. Segre, M. Delfini, G. Conti, and A. Boicelli, *Polymer* **16**, 338 (1975).
101 M. Delfini, A. L. Segre, and F. Conti, *Macromolecules* **6**, 456 (1973).
102 E. Klesper, A. Johnson, W. Gronski, and F. W. Wehrli *Makromol. Chem.* **176**, 1071 (1975).
103 For a comprehensive review, see *Ring-Opening Polymerization* (Eds K. J. Ivin and T. Saegusa), Vols 1–3, Elsevier, New York, 1984.
104 See K. J. Ivin, *Olefin Metathesis*, Chap. 3, Academic Press, New York, 1983.
105 *Ibid.*, Chapters 11 and 13.
106 K. J. Ivin, D. T. Laverty, and J. J. Rooney, *Makromol. Chem.* **178**, 1545 (1977).
107 K. J. Ivin, D. T. Laverty, J. H. O'Donnell, J. J. Rooney, and C. D. Stewart, *Makromol. Chem.* **180**, 1989 (1979).

108 K. J. Ivin, D. T. Laverty, J. J. Rooney, and P. Watt, *Rec. trav. chim.* **96**, M54 (1977).
109 K. J. Ivin, J. J. Rooney, L. Bencze, J. G. Hamilton, L.-M. Lam, G. Lapienis, B. S. R. Reddy, and H. H. Thoi, *Pure Appl. Chem.* **54**, 447 (1982).
110 K. J. Ivin, G. Lapienis, and J. J. Rooney, *J. Chem. Soc. Chem. Commun.* 1068 (1979).
111 K. J. Ivin, G. Lapienis, and J. J. Rooney, *Polymer* **21**, 436 (1980).
112 G. A. Gray, *CRC Critical Reviews in Biochemistry*, 247 (1973).
113 W. Horsley, H. Sternlicht, and J. S. Cohen, *Biochem. Biophys. Res. Commun.* **37**, 47 (1969).
114 W. Horsley, H. Sternlicht, and J. S. Cohen, *J. Am. Chem. Soc.* **92**, 680 (1970).
115 F. R. N. Gurd, P. J. Lawson, D. W. Cochran, and E. Wenkert, *J. Biol. Chem.* **246**, 3725 (1971).
116 P. Keim, R. A. Vigna, J. S. Morrow, R. C. Marshall, and F. R. N. Guard, *J. Biol. Chem.* **248**, 7811 (1973).
117 R. Walter, K. U. M. Prasad, R. Deslauriers, and I. C. P. Smith, *Proc. Nat. Acad. Sci. US* **70**, 2086 (1973).
118 R. Deslauriers, R. Walter, and I. C. P. Smith, *Proc. Nat. Acad. Sci. US* **71**, 365 (1974).
119 I. C. P. Smith, H. J. Jennings, and R. Deslauriers, *Accounts Chem. Research* **8**, 306 (1975).
120 H. Kessler, H. R. Loosli, and H. Oschkinat, *Helv. Chim. Acta* **68**, 661 (1985).
121 (a) C. Wynants, K. Hallenga, G. Van Binst, A. Michel, and J. Zanen, *J. Magn. Reson.* **57**, 93 (1984); (b) K. Hallenga, G. Van Binst, M. Knappenberg, J. Brison, A. Michel, and J. Dirkx, *Biochim. Biophys. Acta* **577**, 82 (1979).
122 G. Bodenhausen and R. Freeman, *J. Magn. Reson.* **28**, 471 (1977).
123 (a) R. Freeman, *Proc. Roy. Soc. London Ser. A* **373**, 149 (1980); (b) R. Freeman and G. A. Morris, *Bull. Magn. Reson.* **1**, 5 (1979).
124 K. Hallenga and G. Van Binst, *Bull. Magn. Reson.* **2**, 343 (1980).
125 H. Kessler and W. Bermel in *Applications of NMR Spectroscopy to Problems in Stereochemistry and Conformational Analysis* (Eds Y. Takeuchi and A. P. Marchand) VCH Publishers, Deerfield Beach, Florida, 1986, pp. 179–205.
126 L. Paolillo, T. Tancredi, P. A. Temussi, E. Trivellone, E. M. Bradbury, and C. Crane-Robinson, *Chem. Commun.* 335 (1972).
127 S. Tadokoro, S. Fujiwara, and Y. Ichihara, *Chem. Lett.* 849 (1973).
128 H. Saitô and I. C. P. Smith, *Arch. Biochem. Biophys.* **158**, 154 (1973).
129 A. Allerhand, D. Doddrell, V. Glushko, D. W. Cochran, P. J. Lawson, and F. R. N. Gurd, *J. Am. Chem. Soc.* **93**, 544 (1971).
130 V. Glushko, P. J. Lawson, and F. N. R. Gurd, *J. Biol. Chem.* **247**, 3176 (1972).
131. A. Allerhand, R. F. Childers, and E. Oldfield, *Biochemistry* **12**, 1335 (1973).
132 See, for example, E. Oldfield and A. Allerhand, *Proc. Nat. Acad. Sci. US* **70**, 3531 (1973).
133 T.-M. Chan and J. L. Markley, *J. Am. Chem. Soc.* **104**, 4010 (1982).
134 C. L. Kojiro and J. L. Markley, *FEBS Lett.* **162**, 52 (1983).
135 W. M. Westler, G. Ortiz-Polo, and J. L. Markley, *J. Magn. Res.* **58**, 352 (1984).
136 For a review of the earlier work see, for example, J. B. Stothers, *Carbon-13 Nuclear Magnetic Resonance Spectroscopy*, Chap. 7, Academic Press, New York, 1972.
137 H. H. Mantsch and I. C. P. Smith, *Biochem. Biophys, Res. Commun.* **46**, 808 (1972).
138 R. A. Komoroski and A. Allerhand, *Biochemistry* **13**, 369 (1974).
139 E. T. Lippmaa, M. A. Alla, T. J. Pehk, and G. Engelhardt, *J. Am. Chem. Soc.* **100**, 1929 (1978).
140 W.-I. Shau, E. N. Duesler, I. C. Paul, D. Y. Curtin, W. G. Blann, and C. A. Fyfe, *J. Am. Chem. Soc.* **102**, 4546 (1980).
141 (a) A. K. Cheng, F. A. L. Anet, J. Mioduski, and J. Meinwald, *J. Am. Chem. Soc.* **94**,

2887 (1974); (b) V. Macho, R. D. Miller, and C. S. Yannoni, *J. Am. Chem. Soc.* **105**, 3735 (1983).

142 C. S. Yannoni, *Acc. Chem. Res.* **15**, 201 (1982).

143 (a) C. S. Yannoni and R. D. Kendrick, *J. Chem. Phys.* **74**, 747 (1981); (b) D. Horne, R. D. Kendrick, and C. S. Yannoni, *J. Magn. Reson.* **52**, 299 (1983); (c) T. C. Clarke, C. S. Yannoni, and T. J. Katz, *J. Am. Chem. Soc.* **105**, 7787 (1983).

144 T. J. Katz, S. M. Hacker, R. D. Kendrick, and C. S. Yannoni, *J. Am. Chem. Soc.* **105**, 7787 (1983).

145 For a general review see J. B. Stothers, *Topics in ^{13}C Nuclear Magnetic Resonance Spectroscopy* (Ed. G. C. Levy), Vol. 1, Chap. 6, Wiley-Interscience, New York, 1974.

146 W. D. Crown and M. N. Paddon-Row, *J. Am. Chem. Soc.* **94**, 4746 (1972).

147 (a) W. J. Baron, M. Jones, Jr., and P. P. Gaspar, *J. Am. Chem. Soc.* **92**, 4739 (1970); (b) E. Hedaya and M. E. Kent, *J. Am. Chem. Soc.* **93**, 3283 (1971); (c) G. G. Vander Stouw, A. R. Kraska, and H. Shechter, *J. Am. Chem. Soc.* **94**, 1655 (1972).

148 P. P. Gaspar, J. P. Hsu, S. Chai, and M. Jones, Jr, *Tetrahedron* **41**, 1479 (1985).

149 J. B. Stothers, I. S. Y. Wang, D. Ouchi, and E. W. Warnhoff, *J. Am. Chem. Soc.* **93**, 6702 (1971).

150 J. B. Stothers, C. T. Tan, A. Nickon, F. Huang, R. Sridhar, and R. Weglein, *J. Am. Chem. Soc.* **94**, 8581 (1972).

151 For references see, for example, (a) A. I. Scott, *Science, NY* **184**, 760 (1974); (b) U. Sequin and A. I. Scott, *Science, NY* **186**, 101 (1974); (c) A. G. McInnes, J. A. Walter, J. L. C. Wright, and L. C. Vining, *Topics in ^{13}C Nuclear Magnetic Resonance Spectroscopy* (Ed. G. C. Levy), Vol. 2, Chap. 3, Wiley-Interscience, New York, 1976.

152 M. Tanabe, T. Hamasaki, D. Thomas, and L. F. Johnson, *J. Am. Chem. Soc.* **93**, 273 (1971).

153 F. Aragozzini, M. G. Beretta, G. Severini Ricca, C. Scolastico, and F. W. Wehrli, *JCS, Chem. Commun.* 788 (1973).

154 M. Tanabe, K. T. Suzuki, and W. C. Jankowski, *Tetrahedron Lett.* 2271 (1974).

155 J. W. Westley, D. L. Pruess, and R. G. Pitcher, *JCS, Chem. Commun.* 161 (1972).

156 (a) N. Neuss, C. H. Nash, P. A. Lempke, and J. B. Grutzner, *J. Am. Chem. Soc.* **93**, 2337 (1971); (b) N. Neuss, C. H. Nash, P. A. Lempke, and J. B. Grutzner, *Proc. Roy. Soc., Ser. B* **179**, 335 (1971).

157 D. E. Cane, T. Rossi, A. M. Tilman, and J. P. Pachlatko, *J. Am. Chem. Soc.* **103**, 1838 (1981).

158 S. J. Gould and D. E. Cane, *J. Am. Chem. Soc.* **104**, 343 (1982).

159 J. A. Findlay, J. -S. Liu, and L. Radics, *Can. J. Chem.* **61**, 323 (1983).

160 G. T. Carter, A. A. Fantini, J. C. James, D. B. Borders, and R. J. White, *Tetrahedron Lett.* **25**, 255 (1984).

161 T. T. Nakashima and J. C. Vederas, *J. Am. Chem. Soc.* **107**, 3694 (1985).

162 J. R. Lyerla, Jr, and D. M. Grant, *International Review of Science, Physical Chemistry Series* (Ed. C. A. McDowell), Vol. 4, Chap. 5, Medical and Technical Publishing Co., Chicago, 1972.

163 See for example D. A. Wright, D. E. Axelson and G. C. Levy, *^{13}C Nuclear Magnetic Resonance, Spectroscopy* (Ed. G. C. Levy), Vol. 3, Chap. 2, Wiley-Interscience, New York, 1979.

164 For a review of structural aspects of ^{13}C relaxation see, for example, F. W. Wehrli, *Topics in ^{13}C Nuclear Magnetic Resonance Spectroscopy* (Ed. G. C. Levy), Vol. 2, Chap. 6, Wiley-Interscience, New York, 1976.

165 A. Allerhand, D. Doddrell, and R. Komoroski, *J. Chem. Phys.* **55**, 189 (1971).

166 F. W. Wehrli, *Advanc. Mol. Relax. Processes* **6**, 139 (1974).

167 C. Yu and G. C. Levy, *J. Am. Chem. Soc.* **106**, 6533 (1984).

168 P. L. Rinaldi, *J. Am. Chem. Soc.* **105**, 5167 (1983).

169 N. Niccolai, C. Rossi, P. Mascagni, W. A. Gibbons, and V. Brizzi, *JCS, Perkin Trans. I*, 239 (1985).
170 N. Niccolai, C. Rossi, V. Brizzi, and W. A. Gibbons, *J. Am. Chem. Soc.* **106**, 5732 (1984).
171 M. A. Khaled and C. L. Watkins, *J. Am. Chem. Soc.* **105**, 3363 (1983).
172 P. Mascagni, M. Pope, W. A. Gibbons, L. M. Ciuffetti, and H. W. Knoche, *Biochem. Biophys. Res. Commun.* **113**, 10 (1983).
173 N. Niccolai, C. Rossi, P. Mascagni, P. Neri, and W. A. Gibbons, *Biochem. Biophys. Res. Commun.* **124**, 739 (1984).
174 J. J. Ford, W. A. Gibbons, and N. Niccolai, *J. Magn. Reson.* **47**, 522 (1982).
175 P. Mascagni, A. Prugnola, W. A. Gibbons, and N. Niccolai, *JCS, Perkin Trans. I*, 1015 (1986).
176 H. Kessler and W. Bermel, *Applications of NMR Spectroscopy to Problems in Stereochemistry and Conformational Analysis* (Eds Y. Takeuchi and A. P. Marchand), Chap. 6, p. 188, VCH Publishers, Inc., Deerfield Beach, Florida, 1986.
177 W. T. Huntress, Jr., *Advan. Magn. Reson.* **4**, 1 (1970).
178 K. F. Kuhlmann and D. M. Grant, *J. Chem. Phys.* **55**, 2998 (1971).
179 G. C. Levy, J. D. Cargioli, and F. A. L. Anet, *J. Am. Chem. Soc.* **95**, 1527 (1973).
180 T. D. Alger, D. M. Grant, and R. K. Harris, *J. Phys. Chem.* **76**, 281 (1972).
181 G. C. Levy, *Acc. Chem. Res.* **6**, 161 (1973).
182 J. W. ApSimon, H. Beierbeck, and J. K. Saunders, *Can. J. Chem.* **53**, 338 (1975).
183 G. C. Levy, D. M. White, and F. A. L. Anet, *J. Magn. Res.* **6**, 453 (1972).
184 F. W. Wehrli, Paper presented at the 3rd International Symposium on Nuclear Magnetic Resonance, held at St Andrews, Scotland, 1975.
185 T. D. Alger, D. M. Grant, and J. R. Lyerla, Jr, *J. Phys. Chem.* **75**, 2539 (1971).
186 D. Doddrell and A. Allerhand, *J. Am. Chem. Soc.* **93**, 1558 (1971).
187 D. E. Woessner, *J. Chem. Phys.* **37**, 647 (1962).
188 D. M. Grant, R. J. Pugmire, E. P. Black, and K. A. Christensen, *J. Am. Chem. Soc.* **95**, 8465 (1973).
189 S. Berger, F. R. Kreissl, D. M. Grant, and J. D. Roberts, *J. Am. Chem. Soc.* **97**, 1805 (1975).
190 D. Doddrell, V. Glushko and A. Allerhand, *J. Chem. Phys.* **56**, 3683 (1972).
191 A. Allerhand and E. Oldfield, *Biochemistry*, **12**, 3428 (1973).
192 J. Schaefer and D. F. S. Natusch, *Macromolecules* **5**, 416 (1972).
193 J. Schaefer, *Macromolecules* **5**, 590 (1972).
194 J. Schaefer, *Macromolecules* **5**, 427 (1972).
195 F. W. Wehrli, T. Wirthlin, E. Klesper, and A. Johnsen, Unpublished data.
196 G. C. Levy and R. A. Komoroski, *J. Am. Chem. Soc.* **96**, 678 (1974).
197 (a) O. A. Gansow, A. R. Burke, and G. N. La Mar, *Chem. Commun.* 456 (1972); (b) G. C. Levy, J. D. Cargioli, P. C. Juliano, and T. D. Mitchell, *J. Am. Chem. Soc.* **95**, 3445 (1973); (c) L. F. Farnell, E. W. Randall, and A. I. White, *Chem. Commun.* 1159 (1972).
198 J. W. Faller, M. A. Adams, and G. N. La Mar, *Tetrahedron Lett.* 699 (1974).
199 D. Welti, PhD Dissertation, ETH, Zürich, 1975.
200 J. N. Shoolery, *Progress in NMR Spectroscopy* **11**, 79 (1977).
201 J. N. Shoolery and W. C. Jankowski, *Applications Note*, No. 4, Varian Assoc., Palo Alto, California, 1973.
202 W. Bremser, H. D. W. Hill, and R. Freeman, *Messtechnik* **79**, 14 (1971).
203 S. Gillet and J. -J. Delpuech, *J. Magn. Reson.* **38**, 433 (1980).
204 S. Gillet, J. -J. Delpuech, P. Valentin, and J. -C. Escalier, *Anal. Chem.* **52**, 813 (1980).
205 G. E. Maciel and M. J. Sullivan, *ACS Symp. Ser.*, No. 191, 319 (1982).

206 See, for example, *Magnetic Resonance-Introduction, Advanced Topics and Applications to Fossil Energy* (Eds. L. Petrakis and J. P. Fraissard), Reidel, Dordrecht, 1984.
207 J. N. Shoolery, *Application Note*, No. 3, Varian Assoc., Palo Alto, California, 1975.
208 E. Hofer, R. Keuper, and H. Renken, *Tetrahedron Lett.* **25**, 1141 (1984).
209 E. Hofer and R. Keuper, *Tetrahedron Lett.* **25**, 5631 (1985).
210 J. F. Haw, G. E. Maciel, and H. A. Schroder, *Anal. Chem.* **56**, 1323 (1984).
211 D. S. Himmelsbach, F. E. Barton, II, and W. R. Windham, *J. Agric. Food Chem.* **31**, 401 (1983).
212 M. Alla and E. Lippmaa, *Chem. Phys. Lett.* **37**, 260 (1976).
213 P. D. Murphy, T. J. Cassady, and B. C. Gerstein, *Fuel* **61**, 1233 (1982).
214 D. L. VanderHart and H. L. Retcofsky, *Fuel* **55**, 202 (1976).
215 H. Sfihi, M.F. Quinton, A. Legrand, S. Pregermain, D. Carson, and P. Chiche, *Fuel* **65**, 1006 (1986).
216 L. B. Alemany, D. M. Grant, R. J. Pugmire, and L. M. Stock, *Fuel* **63**, 513 (1984).
217 R. L. Dudley and C. A. Fyfe, *Fuel* **61**, 651 (1982).
218 M. A. Wilson, A. M. Vassallo, P. J. Colling, and H. Rottendorf, *Anal. Chem.* **56**, 433 (1984)
219 D. D. Whitehurst, *Discuss. Fuel* **61**, 967 (1982).
220 K. W. Zilm and G. G. Webb, *Fuel* **65**, 721 (1986).
221 J. J. H. Ackerman, T. H. Grove, G. C. Wong, G. D. Gadian, and G. K. Radda, *Nature* **283**, 167 (1980).
222 F. W. Wehrli, D. Shaw and B. Kneeland, *Principles, Methodology and Applications of Biomedical Magnetic Resonance*, VCH Publishers, Inc., New York, 1988.
223 G. D. Gadian, *Nuclear Magnetic Resonance and its Applications to Living Systems*, The Clarendon Press, Oxford, 1982.
224 T. L. James and A. R. Margulis, *Biomedical Magnetic Resonance*, Radiology Research and Education Foundation, San Francisco, 1984.
225 P. A. Bottomley, H. R. Hart Jr, W. A. Edelstein, J. F. Schenck, L. S. Smith, W. M. Leue, O. M. Mueller, and R. W. Redington, *Radiology* **150**, 441 (1984).
226 F. W. Wehrli, *Localization Methods for Nuclear Magnetic Resonance Spectroscopy In Vivo*, Circulation 72 (Supplement IV), IV-97 (1985).
227 P. A. Bottomley, T. B. Foster, and R. D. Darrow, *J. Magn. Reson.* **59**, 338 (1984).
228 W. P. Aue, S. Muller, T. A. Cross, and J. Seelig, *J. Magn. Reson.* **56**, 350 (1984).
229 D. I. Hoult and P. C. Lauterbur, *J. Magn. Reson.* **34**, 425 (1979).
230 C. J. Unkefer, R. M. Blazer, and R. E. London, *Science* **222**, 62 (1983).
231 J. R. Alger, L. O. Sillerud, K. L. Behar, R. J. Gillies, R. G. Shulman, R. E. Gordon, D. Shaw, and P. E. Hanley, *Science* **214**, 660 (1981).
232 D. L. Rothman, K. L. Behar, H. P. Hetherington, J. A. Den Hollander, M. R. Bendall, O. A. C. Petroff, and R. G. Shulman, *Proc. Nat. Acad. Sci. US* **82**, 1633 (1985).
233 T. Jue, J. A. B. Lohman, R. J. Ordidge, and R. G. Shulman, *Fifth Annual Meeting of the Society of Magnetic Resonance in Medicine*, Montreal, 1986, Book of Abstracts, p. 571.
234 R. A. Heinlein, *By His Bootstraps*, Street & Smith Publications, Inc., New York, 1941.

APPENDIX

Solutions to Problems

PROBLEM 1

According to the general scheme outlined in Table 5.1, the following can be immediately concluded: (1) the molecule has no double bonds, and (2) it is likely to have symmetry since the eight carbons give rise to only five lines. From the off-resonance spectrum, we learn that lines 1 to 5 have t, s, q, q, and d multiplicity. The splitting pattern for line 5 is difficult to identify because it is highly second order. However, if it is assumed that the five lines 1 to 5 correspond to 1, 1, 3, 2, 1 carbons as suggested by their intensities, signal 5 must be due to a methine carbon. The large deshielding of the carbon resonating at 53.4 ppm points to heavy α-substitution of this methylene carbon. Three equivalent methyl carbons can only be rationalized by the presence of a tertiary butyl group unless the molecule has threefold symmetry, which can be excluded. The following structure elements can immediately be derived:

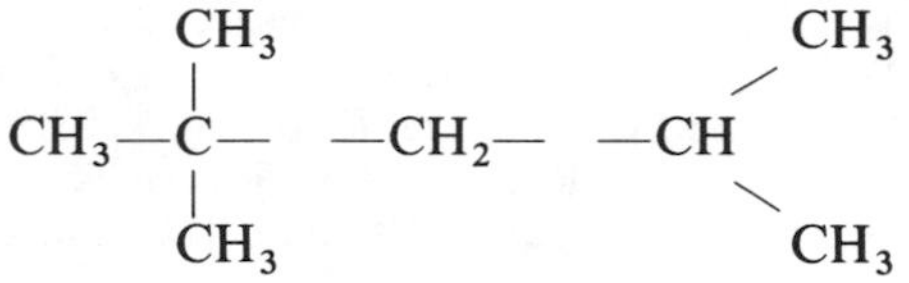

$$
\begin{array}{cc}
\mathrm{CH_3} & \mathrm{CH_3} \\
| & \diagup \\
\mathrm{CH_3-C-} \quad -\mathrm{CH_2-} \quad -\mathrm{CH} & \\
| & \diagdown \\
\mathrm{CH_3} & \mathrm{CH_3}
\end{array}
$$

These can be linked in only one way, to give 2, 2, 4-trimethylpentane,

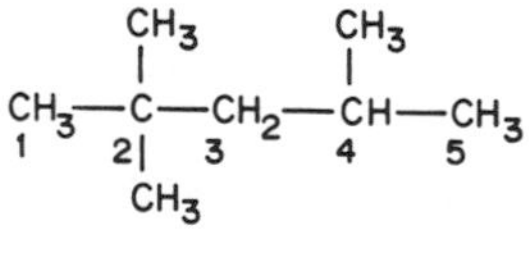

$$
\begin{array}{c}
\underset{1}{\mathrm{CH_3}} \overset{\mathrm{CH_3}}{\underset{2|}{-\mathrm{C}}} \underset{3}{-\mathrm{CH_2}} \overset{\mathrm{CH_3}}{\underset{4}{-\mathrm{CH}}} \underset{5}{-\mathrm{CH_3}} \\
\mathrm{CH_3}
\end{array}
$$

Let us now calculate the chemical shifts on the basis of Lindeman and Adams' additivity rules for open-chain aliphatic hydrocarbons using the parameters listed in Table 2.9. As an example we may choose C-3, for which $n = 2$, $N_0 = 1$, $N_1 = 1$, $N_2 = 0$, $N_\gamma = N_\delta = 0$.

The following increments are therefore needed: $A_2 = 15.34$, $\alpha_{20} = 21.43$, $\alpha_{21} = 16.70$. Thus for $\delta_c(3)$ one obtains

$$\delta_c(3) = 15.34 + 21.43 + 16.70 = 53.47$$

The predicted chemical shifts for the remaining carbons are listed below.

Carbon	δ_c(calc)	δ_c(obs)
1	30.27	30.21
2	31.75	31.13
3	53.47	53.44
4	23.85	24.87
5	23.11	25.59

From this table it is recognized that satisfactory agreement between experimental and predicted values is obtained except for C-5, where the deviation exceeds 2 ppm. It should be borne in mind, however, that the procedure used is based on rules which are purely empirical.

PROBLEM 2

The spectrum contains ten lines of comparable intensities, two between 110 and 140 ppm and the remaining eight in the aliphatic region between 10 and 35 ppm. These facts permit the conclusion that the molecule has ten carbons, two being olefinic and eight aliphatic. The absence of resonances in the lower field aliphatic region excludes the presence of electronegative heteroatoms. From the signal multiplicities a hydrogen number of 20 is calculated corresponding to the molecular formula $C_{10}H_{20}$ and $F = 1$, i.e. there are no rings present. The signal multiplicities for lines 1 and 2 indicate that the double bond is terminal. Since there is only one methyl carbon (line 10), the molecule is unbranched: 1-decene,

$$CH_2{=}CH{-}CH_2{-}CH_2{-}CH_2{-}CH_2{-}CH_2{-}CH_2{-}CH_2{-}CH_3$$

Although no further corroborative arguments are required in principle, the use of chemical shift increments for the prediction of olefinic carbon shieldings will be shown in this example (cf. Table 2.10). For C-1, the parameters required are $A_{1\alpha} = 10.6$, $A_{1\beta} = 7.2$, $A_{1\gamma} = -1.5$ and one obtains for $\delta_c(1)$

$$\delta_c(1) = 123.3 + 10.6 + 7.2 - 1.5 = 139.6$$

and in analogous manner for $\delta_c(2)$

$$\delta_c(2) = 123.3 - 7.9 - 1.8 + 1.5 = 115.1$$

Both values are in satisfactory agreement with the experimental data.

PROBLEM 3

$F = 5$ is determined from the molecular formula, i.e. the sum of double bonds and rings in this molecule is 5. The total line number in the ^{13}C spectrum is 11, pointing to symmetry within a group of the system. Line 10, which is suspected to be degenerate and appears to accommodate three carbons, indicates a *t*-butyl group. Line 1 at 205.2 ppm is comparable with an aldehydic, ketonic, or the centre carbon of an allene. An allenic function would also give rise to resonances between 80 and 100 ppm which, however, do not occur. The low intensity of this signal is characteristic of a quaternary carbon and therefore strongly implies a ketone. Lines 2 to 7, between 160 and 110 ppm, are typical of a substituted benzene with the two low-intensity lines indicating disubstitution.

Since all carbons are anisochronous, 1, 4-disubstitution is not possible. The large deshielding of one of the quaternary carbons (line 2:160 ppm) suggests attachment of an electronegative substituent. Since the only oxygen atom pertains to the ketone function, we conclude that nitrogen is bonded to this carbon. Proceeding to the aliphatic region, one finds a quaternary carbon resonance at 70.8 ppm (line 8) whose deshielding requires it to be next to an electronegative atom, i.e. nitrogen. A further quaternary carbon signal is found at 36.9 ppm (line 9) followed by the high-intensity signal at 24.90 assignable to the *t*-butyl methyls and a further line at 19.2 ppm which is likely to belong to another methyl carbon. Returning to the unsaturation number of 5, one recognizes that, in addition to the benzene ring, a further ring must be present.

A glance at the proton spectrum is appropriate at this stage. The splitting pattern observed for the four non-equivalent aromatic protons is compatible only with *ortho* substitution. The deshielding of the two low-field protons tells us that the second substituent must be a carbonyl. The broad singlet at around 4.8 ppm must belong to a labile proton, in this case NH. The two remaining signals at highest field corresponding to 3 and 9 protons, respectively, can only be explained by quaternary methyl groups whose counterparts in the ^{13}C spectrum are peaks numbered 10 and 11. The following structure elements can now be formulated:

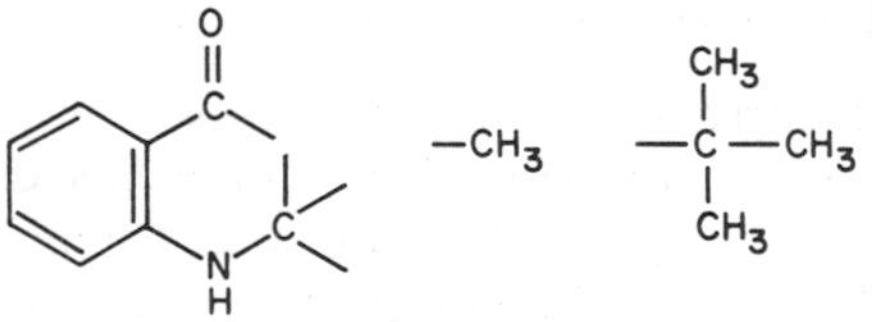

These three elements can be linked in only one way to yield

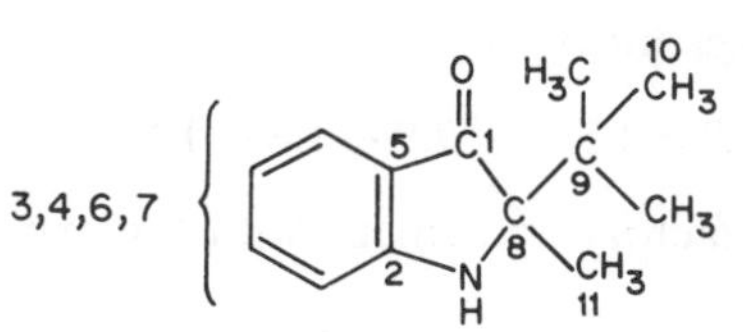

The chemical shift found for the carbonyl carbon may at first sight appear too large for an unsaturated ketone. It is known, however, that increased α-substitution is accompanied by a deshielding of several parts per million.

The assignment of the eleven lines is as given in the formula.

PROBLEM 4

Depending on absolute configuration, the proposed structure possesses either a centre of symmetry (phenyl groups *trans*) or a twofold axis (phenyl groups *cis*). Assuming free rotation of the phenyls in both cases, eleven signals are expected which, *eo ipso*, rule out the structure displayed. If the molecule had no symmetry, then a total of twenty-two lines should be obtained. The lower number of lines, namely sixteen, thus indicates overall symmetry.

It is noteworthy that the low-field region of the spectrum between 105 and 160 ppm contains an odd number of lines, namely fifteen, whereas non-equivalence of the phenyls and equivalence of the backbone aromatic rings would result in fourteen lines. The reverse situation with the phenyls being equivalent but non-equivalence of the backbone benzene rings is unlikely and can be ruled out because this would give sixteen lines in this region. It can thus be inferred that one of the resonances in question is due to an sp^3-type carbon. This is a fairly rare situation, and its occurrence requires attachment to more than one electronegative substituent. In acetals, ketals and *ortho* esters, for example, where two and three oxygens are bound to the same carbon, shifts as large as 120 ppm may be encountered. It is therefore likely that the two oxygens present in the molecule are adjacent to the same carbon since no other hetero atoms are present. The second sp^3 carbon, which is found at about 70 ppm and which evidently is quaternary, is too highly shielded to be compatible with an ether-type carbon. This unusual deshielding of a quaternary aliphatic carbon can be explained by the high degree of α substitution. The structure nearest to the one proposed but in accordance with the n.m.r. data is

This molecule has a plane of symmetry which bisects the two sp^3 carbons. Independent of absolute configuration of these carbons, sixteen lines are expected, with six being due to singly occurring carbons, namely C-7 and C-8 (lines 12 and 16), C-9 and C-13 (lines 2/3), and C-12 and C-16 (lines 7/10).

PROBLEM 5:[1]

(1) The four possible structures, **A–D**, are shown below:

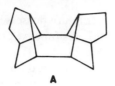

A

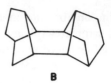

B

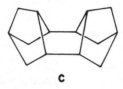

C

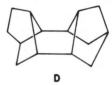

D

The following structures can be eliminated on the basis of the given ^{13}C n.m.r. spectrum of **2**:

(i) We can rule out **C**, which has C_{2h} symmetry and for which we expect to observe only five ^{13}C resonances.

(ii) We can also rule out **D**, which has C_1 symmetry and for which we expect to observe fourteen ^{13}C resonances.

However, both **A** (C_s symmetry) and **B** (C_2 symmetry) fit the given ^{13}C n.m.r. spectrum for **2** (i.e. we expect to observe seven ^{13}C resonances for either compound).

(2) From Problem Fig. 5(b), we obtain the following connectivities:

(a, b) (b, d) (c, e) (d, g)
(a, c) (b, e) (c, f) (f, g)
(a, d)

Note in particular the three connectivities to a, i.e. (a, b), (a, c), and (a, d). All four resonances a–d correspond to *methine* carbons. Inspection of structure **A** reveals that there exists no methine carbon which is connected to three other *non-equivalent* methine carbons. (Note that connectivities $(1, 2)$ and $(8, 9)$ in structure **A** involve *equivalent* carbons.) Hence, structure **A** cannot possibly correspond to **2**; we need only demonstrate that structure **B** is the correct choice for **2**.

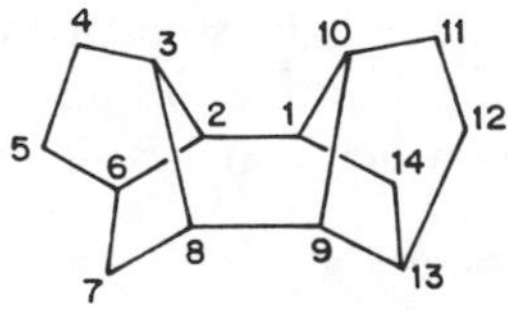

B

In preparing the following arguments, we make use of the fact that **B** possesses C_2 symmetry:

(i) Peak a is connected to b, c, and d. Hence, $\mathbf{a} = (\mathbf{2,9})$ and $[b, c,$ and $d]$ correspond to $[(1, 8), (3, 10),$ and $(6, 13)]$.

(ii) Of the three methylene carbon resonances, only 7 is connected to two methine carbons (4 and 5 are each connected to one methine and to one methylene carbon). Hence, $\mathbf{e} = (\mathbf{7, 14})$.

(iii) b(CH) is connected to $a(2, 9)$, d(CH), and $e(7, 14)$. Hence $\mathbf{b} = (\mathbf{1, 8})$ and $\mathbf{d} = (\mathbf{3, 10})$.

(iv) Recall from (i), above, that $[b, c,$ and $d]$ correspond to $[(1, 8), (3, 10),$ and $(6, 13)]$. Since we have already assigned $b = (1, 8)$ and $d = (3, 10)$, therefore $\mathbf{c} = (\mathbf{6, 13})$.

(v) Peak $c(6, 13)$ is connected to $e(7, 14)$ and $f(\mathrm{CH_2})$. Therefore, $\mathbf{f} = (\mathbf{5, 12})$.

(vi) The only remaining signal is $\mathbf{g} = (\mathbf{4, 11})$.

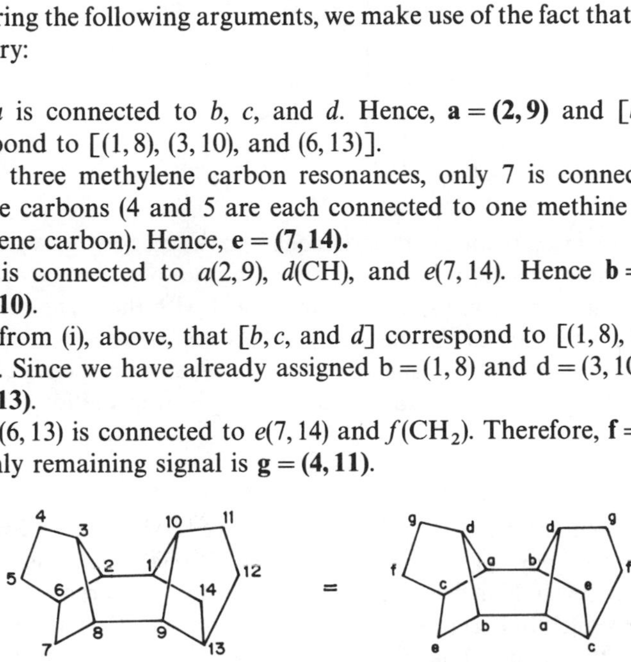

B = 2

PROBLEM 6

The symmetry of the proposed structure allows one to predict a maximum of eleven resonances with only two different types of phenyl rings. The spectrum exhibits seventeen distinct lines but further accidental equivalences of some aromatic resonances must be expected.

A conspicuous feature of this spectrum is line 1 corresponding to a chemical shift of 210 ppm. In a hydrocarbon there is only one carbon type matching this extremely low shielding: the centre carbon of an allene. By contrast, the outer carbons resonate at considerably higher field and a brief check of allenic carbon shifts shows that in phenyl-substituted systems these resonate between 100 and 120 ppm ($\rightarrow$ lines 13, 14). The positions of lines 15 and 16, on the other hand, are typical of acetylenic carbons. Hence both an allenic and an acetylenic function are present. The low intensity of the signals associated with these groupings indicate full substitution. Since a line pertaining to an sp^3 carbon is also present, the following structure is derived which matches the previously discussed data.

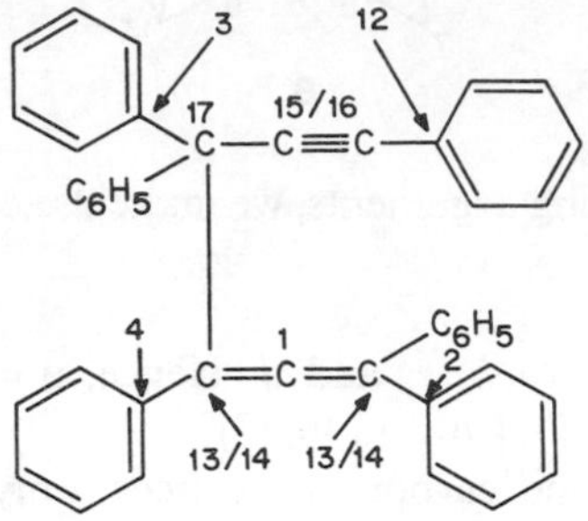

All proton-bearing phenyl carbons resonate in the narrow region between 125 and 129 ppm and thus cannot be assigned on a one-to-one basis. However, the substituted ring carbons are shifted to lower field with the exception of the C-1 phenyl carbon attached to the acetylenic moiety (cf. Table 2.15). Since there are four observationally distinguishable phenyl groups, four such signals can be detected (lines 2–4, 12). A partial assignment is given in the formula.

PROBLEM 7

From the molecular formula $C_{20}H_{16}O_6$, an unsaturation number $F = 13$ follows and the line number $l = 9$ indicates symmetry. When scanning through the spectrum from low to high field, we first encounter two quaternary carbons which have to be assigned to carbonyls (lines 1 and 2). The very low intensity of line 2 suggests that this carbon relaxes particularly slowly. This is typical of carbons which are spatially isolated from protons. Lines 3, 4, and 5, all quaternaries, belong to sp^2 carbons with at least two of them (3 and 4) adjacent to oxygen. Lines 6, 7, and 8 having doublet multiplicity (121.8, 115.3, and 110.4) are due to proton-bearing aromatic or olefinic carbons. The intensity ratio furthermore points to symmetric substitution. The only aliphatic carbon resonance having quartet multiplicity is attributed to OCH_3 (line 9).

The proton spectrum exhibits resonances at around 7.0, 5.6, and 3.75 ppm with an intensity ratio of 4:1:3, thus corresponding to 8, 2, and 6 protons, respectively. The splitting pattern of the lowest field signal is characteristic of AA′BB′ systems. The almost identical chemical shift for both types of protons and their relative shielding compared to benzene suggests unsymmetric 1,4-substitution with both substituents being oxygen. The two olefinic protons at 5.7 ppm are difficult to allocate at this stage, while the highest field resonance confirms the presence of two methoxyl groups. We know at this stage that line 8 in the ^{13}C spectrum is not aromatic, it must belong to two olefinic carbons. From the knowledge of the presence of two 1,4-dioxo-substituted benzene rings, two carbonyls and two olefinic double bonds, the signals 3, 4, and 5 can readily be explained: 3 and 4 must be assigned to the two types of oxygen-bearing aromatic carbons, while 5 originates from the second olefinic carbon. The structure elements thus far established accommodate twelve double bonds and rings with all carbons being assigned. It follows, therefore, that the structure must contain one further ring. Since four of the six oxygens are substituted to the two aromatic rings and the remaining two are in carbonyl functions, it can be inferred that the additional ring must be homocyclic. When attempting to integrate the four olefinic and the two carbonyls into a ring, we find that only one isomeric species leads to an arrangement having four observationally different carbons:

The structure also explains the large intensity differences of the two carbonyl carbon resonances.

PROBLEM 8[2]

The following carbon–carbon connectivities are established by examination of the double quantum coherence data presented in Problem Fig. 8: $(d, e), (d,f), (a, e),$ $(b, e), (b, g), (c, g),$ and (c, f).

(i) It seems reasonable to expect that the peak situated farthest downfield (i.e. the resonance associated with the most highly deshielded aliphatic carbon atom in **I**) corresponds to the benzylic carbon, α. Hence, we can tentatively assign $\mathbf{a} = \alpha$.

(ii) Peak a is connected to e; hence, $\mathbf{e} = \beta$.

(iii) In addition to its connectivity to a, peak e is connected also to b, and d. These latter two peaks, therefore, correspond to the two γ carbon atoms (γ_s, γ_a).

(iv) The difference in chemical shift values between γ_s and γ_a is substantial (*ca* 7.3 ppm). If we consider the 2-phenyl group to be the '*gauche* γ-substituent' referred to by Beierbeck and Saunders,[3] then we expect that γ_s will be shielded relative to γ_a. Therefore, we can assign $\mathbf{d} = \gamma_s$ and $\mathbf{b} = \gamma_a$.

(v) The observed connectivities (b, g) and (d,f) then permit the following assignments to be made: $\mathbf{g} = \delta_s$ and $\mathbf{f} = \delta_a$.

(vi) Finally, c is connected only to f and g; hence, $\mathbf{c} = \varepsilon$.

PROBLEM 9

This compound's double bond equivalence is 5. Its ^{13}C spectrum affords twelve lines with each of the two resonances at 128.0 and 125.7 ppm (3 and 4) being-degenerate. Line 1 at 169.8 ppm of singlet multiplicity points to a carbonyl, most probably in an acid or ester function, as suspected from the chemical shift. The four lines at next higher field are characteristic of a monosubstituted benzene. The chemical shifts of the proton-bearing carbons (lines 3 to 5) furthermore suggest that the substituent is neither electron-withdrawing nor donating. This is corroborated by the featureless signal found in the proton spectrum at *ca* 7.2 ppm corresponding to 5 protons. Since lines 1 to 5 account for four double bonds and one ring, we can also conclude that neither additional rings nor double bonds are present. Line 6 at 68.0 ppm conveys that a methine carbon is next to oxygen. Hence the carboxy group belongs to an ester function. Lines 7 to 12 in the range between 37 and 20 ppm originate from one methylene, one quaternary and four methyl carbons. In the proton spectrum we find the correct number of protons in the high-field region, namely 14. The single proton resonance centred at *ca* 4.9 ppm splits into a complex multiplet and is likely to be associated with carbon line 6. While three of the methyl proton signals are singlets, the one at highest field is a doublet, which must arise from vicinal coupling to the only occurring methine proton. The two geminal protons are evidently non-equivalent, as seen from the six lines at around 2 ppm. With a little experience, this is identified as the AB part of an ABX spin system with two lines probably buried under the lowest field methyl resonance. These facts allow us to derive the subsequent partial structures:

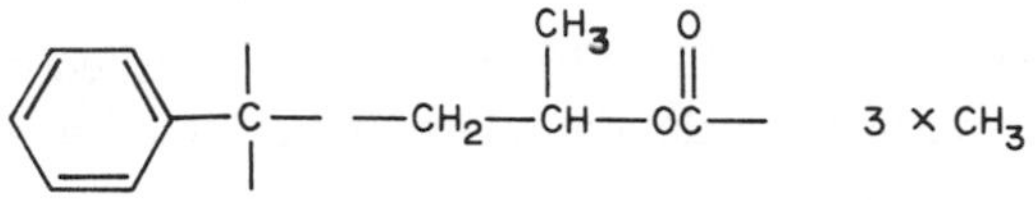

Since we know that the three so far unassigned methyl groups must be bonded to quaternary carbons, this unavoidably leads to the structure:

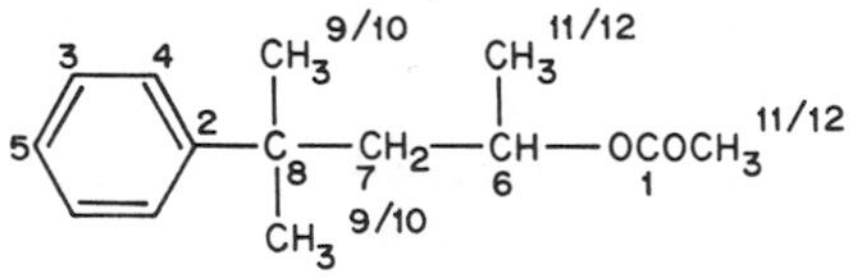

The assignments are as given in the formula.

PROBLEM 10

This molecule has an unsaturation number of 3. The two resonances at 144 and 116 ppm, respectively, which have to be assigned to sp^2 carbons, further tell us that the molecule contains one double bond. Hence we have to accommodate two rings in the structure. The off-resonance decoupled spectrum of the aliphatic region indicates the presence of three CH_3's, two CH_2's, two CH's, and one quaternary carbon. This leaves only seven carbons for the formation of the two rings. The aliphatic region of the proton spectrum exhibits two sharp methyl resonances, thus proving that two of the methyls at least are adjacent to a quaternary carbon. The third methyl proton signal has quartet multiplicity with a splitting of ca 2.5 Hz, typical of an allylic coupling. However, since there is only one olefinic proton, other couplings must be involved. This is corroborated by irradiating the vinyl proton resonance at 5.25 ppm, which converts the quartet into a somewhat distorted triplet. Coupling constants of the same order of magnitude are generally found in homoallylic arrangements. This allows us to deduce the following structure element:

$$-CH_2-CH=\overset{|}{C}-CH_3$$

Since there is only one quaternary sp^3 carbon and the two remaining methyls are known to be adjacent to a fully substituted carbon, we further conclude that these two methyls are in geminal position, i.e.

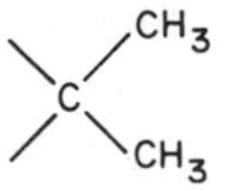

We should now examine all possibilities for bicyclic structures compatible with the above-listed structure elements consisting of seven carbons. A spirocyclic structure can *a priori* be excluded since the only quaternary sp^3 carbon is known to carry two methyl groups. The following bicycloheptanes need be considered: (a) [4.1.0], (b) [3.2.0], (c) [3.1.1], (d) [2.2.1].

(a) Such a structure contains a cyclopropane moiety, which can be ruled out on the basis of the chemical shifts observed for methylene, methine, and quaternary, carbons since in cyclopropyl rings these are characteristically shielded (between 20 and 0 ppm). The same reasoning applies to the proton chemical shifts.

(b) Two structures of this type are conceivable:

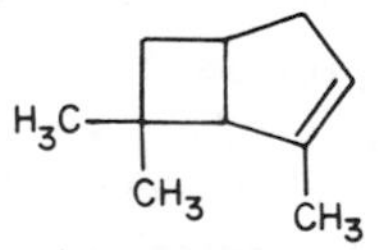

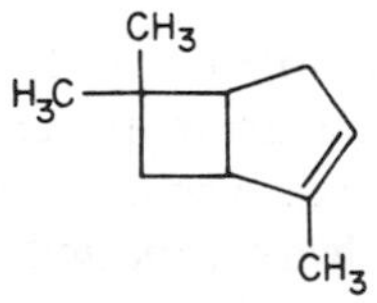

A strong argument against both structures concerns the chemical shifts found for the methylene carbons. A CH_2 carbon adjacent to a quaternary and tertiary carbon would certainly be less shielded than the resonance observed (number of α and β interactions).

(c)

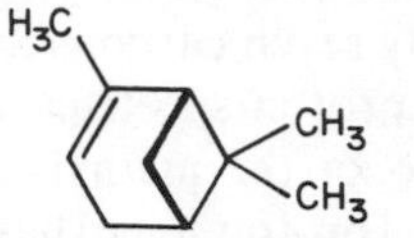

α-pinene

This structure is compatible with the chemical shifts of methylene and methine carbons. The methine carbons, being α to a quaternary carbon, are anticipated to be the least shielded in accordance with the data. We need not consider (4) since a [2.2.1]bicycloheptane cannot accommodate one of the partial structures that are requisite.

PROBLEM 11

The region displayed shows the resonances of the aromatic carbons. The two proton-bearing carbons resonate at *ca* 141 ppm in either isomer and both exhibit a large splitting of 181 Hz due to the one-bond C—H coupling. However, there is a striking difference in that in the proton-coupled spectrum of **I** no further multiplet structure is present, whereas for **II** each branch consists of a six-line pattern which simplifies to a doublet upon irradiation of the methyl protons. In Chapter 3 we have seen that in mesitylene, for example, the aromatic carbons couple with the methyl protons across two and three bonds ($^2J = 6\,\text{Hz}$ and $^3J = 4\,\text{Hz}$) but only to a negligible extent across four bonds. This allows us to assign **I** to 2, 3- and **II** to 2, 6-dimethylpyrazine:

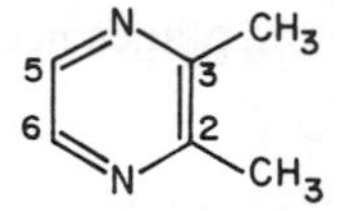

I

$^1J_{C_5H_5} = 181\,\text{Hz}$
$^2J_{C_5H_6} = 10.8\,\text{Hz}$
$^4J_{C_5Me} \approx 0$

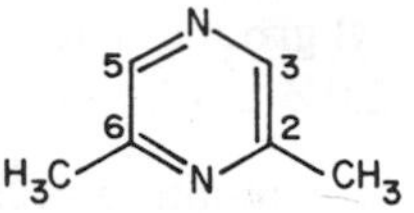

II

$^1J_{C_3H_3} = 181\,\text{Hz}$
$^3J_{C_3H_3} = 10.5\,\text{Hz}$
$^2J_{C_3Me} = 4.8\,\text{Hz}$
$^2J_{C_2H_3} = 9.5\,\text{Hz}$

It is interesting to note that the geminal and vicinal ^{13}C—H coupling constants involving aromatic carbons and protons are of equal order of magnitude ($\sim 10\,\text{Hz}$), whereas normally $^3J_{CCCH} \gg {}^2J_{CCH}$. However, deviations from this rule are always expected when the ring is heterocyclic. In these cases the geminal ^{13}C—H coupling constant involving the proton adjacent to the hetero atom increases considerably in magnitude (cf. Table 2.21). It should be noted that the apparent coupling constants obtained from the selectively decoupled spectra are reduced, as a comparison between the splittings observed in the single and double resonance spectra proves.

PROBLEM 12[4]

(1) Note that f is the only doublet among the six methyl resonances. Hence, we can readily assign $f = (CH_3)_\gamma$

(2) The two low-field carbon resonances, i and ii, must correspond to the two carbonyl carbons (C-1 and C-7). Carbon ii is coupled only to methyl group a; this methyl group is coupled to no other carbon atom. In contrast, carbon i is coupled only to methyl group c, which in turn is coupled to carbon atoms iv and v. Since the acetate carbonyl is insulated from the rest of the molecule by the intervening oxygen atom, it follows that $ii = $ **C-7** (the acetate carbonyl carbon atom) and, hence, $i = $ **C-1** (the amide carbonyl carbon atom).

(3) Since i is coupled only to methyl group c, we can assign $c = (CH_3)_\alpha$.

(4) We noted above that methyl group c is coupled to carbon atoms iv and v; accordingly, iv and v must correspond to C-2 and C-3. Carbon iv is coupled to methyl groups c, d, and f, whereas carbon v is coupled to methyl groups c and d. The fact that iv and not v is coupled to f permits the assignment $iv = $ **C-3**; hence, $v = $ **C-2**.

(5) Methyl group f is coupled to both carbons iv and vi. Since we have already assigned $iv = $ C-3, it follows that $vi = $ **C-5**.

(6) Only one carbon atom remains unassigned; hence, $iii = $ **C-6**.

(7) Carbon vi is coupled to three methyl groups, i.e. b, e, and f. It was noted in the statement of the problem that the coupling between vi and e is especially intense, suggesting that these groups are geminal; hence, $e = (CH_3)_\delta$.

(8) Three methyl groups, a, b, and d, remain unassigned. Methyl group a is coupled only to carbon ii (i.e. to C-7); hence, $a = (CH_3)_\zeta$. Methyl group b is coupled only to carbons iii and vi (i.e. to C-5 and C-6); hence, $b = (CH_3)_\varepsilon$.

(9) The lone remaining methyl group, d, must be $d = (CH_3)_\beta$. Indeed, d is coupled only to C-2 and C-3, a fact that is consistent with this assignment.

PROBLEM 13

From the molecular formula one obtains $F = 6$. The ^{13}C spectrum contains thirteen separate lines with two of them being degenerate. In addition to the seven resonances expected from the formyl carbon, the unsymmetrically disubstituted furan ring and two additional sp^2 carbon resonances originating from the side-chain double bond (lines 1–6), there is no evidence for other multiple bonds. Together these account for $F = 5$.

In order to account for the known partial structure we have

$$-CH_2-CH-\overset{\displaystyle |}{C}=CH-$$
$$\underset{\displaystyle CH_3}{|}$$

to subtract one CH_3, CH_2, and CH peak from the high-field resonances. This leaves us with the following unassigned carbons:

$$2 \times -CH_3$$
$$2 \times -CH_2-$$
$$1 \times -\overset{|}{\underset{|}{C}}-$$

Incorporating the olefinic double bond into a cyclic partial structure leads to a cyclopentene

Alternative structures with one of the methyls attached to an olefinic carbon are

Although these two structures would account well for the equivalence of two methylene resonances, they are ruled out on the basis of the chemical shifts since cyclopropyl carbons are more highly shielded (between 20 and 0 ppm, depending upon substitution). Chemical shifts of aliphatic methylene carbons near 50 ppm require heavy α-substitution. This favours the cyclopentene structure with the geminal methyl group β to the double bond.

The substitution of the furan ring remains to be established. Among the four possibilities 2, 3; 2, 4; 2, 5, and 3, 4 only the last is compatible with the observed chemical shifts. The doublet resonances at 148 and 141 ppm require carbons 2 and 5 of the furan moiety to be unsubstituted. The correct structure is therefore

PROBLEM 14

Except for an overlap of two high-field lines occurring in the spectrum of **II** (line 17 at 20.15 ppm is composed of a quartet and a triplet carbon), the total carbon number is the same for the two species. However, the multiplicities in the spectrum of **II** indicate 24 as compared to 26 hydrogens for **I**. Hence structure **II** differs from **I** by possessing one additional double bond or ring. In fact, two more olefinic resonances appear in the low-field region of **II**, thus pointing to a further double bond. From their multiplicities (both are doublets) we can rule out positions which would lead to quaternary carbons. If we assume that the Δ_{45} double bond is retained, the second olefinic double bond can be in position 1, 2; 6, 7; 11, 12; or 15, 16. From a comparison of the chemical shift of line 1 assignable to the ketonic carbonyl carbon, it is recognized that this is displaced upfield in **II** (186.2 against 199.2 ppm). This requires this double bond to be conjugated with the enone system (cf. Table 2.6). Consequently we deal with an $\alpha, \beta, \gamma, \delta$, or a cross-conjugated dienone. There are various ways of differentiating the two possibilities but only one will be outlined. One line that is readily identified pertains to the quaternary carbon 10 (line 9 in **I** and 10 in **II**). In the case of the double bonds being in position 4, 5, 6, 7, the chemical shift should be little affected. A chemical shift comparison in the two spectra shows that C-10 experiences a 5 ppm deshielding in **II**. This behaviour favours assignment of the spectra of **II** to an $\alpha, \beta, \alpha', \beta'$-dienone:

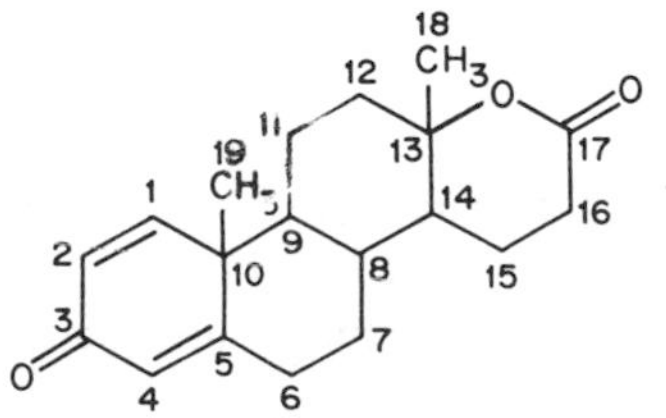

The sp^2 carbon resonances are readily assigned using chemical shift correlation arguments. It is to be further noted that olefinic carbons β to a carbonyl group are considerably deshielded (Fig. 2.4(c)).

	Line number	
Carbon	I	II
1		4
2		5
3	1	1
4	4	6
5	3	3
17	2	2

PROBLEM 15

Apart from the carbonyl function, the molecule has no sites of unsaturation. For the ten lines (2–11; the carbonyl resonance is not shown) the signal multiplicities are found to be: 2d, 3d, 4t, 5d, 6d, 7t, 8q, 9q, 10q, 11q. The presence of four methyl carbons points to a highly branched carbon chain. Line 2 at *ca* 76 ppm is the only resonance which can immediately be assigned (carbon bearing the hydroxyl group). Among the remaining lines, 2 exhibits the largest residual splitting. Since, as usual, the decoupler has been placed at high field from TMS, a large residual splitting is equivalent to relative deshielding of the corresponding proton resonance. Line 2 is therefore identified as belonging to a CH carbon adjacent to the carboxyl function. The single-proton quintuplet at 2.65 ppm in the 100 MHz spectrum is suspected to arise from the methine proton next to COOH. Its multiplicity could originate from nearly equal vicinal coupling to methyl protons and another CH proton in β position. The double resonance spectrum confirms this: irradiation at 2.65 ppm converts the doublet at 1.2 ppm to a singlet. At the same time the double doublet at 3.7 ppm assignable to the CH—O proton collapses to a broadened singlet. These arguments allow one to deduce the partial structure

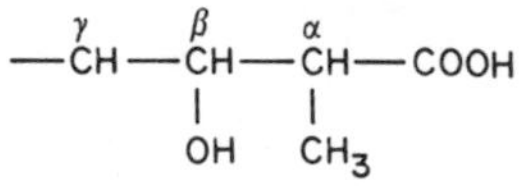

$$-\overset{\gamma}{CH}-\overset{\beta}{\underset{\underset{OH}{|}}{CH}}-\overset{\alpha}{\underset{\underset{CH_3}{|}}{CH}}-COOH$$

The multiplicity of the CH proton resonance tells us that the γ carbon carries only one proton. For the unassigned part of the molecule (C_6H_{13}), the following structures are compatible with the signal multiplicities observed:

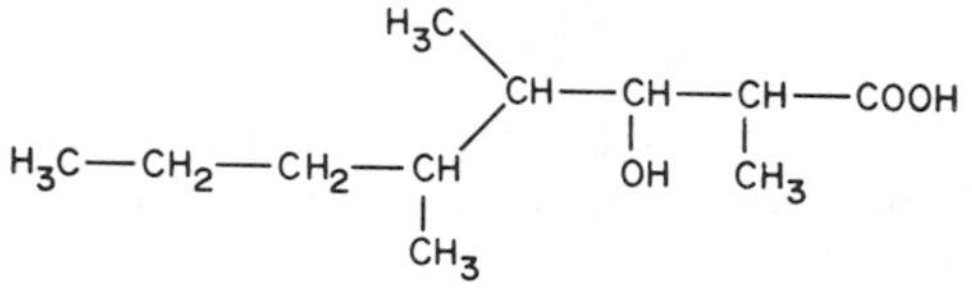

$$H_3C-CH_2-CH_2-\underset{\underset{CH_3}{|}}{CH}\diagup\overset{H_3C}{\diagdown}CH-\underset{\underset{OH}{|}}{CH}-\underset{\underset{CH_3}{|}}{CH}-COOH \quad (i)$$

$$CH_3-CH_2-\underset{\underset{CH_3}{|}}{CH}-CH_2\diagup\overset{H_3C}{\diagdown}CH-R \quad (ii)$$

$$\overset{CH_3}{\underset{CH_3}{\diagdown}}CH-CH_2-CH_2\diagup\overset{H_3C}{\diagdown}CH-R \quad (iii)$$

$$CH_3-CH_2-\underset{\underset{CH_3}{|}}{CH}\diagup\overset{H_3C-CH_2}{\diagdown}CH-R \quad (iv)$$

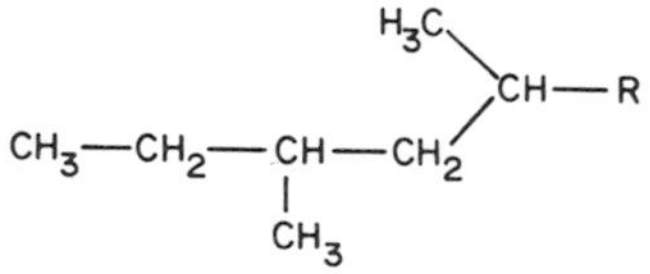

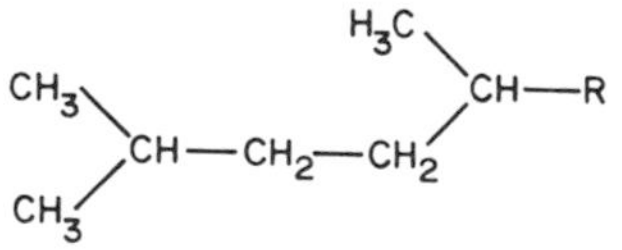

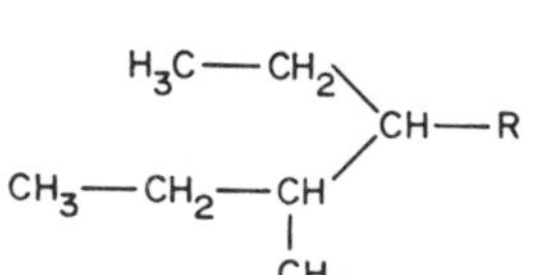

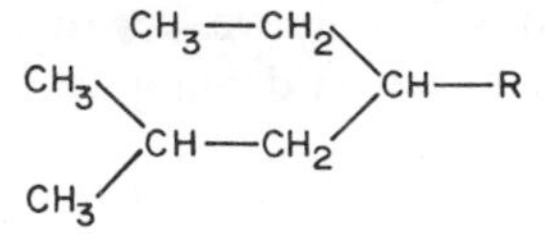

(v)

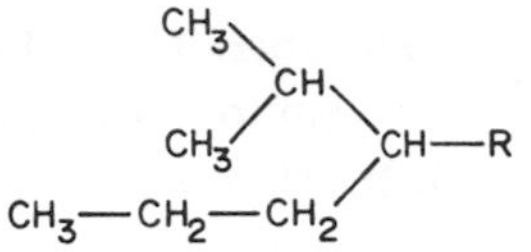

(vi)

Although geminal methyl carbons are non-equivalent in this molecule, the large chemical shift differences observed for the four methyl carbons render structures containing isopropyl groups very unlikely. This lowers the number of substructures to be considered to three (**i, ii, iv**). From a calculation of the CH_2 chemical shifts below (Chapter 2), it is recognized that these match only partial structure (**ii**).

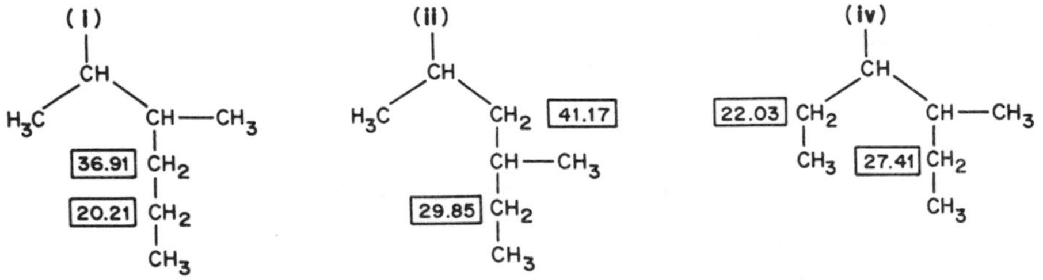

(I)

(II)

(iv)

The correct structure is therefore

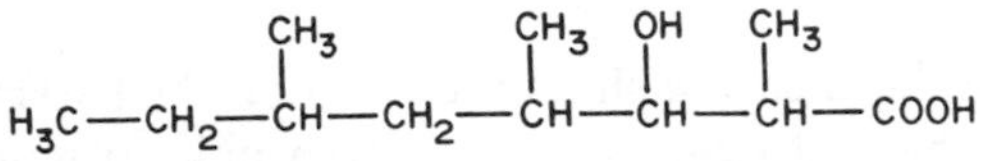

PROBLEM 16

The number of observationally different carbons is clearly inconsistent with the molecular mass of 268 unless line degeneracies are allowed for. Both the 25.2 MHz ^{13}C and the 300 MHz proton spectrum indicate the presence of a saturated hydrocarbon. The integral of the three distinguishable regions in the proton spectrum normalizes to 20 protons. Since 20 is too low a value to account for the molecular mass of 268, we are probably dealing with a molecule containing 40 hydrogens corresponding to $C_{19}H_{40}$. The molecule thus has twofold symmetry with the symmetry operation affecting all but one of the carbons. This means that in the ^{13}C spectrum all except one of the lines represent two-carbon resonances. Although ^{13}C spectra are generally not fully quantitative, it can be seen that line 5 is considerably weaker than the remainder of the resonances. It appears, therefore, that we are dealing with a symmetric structure of the type

$$R—CH_2—R$$

R contains three CH_3s, four CH_2s, and two CHs. The proton spectrum, furthermore, indicates the presence of one isopropyl group in R as shown from the multiplet at 1.53 ppm. This limits the possibilities for the partial structures of R to five:

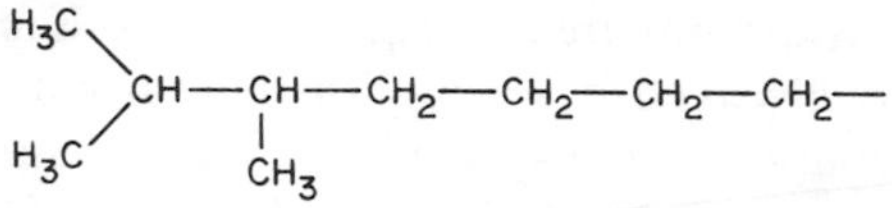

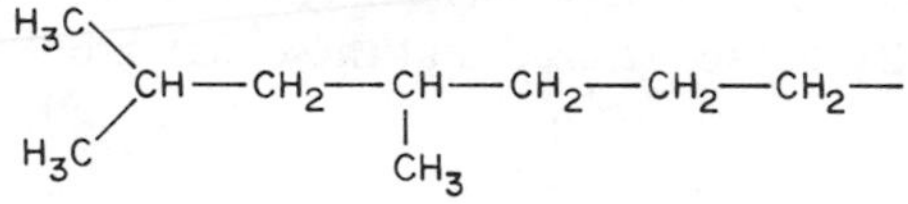

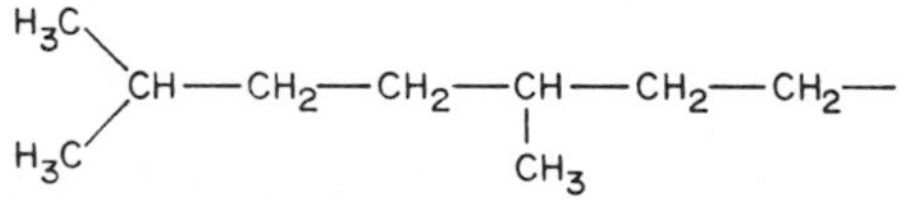

A comparison of the chemical shifts predicted on the basis of Lindeman and Adams's additivity rules (cf. Chapter 2) fit the terpenoid structure depicted below

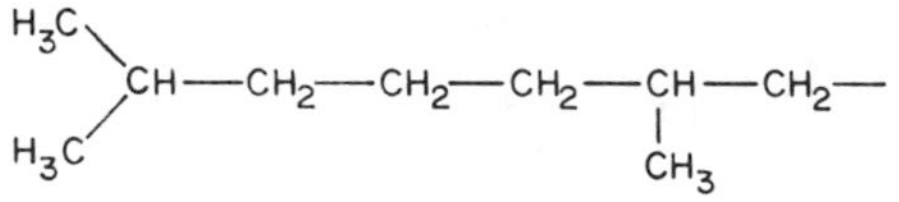

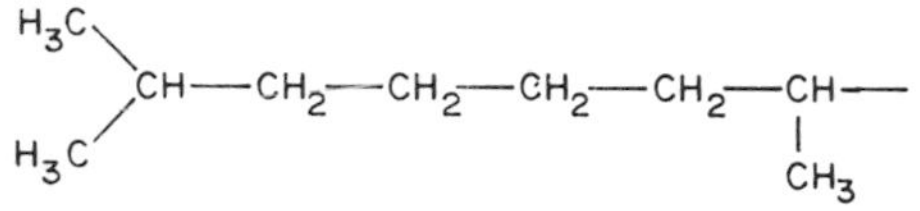

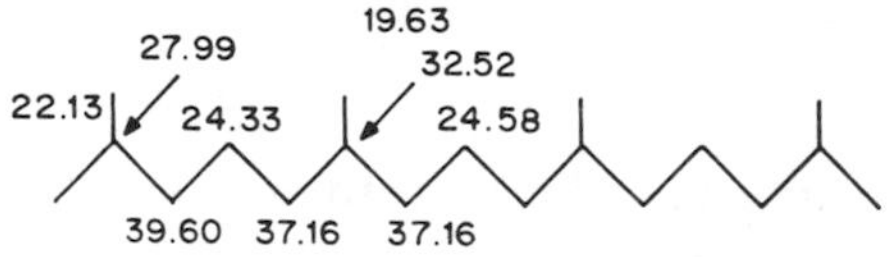

The calculated chemical shifts are indicated in the formula.

PROBLEM 17[5]

(1) From examination of the ^{13}C n.m.r. spectra edited with the DEPT pulse sequence (Problem Fig. 17(a)) we can conclude the following: (a) carbons h, i,

and j are methyl group carbons; (b) carbons e, f, k, l, and m are methylene group carbons; (c) carbons a, b, and c are methine group carbons; (d) carbons d and g are quaternary carbons. In addition, we note that eleven carbon–carbon connectivities have been established via double quantum coherence measurements (Problem Fig. 17(b)). The data in Problem Fig. 17(b) afford the following conclusions:

(i) Peaks h, i correspond to two methyl groups, both of which are connected to g (a quaternary carbon atom trace I). Hence, $\mathbf{g = 8, h, i = 11, 12}$ (degenerate carbon resonances).

(ii) The remaining methyl group must correspond to $\mathbf{j = 10}$.

(iii) Peak g is also connected to peaks a (CH, trace C) and f (CH$_2$, trace G); therefore, $\mathbf{a = 8a}$ and $\mathbf{f = 7}$.

(iv) Peak f is connected to m (CH$_2$, trace J); hence, $\mathbf{m = 6}$.

(v) Peak m is connected to e (CH$_2$, trace H); hence, $\mathbf{e = 5}$.

(vi) Since there are only two quaternary carbons (i.e. d and g), and we have already established that $g = 8$, it follows that $\mathbf{d = 4}$.

(vii) Peak d is connected to c (CH, trace D); hence, $\mathbf{c = 3a}$.

(viii) Peak c is connected to l (CH$_2$, trace F); hence, $\mathbf{l = 3}$.

(ix) Peak l is connected to k (CH$_2$, trace K); hence, $\mathbf{k = 2}$.

(x) Peak k is connected to b (CH, trace E); hence, $\mathbf{b = 1}$.

(xi) The data given from the off-resonance decoupled ^{13}C n.m.r. spectrum of

TABLE 1 ^{1}H and ^{13}C n.m.r. chemical shifts of longifolene

$\delta_{C_6D_6}$	^{13}C Group	Assignment	^{1}H: $\delta_{C_6D_6}$
167.58	C	9	
99.80	CH$_2$	13	4.77 and 4.52
62.67	CH	8a	1.36
48.49	CH	1	2.59
45.67	CH	3a	2.01
44.38	C	4	—
43.85	CH$_2$	5	1.69 and 1.54
36.99	CH$_2$	7	1.73 and 1.02
33.94	C	8	—
30.85	CH$_3$, CH$_3$	11, 12	0.91, 0.95
30.91			
30.42	CH$_3$	10	0.99
30.19	CH$_2$	2	1.68 and 1.18
26.00	CH$_2$	3	1.65 and 1.36
21.67	CH$_2$	6	1.56 and 1.43

longifolene requires that the vinyl carbon peak at $\delta 167.59$ corresponds to C_9 and the vinyl carbon peak at $\delta 98.93$ corresponds to C_{13}.

(2) The ^1H n.m.r. peak assignments for longifolene deduced from the data in Problem Fig. 17(c), together with the corresponding ^{13}C n.m.r. spectral assignments deduced in part (1), above, are summarized in Table 1.

PROBLEM 18

From the number and relative intensity of the lines, it can readily be concluded that we are dealing with a mixture consisting of the two stereoisomers cis- and trans-1, 3-dimethylcyclohexane. Since cyclohexane rings are known to undergo rapid chair–chair interconversion, we have to consider the two equilibria (cf. Chapter 5).

Intercoversion in the case of the trans isomer occurs between equal energy conformers; thus the chemical shifts observed for carbons 1 and 3, 4 and 6, and the axial and equatorial methyls are observed as their averages; for example,

$$\delta_{C_1}(\text{obs}) = \delta_{C_3}(\text{obs}) = 0.5(\delta_{C_1} + \delta_{C_3}) \text{ etc.}$$

Consequently, one expects to observe five resonances for this species.

Cis-1, 3-dimethylcyclohexane also interconverts, but in contrast to the trans isomer the relative energies of the two conformers are very different with a negligible population of the diaxial form. Since the diequatorial form itself possesses a plane of symmetry, five resonances result also. The calculated chemical shifts (using the increments of Table 2.11), together with the measured values, are tabulated below.

Carbon	trans		cis	
	$\delta_C(\text{pred})$	$\delta_C(\text{obs})$	$\delta_C(\text{pred})$	$\delta_C(\text{obs})$
1/3	27.8	27.44	33.3	33.08
2	41.7	41.40	45.3	44.74
4/6	34.4	33.94	36.1	35.29
5	20.9	21.01	27.3	26.69
CH_3	—	20.71	—	22.93

PROBLEM 19

When checking the chemical shift tables provided for the ^{13}C spectra of isomers A and B, a striking similarity of the data is noticed. The 1:1 correspondence of shift values is violated only for two resonances: 111.4 (A), 107.2 ppm (B) and 107.6 (A), 110.2 ppm (B). The signal multiplicities are found to be compatible only with structure **III**. The unusual shielding observed for one of the quaternary sp^2 carbons (107.6 and 110.2 ppm, respectively) is characteristic of C-3 in a

substituted indole which confirms the presence of an indole moiety. A comparison of the chemical shifts with suitable model compounds did not reveal any controversy. However, one question mark remains. The single chiral atom by itself cannot account for the existence of diastereomers. If, on the other hand, rotation about the single bond connecting the two heterocycles via the N—C bond is hindered, then the combination of what is referred to as atropisomerism and one chiral carbon atom allows the existence of stable diastereomers. The presence of atropisomerism as the cause for two observationally different species was verified by heating the pure components, which converted each of them into the equilibrium mixture.

PROBLEM 20[6]

I: $J_{34} = 7.3$ Hz; therefore, H$_4$ in **I** showing the large coupling (double quartet centred at $\delta 2.05$), correlating with C$_4$ in **I** (coupled to the directly attached proton H$_4$) can be assigned to the resonance at $\delta 15.93$, and C$_5$ must therefore correspond to the peak at $\delta 20.22$.

II: Similarly, $J_{34} = 7.0$ Hz; therefore, H$_4$ in **II** shows the large coupling (double quartet centred at $\delta 1.76$). From the correlation diagram, C$_4$ in **II** can be assigned to the resonance at $\delta 13.83$, and C$_5$ must therefore correspond to the peak at $\delta 10.95$.

These conclusions are in agreement with the results of Fraser's earlier study[7] (deduced from 40 MHz ^{1}H n.m.r. measurements on **I** and **II**).

PROBLEM 21[8]

On the basis of chemical shift arguments, the peak at 182 can be assigned to the enone carbonyl resonance (C-1), whereas the peak at 166 corresponds to the amido carbonyl group in **I**. A substantial decrease in the intensity of signals corresponding to C-1, C-2, and C-3 occurs upon H–D exchange in CH_3OD, while carbon C-4 remains relatively unaffected; (compare the middle and bottom traces in Problem Fig. 21). In addition, a substantial NOE is observed for C-1, while none is observed for C-4 in this compound (compare the top and bottom traces in Problem Fig. 21). Both sets of results are consistent with structure **Ia** (*E*-isomer) and not with structure **Ib** (*Z*-isomer).

PROBLEM 22

It is well known that the stability of carbenium ions increases with the ability of the ion to delocalize its positive charge. For a classical carbenium ion increased substitution of hydrogen by alkyl stabilizes the ion. Parallel to this trend, the shielding of the carbon bearing the positive charge increases. Since aromatic substituents delocalize the charge more effectively, considerably increased shielding is to be expected in such systems. Hence the shielding sequence predicted for the six ions is

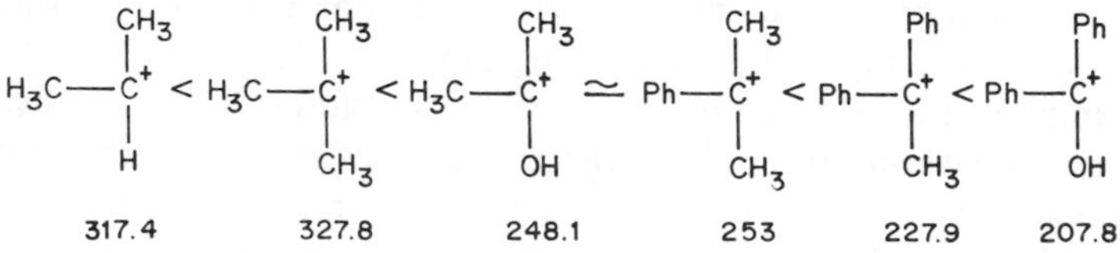

$$
\underset{317.4}{H_3C\!-\!\overset{CH_3}{\underset{H}{\overset{|}{\underset{|}{C^+}}}}}
<
\underset{327.8}{H_3C\!-\!\overset{CH_3}{\underset{CH_3}{\overset{|}{\underset{|}{C^+}}}}}
<
\underset{248.1}{H_3C\!-\!\overset{CH_3}{\underset{OH}{\overset{|}{\underset{|}{C^+}}}}}
\simeq
\underset{253}{Ph\!-\!\overset{CH_3}{\underset{CH_3}{\overset{|}{\underset{|}{C^+}}}}}
<
\underset{227.9}{Ph\!-\!\overset{Ph}{\underset{CH_3}{\overset{|}{\underset{|}{C^+}}}}}
<
\underset{207.8}{Ph\!-\!\overset{Ph}{\underset{OH}{\overset{|}{\underset{|}{C^+}}}}}
$$

The experimental chemical shifts δ_{c^+} are indicated in the formulae. The simple charge delocalization model cannot explain the inverse order of the shieldings found in dimethyl- and trimethylcarbenium ion.

PROBLEM 23

If the dimethylisopropyl carbenium ion were a static species, it would have four non-equivalent carbons.

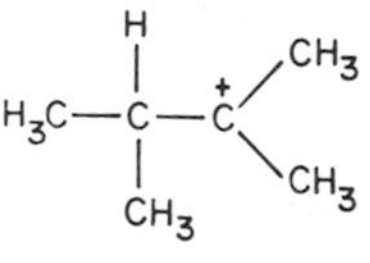

$$
H_3C\!-\!\underset{CH_3}{\overset{H}{\underset{|}{\overset{|}{C}}}}\!-\!\overset{+}{C}\!\underset{CH_3}{\overset{CH_3}{<}}
$$

The chemical shifts of these four types of carbons can be estimated on the basis of the known chemical shifts in the trimethylcarbenium ion. Replacing two hydrogens of a methyl carbon leads to a deshielding of about 15 ppm at the resulting isopropyl carbon when the empirical substituent effects listed in Table 2.7 are used. Accordingly, the neighbouring cationic centre experiences a deshielding of 18 ppm. Hence the two carbons should have shifts of 63 and 346 ppm, respectively. If the system undergoes rapid degenerate equilibration according to chemical shift averaging occurs and the predicted chemical shift of the ionic carbon is $0.5(63 + 346) = 204.5$ ppm, a value which is in good agreement with the experimentally observed shielding of 198 ppm. The coupling constant of 65 Hz is also a result of the averaging; it is 130 Hz and 0 for the two ionic sites.

PROBLEM 24

The fact that five of the six methyl carbons become equivalent in the dication points to a fivefold symmetry of this species. Two structures can in principle explain the degeneracies observed in the spectrum: (1) the non-classical carbonium ion **I** and (2) a species **II** undergoing rapid degenerate rearrangement

I   II

Chemical shift calculations using suitable model compounds strongly favour the non-classical ion **I** over a rapidly equilibrating species **II**. For example, it is difficult to explain the unusual shielding of the apical methyl (-2 ppm) on the basis of **II**.

PROBLEM 25[9]

The answer to this question can be arrived at readily via inspection of the symmetry properties of each of the three possible structural situations for cation **I**:

(1) First consider the possibility that **I** exists as a static structure (for example, **Ia**). Since **Ia** possesses twofold symmetry, we expect to observe four peaks, relative intensities 4:2:2:1, in either its ^1H or its ^{13}C n.m.r. spectrum:

$$C-2 = C-3 = C-6 = C-7$$

$$C-1 = C-8$$

$$C-4 = C-5$$

$$C-9$$

Ia

(2) Next, consider the possibility that **I** exists as a mixture of rapidly equilibrating classical ions (for example, **Ia⇌Ib⇌**, etc.). If this were indeed the case, it can readily be shown that the operation of successive Wagner–Meerwein rearrangements will equilibrate all nine carbon atoms and all nine hydrogen atoms in **I**. Since this situation leads to ninefold symmetry in **I**, we would expect to observe a single peak in either the ^{1}H or the ^{13}C n.m.r. spectrum of **I**.

(3) Finally, if we redraw structure **Ic** in such a way as to emphasize its threefold symmetry, we conclude that we should observe three peaks, relative intensities 1:1:1, in either the ^{1}H or the ^{13}C n.m.r. spectrum of **I** if this cation indeed possesses a non-classical structure.

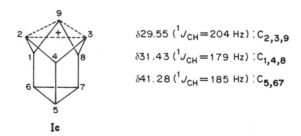

$$C-1 \equiv C-4 \equiv C-8$$
$$C-2 \equiv C-3 \equiv C-9$$
$$C-5 \equiv C-6 \equiv C-7$$

Ic

Since both the ^{1}H and ^{13}C n.m.r. spectra of **I** display three peaks of equal intensity, we conclude that this cation possesses non-classical structure **Ic**. Furthermore, we can make use of the given J_{CH} values to assign the ^{1}H and ^{13}C n.m.r. chemical shifts in **Ic**. It is well known that the magnitude of directly bonded ^{13}C—^{1}H coupling constants varies directly with the hybridization of the carbon atom (i.e. the '%s-character' in the carbon orbital that is involved in bonding to hydrogen).[10] From examination of structure **Ic** we expect the carbon hybridization, expressed as %s-character, to decrease in the order (C-2, C-3, C-9) > (C-5, C-6, C-7) > (C-1, C-4, C-8); hence, the following spectral assignments can be made for cation **I**:

$\delta 29.55$ ($^1J_{CH}=204$ Hz) : $C_{2,3,9}$

$\delta 31.43$ ($^1J_{CH}=179$ Hz) : $C_{1,4,8}$

$\delta 41.28$ ($^1J_{CH}=185$ Hz) : $C_{5,67}$

Ic

PROBLEM 26[11]

(1) Since the product is known to consist of only three deuterated bicyclo[2.1.1]hexan-2-ols, and since only three signals are observed in the

one-dimensional $\{^2H\}^{13}C$ polarization transfer experiment (Problem Fig. 26a), it follows that the CD_2 units must survive intact in all three of the products.

(2) Comparison of the chemical shifts of each of the three peaks in Problem Fig. 26(a) with the given ^{13}C n.m.r. chemical shift data suggests that **I, II**, and **III** possess the structures indicated below:

(3) Inspection of Problem Fig. 26(c) permits the following assignments to be made:

(i) The ^{13}C peak that corresponds to the CD_2 group in **I** (Problem Fig. 26(c)) correlates with ^2H resonances 1 and 5.
(ii) The ^{13}C peak that corresponds to the CD_2 group in **II** (Problem Fig. 26(c)) correlates with ^2H resonances 3 and 4.
(iii) The ^{13}C peak that corresponds to the CD_2 group in **III** (Problem Fig. 26(c)) correlates with ^2H resonances 2 and 6.

Integration of the ^2H n.m.r. spectrum revealed that **I, II**, and **III** were formed in 47.1, 44.9, and 8.0% yield, respectively. A detailed study of the degenerate rearrangement of the 2-bicyclo[2.1.1]hexyl cation in aqueous media by using optically active, deuterium-labelled and ^{13}C-labelled substrates has been reported recently.[12]

PROBLEM 27

The first step involved in the line shape analysis is the measurement of the line widths at different temperatures. Since the approximations to be used for the calculation of the rate constants (Eqns (5.6), (5.7), and (5.9)) are applicable only on condition that the lines do not overlap, the coalescence region (traces recorded at $0°$ and $5°C$) must be ignored. Below coalescence we furthermore confine ourselves to lines 1/5 and 2/6. Since artificial line-broadening effects are within experimental error of the line width measurements, no corrections are required. Following the procedure outlined in Chapter 5, the subsequently listed data set is obtained (Problem Solution Fig. 27).

The figure represents a plot of $-R \ln (2hk_r/kT)$ against $1/T$. The straight line was obtained by regressional analysis ignoring the first two points which fall off

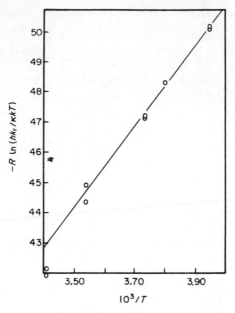

Problem Solution Fig. 27.

the line probably on account of inaccurate temperature determination. From the least-squares fit the following kinetic parameters were obtained:

$$\Delta H^{\neq} = 13.3 \, \text{kcal mol}^{-1}$$
$$\Delta S^{\neq} = 2.4 \, \text{e.u.}$$

$10^3/T$	Δv_{3726}[a]	Δv_{1548}[a]	k_τ[b]	$-R\ln\dfrac{2hk_\tau}{kT}$
3.41	12.6	14.2	1880, 2125	4.21×10^{-2}
				4.19×10^{-2}
3.53	50.4	51.4	470, 639	4.49×10^{-2}
				4.43×10^{-2}
3.73	44.1	47.2	138.5, 148.3	4.72×10^{-2}
				4.71×10^{-2}
3.80	24.6	25.2	77.3, 79.2	4.83×10^{-2}
				4.83×10^{-2}
3.95	9.1	9.4	28.6, 29.5	5.02×10^{-2}
				5.01×10^{-2}

[a] Width of lines due to carbons 1, 5, 4, 8 and 3, 7, 2, 6 above and below coalescence (in Hertz).
[b] $\ln s^{-1}$.

PROBLEM 28[13]

(1) The integrals and chemical shifts for the two signals in the ^{13}C n.m.r. spectrum obtained at $-100°C$ suggest that the low-field signal corresponds to the methine carbons in 2, 3-dimethylbutane. The observation of non-equivalence in the spectrum obtained at $-180°C$ suggests the presence of both the anti and *gauche* rotational isomers in the equilibrium mixture and slow exchange in the MRI regime.

    Whereas all four methyl groups in the anti rotamer are chemically and magnetically equivalent, symmetry considerations require that there be two sets of two non-equivalent methyl groups in the *gauche* rotamer. Therefore, we expect to see a total of five signals in the low-temperature ^{13}C n.m.r. spectrum of 2, 3-dimethylbutane (i.e. two low-field signals, 1 and 2, that correspond to the methine carbon atoms in the anti and *gauche* rotamers, respectively, and three signals at higher field, one for the four equivalent methyl carbon atoms in the anti rotamer and two additional signals corresponding to the two non-equivalent methyl carbon atoms in the *gauche* rotamer). However, only *four* signals (i.e. 1–4) appear in the low-temperature spectrum of this compound; it therefore seems likely that two of the three non-equivalent methyl groups may have very similar chemical shift values (See below).

    Inspection of Newman projections (below) reveals that two of the methyl groups in *gauche* rotamers (i.e. Me_2 and Me_3 in **Gauche I** and Me_1 and Me_4 in **Gauche II**) are *gauche* with respect to both a hydrogen and a methyl group. An analogous situation exists for all four methyl groups in the anti rotamer. Hence, we expect the chemical shift values for the four methyl groups in the anti rotamer to be similar to the corresponding chemical shift values of those methyl groups identified above that are *gauche* with respect to both a

Anti

Gauche I

Gauche II

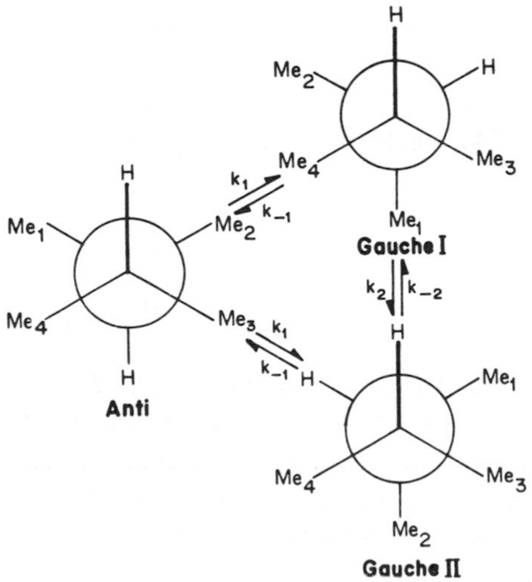

hydrogen and a methyl group in the *gauche* rotamers. Therefore, we can safely assign signal number **3**, the most intense signal in the spectrum, to these methyl carbons in the anti rotamer and to the methyls that are *gauche* to hydrogen in the two *gauche* rotamers. It is important to note that the existence of two *gauche* rotamers statistically confers a 2:1 intensity ratio for signals due to the *gauche* rotamer *vis-à-vis* those due to the anti rotamer (cf. answer to part (2), below).

Signal number 4, then, must be due to the remaining two methyl groups in the *gauche* rotamers (i.e. Me_1 and Me_4 in **Gauche I** and Me_2 and Me_3 in **Gauche II**). Since these methyl groups each experience *two* γ-steric effects due to non-bonded interactions with neighbouring methyl groups (versus one such γ-steric effect for each of the four equivalent methyl groups in the anti rotamer), it is reasonable that signal number 4, i.e. the most highly shielded resonance in the spectrum, should correspond to those methyl groups in the *gauche* rotamers that experience that greatest number of γ-steric interactions.

Finally, we note that the intensities of signal numbers 2 and 4 are nearly equal, an observation that implies that signal number 2 must correspond to the methine carbon atoms in the *gauche* rotamers. The remaining signal, 1, then must correspond to the methine carbon atoms in the anti rotamer, and the task of spectral assignment is thus completed.

(2) Once spectral assignments have been made, it is clear that the observed 1:2 ratio of 1 to 2 is indicative of the ratio of anti rotamer to *gauche* rotamer. It is noteworthy that the ratio of these rotamers in solution is statistically determined. This has also been shown to be the case for 2, 3-dimethylbutane in the gas phase.[14]

PROBLEM 29

The stereochemical sequences in PVC can be shown to obey Bernoullian statistics. For mm-centred pentad placements the probabilities are found to be (Table 5.3):

Designation	Probability
mmmm	P_m^4
mmmr	$2P_m^3(1 - P_m)$
rmmr	$P_m^2(1 - P_m)^2$

From the normalized triad probabilities (Fig. 5.23) $P_m = 0.45$ was determined (see page 332). Hence for the above listed pentad probabilities values of 0.041, 0.100, 0.061 are computed. The three lines are therefore designated in order of increasing field:

mmmm, mmmr, rmmr

PROBLEM 30

The very intense peak at 29.9 ppm is indicative of a large predominance of ethylene in this ethylene–propylene copolymer. Hence we can neglect all types of the possible sequences listed in Table 5.4 except A and B. A corresponds to pure polyethylene and the chemical shift based on Lindeman and Adams's additivity rules is 29.96 ppm in excellent agreement with the observed shift value. A comparison of the spectrum with the predicted shifts for the carbons in a sequence of type B allows easy identification of lines 1 to 6:

$$
\begin{array}{cccccccc}
& & & & \overset{\displaystyle C^m}{\underset{|}{}} & & & \\
-C-C-C-C- & \!\!\!\! C\!\!\!\! & -C- & C- & C-C \\
& & & t & \alpha & \beta & \gamma \\
\end{array}
$$

Carbon	δ_C^{pred}	δ_C^{obs}
Polyethylene	29.96	29.92
m	19.63	19.90
t	32.52	33.15
α	36.91	37.47
β	27.27	27.37
γ	30.21	30.30

Assuming the areas in the spectrum to be proportional to the number of carbons* the mole fractions of propylene and ethylene, P and E, are calculated on the basis of Eqn. (5.20):

$$I(\,CH_3)/\sum_i I_i = P/(2E + 3P)$$

Substituting $E = 1 - P$, the equation can be reformulated

$$P = \frac{2I(CH_3)}{\sum_i I_i - I(CH_3)}$$

Improved accuracy is obtained if the average of the integrals of peaks 1, 2, 5, 6 is taken in lieu of the methyl signal area according to

$$I(CH_3) = \frac{I(C - \alpha) + I(C - t) + I(C - \beta) + I(CH_3)}{6}$$

Division by 6 is required because the statistical weight of C-α and C-β is 2. In this way one obtains $P = 0.06$.

*This is a valid assumption since relaxation in this polymer is known to be governed by segmental motion so that the NOE should be complete for all carbons (cf. Chapter 5).

PROBLEM 31

The two groups of carbonyl resonaces in methylmethacrylate–methacrylic acid copolymer (lines 1–3 and 4–6) are assigned to the acid and ester carbonyls, respectively, on the basis of shielding arguments. Although the integral has not been recorded, it is recognized that the higher field group of lines has a larger area which corroborates the assignment since it is known that the copolymer contains 60% ester and 40% acid functions. If we denote an acid repeating unit by A and one containing an ester function by B, the following compositional triads can be formulated:

$$\begin{array}{ccc} \text{AAA} & & \text{BBB} \\ \text{AAB/BAA} & \text{and} & \text{ABB/BBA} \\ \text{BAB} & & \text{ABA} \end{array}$$

The probability of the carbonyl being in an acid function is designated P_A. Then one calculates for the triad probabilities:

Line	Designation	Triad probability	
1	AAA	p_A^3	0.064
2	AAB/BAA	$2p_A^2(1 - p_A)$	0.192
3	BAB	$p_A(1 - p_A)^2$	0.144
6	BBB	$(1 - p_A)^3$	0.216
5	ABB/BBA	$2p_A(1 - p_A)^2$	0.288
4	ABA	$p_A^2(1 - p_A)$	0.096
			1.000

From the numerical values which have been obtained by inserting $p_A = 0.4$, the line assignment in the left-hand column follows immediately.

PROBLEM 32[15]

Carbon-13 n.m.r. spectra of fumaric acid in solution and in the solid state are very similar. Both solution and solid-state spectra indicate the presence of a symmetry element in fumaric acid (note the ^{13}C n.m.r. chemical shift equivalences: $C_1 = C_4$ and $C_2 = C_3$). In addition, differences between isotropic ^{13}C n.m.r. chemical shift values for fumaric acid in solution and in the solid state are very small, thereby indicating the absence of specific interactions in the solid state.

In contrast to this result, ^{13}C n.m.r. spectra of maleic acid in solution and in the solid state differ markedly. Whereas the solution ^{13}C n.m.r. spectrum of maleic acid indicates the presence of only two non-equivalent carbon atoms, degeneracy is removed in the corresponding solid state ^{13}C n.m.r. spectrum, and four distinct resonances are observed.

Intramolecular hydrogen bonding is expected to be important in the case of maleic acid but not for fumaric acid:

Maleic acid   Fumaric acid

Hence, this problem illustrates the fact that intramolecular hydrogen bonding in maleic acid in the solid state results in loss of molecular symmetry, as expected.[15]

PROBLEM 33[16]

Only three peaks are observed in the ^{13}C n.m.r. spectrum of **I** at room temperature; *no* peak appears at low field in the region expected for a carbonyl group. These observations suggest that a fast chemical exchange (dynamic) process occurs in **I** (in the solid state!) that can be frozen out at $-160°$C. At this low temperature, the peak at $\delta112$ remains unaffected, whereas the two low-field peaks (δ 134 and 173, respectively) become split into two lines. A reasonable explanation for this behaviour is that intramolecular hydrogen bonding occurs in **I** in such a way as to produce two tautomers, **Ia** and **Ib**, that equilibrate rapidly on the n.m.r. time scale at room temperature via movement of two hydrogen atoms:

Ia   Ib

This process results in equilibration of the carbonyl (5,8) and hydroxylic (1,4) carbons, and the three-line ^{13}C n.m.r. spectrum is observed accordingly. At low temperature, a single tautomer is frozen out, and the degeneracy between carbon atoms 1,4 and 5,8 is removed, resulting in the observed splitting of the $\delta173$ resonance into two peaks. Similarly, degeneracy between carbon atoms 2,3 and 6,7 that exists at room temperature is removed at low temperature, and splitting of the $\delta134$ resonance into two peaks occurs as well. (Carbon atoms 9 and 10 remain magnetically equivalent both in the presence and in the absence of the temperature-dependent dynamic process; hence, no splitting of the $\delta 112$ resonance occurs upon cooling.)

PROBLEM 34[17,18]

(1) Semibullvalene (**I**) is a fluxional molecule. It undergoes a degenerate Cope rearrangement (i.e. [3, 3] sigmatropic shift) in solution at $-95°C$ which is rapid on the n.m.r. time scale. In this way, structures **Ia** and **Ib** become 'averaged' to **Ic** as the temperature is raised from $-160°C$ to $-95°C$. The ^{13}C n.m.r. spectrum of **I** at $-160°C$ (at which temperature the degenerate Cope rearrangement becomes slow on the n.m.r. time scale) corresponds to a single structure, **Ia** (or **Ib**), which has been 'frozen out'; hence, the spectrum consists of five distinct resonances as indicated in the table. However, at $-95°C$, averaging occurs in the way suggested by structure **Ic**, and a three-resonance ^{13}C n.m.r. spectrum is obtained accordingly.

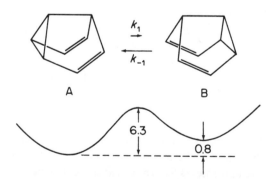

| **Ia** | **Ib** | **Ic** |

$\delta 50.7$

$\delta 87.2$

$\delta 121.7$

(2) The ^{13}C CP/MAS n.m.r. spectrum of solid **I** at $-95°C$ is considerably different from the corresponding spectrum of **I** in solution at the same temperature. In the former case, individual aliphatic and vinyl ^{13}C resonances

A $\xrightarrow{k_1}$ $\xleftarrow{k_{-1}}$ B

6.3   0.8

Problem Solution Fig. 34 Free-energy profile (kilocalories per mole) for the non-degenerate Cope rearrangement of semibullvalene obtained from line-shape analysis of the ^{13}C CPMAS spectra shown in Problem Fig. 34(a).

have merged essentially into two singlets, whereas in the latter case, three well-separated resonances were observed (cf. data in table in part (1), above). The results were considered to reflect the operation of a rapid, *non-degenerate* Cope rearrangement in solid **I**. The results of line-shape analysis of the ^{13}C CP/MAS spectra of **I** (shown in Problem Solution Fig. 34) confirmed the fact that degeneracy was removed during Cope rearrangement of solid **I**.

Further evidence which suggests that the behaviour of **I** in the solid state is different from that in solution is provided by the data in Problem Fig. 34(b). After solid **I** had been annealed for several hours, irreversible changes occurred in the sample that were reflected in the corresponding ^{13}C CP/MAS n.m.r. spectra (i.e. the appearance of the spectrum was no longer temperature-dependent after annealing). The small splitting of the vinyl carbon peaks (Problem Fig. 34(b)) suggests that annealing results in a loss of molecular symmetry in **I**. The authors concluded that 'solid-state morphology can have a significant influence on intramolecular rearrangements'.[18]

PROBLEM 35

In thermozymocidin it was suspected that the C_{3-20} chain was polyacetate-derived. In order to establish the labelling pattern at least a partial assignment is required. Carbons which lend themselves for unequivocal assignment are C-6, 7, 13, 15, 19, 20. The olefinic carbons 6 and 7, for example, are predicted to resonate at 129.9 and 133.7 ppm, respectively (Table 2.12). Although the deviations from the observed values are relatively large, we can assign line 6 to C-7 and line 7 to C-6. The spectrum of the labelled species indicates enrichment at site 7. Hence ^{13}C incorporation should have taken place at positions 3, 5, 7, 9, 11, 13, 15, 17, and 19. We will not verify this for every single position. Let us choose C-13 and 15, which should both be labelled. A chemical shift estimate can be made by calculating the shielding of a methyl carbon in a straight-chain hydrocarbon and subsequent replacement of a hydrogen by —COR (Tables 2.7 and 2.11). The value thus predicted is 43.9 ppm. A two-carbon resonance is indeed observed at 42.9 ppm, well separated from other lines. In the spectrum of biosynthesized thermozymocidintriacetyl-γ-lactone the corresponding line exhibits an intensity which is approximately twice that found for a single enriched carbon. Line 20 belonging to C-20 may serve as additional confirmation. This site should be unlabelled, which is obviously the case.

In the critical region where carbons bonded to electronegative heteroatoms resonate, only one line exhibits enhanced intensity in the spectrum of the biosynthesized product (line 9). This is assigned to C-3. Lines 10 and 11 on the basis of their unique multiplicity are assigned to C-21 and C-2, respectively. These, together with one of the carbonyl resonances, originate from the serine moiety. None of them exhibits enhanced intensity, which proves that acetate does not contribute appreciably to the biosynthesis of serine.

Thermozymocidin grown on [1-^{13}C]acetate yields the following labelling:

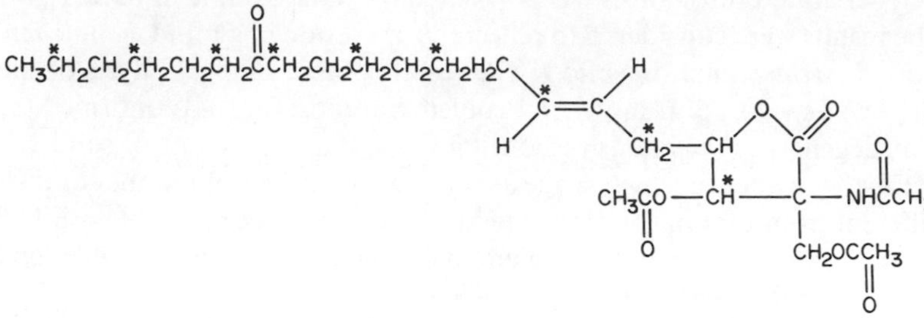

PROBLEM 36

The time dependence of the longitudinal magnetization M_z in an inversion-recovery experiment is given by Eqn. (4.35). Remembering that the signal amplitudes S are proportional to the corresponding individual magnetizations, Eqn. (4.35) may be expressed as

$$S_\infty - S(t) = 2S_\infty e^{-t/T_1}$$

In the spectra given, the signal intensities are equal to $S_\infty - S(t)$. Consequently a plot of $\ln(S_\infty - S(t))/2S_\infty$ against t, the delay time between $180°$ and $90°$ pulse, results in a straight line of slope $-1/T_1$. The simplest way of extracting T_1 consists of plotting the intensities on semilogarithmic paper, followed by visually drawing a straight line through the points. More accurate values are derived from least-squares regression. This will be done in the present case by using the following data set:

$\ln\dfrac{S_\infty - S(t)}{2S_\infty}$		
CN	C-1	t/s
− 0.693	− 0.693	1
− 0.801	− 0.847	10
− 0.922	− 0.965	20
− 1.098	− 1.037	30
− 1.199	− 1.201	40
− 1.221	− 1.273	50
− 1.386	− 1.410	60
− 1.553	− 1.513	70
− 1.717	− 1.658	80
− 1.791	− 1.739	90

This yields the following values:

$$T_1(CN) = 81 \pm 3\,s$$
$$T_1(C-1) = 86 \pm 3\,s$$

From a comparison of the peak areas in the proton-coupled and decoupled spectra the following NOE enhancement factors can be determined:

$$\eta(CN) = 0.7 \pm 0.2$$
$$\eta(C\text{-}1) = 1.2 \pm 0.2$$

With the aid of Eqn. (4.40) one obtains for the dipolar relaxation times T_1^{DD}

$$T_1^{DD}(CN) = 230 \pm 50\,s$$
$$T_1^{DD}(C\text{-}1) = 143 \pm 20\,s$$

Although the experimental T_1 s are approximately equal, T_1^{DD} is significantly different for the two carbons. The shorter value found for C-1 is due to the shorter ^{13}C—H internuclear distance of this nucleus to the protons in *ortho* position.

PROBLEM 37

Since the molecule is known to reorient in an isotropic fashion and the relaxation mechanism is dipolar, the relaxation rate, $1/T_1$, obeys Eqn. (4.15). If we interpret Eqn. (4.15) qualitatively we may state that the relaxation time of a quaternary carbon is expected to decrease with increasing number of geminal hydrogens. If, on the other hand, the number of geminal hydrogens is equal for two carbons, the vicinal hydrogens are important. In this case, however, not only the number but also their geometry will be relevant, depending on whether the carbon in question is in a cisoid or transoid relationship to the neighbouring proton. Using combined chemical shift and relaxation arguments, an unambiguous assignment can be made.

Beginning at the low-field end of the spectrum, we first encounter the carbonyl resonances at 173.2 and 165.9 ppm which are assigned to C-33 and C-22, respectively. The former possesses one geminal proton, whereas its higher field analogue has only vicinal protons and is thus less efficiently relaxed. Moreover, it is known that carbonyl carbons which are conjugated with a π-electron system are more heavily shielded. The carbons at next higher field are those adjacent to oxygen. Among these, C-3 should relax fastest. The resonance at 156.5 ppm (2.9 s) is thus assigned to C-3, the two-carbon resonance at 153.3 ppm (4.9 s) to C-25 and 27, and the one at 142.8 ppm (12.8 s) to C-26. This carbon relaxes very slowly because it has no geminal hydrogens while those in vicinal position are either *trans* (C_{24}—H and C_{28}—H) or belong to rapidly rotating groups (OCH_3). Among the remaining carbons, C-6 should exhibit the longest T_1 since it has only one α-carbon bearing a hydrogen. Hence the signal at 125 ppm (7.5 s) is assigned to C-6. The four remaining carbons are all relaxed by two geminal protons. Whereas relaxation of C-5, 8 and 9 is further enhanced by contributions from vicinal protons, that of C-23 is not. The longest relaxation time of the four unassigned carbons is therefore associated with C-23 ($T_1 = 3.8$ s). The resonances at 130.9 and 136.8 ppm are assigned to the two nitrogen-bearing carbons 8 and 5. Since relaxation of C-8 benefits from contributions due to the spatially proximate hydrogens in position 14, the shorter T_1 of 1.6 s must belong to this carbon. The characteristically large shielding of 108.2 ppm confirms the assignment of this resonance to C-9 (cf. Problem 19). A total assignment of all quaternary carbons in reserpine is summarized in the table.

Carbon	T_1/s	Carbon	T_1/s
3	2.9	6	7.5
5	2.5	25	4.9
8	1.6	27	4.9
9	1.6	33	2.4
23	3.8	22	5.6
		26	12.8

PROBLEM 38[19]

Let us first examine the nine ^{13}C T_1 values given in the problem for fluoranthene. These fall into two distinct groups: five values lie in the range 2.1–2.5 s and four lie in the range 11.6–16.1 s. The four ^{13}C atoms that display the longest T_1 values are most likely to be the four quaternary carbons (i.e. C-11, C-13, C-15, and C-16). Two of these, C-16 and C-15, can be assigned readily, as follows. Carbon atom C-15 is the only quaternary carbon atom that has no neighbouring (α-) proton; hence, it is expected to show the smallest NOE of any of the four quaternary carbon atoms in fluoranthene. Inspection of the HOESY spectrum of fluoranthene reveals that only one carbon resonance shows no cross peak; hence, this carbon resonance can be assigned to C-15. Carbon atom C-16 has two neighbouring α-protons whereas the remaining two quaternary carbons (C-11 and C-13) have only one; hence, dipole–dipole spin–lattice relaxation of C-16 is expected to be the most efficient of these three quaternary carbon atoms. Accordingly, the quaternary carbon atom that possesses the shortest relaxation time (11.6 s) can be assigned to C-16.

Inspection of the HOESY spectrum of fluoranthene reveals that C-16 has one cross peak; this cross peak must arise via dipolar interaction of C-16 with proton H-3 (a doublet in the one-dimensional proton n.m.r. spectrum of fluoranthene; see F_2 dimension in the HOESY spectrum). Proton H-3 also gives a strong cross peak with its attached carbon atom, C-3, thereby permitting assignment of the C-3 ^{13}C signal.

Protons H-7 and H-8 comprise half of an AA'BB' signal. Such patterns can be quite complex, but they are always centrosymmetric. The highest field and lowest field multiplets in the one-dimensional proton n.m.r. spectrum of fluoranthene correspond to this pattern; hence, these multiplets must correspond to H-7 and H-8, or vice versa. (Firm assignment of these proton signals will be made below.)

Three proton n.m.r. patterns remain: two doublets and a double doublet. Note that H-1 and H-3 are coupled only to H-2, while H-2 is coupled to both H-1 and H-3. The two doublets in the one-dimensional proton n.m.r. spectrum therefore correspond to H-1 and H-3; since H-3 has already been assigned, the other doublet (downfield from H-3) must be due to H-1. The remaining pattern (double doublet) can be assigned to H-2. Once these protons have been assigned,

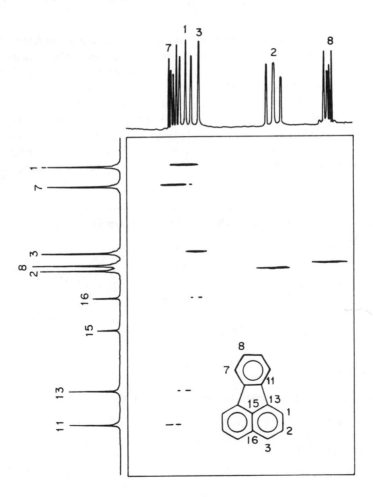

the corresponding directly bonded carbon atoms (C-1 and C-2) can be assigned by noting the appropriate cross peaks in the HOESY spectrum.

Finally, we expect to see cross peaks corresponding to dipolar interactions between C-11 and H-7 and between C-13 and H-1. Both are present in the HOESY spectrum of fluoranthene. Since the resonance that corresponds to H-1 has already been assigned, this observation permits concomitant assignment of C-13, C-11, and H-7. The observation of the strong cross peak with H-7 in the HOESY spectrum permits assignment of C-7. The lone remaining ^{13}C and proton resonances must be due to C-8 and H-8, respectively. Assignments of the nine ^{13}C resonances and the five proton resonances in the one-dimensional ^{13}C and proton n.m.r. spectra, respectively, of fluoranthene are summarized in Problem Solution Fig. 38.

PROBLEM 39

The lines pertaining to the carbons in *ortho* and *meta* position in the unsubstituted phenyl ring (C-2′, 3′) are readily identified on the basis of shielding and intensity arguments (C-2′:127.0 ppm, C-3′: 129.1 ppm). C-4 and C-4′, whose C—H bond vectors are aligned along the biphenyl axis, are expected to have the shortest T_1 s. In conjunction with chemical shift arguments (Table 2.15) this allows us to assign the lines at 128.1 and 130.3 ppm to C-4′ and C-4, respectively. Similarly, C-6 should have a shortened T_1 because of preferential rotation of the molecule about the C-3—C-6 axis. Hence the resonance at 125.8 ppm corresponding to $T_1 = 0.96$ s is assigned to C-6. The remaining lines are assigned on the basis of shielding arguments.

Line	Carbon	Line	Carbon
1	1	6	3′
2	1′	7	4′
3	2	8	2′
4	4	9	6
5	5	10	3

PROBLEM 40

Let us first introduce the designations A to D for the four esters according to

$CH_3COOCH_2CH_3$	: A
$BrCH_2COOCH_3$	: B
$HCOOCH_2CH=CH_2$	: C
$PhCOOCH_3$	: D

It is easily realized that only A and B give rise to absorptions in the high-field sp^3 carbon region. Lines 16 and 17 are thus immediately identified as the CH_2 and CH_3 carbon resonance in A while the weaker resonance 15 must belong to CH_2 in B. Lines 13 and 14, in the region where carbons next to singly bonded oxygen are known to resonate, are assigned to OCH_3 in B and D. Using intensity and chemical shift criteria, the assignments summarized in the subsequent table are obtained.

Line	Carbon	Line	Carbon
1	CO(A)	10	$=CH_2(C)$
2	CO(B)	11	$OCH_2(C)$
3	CO(D)	12	$OCH_2(A)$
4	CO(C)	13	$OCH_3(B)$
5	p-phenyl(D)	14	$OCH_3(D)$
6	$=CH-(C)$	15	$CH_2Br(B)$
7	C_1-phenyl(D)	16	$*CH_3CO(A)$
8	o/m-phenyl(D)	17	$*CH_3CH_2(A)$
9	o/m-phenyl(D)		

The composition in mole per cent is calculated from the normalized carbonyl peak areas, giving

$$A: 48.2$$
$$B: 12.1$$
$$C: 15.0$$
$$D: 24.8$$

PROBLEM 41[20,21]

The kinetic approach is analogous to the first-order reactions of two compounds, A and B, that produce a common product:[22]

$$A \xrightarrow{\; k_A \;} C$$

$$B \xrightarrow{\; k_B \;} C$$

The integrated rate law is given by $\log([C_\infty] - [C]) = \log([A_0]e^{-k_A t} + [B_0]e^{-k_B t})$. In the present instance, the intensity I_{R+S} represents the common product, C. A plot of $\log(I_\infty - I_t)$ versus time will be non-linear if $k_R \neq k_S$. However, if the T_1 values of the enantiomeric carbon atoms are quite different in magnitude, a straight line will be obtained after the carbon atom with the faster relaxation rate has fully relaxed. This is analogous to the above kinetic situation where the reaction rates are very dissimilar, for example, $k_A \gg k_B$; the plot of $\log([C_\infty] - [C])$ versus t will become linear after the more reactive component (in this case, A has disappeared. In the present instance, the semilogarithmic plot will then afford a straight line which can be extrapolated to $t = 0$. Upon extrapolation, the y-intercept will afford the value of I_0 for one enantiomer. The enantiomeric excess can then be calculated from the values of I_∞ and I_0.

Let us assume that for the two enantiomeric methylephedrines $k_R \gg k_S$. Then, extrapolation of the straight line shown in Problem Fig. 41 to $t = 0$ affords $\log I_{OS} = 0.59$, which corresponds to $I_{OS} = 3.9$. I_{OR} can be calculated by difference: $I_{OR} = I_\infty - I_{OS} = 5.65 - 3.9 = 1.75$. Then $I_{OS}/(I_{OR} + I_{OS}) = 3.9/5.65 = 0.69$. Hence, the composition of the mixture of enantiomeric methylephedrines is 69% S-enantiomer and 31% R-enantiomer; the enantiomeric excess (ee) of S-enantiomer is thus 38%.

An advantage to this approach is that it is not necessary to know the T_1 values of the pure enantiomers in order to determine the enantiomeric composition of a given sample. However, it should be noted that the relaxation rates k_R and k_S for the two enantiomers must differ substantially in order to achieve accurate results via application of this method.[21]

References

1   V. V. Krishnamurthy, J. G. Shih, and G. A. Olah, *J. Org. Chem.* **50**, 3005 (1985).
2   A. N. Abdel-Sayed and L. Bauer, *Tetrahedron Lett.* **26**, 2841 (1985).
3   H. Beierbeck and J. K. Saunders, *Can. J. Chem.* **54**, 2985 (1978).
4   H. Kessler, C. Griesinger, and J. Lautz, *Angew. Chem., Internat. Edit.* **23**, 444 (1984).
5   P. Joseph-Nathan, R. L. Santillan, P. Schmitt, and H. Günther, *Org. Magn. Reson.* **22**, 450 (1984).
6   P. Joseph-Nathan, J. R. Wesener, and H. Günther, *Org. Magn. Reson.* **22**, 190 (1984).
7   R. R. Fraser, *Can. J. Chem.* **38**, 549 (1960).
8   M. J. Shapiro, M. X. Kolpak, and T. L. Lemke, *J. Org. Chem.* **49**, 187 (1984).
9   R. M. Coates and E. R. Fretz, *J. Am. Chem. Soc.* **97**, 2538 (1975).
10 N. Muller and D. E. Pritchard, *J. Chem. Phys.* **31**, 768, 1471 (1959); See also Chapter 2, p. 66.
11 J. R. Wesener and H. Günther, *J. Am. Chem. Soc.* **107**, 1537 (1985).
12 W. Kirmse, V. Zellmer, and B. Goer, *J. Am. Chem. Soc.* **108**, 4912 (1986).
13 L. Lunazzi, D. Macciantelli, F. Bernardi, and K. U. Ingold, *J. Am. Chem. Soc.* **99**, 4573 (1977).
14 A. L. Verma, W. F. Murphy, and H. J. Bernstein, *J. Chem. Phys.* **60**, 1540 (1974).
15 E. T. Lippmaa, M. A. Alla, T. J. Pehk, and G. Engelhardt, *J. Am. Chem. Soc.* **100**, 1929 (1978).
16 W.-I. Shiau, E. N. Duesler, I. C. Paul, D. Y. Curtin, W. G. Blann, and C. A. Fyfe, *J. Am. Chem. Soc.* **102**, 1546 (1980).
17 A. K. Cheng, F. A. L. Anet, J. Mioduski, and J. Meinwald, *J. Am. Chem. Soc.* **96**, 2887 (1974).

18 V. Macho, R. D. Miller, and C. S. Yannoni, *J. Am. Chem. Soc.* **105**, 3735 (1983).
19 C. Yu and G. C. Levy, *J. Am. Chem. Soc.* **106**, 6533 (1984).
20 E. Hofer, R. Keuper, and H. Renken, *Tetrahedron Lett.* **25**, 1141 (1984).
21 E. Hofer and R. Keuper, *Tetrahedron Lett.* **25**, 5631 (1984).
22 (a) H. C. Brown and R. S. Fletcher, *J. Am. Chem. Soc.* **71**, 1845 (1949); (b) A. A. Frost and R. G. Pearson, *Kinetics and Mechanism*, John Wiley, New York, 2nd edn, 1961, pp. 162–4.

Compound Index

Subject Index